POLYMER
CHEMISTRY

THIRD EDITION

POLYMER
CHEMISTRY

AN INTRODUCTION

Malcolm P. Stevens
University of Hartford

New York   Oxford

OXFORD UNIVERSITY PRESS

1999

Oxford University Press

Oxford New York
Athens Auckland Bangkok Bogotá Buenos Aires Calcutta
Cape Town Chennai Dar es Salaam Delhi Florence Hong Kong Istanbul
Karachi Kuala Lumpur Madrid Melbourne Mexico City Mumbai
Nairobi Paris São Paulo Singapore Taipei Tokyo Toronto Warsaw

and associated companies in
Berlin Ibadan

Copyright © 1999 by Oxford University Press, Inc.

Published by Oxford University Press, Inc.
198 Madison Avenue, New York, New York 10016
http://www.oup-usa.org

Oxford is a registered trademark of Oxford University Press

Library of Congress Cataloging-in-Publication Data
Stevens, Malcolm P., 1934–
    Polymer chemistry : an introduction / Malcolm P. Stevens. — 3rd
ed.
        p. cm.
    Includes bibliographical references and index.
    ISBN 0-19-512444-8 (hardcover)
    1. Polymers. 2. Polymerization. I. Title.
QD381.S73   1999   98-23083
547'.7—dc21   CIP

Printing (last digit): 9 8 7 6 5 4 3 2

Printed in the United States of America
on acid-free paper.

To Marcia,
Jeff, Mandy, Megan and Matthew
Phil and Laurissa

CONTENTS

PART II   VINYL POLYMERS

6. Free radical polymerization,   167

PREFACE

Polymer chemistry is a marvelous subject. It draws upon all the traditional subdivisions of chemistry: organic, inorganic, physical, analytical, and biochemistry. And it includes occasional side trips into physics, engineering, and even business economics. What makes the subject particularly appealing to the student is that it deals with materials that not only are commercially important, but are a part of the everyday experience. Polymers also represent the area of chemistry in which the student is most likely to find employment on graduation. Yet paradoxically, polymers have traditionally been among the most neglected topics in both the undergraduate and graduate curriculum.

Some years ago the renowned polymer chemist Carl S. Marvel remarked: "Polymer chemistry has become such an important part of chemical technology, and polymers have come to play such a role in everyday living, that no chemist can consider himself adequately trained in his science without some introduction to this field" (*Journal of Chemical Education*, January 1965). When the late Professor Marvel made that comment—indeed, when the first edition of this book was published in 1975—few chemistry departments offered an elective in polymer chemistry. Much has changed in the intervening years. The American Chemical Society's Committee on Professional Training now includes polymer chemistry among the electives that students should be "strongly encouraged" to take as part of their undergraduate training; and guidelines have been published (*Journal of Chemical Education*, **59**, 652, 1982) outlining topics that should be included in such courses. The ACS also offers a standardized examination in polymer chemistry. Paralleling this recognition of the importance of polymer education, a number of polymer research centers have been established on university campuses.

This book is intended as an introductory text for a course in polymer chemistry at the advanced undergraduate or beginning graduate level. It should also serve as an introduction to the field for the industrial chemist with no prior training in polymer chemistry, or as a reference source for the practicing polymer chemist. I assume that students using this book will have completed undergraduate courses in organic and physical chemistry and will be familiar with the more commonly used spectroscopic and chromatographic methods of analysis and characterization.

This new edition was prompted by significant new developments in polymer chemistry. Advances in "single-site" metallocene catalysts now allow manufacturers to tailor polyolefin structure in ways heretofore impossible, and will undoubtedly bring a variety of new plastics and elastomers to the marketplace. New soft ionization techniques coupled with time-of-flight mass spectrometers have added a powerful new technique for measuring the molecular weights and molecular weight distributions of polymers. Supramolecular assemblies such as polyrotaxanes are poised to move from the realm of laboratory curiosity to applications as molecular devices. Dendritic and hyperbranched polymers having molecular dimensions in the nanometer range are being developed commercially because of their unique properties. Polymers are also taking an increasingly important role in the area of photonics.

The organization of this edition follows that of the previous one. Polymer properties, morphology, and methods of characterization are covered in Part I. Carbon-chain polymers prepared from olefinic and vinyl monomers are discussed in Part II; and nonvinyl synthetic and natural polymers are in Part III. The only significant organizational change is that heterocyclic polymers are now gathered into a separate chapter (Chapter 15), whereas previously they were divided between two chapters. This posed the question of whether to include polyimides in the new chapter or to leave them with the structurally related polyamides in Chapter 13. I chose the latter. Similarly microbial polyesters were included with other polyesters in Chapter 12 rather than with natural polymers in Chapter 18.

As before, end-of-chapter references are based in many instances on review articles or monographs to allow the student a quick entry into more detailed information on any given topic. Original sources may, of course, be obtained from the reviews. I believe that students at the advanced undergraduate or beginning graduate level should be strongly encouraged to read the journals available to them, which is why I have often included references with end-of-chapter review exercises as an enticement. A solutions manual is available, however, for faculty members who choose to use it.

Three appendices have been continued from the previous edition. Appendix A lists abbreviations commonly used by polymer scientists. Appendix B contains a compilation of the literature—encyclopedias, handbooks, continuing series, and journals—devoted exclusively or in large measure to polymers. The list of specialty handbooks should be particularly useful for industrial chemists. Appendix C contains a rich assortment of laboratory experiments and sources for instructors who wish to design a polymer laboratory course or who wish to integrate polymers into existing laboratory courses.

I would again like to acknowledge individuals who provided me with figures for inclusion in this edition: David Burge of Wescan Instruments (Figure 2.3); Dr. Frank Bovey of Bell Laboratories (Figures 5.3 and 5.4); James Martin and Alvin Tanner of Instron Corporation (Figure 5.22); Professor Harry Workman of the University of Hartford (Figure 6.2); the late Professor Jerome Vinograd of California Institute of Technology (Figure 18.12); and Dr. Leon Barstow of Protein Technology (Figure 18.14).

In the past I have been grateful to individuals, including students, who have pointed out errors in the text or offered suggestions for improvement. I hope readers of the present edition will do likewise. In particular I wish to acknowledge the following individuals who reviewed the proposal for this edition or who offered suggestions or constructive criticism:

Professor Georgia A. Arbuckle, Rutgers University

Dr. David R. Fagerburg, Eastman Chemical Company

Professor Edwin J. Goller, Virginia Military Institute

Professor R. H. Marchessault, McGill University

Professor Lon J. Mathias, University of Southern Mississippi

Professor Eli M. Pearce, Polytechnic University

My thanks also to Mary Dowst, University of Hartford, and my wife Marcia for assistance with manuscript preparation. As always I owe a debt of gratitude to Marcia, who actually puts up with this business of writing a textbook without losing her sense of humor.

West Hartford, Connecticut   M.P.S.
March, 1998

POLYMER
CHEMISTRY

P A R T

POLYMER STRUCTURE AND PROPERTIES

1

BASIC PRINCIPLES

1.1 Introduction and Historical Development

We live in a polymer age. Plastics, fibers, elastomers, coatings, adhesives, rubber, protein, cellulose—these are all common terms in our modern vocabulary, and all a part of the fascinating world of polymer chemistry. Innumerable examples of synthetic polymers may be cited, some everyday ones, others esoteric: polyester and nylon textile fibers; high-strength polyamide fibers for lightweight bulletproof vests; polyethylene plastic for milk bottles; polyurethane plastic for an artificial heart; rubber for automobile tires; fluorinated phosphazene elastomers that remain flexible in arctic environments. Whatever example or application one might select for purposes of illustration, an underlying consideration is that the particular polymer, for reasons of its unique properties or its economy, or both, is used because it does the job better than other available materials.

The purpose of this book is to provide an understanding of the chemistry of polymeric materials—how these materials differ from nonpolymers, how they are synthesized, and how they may be modified to assume a range of chemical and physical properties. While the experimental techniques for handling polymers may be somewhat different from those used with low-molecular-weight compounds, the chemistry of polymers will, in most cases, be familiar to the student who has completed introductory courses in organic and physical chemistry. The major adjustment the student has to make in beginning a study of polymers is to recognize that polymers exhibit certain properties, especially macroscopic ones, that differ markedly from those of the low-molecular-weight compounds usually encountered in undergraduate courses.

Polymers are large molecules made up of simple repeating units. The name is derived from the Greek *poly*, meaning "many," and *mer*, meaning "part." *Macromolecule* is a term synonymous with polymer. Polymers are synthesized from simple molecules called *monomers* ("single part") by a process called *polymerization*. Given the fact that most synthetic polymers have molecular weights in the range of several thousand up to several million atomic mass units, it is obviously impractical to attempt to write a definitive polymer structure. Instead a structural unit of the polymer is enclosed by brackets (or parentheses). Following are the structures of five representative monomers and their polymers:

Monomer	*Polymer*
$CH_2{=}CH_2$	$+CH_2CH_2+$
$CH_2{=}CHCl$	$+CH_2CH+$
	$\quad\quad\quad\ \mid$
	$\quad\quad\quad\ Cl$

$$\overset{\displaystyle O}{\overset{\diagup\ \diagdown}{CH_2{-}CH_2}}\qquad\qquad +CH_2CH_2O+$$

$$HOCH_2CH_2OH\qquad\qquad +CH_2CH_2O+$$

$$HO{-}\hexagon{-}CO_2H \qquad\qquad \overset{\displaystyle O}{\underset{}{}}$$

The structural units shown are understood to be repeated multifold to make up the polymer. To accent the repetition, a subscripted n is frequently placed after the closing bracket, for example, $+CH_2CH_2+_n$.

It may be noted that the first three polymers listed contain the same atoms as their monomers, whereas the last two contain fewer atoms because a byproduct (water) is formed during polymerization. Furthermore, the third and fourth polymers have the same structural unit, which illustrates that a particular polymer may be synthesized in different ways.

The reactions by which the monomers are converted to their respective polymers are described later in the book, but certain aspects of the transformations should be familiar. The first two involve alkene addition reactions. The third involves the opening of a strained three-membered ring cyclic ether (ethylene oxide) to form a more stable open-chain *polyether*. The fourth monomer, ethylene glycol, undergoes intermolecular dehydration, also to form a polyether; and the fifth, *p*-hydroxybenzoic acid, undergoes intermolecular esterification to yield a *polyester*. It should be emphasized at this point that although the third and fourth polymers have identical structural units, the two polymers may be quite different. Ethylene glycol, for example, yields a relatively low-molecular-weight polyether having little practical utility because side reactions compete with polymerization. Ethylene oxide, on the other hand, yields a high-molecular-weight, useful polymer.

Monomers such as ethylene glycol and *p*-hydroxybenzoic acid are *difunctional*; that is, each contains two functional groups. Monofunctional compounds such as ethanol and benzoic acid react together to form simple (nonpolymeric) compounds (reactions 1.1 and 1.2). Difunctional compounds such as *p*-hydroxybenzoic acid or ethylene glycol and terephthalic acid react to form polymers (reactions 1.3 and 1.4). Thus a functionality of at least two is necessary to form polymers by other than alkene addition or ring opening.

$$2CH_3CH_2OH \longrightarrow CH_3CH_2OCH_2CH_3 + H_2O \qquad (1.1)$$

$$\hexagon{-}CO_2H + CH_3CH_2OH \longrightarrow \hexagon{-}\overset{\displaystyle O}{\overset{\|}{C}}OCH_2CH_3 + H_2O \qquad (1.2)$$

$$n \; HO-\bigcirc-CO_2H \longrightarrow \left[O-\bigcirc-\overset{\overset{\textstyle O}{\|}}{C} \right]_n + n \; H_2O \qquad (1.3)$$

$$n \; HO_2C-\bigcirc-CO_2H + n \; HOCH_2CH_2OH \longrightarrow$$

$$\left[\overset{\overset{\textstyle O}{\|}}{C}-\bigcirc-\overset{\overset{\textstyle O}{\|}}{C}OCH_2CH_2O \right]_n + 2n \; H_2O \qquad (1.4)$$

Polymers of the preceding type are well characterized, even to the point of understanding their stereochemistry, but this accumulation of knowledge is the result of a considerable research effort spanning more than a century.[1–3] The word *polymer* was first used by the Swedish chemist Berzelius in 1833.[4] Throughout the nineteenth century chemists worked with macromolecules without having any clear understanding of their structure. Some modified natural polymers were, in fact, commercialized. Nitrated cellulose (known by the misnomer of nitrocellulose), for example, was marketed under such names as Celluloid and guncotton. As long ago as 1839 the polymerization of styrene was reported,[5] and during the 1860s the synthesis of poly(ethylene glycol) and poly(ethylene succinate) was published even with the correct structures.[6] (Nomenclature of polymers will be discussed later.)

$$\left[C_2H_4O \right] \qquad \text{Poly(ethylene glycol)}$$

$$\left[OCH_2CH_2O\overset{\overset{\textstyle O}{\|}}{C}CH_2CH_2\overset{\overset{\textstyle O}{\|}}{C} \right] \qquad \text{Poly(ethylene succinate)}$$

About the same time, isoprene was isolated as a degradation product of rubber,[7] although how the isoprene was incorporated into the polymer was not understood at the time. Numerous other examples of macromolecular chemistry may be found in the nineteenth-century chemical literature.[8]

The first truly synthetic polymer to be used on a commercial scale was a phenol–formaldehyde resin. Developed in the early 1900s by the Belgian-born chemist Leo Baekeland[9] (who had already earned considerable success with his invention of light-sensitive photographic paper), it was known commercially as Bakelite. By the decade of the 1920s, Bakelite had found its way into a wide spectrum of consumer products, and its inventor had achieved the ultimate in visibility—he was featured on a cover of *Time*. Other polymers, notably alkyd (polyester) paints and polybutadiene rubber, were introduced about the same time. Yet despite such commercial successes, most scientists had no clear concept of polymer structure. The prevailing theory was that polymers were aggregates of small molecules, much like colloids, but were held together by some mysterious secondary force.

This aggregation or association theory eventually gave way, with no small amount of resistance, to the theories of the German chemist Hermann Staudinger, who attributed the remarkable properties of polymers to ordinary intermolecular forces between molecules of very high molecular weight. Staudinger also introduced the term *makromolekül*.[10] In recognition of his contributions, Staudinger was awarded the Nobel Prize in Chemistry in 1953. In the 1930s, the brilliant work of the American chemist Wallace Hume Carothers placed the theories of Staudinger on a firm experimental basis and led to the commercial development of neoprene rubber and polyamide (nylon) fibers.[11]

World War II led to significant advances in polymer chemistry, particularly with the development of synthetic rubber when the natural rubber-growing regions of the Far East became inaccessible to the Allies. Among the more significant developments of the postwar years was the discovery by Karl Ziegler[12] in Germany of new coordination catalysts for initiating polymerization reactions and the application by Giulio Natta in Italy of these new systems to development of polymers having controlled stereochemistry.[13] Their work has revolutionized the polymer industry, for these so-called stereoregular polymers have mechanical properties superior in most instances to those of nonstereoregular polymers. The importance of their discoveries was recognized by the award of the Nobel Prize in Chemistry jointly to Ziegler and Natta in 1963. Equally significant was the work of Paul Flory[14] (Nobel Prize 1974), who established a quantitative basis for polymer behavior, whether it be the physical properties of macromolecules in solution or in bulk or such chemical phenomena as crosslinking and chain transfer (concepts to be encountered later in this text).

More recent years have seen a number of important advances in polymer science, which will be elaborated on in this and later chapters. Examples include:

Polymers having excellent thermal and oxidative stability, for use in high-performance aerospace applications

Engineering plastics—polymers designed to replace metals

High-strength aromatic fibers, some based on liquid crystal technology, for use in a variety of applications from tire cord to cables for anchoring oceanic oil-drilling platforms

Nonflammable polymers, including some that emit a minimum of smoke or toxic fumes

Degradable polymers, which not only help reduce the volume of unsightly plastics waste but also allow controlled release of drugs or agricultural chemicals

Polymers for a broad spectrum of medical applications, from degradable sutures to artificial organs

Conducting polymers—polymers that exhibit electrical conductivities comparable to those of metals

Polymers that serve as insoluble supports for catalysts or for automated protein or nucleic acid synthesis (Bruce Merrifield, who originated solid-phase protein synthesis, was awarded the Nobel Prize in Chemistry in 1984)

This list, by no means exhaustive, clearly illustrates that polymer chemistry is an exciting field with almost limitless possibilities.

1.2 Definitions

As already mentioned, the term *polymer* refers to large molecules—macromolecules—whose structure depends on the monomer or monomers used in their preparation. If only a few monomer units are joined together, the resulting low-molecular-weight polymer is called an *oligomer* (Greek *oligos*, "few"). The structural unit enclosed by brackets or parentheses is referred to as the *repeating unit* (or *monomeric unit*). One might reasonably argue that the first repeating unit shown previously is $+CH_2+$ rather than $+CH_2CH_2+$; however, it is more conventional to define repeating units in terms of monomer structure. The *smallest* possible repeating unit ($+CH_2+$ in this instance) is referred to as the *base unit*.

End groups are the structural units that terminate polymer chains. Where end groups are specified, they are shown outside the brackets, for example,

$$CH_3CH_2+CH_2CH_2+CH=CH_2$$

Some polymers are deliberately synthesized with reactive end groups for undergoing further reaction. Polymers containing reactive end groups are called *telechelic polymers* (from the Greek *tele*, far, and *chele*, claw). Related to telechelic polymers are the commercially important *reactive oligomers*, which are oligomers containing end groups capable of undergoing polymerization, usually by heating, to form network polymers.

Polymers formed by alkene addition reactions are called *homochain* polymers because the polymer chain, or *backbone*, as it is commonly called, consists of a single atom type— carbon—with other atoms or groups of atoms attached. *Heterochain* polymers such as polyethers or polyesters contain more than one atom type in the backbone.

The *degree of polymerization* (DP) refers to the total number of structural units, including end groups, and hence is related to both chain length and molecular weight. Consider, for example, the polymerization of vinyl acetate (an important industrial monomer) in reaction (1.5):

$$
n CH_2{=}CH \atop \underset{\underset{O}{\|}}{\overset{|}{CH_3CO}} \longrightarrow
\left[CH_2{-}CH \atop \underset{\underset{O}{\|}}{\overset{|}{CH_3CO}} \right]_{n-2}
\tag{1.5}
$$

DP in this case is equivalent to n (note that two monomer units are at the chain ends), and the molecular weight of the macromolecule is the product of DP and the molecular weight of the structural unit. For a DP of 500, for example,

$$\text{Molecular weight} = 500 \times 86 = 43,000$$

Because polymer chains within a given polymer sample are almost always of varying lengths (except for certain natural polymers like proteins), we normally refer to the *average degree of polymerization* $(\overline{DP})$.

A polymer prepared from a single monomer is called a *homopolymer*. If two or more monomers are employed, the product is a *copolymer*. In copolymers the monomeric units may be distributed randomly (*random copolymer*), in alternating fashion (*alternating copolymer*), or in blocks (*block copolymer*). A *graft copolymer* consists of one polymer branching from the backbone of the other. These various possibilities are illustrated schematically in Figure 1.1 for hypothetical monomers A and B. There are different kinds of block copolymers. Where blocks of A and B alternate in the backbone, the polymer is designated an $+AB+$ *multiblock* copolymer. If the backbone consists of a single block of each, it is an AB *diblock* copolymer. Other possibilities include ABA (*triblock*: a central B block with terminal A blocks) and ABC (*triblock*: one each of three different blocks). As will be seen later, certain monomer combinations display a tendency toward alternation during copolymerization, whereas formation of block and graft copolymers requires special techniques. Polyesters

```
 -A-A-A-A-A-A-A-A-         Homopolymer


 -A-B-B-A-B-A-A-B-         Random copolymer


 -A-B-A-B-A-B-A-B-         Alternating copolymer


 -A-A-A-A-B-B-B-B-         Block copolymer


 -A-A-A-A-A-A-A-A-         Graft copolymer
        |
      B-B-B-B-B-B-
```

FIGURE 1.1. Representations of homopolymer and copolymers.

of the type shown in equation (1.4) may also be considered copolymers, since two monomers, dibasic acid and glycol, are employed in the synthesis. More commonly, however, the term *copolymer* is reserved for polymers having more than one kind of repeating unit—for example, a *copolyester* prepared from two different dibasic acids and a glycol.

One can also describe polymers as *linear, branched,* and *network* (Figure 1.2). A linear polymer has no branching other than the pendant groups associated with the monomer. Graft copolymers, on the other hand, are examples of branched polymers. It should be stressed, however, that a branched polymer is not necessarily a graft copolymer. Low-density polyethylene (discussed later) is a common example of a branched homopolymer whereby chain branching arises as a result of side reactions during the polymerization process.

Some more unusual polymer architectures are represented in Figure 1.3. *Star polymers* contain three or more polymer chains emanating from a core structural unit. *Comb polymers*

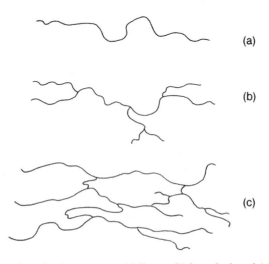

FIGURE 1.2. Representation of polymer types: (a) linear, (b) branched, and (c) network.

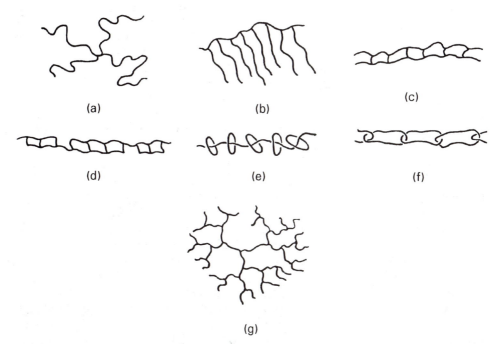

FIGURE 1.3. Representations of (a) star polymer; (b) comb polymer; (c) ladder polymer; (d) semi-ladder (or stepladder) polymer; (e) polyrotaxane; (f) polycatenane; (g) dendrimer.

contain pendant chains (which may or may not be of equal length) and are related structurally to graft copolymers. Such a polymer may be synthesized by polymerizing a long-chain vinyl monomer, or even by polymerizing an existing polymer (or oligomer) containing a polymerizable double bond at the end of the chain (reaction 1.6). Such polymeric or oligomeric monomers are termed *macromonomers* or *macromers*.

$$(1.6)$$

Ladder polymers are made up of recurring fused-ring structures, whereas *semiladder* or *step-ladder polymers* have fused-ring sections interspersed with open-chain units. Poly-rotaxanes and polycatenanes[15–17] are examples of *supramolecular assemblies*—molecules linked intermolecularly by noncovalent bonding. The former consist of polymer chains (or polymer side chains) threaded through cyclic compounds,[18] while the latter are made up of interlocking rings.[19] *Dendrimers*,[20–22] also known as dendritic, Starburst (a Michigan Molecular Institute trademark), or cascade polymers, resemble star polymers except that each leg of the star exhibits repetitive branching in the manner of a tree (Greek *dendron*, tree). Certain extremely branched polymers, termed *hyperbranched polymers*, are related to dendrimers in that they exhibit dendritic branching, but the branches do not emanate from a central core, nor is the branching necessarily regular as it is in dendrimers. Dendrimers, hyperbranched polymers, and supramolecular assemblies represent new and rapidly developing areas of polymer chemistry with potentially useful industrial applications.[23, 24]

Network polymers are formed when linear or branched polymer chains are joined together by covalent bonds, a process called *crosslinking*. Vulcanization of rubber (discussed in Chapter 9) is an example of crosslinking. Network polymers are also formed from polyfunctional monomers. If ethylene glycol in reaction (1.4) is replaced with the triol glycerol, for example, a network polyester is formed. Because of crosslinking, the polymer chains of network polymers lose their ability to flow past one another. As a result the polymer will not melt or flow and cannot, therefore, be molded. Such polymers are said to be *thermosetting* or *thermoset*. To manufacture useful articles out of thermosetting polymers, one must accomplish the crosslinking reaction in place or temporarily disrupt the crosslinking to allow the polymer to flow. Thermosetting polymers are also insoluble because the crosslinking causes a tremendous increase in molecular weight. At most, thermosetting polymers only swell in the presence of solvent, as solvent molecules penetrate the network. It is interesting to reflect that an article made with a thermosetting polymer may be considered one gigantic molecule if one can believe that all polymer chains present are linked together!

Polymers that are not crosslinked (linear or branched) can usually be dissolved in some solvent, and in most instances they will melt and flow. Such materials are said to be *thermoplastic*.

Polymers may also be categorized according to physical or mechanical properties or end use into *plastics*, *fibers*, *rubbers* (*elastomers*), *coatings*, and *adhesives*. These classifications are discussed in Section 1.8.

1.3 Polymerization Processes

Traditionally, polymers have been classified into two main groups, *addition polymers* and *condensation polymers*. This classification, first proposed by Carothers,[25] is based on whether the repeating unit of the polymer contains the same atoms as the monomer. An addition polymer has the same atoms as the monomer in its repeating unit, whereas condensation polymers contain fewer because of the formation of byproducts during the polymerization process. The corresponding polymerization processes would then be called addition polymerization and condensation polymerization. Labeling a given polymer as the addition or condensation type is complicated, however, by the fact that the polymer might well be synthesized by either addition or condensation polymerization or by ring opening. The formation of polyether from both ethylene oxide and ethylene glycol, described earlier, is one example. Other examples are shown in reactions (1.7)–(1.14):

1. Polyester from lactone (1.7), and from ω-hydroxycarboxylic acid (1.8):

$$\underset{R}{\overset{O}{\underset{\|}{\overset{\|}{C}}}} \longrightarrow \left[O-R-\overset{\overset{O}{\|}}{C} \right] \tag{1.7}$$

$$HO-R-CO_2H \longrightarrow \left[O-R-\overset{\overset{O}{\|}}{C} \right] + H_2O \tag{1.8}$$

2. Polyamide from lactam (1.9), and from ω-aminocarboxylic acid (1.10):

$$\underset{R}{\underbrace{(NH-\overset{\overset{\displaystyle O}{\|}}{C})}} \longrightarrow \left[NH-R-\overset{\overset{\displaystyle O}{\|}}{C}\right] \tag{1.9}$$

$$H_2N-R-CO_2H \longrightarrow \left[NH-R-\overset{\overset{\displaystyle O}{\|}}{C}\right] + H_2O \tag{1.10}$$

3. Polyurethane from diisocyanate and dialcohol (1.11), and from diamine and bischloroformate (1.12):

$$OCN-R-NCO + HO-R'-OH \longrightarrow$$

$$\left[\overset{\overset{\displaystyle O}{\|}}{C}NH-R-NH\overset{\overset{\displaystyle O}{\|}}{C}O-R'-O\right] \tag{1.11}$$

$$H_2N-R-NH_2 + \overset{\overset{\displaystyle O}{\|}}{Cl C}O-R'-O\overset{\overset{\displaystyle O}{\|}}{C}Cl \longrightarrow$$

$$\left[\overset{\overset{\displaystyle O}{\|}}{C}NH-R-NH\overset{\overset{\displaystyle O}{\|}}{C}O-R'-O\right] + 2HCl \tag{1.12}$$

4. Hydrocarbon polymer from ethylene (1.13), and from α,ω-dibromide (1.14):

$$CH_2{=}CH_2 \xrightarrow{\text{initiator}} \left[CH_2CH_2\right] \tag{1.13}$$

$$BrCH_2(CH_2)_8CH_2Br \xrightarrow{2Na} \left[CH_2CH_2\right]_5 + 2NaBr \tag{1.14}$$

To avoid ambiguity, polymerization reactions are now more commonly characterized according to the polymerization mechanism. If the polymer chains are built up in a stepwise fashion by the random union of monomer molecules to form dimers, trimers, and higher species throughout the monomer matrix, the process is called *step-reaction* or *step-growth* polymerization. If, on the other hand, the molecular weight increases by the successive linking of monomer molecules to the end of a growing chain, the process is called *chain-reaction* or *chain-growth* polymerization. Ring-opening polymerization reactions may be either step-reaction or chain-reaction processes.

Before we consider the mechanistic implications of step-reaction and chain-reaction polymerization, it should be noted that polymers having identical repeating units but formed by entirely different reactions do not necessarily have identical properties. On the contrary, physical and mechanical properties may differ markedly because different polymerization processes may give rise to differences in molecular weight, end groups, stereochemistry, or possibly chain branching. (How structural variations influence properties is considered in later chapters.) Spectroscopic properties and elemental analyses will, however, be the same, apart from small differences arising from different end groups, such differences becoming less significant as the molecular weight increases.

1.4 Step-reaction Polymerization

Consider hypothetical functional groups A and B that react together to form group X. There are two approaches to preparing linear step-reaction polymers, one having both reactive functional groups in one molecule:

$$A—R—B \rightarrow +R—X+$$

and the other having two difunctional monomers:

$$A—R—A + B—R'—B \rightarrow +R—X—R'—X+$$

Reactions (1.8) and (1.10) illustrate the former, and (1.11) and (1.12) the latter.

Polyesterification, whether between diol and dibasic acid (1.4) or intermolecularly between hydroxy acid molecules (1.3 or 1.8), is an example of a step-reaction process. The esterification reaction occurs anywhere in the monomer matrix where two monomer molecules possessing the proper orientation and energy collide. Once the ester has formed (with loss of a water molecule), it, too, can react further by virtue of its still-reactive hydroxyl and carboxyl groups. The net effect is that monomer molecules are consumed rapidly without any concomitant large increase in molecular weight. Figure 1.4 schematically illustrates this phenomenon for a system of 12 monomer molecules containing functional groups A and B (e.g., a hydroxy acid). When half the monomer molecules are consumed in the manner shown (Figure 1.4b), the degree of polymerization (the number of molecules initially divided by the number remaining) is 1.3. Figure 1.4c shows 75% of the monomer molecules consumed with $\overline{DP}$ still only 1.7. With all the monomer molecules reacted (Figure 1.4d), $\overline{DP}$ is still low. But each polymer (or oligomer) molecule still has reactive end groups; hence the polymerization reaction will continue in stepwise fashion with each step being mechanistically the same. The polymerization rate will decrease as the functional groups are consumed. Thus the molecular weight of the polymer increases slowly even at high levels of monomer conversion, and it will continue to increase until the concentration of reactive end groups becomes too low for the reaction to proceed at a significant rate. Other factors, such as increasing vis-

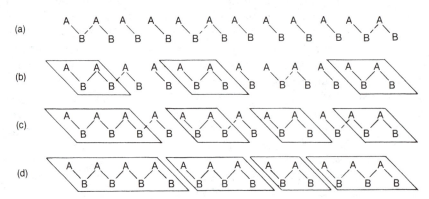

FIGURE 1.4. Step-reaction polymerization: (a) unreacted monomer; (b) 50% reacted, $\overline{DP} = 1.3$; (c) 75% reacted, $\overline{DP} = 1.7$; (d) 100% reacted, $\overline{DP} = 3$. (Broken lines represent reacting species.)

cosity of the reaction medium, which reduces the mobility of the reactive chain ends and increases the difficulty of removing byproducts, may also contribute to a rate reduction.

Carothers developed a simple equation for relating $\overline{DP}$ to percent conversion of monomer. If one assumes that there are N_0 molecules initially and N molecules (total) after a given reaction period, then the amount reacted is $N_0 - N$. The reaction conversion, p, is then given by the expression

$$p = \frac{N_0 - N}{N_0}$$

or

$$N = N_0(1 - p)$$

The average number of repeating units of all molecules present, $\overline{DP}$, is equal to N_0/N. Substituting this in the above expression reduces the latter to

$$\overline{DP} = \frac{1}{1 - p}$$

This simple equation demonstrates one fundamental aspect of step-reaction polymerization—that very high conversions are necessary to achieve practical molecular weights. At 98% conversion, for example,

$$\overline{DP} = \frac{1}{1 - 0.98}$$

and $\overline{DP}$ is still only 50. Thus very high $\overline{DP}$s are generally not obtainable for linear polymers by step polymerization.

It can also be shown, by using similar types of relationships, that in the A—R—A + B—R'—B type of polymerization, an exact stoichiometric balance and very pure monomers are necessary to achieve high molecular weights. If some monofunctional impurity is present, its reaction will limit the molecular weight by rendering a chain end inactive. Similarly, high-purity monomers are necessary in the A—R—B type of polymerization, and it follows that high-yield reactions are the only practical ones for polymer formation because side reactions will upset the stoichiometric balance. These topics will be considered in more detail in Chapter 10.

1.5 Chain-reaction Polymerization

Chain-reaction polymerization involves two distinct kinetic steps, *initiation* and *propagation*. The first requires an *initiator* to start the reaction. The initiator might, for example, be a free radical ($R\cdot$) to initiate polymerization of a monomer such as ethylene:

Initiation
$$R\cdot + CH_2 = CH_2 \rightarrow RCH_2CH_2\cdot$$

Propagation
$$RCH_2CH_2\cdot + CH_2 = CH_2 \rightarrow RCH_2CH_2CH_2CH_2\cdot$$

Alternatively, it might be an anion (B:$^-$) to initiate ring-opening polymerization of a cyclic monomer such as ethylene oxide:

Initiation

$$B:^- + CH_2{-}CH_2 \longrightarrow BCH_2CH_2O^-$$

Propagation

$$BCH_2CH_2O^- + CH_2{-}CH_2 \longrightarrow BCH_2CH_2OCH_2CH_2O^-$$

Other types of initiators (that are discussed in later chapters) include cationic and complex coordination compounds. (Note that it is incorrect to refer to initiating species as catalysts if they are consumed in the reaction.) In chain-reaction polymerization, the reaction propagates at a reactive chain end and continues until a *termination* reaction renders the chain end inactive (e.g., combination of radicals), or until monomer is completely consumed. The probability of termination reactions occurring increases as monomer concentration decreases. Because polymerization occurs at the chain end, the chain length increases rapidly even though relatively large amounts of monomer remain unreacted. This is shown schematically for a vinyl monomer in Figure 1.5. Kinetics and mechanisms of chain-reaction polymerization are discussed in Chapter 6.

Figures 1.3 and 1.4 illustrate one of the fundamental differences between step-reaction and chain-reaction polymerization. In the former, molecular weight increases slowly while monomer is consumed rapidly; in the latter, high molecular weights are reached rapidly at low monomer conversions. Furthermore, in chain-reaction processes, initiation, propagation, and termination are all different reactions occurring at different rates, and, once termination has occurred, the polymerization cannot proceed further. If, on the other hand, monomer is completely consumed but *no* termination has taken place, polymerization will continue if more monomer is added. By contrast, in step-reaction polymerization, initiation and propagation reactions are for all practical purposes the same, and once the polymeriza-

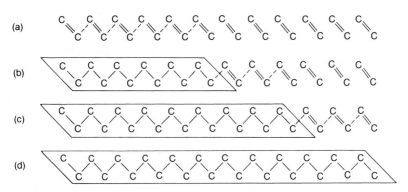

FIGURE 1.5. Chain-reaction polymerization: (a) unreacted monomer; (b) 50% reacted, $\overline{DP} = 1.7$; (c) 75% reacted, $\overline{DP} = 3$; (d) 100% reacted, $\overline{DP} = 12$. (Broken lines represent reacting species.)

TABLE 1.1. Comparison of Step-Reaction and Chain-Reaction Polymerization

Step Reaction	*Chain Reaction*
Growth occurs throughout matrix by reaction between monomers, oligomers, and polymers	Growth occurs by successive addition of monomer units to limited number of growing chains
DP[a] low to moderate	DP can be very high
Monomer consumed rapidly while molecular weight increases slowly	Monomer consumed relatively slowly, but molecular weight increases rapidly
No initiator needed; same reaction mechanism throughout	Initiation and propagation mechanisms different
No termination step; end groups still reactive	Usually chain-terminating step involved
Polymerization rate decreases steadily as functional groups consumed	Polymerization rate increases initially as initiator units generated; remains relatively constant until monomer depleted

[a]DP, average degree of polymerization.

tion reaction ends, the polymer still has reactive end groups. Major differences between the two polymerization processes are summarized in Table 1.1.

1.6 Step-reaction Addition and Chain-reaction Condensation

Most step-reaction polymerizations are condensation processes, and most chain-reaction polymerizations are addition processes, but there are exceptions. Reaction (1.11), for example, illustrates a step-growth process that is also an addition reaction. Another example, a polymer formed by the Diels–Alder reaction between a bisdiene, 1,6-bis(cyclopentadienyl)hexane, and benzoquinone, is shown in reaction (1.15). The boron trifluoride-catalyzed polymerization of diazomethane (1.16), on the other hand, is an example of a chain-growth condensation reaction.

$$\text{(structure)} \quad (1.15)$$

$$CH_2N_2 \xrightarrow{BF_3} -[CH_2]- + N_2 \quad (1.16)$$

In 1994 the International Union of Pure and Applied Chemistry (IUPAC) proposed four new polymerization classifications:[26]

> *Polycondensation*—the same as step-reaction polymerization with concurrent formation of low-molecular-weight byproducts (e.g., reaction 1.12 or 1.14).

Polyaddition—the same as step-reaction polymerization without formation of byproducts (e.g., 1.11 or 1.15).

Chain polymerization—the same as chain-reaction polymerization without formation of byproducts (e.g., 1.13).

Condensative chain polymerization—the same as chain-reaction polymerization with the formation of low-molar-mass byproducts (e.g., 1.16).

Because the IUPAC classifications have yet to be widely accepted, we shall continue to use the more familiar step-reaction (or step-growth) and chain-reaction (or chain-growth) terminology.

1.7 Nomenclature

Polymer nomenclature is complicated for a variety of reasons. First, most polymer names are based on names of the corresponding monomers, and although this *source-based* system is widely accepted, one frequently encounters variations in format. Second, although the Macromolecular Nomenclature Commission of the IUPAC has grappled valiantly with the complexities of polymer structure and has proposed perfectly logical rules,[27] the IUPAC system is not widely used except in reference works. Third, some polymer structures are so complex, particularly where branching or crosslinking is involved, that naming them defies all but the most persistent of nomenclature purists. And finally, polymer science has some fairly sharply defined boundaries; rubber chemists, for example, employ terminology that might be unintelligible to a plastics or textiles chemist. All this takes on added significance in this age of computer-based information storage and retrieval, where a standardized set of abbreviations is a necessity.[28,29] Given the magnitude of the problem, therefore, we will attempt in this section to clarify the most generally accepted terminology.

As should be apparent from the previous sections, polymer types, or families, are named according to the functional group present in the repeating unit, with the prefix *poly*: polyesters, polyamides, polyethers, and so on. Polyamides are unusual in that they are also called *nylons*, a term that originated as a trade name but then evolved over the years into general use. If more than one functional group is present, the polymer is named accordingly, for example, polyetherimide.

Where polymer structure is complex and not easily definable, the family is usually named for the monomers employed, as with phenol–formaldehyde polymers. Part III of this book discusses a variety of polymers according to the family classification. Polymers derived from monomers containing carbon–carbon double bonds are often referred to as *vinyl polymers*, although it would be more correct to reserve that term for monomers containing the vinyl group, CH_2=CH—, such as vinyl chloride, CH_2=$CHCl$, or acrylonitrile, CH_2=$CHCN$. The term *vinylidene* is used with 1,1-disubstituted ethylenes, for example, vinylidene chloride, CH_2=CCl_2. Polymers derived from simple alkenes such as ethylene or propylene are often called *polyolefins*. For the sake of convenience, *we shall refer to all polymers prepared by chain-reaction polymerization of alkenyl monomers as vinyl polymers*.

1.7.1 Vinyl Polymers

The most widely accepted method of naming vinyl polymers is to place the prefix *poly* before the name of the corresponding monomer; for example,

$-[CH_2CH_2]-$ Polyethylene

$-[CF_2CF_2]-$ Polytetrafluoroethylene

$-[CH_2CH]-$ Polystyrene

(Note that common names of the monomers are used, not ethene or phenylethene.) By this system, the polymer $-[CH_2]-$ prepared according to reaction (1.16) is called polymethylene to distinguish it from the commercial polymer prepared from ethylene. (In the British literature, polyethylene is frequently called *polythene*.) Occasionally a trade name may be employed in casual usage, for example, Teflon (du Pont) for polytetrafluoroethylene, but this occurs only where such names are widely recognized.

If the monomer name consists of more than one word, or if the name is preceded by a letter or number, the monomer name is enclosed in parentheses and prefixed with *poly*:

$-[CH_2CH]-$
$\quad CO_2H$
Poly(acrylic acid)

$\quad CH_3$
$-[CH_2C]-$
Poly(α-methylstyrene)

$-[CH_2CH]-$
$\quad CH_2CH_2CH_3$
Poly(1-pentene)

Unfortunately, one still sees parentheses omitted—polyacrylic acid, for example—especially in trade literature. This nomenclature is no longer considered acceptable and should be avoided.

The IUPAC recommends that names be derived from the structure of the base unit, or *constitutional repeating unit* (CRU) as the IUPAC prefers to call it, according to the following steps:

1. The smallest structural unit (CRU) is identified.
2. Substituent groups on the backbone are assigned the lowest possible numbers.
3. The name is placed in parentheses (or brackets and parentheses, where necessary), and prefixed with *poly*.

The preceding polymers would thus be named as follows:

Source name	IUPAC name
Polyethylene	Poly(methylene)
Polytetrafluoroethylene	Poly(difluoromethylene)
Polystyrene	Poly(1-phenylethylene)
Poly(acrylic acid)	Poly(1-carboxylatoethylene)
Poly(α-methylstyrene)	Poly(1-methyl-1-phenylethylene)
Poly(1-pentene)	Poly[1-(1-propyl)ethylene]

In the IUPAC system, parentheses are *always* used following the prefix; also ethylene (rather than ethene) is the acceptable structural name. Furthermore, the IUPAC system makes no distinction between polyethylene and polymethylene.

Diene monomers such as 1,3-butadiene (CH_2=CH—CH=CH_2) create special difficulties because they can undergo both 1,2- and 1,4-addition. Thus poly(1,3-butadiene) may have two possible repeating units:

$$\left[CH_2CH\underset{\displaystyle CH=CH_2}{}\right] \qquad -[CH_2CH=CHCH_2]-$$

1,2-addition   1,4-addition

with the latter capable of exhibiting *cis–trans* isomerism. To name the polymer by IUPAC rules, the CRU of the 1,4-addition product is best written in the less familiar format:

$$-[CH=CHCH_2CH_2]-$$

Assuming each butadiene polymer were to consist of a single repeating unit, the names would be as follows:

	Source name	IUPAC name
1,2-addition	1,2-Poly(1,3-butadiene)	Poly(1-vinylethylene)
1,4-addition	1,4-Poly(1,3-butadiene)	Poly(1-butene-1,4-diyl)

The prefixes *cis* and *trans* are placed in front of the name, as appropriate, for the 1,4 polymer. Until recently the IUPAC had recommended the name poly(1-butenylene) for the 1,4 polymer, but this has been changed to identify (by "1,4-diyl") the bonds linking the repeating units. The source-based names are commonly simplified to 1,2-polybutadiene and 1,4-polybutadiene.

In polydiene synthesis, both repeating units and stereochemistries are invariably present in the final product, with one predominating depending on the reaction conditions. This is explored more fully in Part II. Other aspects of polymer stereochemistry are discussed in Chapter 3. Some additional examples of vinyl polymer nomenclature are given in Table 1.2

TABLE 1.2. Representative Nomenclature of Vinyl Polymers

Monomer Structure	Monomer Name	Polymer Repeating Unit	Source Name	IUPAC[a] Name
$CH_3CH{=}CH_2$	Propylene	$\left[CH_2CH\text{—}\right]$ CH_3	Polypropylene	Poly(propylene)
$CH_2{=}CHCl$	Vinyl chloride	$\left[CH_2CH\text{—}\right]$ Cl	Poly(vinyl chloride)	Poly(1-chloro ethylene)
$CH_2{=}C(CH_3)_2$	Isobutylene	CH_3 $\left[\text{—}CH_2C\text{—}\right]$ CH_3	Polyisobutylene	Poly(1,1-dimethyl-ethylene)
$CH_2{=}CHCN$	Acrylonitrile	$\left[CH_2CH\text{—}\right]$ CN	Polyacrylonitrile	Poly(1-cyano ethylene)
$CH_2{=}CCO_2CH_3$ (CH_3)	Methyl methacry-late	CH_3 $\left[\text{—}CH_2C\text{—}\right]$ CO_2CH_3	Poly(methyl methacrylate)	Poly[1-(methoxy-carbonyl)-1-meth-ylethylene]
$CH_2{=}CHOCCH_3$ (O)	Vinyl acetate	$\left[CH_2CH\text{—}\right]$ $OCCH_3$ O	Poly(vinyl acetate)	Poly(1-acetoxyeth-ylene)
2-pyridyl-$CH{=}CH_2$	2-Vinyl-pyridine	$\left[CH_2CH\text{—}\right]$ (pyridyl)	Poly(2-vinyl-pyridine)	Poly[1-(2-pyridinyl)-ethylene]
CH_3 $CH_2{=}C\text{—}CH{=}CH_2$	Isoprene	$\left[CH_2\quad CH_2\right]$ C=C $CH_3\quad H$	*cis*-1,4-Poly-isoprene	*cis*-Poly(1-methyl-1-butene-1,4-diyl)
Cl $CH_2{=}C\text{—}CH{=}CH_2$	Chloroprene	$\left[CH_2CH\text{—}\right]$ $ClC{=}CH_2$	3,4-Polychloro-prene	Poly[1-(1-chloro-vinyl)ethylene]

[a]IUPAC, International Union of Pure and Applied Chemistry.

Having two such nomenclature systems may be confusing at first. The source-based system is still the one of choice among most polymer chemists, notwithstanding the fact that such important reference works as *Chemical Abstracts* and *Polymer Handbook* have adopted the IUPAC system. The IUPAC is not dogmatic on the point; it recognizes the use of source-based names where no ambiguity arises, but it *advocates* a more systematic approach.

1.7.2 Vinyl Copolymers

By definition, copolymers are polymers derived from more than one species of monomer. (The term *terpolymer* is sometimes used to denote a copolymer prepared from three different monomers.) For example, a copolymer prepared from the monomers styrene and methyl methacrylate quite logically is called styrene–methyl methacrylate copolymer, using source-based terminology. The IUPAC recognizes that defining a particular constitutional repeating unit for copolymers is frequently impossible; even when it is possible, as with alternating, block, and graft copolymers, structure-based nomenclature can be very unwieldy. The IUPAC, therefore, recommends source-based names for copolymers.[30] In the IUPAC system the polymer example just given is called by either of the two names

Poly[styrene-*co*-(methyl methacrylate)]

Copoly(styrene/methyl methacrylate)

The first is more systematic; the second is more concise. For very complex copolymers, the more systematic approach is recommended. In both names *co* implies that no distribution of the monomeric repeating units is specified. In place of *co*, *ran* is used if the distribution is known to be completely random.*

Alternating, block, and graft copolymers of the same monomers are named as follows:

Systematic

Poly[styrene-*alt*-(methyl methacrylate)]

Polystyrene-*block*-poly(methyl methacrylate)

Polystyrene-*graft*-poly(methyl methacrylate)

Concise

Alt-copoly(styrene/methyl methacrylate)

Block-copoly(styrene/methyl methacrylate)

Graft-copoly(styrene/methyl methacrylate)

With graft copolymers, the order of monomer names is important. As written above, the name specifies that poly(methyl methacrylate) is grafted onto a polystyrene backbone. If the reverse were the case, the monomer names would, of course, be reversed. Some additional examples of copolymer names, systematic followed by concise, are the following:

Poly[styrene-*co*-butadiene-*co*-(vinyl acetate)]

Copoly(styrene/butadiene/vinyl acetate)

(a copolymer of the three named monomers with no specified distribution of repeating units)

Polystyrene-*block*-polyisoprene-*block*-polystyrene

Block-copoly(styrene/isoprene/styrene)

*If the monomer sequence is known to obey some specified statistical distribution, for example, Bernoulli, the term *stat* is recommended instead of *co* or *ran*.

(an ABA block copolymer with polystyrene terminal blocks)

Poly[styrene-*alt*-(maleic anhydride)-*block*-poly(vinyl chloride)]

Block-copoly[*alt*-co(styrene/maleic anhydride/vinyl chloride)]

[a block copolymer, one block made up of an alternating copolymer of styrene and maleic anhydride, and the other of poly(vinyl chloride)]

Polybutadiene-*block*-(polystyrene-*graft*-polyacrylonitrile)

Block-copoly[butadiene/*graft*-co(styrene/acrylonitrile)]

(a block copolymer of styrene and butadiene with polyacrylonitrile grafted to the polystyrene block)

The last two copolymers could be correctly named with the blocks in reverse order.

1.7.3 Nonvinyl Polymers

Nomenclature of nonvinyl polymers is much more complicated because of the great variety of possible repeating units. Three of the more important commercial polymer types—polyethers, polyesters, and polyamides—are included in Table 1.3 under the appropriate subheadings.

Polyethers are named according to monomer used (source) or structure (IUPAC). In the latter case, seniority is given to the oxygen atom, and each subunit is named in sequence.

Like polyethers, polyesters can be formed by ring-opening as well as by step-reaction polymerization. Ring opening of cyclic esters (lactones) yields products that are named according to the source or the ester repeating unit in the common nomenclature, as shown for β-propiolactone in Table 1.3. The IUPAC name is based on two subunits of the CRU, with the carbonyl oxygen named as an *oxo* substituent:

$$
\underset{\text{oxy}}{-\text{O}-}\underset{\text{1-oxopropane-1,3-diyl}}{\overset{\overset{\text{O}}{\|}}{\text{C}}\text{CH}_2\text{CH}_2-}
$$

Note that the oxo substituent in the subunit requires that the subunit be enclosed with parentheses; also that for a carbon chain longer than ethylene, the parent alkane (propane) has the location of the bonds between repeating units specified (1,3-diyl).[31] The older IUPAC name for this polyester was poly[oxy(1-oxotrimethylene)], a format still found in reference books and journals. The polyester formed from 10-hydroxydecanoic acid is named in analogous fashion, with the older name being poly[oxy(1-oxodecamethylene)]. Occasionally one may see the name poly(10-hydroxydecanoic acid), but this is less common.

For the polyester prepared from ethylene glycol and terephthalic acid, the IUPAC recognizes the larger terephthaloyl structural unit:

$$
\underset{\text{oxy}}{-\text{O}-}\underset{\text{ethylene}}{\text{CH}_2\text{CH}_2-}\underset{\text{oxy}}{\text{O}-}\underset{\text{terephthaloyl}}{\overset{\overset{\text{O}}{\|}}{\text{C}}-\bigcirc-\overset{\overset{\text{O}}{\|}}{\text{C}}-}
$$

TABLE 1.3. Representative Nomenclature of Nonvinyl Polymers

Monomer Structure	Monomer Name	Polymer Repeating Unit	Source or Common Name	IUPAC Name
Polyethers				
CH_2—CH_2 (O)	Ethylene oxide	$+CH_2CH_2O+$	Poly(ethylene oxide)	Poly(oxyethylene)
$HOCH_2CH_2OH$	Ethylene glycol	$+CH_2CH_2O+$	Poly(ethylene glycol)	Poly(oxyethylene)
CH_2O	Formaldehyde	$+CH_2O+$	Polyformaldehyde	Poly(oxymethylene)
CH_3CHO	Acetaldehyde	$+CHO+$, CH_3	Polyacetaldehyde	Poly(oxyethylidene)
Polyesters				
(β-propiolactone structure)	β-Propiolactone	$[OCH_2CH_2C{=}O]$	Poly(β-propiolactone) or poly(3-propionate)	Poly[oxy(1-oxopropane-1,3-diyl)]
$HO(CH_2)_9CO_2H$	10-Hydroxydecanoic acid	$[O(CH_2)_9C{=}O]$	Poly(10-decanoate)	Poly[oxy(1-oxodecane-1,10-diyl)]
$HOCH_2CH_2OH$ + (terephthalic acid structure with HO_2C — CO_2H)	Ethylene glycol + Terephthalic acid	$[OCH_2CH_2OC({=}O){-}\bigcirc{-}C{=}O]$	Poly(ethylene terephthalate)	Poly(oxyethylene-oxyterephthaloyl)

Polyamides

Monomer	Polymer structure	Common name	Systematic name
Caprolactam	$[NH(CH_2)_5C{=}O]$	Polycaprolactam or nylon 6	Poly[imino(1-oxohexane-1,6-diyl)]
11-Aminoundecanoic acid, $H_2N(CH_2)_{10}CO_2H$	$[NH(CH_2)_{10}C{=}O]$	Poly(11-undecanoamide) or nylon 11	Poly[imino(1-oxoundecane-1,11-diyl)]
Hexamethylenediamine $H_2N(CH_2)_6NH_2$ + Sebacic acid $HO_2C(CH_2)_8CO_2H$	$[NH(CH_2)_6NHC(CH_2)_8C]$ (diketo)	Poly(hexamethylene sebacamide) or nylon 610	Poly(iminohexane-1,6-diyliminosebacoyl)
Terephthaloyl chloride (ClOC–C₆H₄–COCl) + m-Phenylenediamine	Poly(m-phenylene-terephthalamide) structure	Poly(m-phenylene-terephthalamide)	Poly(imino-1,3-phenylene-iminoterephthaloyl)
Tetramethylene diamine $H_2N(CH_2)_4NH_2$ + m-Benzenedisulfonyl chloride (ClO₂S–C₆H₄–SO₂Cl)	$[NH(CH_2)_4NHO_2S{-}C_6H_4{-}SO_2]$	Poly(tetramethylene-m-benzenesulfonamide)	Poly(sulfonyl-1,3-phenylenesulfonylimino-butane-1,4-diylimino)

Where no recognized larger unit is present, the carbonyl group is named separately, as with the commercially important polycarbonate

which bears the IUPAC name poly(oxycarbonyloxy-1,4-phenyleneisopropylidene-1,4-phenylene). More commonly this polymer is called bisphenol A polycarbonate after the trivial name for the bisphenol monomer used in its synthesis.

Polyamides are named like polyesters with two important distinctions. In the IUPAC system, the term *imino* denotes the —NH— group; and in the common nomenclature, the term *nylon* is applied to aliphatic polyamides (although the generic term *nylon* is often used to denote all polyamides). The first three polyamides in Table 1.3 are illustrative. For polyamides prepared by ring-opening of cyclic amides (lactams) or by step polymerization of amino acids, the number following the nylon name indicates the number of carbon atoms in the repeating unit. For those prepared from diamines and dicarboxylic acids, the number indicates the number of carbons in the diamine and diacid moieties, respectively. Thus nylon 610 (or 6-10) is prepared from 1,6-diaminohexane and 1,10-decanedioic acid (sebacic acid). Where more than one heteroatom appears in the backbone, the CRU is named with the heteroatoms in descending order of seniority: O, S, N, P; thus sulfonyl is named before imino.

1.7.4 Nonvinyl Copolymers

The IUPAC also recommends source-based nomenclature for nonvinyl copolymers. A copolyester formed, for example, from a 2:1:1-molar ratio of the monomers ethylene glycol, terephthalic acid, and isophthalic acid could be named most simply,

Poly(ethylene terephthalate-*co*-ethylene isophthalate)

A copolyamide prepared from a mixture of 6-aminohexanoic acid and 11-aminoundecanoic acid would be called by either of the following names:

Poly[(6-aminohexanoic acid)-*co*-(11-aminoundecanoic acid)]

Poly[(6-aminohexanoamide)-*co*-(11-aminoundecanoamide)]

1.7.5 End Groups

End groups are not normally named unless their structures are known with certainty and there is a compelling reason for naming them. In such cases the alpha and omega symbols are used to designate the groups, for example:

H+OCH₂CH₂+OH

α-Hydro-ω-hydroxypoly(oxyethylene)

1.7.6 Abbreviations

Not surprisingly, given the complexities of polymer structure, polymer scientists are very fond of using abbreviations [PVC for poly(vinyl chloride), HDPE and LDPE for high- and low-density polyethylene, PET for poly(ethylene terephthalate), etc.]. Not only are abbreviations used in journal articles as well as trade literature, but efforts are underway to standardize them. Commonly used abbreviations will be noted where appropriate and are listed in Appendix A.

1.8 Industrial Polymers

One of the greatest challenges confronting students of chemistry is to relate the abstracts of chemical formulas and physical principles to the applied reality of the chemical industry and consumer products. No other branch of chemistry allows this chasm to be bridged more tangibly than that of polymer chemistry. In no other area of chemistry can one relate chemical structure to physical or mechanical properties so readily. This will be a continuing theme throughout the book and will be addressed in some detail in Chapters 3 and 4. In preparation for this most intellectually satisfying aspect of polymer chemistry, let us take a brief look at the polymer industry—which of the myriad of known polymers have achieved commercial success, how they are used, and in which direction the industry is moving. This overview of what is the single biggest division of the chemical industry is intended to serve as an introduction to later chapters, where some of the polymers described here are discussed more fully. An important point for the student to keep in mind is that chemical structures encountered in this book frequently bear direct relation to articles of commerce and are by no means just laboratory curiosities.

As mentioned earlier, there are five major classifications of the polymer industry: plastics, fibers, rubber (elastomers), adhesives and coatings. The differentiation (and resultant end use) of the first three is based to a large degree on a particular mechanical property called *modulus*, which in more common terms means stiffness. Fibers have the highest modulus, rubbers the lowest. Included in all three divisions are synthetic and natural products. The latter include cellulose-based plastics, cellulose (cotton, sisal, etc.) and protein (wool, silk) fibers, and natural rubber. Adhesives and coatings are delineated by end use.

A given polymer may fit into more than one category. Styrene–butadiene copolymer, for example, serves as both a rubber and—in the form of an aqueous emulsion or *latex*—as a paint; and the polyester poly(ethylene terephthalate) is widely used as both plastic and fiber.

World consumption of synthetic polymers is now in the neighborhood of 150 million metric tons per year,[32] approximately 56% of which are plastics, 18% fibers, and 11% synthetic rubber. The balance is made up of coatings and adhesives. Organizations such as the Society of the Plastics Industry (SPI), the Textile Economics Bureau, and the Rubber Manufacturers Association regularly provide detailed statistics on polymer production. Summaries are published frequently in industry-oriented chemistry and engineering journals.

1.8.1 Plastics

In the 1980s the total volume of plastics consumed worldwide overtook that of iron and steel. There are sound economic reasons for the increasing use of plastics. They weigh less and are generally more corrosion resistant than metals. Like metals they are blended (alloyed)

TABLE 1.4. Commodity Plastics

Type	Abbreviation	Major Uses
Low-density polyethylene	LDPE	Packaging film, wire and cable insulation, toys, flexible bottles, housewares, coatings
High-density polyethylene	HDPE	Bottles, drums, pipe, conduit, sheet, film, wire and cable insulation
Polypropylene	PP	Automobile and appliance parts, furniture, cordage, webbing, carpeting, film packaging
Poly(vinyl chloride)	PVC	Construction, rigid pipe, flooring, wire and cable insulation, film and sheet
Polystyrene	PS	Packaging (foam and film), foam insulation, appliances, housewares, toys

to improve their properties. And of particular significance with increasing costs of energy, they can be manufactured and processed with lower energy input than either metals or glass.

Plastics are divided into two major categories on the basis of economic considerations and end use: *commodity* and *engineering*.[33–36] Commodity plastics are characterized by high volume and low cost; they may be compared with steel and aluminum in the metals industry. They are used frequently in the form of disposable items such as packaging film, but also find application in durable goods. Engineering plastics are higher in cost and much lower in volume, but have superior mechanical properties and greater durability. They compete with metals, ceramics, and glass in a variety of applications.

Commodity plastics (Table 1.4) consist principally of four major thermoplastic polymers: polyethylene, polypropylene, poly(vinyl chloride), and polystyrene. Polyethylene is divided into low-density (<0.94 g/cm^3) and high-density (>0.94 g/cm^3) product. The difference in densities arises from structure; high-density polyethylene (HDPE) is essentially a linear polymer, and low-density polyethylene (LDPE) is branched. A less common classification is medium-density polyethylene (MDPE) for the density range 0.926–0.940 g/cm^3. Commodity plastics represent about 90% of all thermoplastics production, with the remainder divided among styrene–butadiene and acrylonitrile–butadiene–styrene (ABS) copolymers, polyamides, and polyesters.

Principal engineering plastics[37] are listed alphabetically in Table 1.5 along with the chapters in which their chemistry is discussed. Polyamide, polycarbonate, acetal, poly(phenylene oxide), and polyester represent about 99% of this market. Not listed are engineering grades of acrylonitrile–butadiene–styrene copolymers, a variety of fluorinated polymers, and an increasing number of copolymers and polymer blends.

There is considerable overlap in markets for engineering plastics, but primarily these plastics find use in transportation (automobiles, trucks, aircraft), construction (housing, plumbing, hardware), electrical and electronic goods (business machines, computers), industrial machinery, and consumer goods. Besides the ones listed, an increasing number of copolymers and polymer blends are especially tailored to improve properties. Engineering plastics are a rapidly growing market, with consumption projected to increase by up to 10% per year in the foreseeable future.

TABLE 1.5. Principal Engineering Plastics

Type	Abbreviation	Chapter Where Discussed
Acetal[a]	POM	11
Polyamide[b]	—	13
Polyamideimide	PAI	13
Polyarylate	—	12
Polybenzimidazole	PBI	17
Polycarbonate	PC	12
Polyester[c]	—	12
Polyetheretherketone	PEEK	11
Polyetherimide	PEI	11
Polyimide	PI	13
Poly(phenylene oxide)	PPO	11
Poly(phenylene sulfide)	PPS	11
Polysulfone[d]	—	11

[a]Common name for polyformaldehyde. Abbreviation refers to poly(oxymethylene).

[b]Principally nylons 6 and 66.

[c]Principally poly(ethylene terephthalate) (PET) and poly(butylene terephthalate) (PBT).

[d]Several types marketed.

Almost all the plastics described thus far are thermoplastic. Table 1.6 lists principal thermosetting plastics in decreasing order of consumption. Of these, only certain grades of epoxy qualify as engineering plastics. Phenol– and urea–formaldehyde polymers and unsaturated polyesters represent about 90% of production. The ratio of thermoplastic to thermosetting plastics production is roughly 6:1.

1.8.2 Fibers

Fibers[38] are characterized by having high strength and modulus, good elongation (stretchability), good thermal stability (enough to withstand ironing, for example), spinnability (the ability to be converted to filaments), and a host of other properties depending on whether they are to be used in textiles, tire cord, rope and cable, and so on. A partial list of properties might include dyeability, chemical resistance, insect and mildew resistance, crease resistance, and luster.

There are two principal natural fibers: cotton and wool, the former the polysaccharide cellulose, and the latter a protein. Silk, another protein fiber, is produced in much smaller quantities. World production of cotton is roughly five times that of wool on a weight basis. Natural fibers are covered in greater detail in Chapter 18. Following the pioneering work of Carothers at the du Pont Corporation, and with the increasing availability of low-cost raw materials, the synthetic fibers industry developed rapidly as demand for textiles outstripped production of cotton and wool. At the present time, world production of all fibers exceeds 30 million metric tons a year, with approximately 50% being synthetic. Production of synthetic fibers overtook that of cotton in the late 1970s.

TABLE 1.6. Principal Thermosetting Plastics

Type	Abbreviation	Typical Uses	Chapter Where Discussed
Phenol–formaldehyde	PF	Electrical and electronic equipment, automobile parts, utensil handles, plywood adhesives, particle board binder	14
Urea–formaldehyde	UF	Similar to PF polymers; also treatment of textiles, coatings	14
Unsaturated polyester	UP	Construction, automobile parts, boat hulls, marine accessories, corrosion-resistant ducting, pipe, tanks, etc., business equipment	12
Epoxy	—	Protective coatings, adhesives, electrical and electronics applications, industrial flooring, highway paving materials, composites	11
Melamine–formaldehyde	MF	Similar to UF polymers; decorative panels, counter and table tops, dinnerware	14

Synthetic fibers are classified as *cellulosic* and *noncellulosic*. The principal types are described in Table 1.7. Besides those listed, some high-performance specialty fibers are produced, but these represent a very small percentage of total fibers production. Polyester and nylon account for about 70% of the total.

1.8.3   Rubber (Elastomers)

Some scientists associate the word *rubber* with the naturally occurring polymer and *elastomer* with the synthetic, but more often than not the two terms are used interchangeably. They refer to polymers that exhibit resilience, or the ability to stretch and retract rapidly.[39] In most commercial applications rubbers are converted by crosslinking to a network structure. The principal form of natural rubber is structurally almost identical to *cis*-1,4-polyisoprene (Table 1.2). More information on natural rubber is provided in Chapter 18. Synthetic rubber, which surpassed natural rubber in production in the 1950s, comes in a variety of forms, the principal ones being given in Table 1.8. Styrene–butadiene, polybutadiene, and ethylene–propylene (EPDM) account for about 70% of the total synthetic rubber consumption.

In recent years non-network elastomers, called *thermoplastic elastomers*, have been developed. These products, which make use of block copolymer technology, are making inroads into markets dominated by natural rubber and the principal synthetic rubbers, as are a number of other specialty rubbers. The chemistry of thermoplastic elastomers is covered in Chapter 3.

TABLE I.7. Principal Synthetic Fibers

Type	Description
Cellulosic	
Acetate rayon	Cellulose acetate
Viscose rayon	Regenerated cellulose
Noncellulosic	
Polyester	Principally poly(ethylene terephthalate)
Nylon	Includes nylon 66, nylon 6, and a variety of other aliphatic and aromatic polyamides
Olefin	Includes polypropylene and copolymers of vinyl chloride, with lesser amounts of acrylonitrile, vinyl acetate, or vinylidene chloride (copolymers consisting of more than 85% vinyl chloride are called *vinyon* fibers)
Acrylic	Contain at least 80% acrylonitrile; included are *modacrylic* fibers comprising acrylonitrile and about 20% vinyl chloride or vinylidene chloride

1.8.4 Coatings and Adhesives

Coatings and adhesives are as old as civilization. Since the first prehistoric artist daubed colored clay on the walls of a cave, the science of coatings[40–42] developed very slowly. Early painters discovered the value of oils for portraits and landscapes, and some experimented with eggs for more durable murals on the exterior walls of churches and monasteries. Later, natural varnishes were employed as decorative coatings, then lacquers such as cellulose nitrate. With the development of synthetic polymers in the twentieth century, the coatings in-

TABLE I.8. Principal Types of Synthetic Rubber

Type	Description
Styrene–butadiene	Copolymer of the two monomers in various proportions depending on properties desired; called SBR for styrene–butadiene rubber
Polybutadiene	Consists almost entirely of the *cis*-1,4 polymer
Ethylene–propylene	Often abbreviated EPDM for ethylene–propylene–diene monomer; made up principally of ethylene and propylene units with small amounts of a diene to provide unsaturation
Polychloroprene	Principally the *trans*-1,4 polymer, but also some *cis*-1,4 and 1,2 polymer; also known as *neoprene* rubber
Polyisoprene	Mainly the *cis*-1,4 polymer; sometimes called "synthetic natural rubber"
Nitrile	Copolymer of acrylonitrile and butadiene, mainly the latter
Butyl	Copolymer of isobutylene and isoprene, with only small amounts of the latter
Silicone	Contains inorganic backbone of alternating oxygen and methylated silicon atoms; also called polysiloxane (Chap. 15)
Urethane	Elastomers prepared by linking polyethers through urethane groups (Chap. 13)

dustry evolved from the earliest polyester (alkyd) synthetic varnishes and paints introduced in the 1920s to the more recent interior latex wall paints using styrene–butadiene copolymers, then the exterior latex paints containing poly(vinyl acetate) and poly(acrylate ester)s.

A similar history can be traced with adhesives,[43,44] from the earliest uses of bitumens and natural gums and resins, even blood. Starch and cellulose nitrate adhesives came much later. Like coatings, adhesives came into their own with the development of synthetic polymers, particularly the phenol–formaldehyde and urea–formaldehyde polymers that are still used extensively, especially in the wood industries (plywood, particle board, etc.). Later came the epoxies and cyanoacrylates. In more recent years the adhesives industry has encompassed most of the major polymer types, including some high-temperature-resistant varieties for use in aerospace applications.

Both the coatings and adhesives industries are far more complex than the polymers used. Formulations include solvents, fillers, stabilizers, pigments, and a variety of other additives depending on application.

1.9 Polymer Recycling

One of the blessings of synthetic polymers is their durability. But in this age of environmental awareness, durability can also be a curse because polymers, particularly plastics and rubber, are a highly visible component of society's waste stream. There are a variety of ways to address the problem of polymer waste: making polymers degradable for composting in sanitary landfills; incinerating combustible polymers for energy recovery; recycling; and finding innovative uses for used polymers. Some used automobile tires, for example, are ground and blended into molded rubber products or asphalt paving materials, while others are used to construct barrier reefs. Such measures, however, do little to reduce the stain of some 200 million scrap tires discarded each year in the United States alone.

While recycling of paper (primarily the natural polymer cellulose) has been in effect for a good many years, recycling of plastics is still in its infancy,[45] with only 2–10% being recycled on a worldwide basis.[46] In 1988, the Society of the Plastics Industry (SPI) adopted the Voluntary Plastic Container Coding System to aid in the identification of various types of plastics for recycling. The code, now widely accepted, consists of a series of symbols imprinted or molded on the bottom of plastic containers. Each symbol consists of a number in the center of a three-arrow triangle, with identifying letters imprinted below. The symbol for high-density polyethylene, for example, is

HDPE

The code is summarized in Table 1.9. A more broadly based system encompassing more than one hundred polymers and polymer blends was issued in 1991 by the American Society for Testing and Materials (ASTM). The ASTM system is based on abbreviations recommended by the International Organization for Standardization (ISO) and is intended for identifying all industrial plastics for handling and disposal. A third system, adopted by the Society of Automotive Engineers (SAE) and also based on the ISO abbreviations, is used to identify plastics used in automotive parts.

TABLE 1.9. Plastics Recycling Code[a]

Number	Letters	Plastic
1	PETE[b]	Poly(ethylene terephthalate)
2	HDPE	High-density polyethylene
3	V *or* PVC	Poly(vinyl chloride)
4	LDPE	Low-density polyethylene
5	PP	Polypropylene
6	PS	Polystyrene
7	OTHER	Others or mixed plastics

[a] Adopted by the Society of the Plastics Industry (SPI).
[b] PET is the more widely accepted abbreviation.

Recycling methodologies vary from remolding, blending, and depolymerization to monomer with subsequent repolymerization. Poly(ethylene terephthalate) bottles, for example, are reduced to the monomers by hydrolysis or alcoholysis, which are then repolymerized to polyester for fiber or film applications, or reformulated into polyurethanes or thermosetting unsaturated polyesters. Poly(ethylene terephthalate) and high-density polyethylene share more than 70% of the demand for recycled plastics. The recycling industry has experienced economic difficulties because most "virgin" plastics are not only of better quality than their recycled counterparts, but are often less expensive.[47] Nevertheless, a considerable research effort is now directed toward recycling,[48] and the demand for recycled polymers will inevitably increase as automated sorting methods are refined and quality is improved.

References

1. H. Staudinger, *From Organic Chemistry to Macromolecules*, Wiley-Interscience, New York, 1970.
2. H. Morawetz, *Polymers—The Origins and Growth of a Science*, Wiley-Interscience, New York, 1985.
3. P. J. T. Morris, *Polymer Pioneers*, Center for the History of Chemistry, Philadelphia, 1986.
4. J. J. Berzelius, *Jahresberichte*, **12**, 63 (1833).
5. E. Simon, *Liebigs Ann. Chem.*, **31**, 265 (1839).
6. A-V. Lourenço, *Compt. Rend.*, **51**, 365 (1860); *Ann. Chim. Phys.*, **3**, 67, 257 (1863).
7. J. C. F. Williams, *J. Chem. Soc.*, **15**, Pt. 10 (1862).
8. G. A. Stahl, in *Polymer Science Overview* (G. A. Stahl, Ed.), ACS Symp. Ser. 175, American Chemical Society, Washington, D.C., 1981, Chap. 3.
9. L. H. Baekeland, *Ind. Eng. Chem.*, **5**, 506 (1913).
10. H. Staudinger, *Chem. Ber.*, **57**, 1203 (1924).
11. H. Mark and G. S. Whitby (Eds.), *Collected Papers of Wallace Hume Carothers on High Polymeric Substances*, Wiley-Interscience, New York, 1940.
12. K. Ziegler, E. Holzkamp, H. Breil, and H. Martin, *Angew. Chem.*, **67**, 541 (1955).
13. G. Natta, P. Pino, P. Corradini, F. Danusso, E. Mantica, G. Mazzanti, and G. Moraglio, *J. Am. Chem. Soc.*, **77**, 1708 (1955).
14. P. J. Flory, *Principles of Polymer Chemistry*, 2nd ed., Cornell University Press, Ithaca, N.Y., 1953.
15. G. Schill, *Catenanes, Rotaxanes and Knots*, Academic, New York, 1971.
16. H. W. Gibson, M. C. Bheda, and P. T. Engen, *Prog. Polym. Sci.*, **19**, 843 (1994), and references therein.
17. P. R. Ashton, J. Huff, S. Menzer, I. W. Parsons, J. A. Preece, J. F. Stoddart, M. S. Tolley, A. J. P. White, and D. J. Williams, *Chem. Eur. J.*, **2**, 31 (1996), and references therein.

18. D. B. Amabilino and J. F. Stoddart, *Pure Appl. Chem.*, **65**, 2351 (1993); D. B. Amabilino, I. W. Parsons, and J. F. Stoddart, *Trends Polym. Chem.*, **2**, 146 (1994).
19. D. B. Amabilino, P. R. Ashton, A. S. Reder, N. Spencer, and J. F. Stoddart, *Angew. Chem. Int. Ed. Engl.*, **33**, 1286 (1994); P. R. Ashton, et al., *Chem. Eur. J.*, **2**, 31 (1996).
20. B. I. Voit, *Acta Polymer.*, **46**, 87 (1995).
21. D. A. Tomalia, A. M. Naylor, and W. A. Goddard, *Angew. Chem. Int. Ed. Engl.*, **29**, 138 (1990).
22. C. N. Moorefield and G. R. Newkome, in *Advances in Dendritic Molecules* (G. R. Newkome, Ed.), JAI Press, Greenwich, CT, 1994, p. 1.
23. J. A. Put, *Macromol. Symp.*, **98**, 1333 (1995).
24. O. W. Webster, *Macromol. Symp.*, **98**, 1361 (1995).
25. W. H. Carothers, *J. Amer. Chem. Soc.*, **51**, 2548 (1928).
26. IUPAC, *Pure Appl. Chem.*, **66**, 2483 (1994).
27. IUPAC, *Compendium of Macromolecular Nomenclature*, Blackwell, Oxford, 1991; *Pure Appl. Chem.*, **65**, 1561 (1993); *ibid.*, **66**, 873 (1994).
28. H. Feuerberg, *J. Polym. Sci., Polym. Lett. Ed.*, **22**, 413 (1984).
29. IUPAC, *Pure Appl. Chem.*, **59**, 691 (1987).
30. IUPAC, *Pure Appl. Chem.*, **57**, 1427 (1985).
31. IUPAC, *Pure Appl. Chem.*, **66**, 2469 (1994).
32. O. Vogl, *J. Macromol. Sci.—Pure Appl. Chem.*, **A33**, 963 (1996).
33. N. Platzer, *J. Appl. Polym. Sci. Symp.*, **36**, v (1981).
34. L. H. Gillespie, in *High Performance Polymers: Their Origin and Development* (R. B. Seymour and G. S. Kirshenbaum, Eds.), Elsevier, New York, 1986, pp. 9–15.
35. H.-G. Elias, *An Introduction to Plastics*, VCH, New York, 1993.
36. S. T. Mossman and P. J. T. Morris, *The Development of Plastics*, CRC Press, Boca Raton, 1994.
37. J. M. Margolis (Ed.), *Engineering Thermoplastics: Properties and Applications*, Marcel Dekker, New York, 1985.
38. R. W. Mancrieff, *Man-Made Fibres*, 6th ed., Wiley, New York, 1975.
39. M. Morton (Ed.), *Rubber Technology*, 2nd ed., Van Nostrand Reinhold, New York, 1973.
40. S. Paul, *Surface Coatings*, 2nd ed., Wiley, New York, 1995.
41. G. P. A. Turner, *Paint Chemistry*, Chapman and Hall, London, 1980.
42. Z. W. Wicks, F. N. Jones, and S. P. Pappas, *Organic Coatings* (2 vols.), Wiley, New York, 1992–93.
43. E. J. Bruno (Ed.), *Adhesives in Modern Manufacturing*, Society of Manufacturing Engineers, Dearborn, 1970.
44. L.-H. Lee (Ed.), *Adhesive Chemistry*, Plenum Press, New York, 1985.
45. R. J. Ehrig (Ed.), *Plastics Recycling*, Hanser–Gardner, Cincinnati, 1992.
46. W. Weizer, *Chem. Ind.*, No. 24, 1009 (1995).
47. E. M. Kirschner, *Chem. Eng. News*, **74**(45), 19 (Nov. 4, 1996).
48. C. F. Prader, S. D. Baldwin, D. D. Cornell, G. D. Sadler, and R. F. Stockel (Eds.), *Plastics, Rubber and Paper Recycling: A Pragmatic Approach, ACS Symp. Ser. 609*, American Chemical Society, Washington, D.C., 1995.

Review Exercises

1. Following is an alphabetical listing of important terms encountered in this chapter. Their meaning must be clear to you before you proceed further into the book. Write a concise definition of each, using examples as appropriate.

(a) addition polymerization

(b) alternating copolymer

(c) block copolymer

(d) branched polymer

(e) comb polymer

(f) commodity plastic

(g) condensation polymerization

(h) copolymerization

(i) crosslinking

(j) degree of polymerization

(k) dendrimer

(l) elastomer

(m) end group

(n) engineering plastic

(o) graft copolymer
(p) heterochain polymer
(q) homochain polymer
(r) homopolymer
(s) hyperbranched polymer
(t) ladder polymer
(u) linear polymer
(v) macromer
(w) macromolecule
(x) monomer
(y) oligomer
(z) polycatenane

(a') polyolefin
(b') polyrotaxane
(c') reactive oligomer
(d') repeating unit
(e') ring-opening polymerization
(f') star polymer
(g') supramolecular assembly
(h') telechelic polymer
(i') thermoplastic
(j') thermosetting
(k') vinyl monomer
(l') vinylidene monomer

2. In your own words, explain the difference between step-reaction polymerization and chain-reaction polymerization. What is the advantage of this terminology over the more traditional addition and condensation polymerization?

3. In most instances end groups contribute very little to the physical properties of polymers. With which types of polymers would you expect end groups to contribute significantly?

4. In general, how do the properties of network polymers differ from those of linear or branched polymers?

5. Write a representative section of a network polymer formed from three moles of terephthalic acid (see reaction 1.4) and two moles of glycerol.

6. What is the $\overline{DP}$ of (a) a sample of poly(methyl methacrylate) (Table 1.2) of average molecular weight 50,000, and (b) a sample of poly(tetramethylene-*m*-benzenesulfonamide) (Table 1.3) of average molecular weight 26,000?

7. What percent conversion is represented by the polymer in question 6(b)?

8. What would be the molecular weight of the polyamide formed from 11-aminoundecanoamide (Table 1.3) if the reaction were carried to 96% conversion? To 99.9% conversion?

9. Write the name and structure of the monomers needed to synthesize the following vinyl polymers:

(a) $-\!\!\left[CH_2CHF\right]\!\!-$ (b) $-\!\!\left[CH_2CH\right]\!\!-$ with $CH_2CH_2CH_3$ (c) $-\!\!\left[CH_2\underset{\underset{CH_3}{|}}{\overset{\overset{CH_3}{|}}{C}}\right]\!\!-$

(d) $-\!\!\left[CH_2\underset{\underset{CN}{|}}{\overset{\overset{CN}{|}}{C}}\right]\!\!-$ (e) $-\!\!\left[CH_2CH\right]\!\!-$ (pyridyl, N) (f) $-\!\!\left[CH_2CH\right]\!\!-$ (chlorophenyl, Cl)

(g) $-\!\!\left[CH_2\underset{\underset{(phenyl)}{|}}{\overset{\overset{CH_3}{|}}{C}}\right]\!\!-$ (h) $-\!\!\left[\underset{\underset{Cl}{|}}{CH}\!-\!\underset{\underset{OCH_3}{|}}{CH}\right]\!\!-$

10. Write the name and structure of the monomers needed to synthesize the following nonvinyl polymers. In some instances, more than one answer may be correct. (You might wish to review organic nomenclature before attempting to name some of these monomers.)

(a) $\left[\begin{array}{c} CH_2CHO \\ | \\ CH_3 \end{array} \right]$
(b) $\left[\begin{array}{c} CHO \\ | \\ CH_2CH_3 \end{array} \right]$
(c) $\left[CH_2CH_2CH_2CH_2\overset{\overset{\displaystyle O}{||}}{C}O \right]$

(d) $\left[\overset{\overset{\displaystyle O}{||}}{C}\text{—}\bigcirc\text{—}O\text{—}\bigcirc\text{—}\overset{\overset{\displaystyle O}{||}}{C}OCH_2CH_2CH_2O \right]$
(e) $\left[NHCH_2CH_2\overset{\overset{\displaystyle O}{||}}{C} \right]$

(f) $\left[SO_2\text{—}\bigcirc\text{—}SO_2NH(CH_2)_6NH \right]$

(g) $\left[\overset{\overset{\displaystyle O}{||}}{C}(CH_2)_4\overset{\overset{\displaystyle O}{||}}{C}NHCH_2\text{—}\bigcirc\text{—}CH_2NH \right]$

11. Write common or source-based names and IUPAC-recommended names for each of the polymers in exercises 9 and 10.

12. Write all possible head-to-tail repeating units for polyisoprene and give each one an acceptable source-based and IUPAC name.

13. An inconsistency arises in using source-based nomenclature with the commercially important poly(vinyl alcohol), $\left[CH_2CH(OH) \right]$. What is the inconsistency? (You can check your answer in Chapter 9.) What is the IUPAC name for poly(vinyl alcohol)?

14. The first commercially available polyamide fiber was poly(iminohexane-1,6-diyliminoadipoyl). Write the structure, and suggest two alternate names.

15. A polyester fiber used for surgical sutures is poly[oxy(1-oxoethylene)]. Write the structure and suggest an alternate name.

16. In the following schematic copolymer diagrams, A = acrylonitrile, B = butadiene, M = methyl methacrylate, S = styrene, and V = vinyl acetate. Write an acceptable name for each.
    (a) MBMMVBBVMVVBVMB   (b) SSSSSVVVVVVVVSSSSS
    (c) MMMMMMMMSVSVSVSV   (d) VVVVVVVVBBBBBBBB

17. Using the same designations as in exercise 16, write schematics for an $\left[AB \right]$ block copolymer of styrene and methyl methacrylate, and an ABC block copolymer of vinyl acetate, butadiene, and acrylonitrile.

18. Using appropriate letters or symbols, draw schematic structures for the following copolymers:
    (a) Poly[propylene-*ran*-(methyl methacrylate)]
    (b) Copoly(isoprene/styrene/acrylonitrile)
    (c) Polystyrene-*block*-[butadiene-*alt*-(vinyl acetate)]
    (d) *Block*-copoly[*graft*-*co*(ethylene/styrene)/α-methylstyrene]
    (e) *Block*-copoly(acrylic acid/1-butene/vinyl chloride)

2

MOLECULAR WEIGHT AND
POLYMER SOLUTIONS

2.1 Number Average and Weight Average Molecular Weight

Molecular weight is an extremely important variable because it relates directly to a polymer's physical properties. In general, the higher the molecular weight, the tougher the polymer; however (as will be shown in the next chapter), too high a molecular weight can lead to processing difficulties. What defines an optimum molecular weight depends in large measure on the chemical structure of the polymer and the application for which it is intended. In this chapter we will be concerned with molecular weight definitions, methods of determining molecular weight, and methods of determining the distribution of molecular weights in a polymer sample. We will reserve our discussion of molecular structure for the following chapter. We will also be concerned with solution properties of polymers, because solubility is prerequisite to most methods for determining molecular weight.

Given that polymers are high-molecular-weight compounds, we must begin with the question, what do we *mean* by high molecular weight? Where does low molecular weight end and high begin? There is no simple answer, because what constitutes a "low" molecular weight for a sample of polyethylene, for example, might be ideal for a sample of polyamide. Furthermore, some polymers are deliberately prepared with low molecular weight (even as oligomers) to facilitate initial processing, and the molecular weight is increased in a subsequent processing step. In general, however, we think of polymers as having molecular weights that run from the low thousands up to the millions, with optimum molecular weight depending on chemical structure and application. Vinyl polymers of any commercial import normally have molecular weights in the range of 10^5 to 10^6. Polymers having very polar functional groups, such as polyamides, may have molecular weights as low as 15,000 to 20,000.

To determine molecular weights of simple (nonpolymeric) compounds, we employ the familiar techniques of mass spectrometry, freezing-point depression (cryoscopy), boiling-point elevation (ebulliometry), and, where suitable functional groups are present, titration (for example, neutralization or saponification equivalents). Determining molecular weights of polymers, however, is considerably more complex for two basic reasons. First, in any poly-merization process, it is virtually impossible for all growing polymer chains to terminate at the same size; hence one must necessarily deal with *average* molecular weights. (Certain natural polymers having discrete molecular weights provide exceptions to this rule.) Second, the tech-niques of cryoscopy, ebulliometry, and titration are effective only with relatively low-molecu-lar-weight polymers; more sophisticated methods must be used for polymers with molecular weights higher than about 40,000. Conventional mass spectrometry has not been used exten-sively in the polymer field beyond the characterization of polymer degradation products because of the requirements for volatilizable samples. Recent years, however, have seen some exciting new developments that have extended mass spectrometry into the macromolecular range.

Techniques more commonly used for determining molecular weights of polymers in-clude osmometry, light scattering, and ultracentrifugation, although titration (end-group analysis), cryoscopy, and ebulliometry are also used in some applications.[1,2] The most con-venient method for routinely determining molecular weights involves measuring solution vis-cosities, but this is not an absolute method and can be used only in conjunction with one of the techniques of measuring absolute molecular weights.

Molecular weight values obtained depend in large measure on the method of measure-ment. Methods that depend on end-group analysis or colligative properties (freezing-point depression, boiling-point elevation, osmotic pressure) give rise to the *number average mo-lecular weight*, $\overline{M}_n$, because the number of molecules of each weight in the sample are counted. This is defined as:

$$\overline{M}_n = \frac{\Sigma N_i M_i}{\Sigma N_i}$$

where N_i is the number of molecules—or the number of moles of those molecules—having molecular weight M_i. Suppose, for example, we have a polymer sample consisting of 9 moles having molecular weight 30,000 and 5 moles having molecular weight 50,000:

$$\overline{M}_n = \frac{(9 \text{ mol} \times 30,000 \text{ g/mol}) + (5 \text{ mol} \times 50,000 \text{ g/mol})}{9 \text{ mol} + 5 \text{ mol}} = 37,000 \text{ g/mol}$$

Suppose, instead, our sample consists of 9 grams having molecular weight 30,000 and 5 grams having molecular weight 50,000. Given that $N_i M_i$ is equal to the weight, w_i, then:

$$\overline{M}_n = \frac{9 \text{ g} + 5 \text{ g}}{(9 \text{ g}/30,000 \text{ g/mol}) + (5 \text{ g}/50,000 \text{ g/mol})} = 35,000 \text{ g/mol}$$

Light scattering and ultracentrifugation, on the other hand, are methods of determining mo-lecular weight based on mass or polarizability of the species present. The greater the mass, the greater is the contribution to the measurement. In contrast to number average molecular weight (which is the summation of the *mole* fraction of each species times its molecular weight), these methods sum the *weight* fraction of each species times its molecular weight. The value thus ob-tained is called the *weight average molecular weight*, $\overline{M}_w$, and is expressed mathematically as

$$\overline{M}_w = \frac{\Sigma W_i M_i}{\Sigma w_i} = \frac{\Sigma N_i M_i^2}{\Sigma N_i M_i}$$

Consider the same two samples described previously. Nine moles having 30,000 molecular weight and 5 moles having 50,000 molecular weight:

$$\overline{M}_w = \frac{9 \text{ mol}(30{,}000 \text{ g/mol})^2 + 5 \text{ mol}(50{,}000 \text{ g/mol})^2}{9 \text{ mol}(30{,}000 \text{ g/mol}) + 5 \text{ mol}(50{,}000 \text{ g/mol})} = 40{,}000 \text{ g/mol}$$

Substituting grams for moles:

$$\overline{M}_w = \frac{9 \text{ g}(30{,}000 \text{ g/mol}) + 5 \text{ g}(50{,}000 \text{ g/mol})}{9 \text{ g} + 5 \text{ g}} = 37{,}000 \text{ g/mol}$$

In each instance, we see that $\overline{M}_w$ is greater than $\overline{M}_n$.

In measurements of colligative properties, each molecule contributes equally regardless of weight, whereas with light scattering, the larger molecules contribute more because they scatter light more effectively. It is for this reason that weight average molecular weights are always greater than number average molecular weights except, of course, when all molecules are of the same weight; then $\overline{M}_w = \overline{M}_n$. The narrower the molecular weight range, the closer are the values of $\overline{M}_w$ and $\overline{M}_n$, and the ratio $\overline{M}_w / \overline{M}_n$ may thus be used as an indication of the breadth of the molecular weight range in a polymer sample. This ratio is called the *polydispersity index*, and any system having a range of molecular weights is said to be *polydisperse*. If all polymer molecules present have the same molecular weight (a protein, for example), the system is *monodisperse*.

There is a general equation for molecular weight averages:

$$\overline{M} = \frac{\sum N_i M_i^{a+1}}{\sum N_i M_i^{a}}$$

For $\overline{M}_n$ and $\overline{M}_w$, a is equal to 0 and 1, respectively. A higher-order molecular weight average, the z average $(\overline{M}_z)$, results when $a = 2$. For a given polydisperse sample, $\overline{M}_z > \overline{M}_w > \overline{M}_n$. Processing characteristics of molten polymer are more dependent on $\overline{M}_z$ than on $\overline{M}_w$ or $\overline{M}_n$.

In our discussions of molecular weight measurement (Section 2.3), detailed descriptions of apparatus and derivations of working equations are not given. Students who wish to explore these areas in more detail are encouraged to consult the references provided. First, however, we will look briefly at *solution* properties of polymers because the various methods of determining molecular weight or molecular weight distribution depend in most instances on solubility and polymer–solvent interactions.

2.2 Polymer Solutions

Dissolving a polymer is unlike dissolving low-molecular-weight compounds because of the vastly different dimensions of solvent and polymer molecules.[3] Solution occurs in two stages. Initially the solvent molecules diffuse through the polymer matrix to form a swollen, solvated mass called a *gel*. In the second stage, the gel breaks up and the molecules are dispersed into a true solution. Dissolution is often a slow process. While some polymers dissolve readily in certain solvents, others may require prolonged periods of heating at temperatures near the melting point of the polymer. Network polymers do not dissolve, but usually swell in the presence of solvent. (Any student who has stoppered a flask containing an organic solvent with a rubber stopper has undoubtedly observed this phenomenon.) Even some linear polymers defy all attempts to dissolve them; hence lack of solubility does not necessarily mean that the polymer is of the network type.

How does one choose a solvent? The simplest way is to consult a polymer handbook, where extensive lists of solvents and nonsolvents for numerous polymers are compiled.[4a] Detailed studies of polymer solubilities using thermodynamic principles have led to semiempirical relationships for predicting solubility. Any solution process is governed by the free-energy relationship.

$$\Delta G = \Delta H - T\Delta S$$

When a polymer dissolves spontaneously, the free energy of solution, ΔG, is negative. The entropy of solution, ΔS, invariably has a positive value arising from increased conformational mobility of the polymer chains. Therefore, the magnitude of the enthalpy of solution, ΔH, determines the sign of ΔG. It has been proposed that the heat of mixing, ΔH_{mix}, for a binary system is related to concentration and energy parameters by the expression

$$\Delta H_{mix} = V_{mix}\left[\left(\frac{\Delta E_1}{V_1}\right)^{1/2} - \left(\frac{\Delta E_2}{V_2}\right)^{1/2}\right]^2 \phi_1\phi_2$$

where V_{mix} is the total volume of the mixture, V_1 and V_2 are molar volumes (molecular weight/density) of the two components, ϕ_1 and ϕ_2 are their volume fractions, and ΔE_1 and ΔE_2 are the energies of vaporization. The terms $\Delta E_1/V_1$ and $\Delta E_2/V_2$ are called the *cohesive energy densities*. If $(\Delta E/V)^{1/2}$ is replaced by the symbol δ, the equation is written more simply:

$$\Delta H_{mix} = V_{mix}(\delta_1 - \delta_2)^2\phi_1\phi_2$$

The symbol δ is called the *solubility parameter*. Clearly, for the polymer to dissolve (negative ΔG), ΔH_{mix} must be small; therefore, $(\delta_1 - \delta_2)^2$ must also be small. In other words, δ_1 and δ_2 should be of about equal magnitude. Where $\delta_1 = \delta_2$, solubility is governed solely by entropy effects. Predictions of solubility are therefore based on finding solvents and polymers with similar solubility parameters, which requires a means of determining cohesive energy densities.

Cohesive energy density is the energy needed to remove a molecule from its nearest neighbors, thus is analogous to heat of vaporization per volume for a volatile compound. For the solvent, δ_1 can be calculated directly from the latent heat of vaporization (ΔH_{vap}) using the relationship

$$\Delta E = \Delta H_{vap} - RT$$

where R is the gas constant, and T is the temperature in kelvins. Thus

$$\delta_1 = \left(\frac{H_{vap} - RT}{V}\right)^{1/2}$$

Since polymers have negligible vapor pressure, the most convenient method of determining δ_2 is to use *group molar attraction constants*. These are constants derived from studies of low-molecular-weight compounds that lead to numerical values for various molecular groupings on the basis of considerations of intermolecular forces. Two sets of values (des-

TABLE 2.1. Representative group molar attraction constants[a]

	$G \ [(cal \ cm^3)^{1/2} \ mol^{-1}]$	
Group	*Small*	*Hoy*
CH$_3$—	214	147.3
—CH$_2$—	133	131.5
$\diagdown$CH—	28	85.99
$\diagdown$C$\diagup\diagup$$\diagdown$	−93	32.03
=CH$_2$	190	126.5
=CH—	19	84.51
—C$_6$H$_5$ (phenyl)	735	—
—CH= (aromatic)	—	117.1
$\diagdown$C=O (ketone)	275	262.7
—CO$_2$— (ester)	310	326.6

[a]Values taken from Refs. 5 and 6.

ignated G) have been suggested, one by Small,[5] derived from heats of vaporization, and the other by Hoy,[6] based on vapor pressure measurements. Typical G values are given in Table 2.1. Clearly there are significant differences between the Small and Hoy values. Which set one uses is normally determined by the method used to determine δ_1 for the solvent.

G values are additive for a given structure, and are related to δ by

$$\delta = \frac{d\Sigma G}{M}$$

where d is density and M is molecular weight. For polystyrene, for example, which has a density (obtained from a handbook[4b]) of 1.05, a repeating unit mass of 104, and structure

$$\left[\begin{array}{c} CH_2-CH \\ | \\ C_6H_5 \end{array}\right]$$

δ is calculated, using Small's G values, as

$$\delta = \frac{1.05(133 + 28 + 735)}{104} = 9.0$$

or using Hoy's data,

$$\delta = \frac{1.05[131.5 + 85.99 + 6(117.1)]}{104} = 9.3$$

Solvent δs may be calculated similarly. A major problem with solubility parameters as discussed above is that they do not take into account strong dipolar forces such as hydrogen bonding. Proposals have been made,[7] however, to separate solubility parameters into components reflecting dispersion forces, dipole–dipole attractions, and hydrogen bonding. Compilations of δ for solvents and polymers are available in the literature.[4c,8]

Once a polymer–solvent system has been selected, another consideration is how the polymer molecules behave in that solvent. Particularly important from the standpoint of molecular weight determinations is the resultant size, or *hydrodynamic volume*, of the polymer molecules in solution. Assuming polymer molecules of a given molecular weight are fully separated from one another by solvent, the hydrodynamic volume will depend on a variety of factors, including interactions between solvent and polymer molecules, chain branching, conformational effects arising from the polarity and steric bulk of the substituent groups, and restricted rotation caused by resonance, for example, of the type common to polyamides,

$$
\begin{array}{ccc}
\overset{\text{O}}{\underset{\|}{}} & & \overset{\text{O}^-}{\underset{|}{}} \\
-\text{C}-\text{NH}- & \longleftrightarrow & -\text{C}{=}\overset{+}{\text{N}}\text{H}-
\end{array}
$$

Because of Brownian motion, molecules are changing shape continuously. Hence any method of trying to predict molecular size must necessarily be based on statistical considerations and average dimensions. If a molecule were fully extended, its size could easily be computed from a knowledge of bond lengths and bond angles. Such is not the case, however, with most common polymers; therefore, size is generally expressed in terms of the mean-square average distance between chain ends, $\bar{r}^2$, for a linear polymer, or the square average radius of gyration about the center of gravity, $\bar{s}^2$, for a branched polymer. Figure 2.1 illustrates the meaning of r and s from the perspective of a coiled structure of an individual polymer molecule having its center of gravity at the origin. The average shape of the coiled molecule is spherical. The greater the affinity of solvent for polymer, the larger will be the sphere, that is, the hydrodynamic volume. As the solvent–polymer interaction decreases, intramolecular interactions become more important, leading to a contraction of the hydrodynamic volume.

It is convenient to express r and s in terms of two factors: an *unperturbed dimension* (r_0 or s_0) and an *expansion factor* (α). Thus,

$$\bar{r}^2 = r_0^2 \alpha^2$$
$$\bar{s}^2 = s_0^2 \alpha^2$$

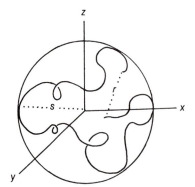

FIGURE 2.1. Schematic representation of a molecular coil, r = end-to-end distance; s = radius of gyration.

The unperturbed dimension refers to the size of the macromolecule exclusive of solvent effects. It arises from a combination of free rotation and intramolecular steric and polar interactions. The expansion factor, on the other hand, arises from interactions between solvent and polymer. For a linear polymer, $\bar{r}^2 = 6\bar{s}^2$. Since

$$\alpha = \frac{(\bar{r}^2)^{1/2}}{(\bar{r}_0^2)^{1/2}}$$

it follows that α will be greater than unity in a "good" solvent, and the actual (perturbed) dimensions will exceed the unperturbed dimensions. The greater the value of α, the "better" the solvent. For the special case where $\alpha = 1$, the polymer assumes its unperturbed dimensions and behaves as an "ideal" statistical coil.

Because solubility properties vary with temperature in a given solvent, α is temperature dependent. For a given polymer in a given solvent, the lowest temperature at which $\alpha = 1$ is called the *theta* (θ) *temperature* (or *Flory temperature*), and the solvent is then called a *theta solvent*. Additionally, the polymer is said to be in a *theta state*. It is convenient to think of the theta state as that in which the polymer is on the brink of becoming insoluble; in other words, the solvent is having a minimal solvation effect on the dissolved molecules. Any further diminution of this effect causes the attractive forces among polymer molecules to predominate, and the polymer precipitates. Extensive compilations of theta solvents and temperatures are available in the literature.[4d]

From the standpoint of molecular weight determinations, the significance of these parameters is that they can be related to dilute solution viscosity according to the Flory–Fox equation,

$$[\eta] = \frac{\phi(\bar{r}^2)^{3/2}}{\overline{M}}$$

where $[\eta]$ is the intrinsic viscosity (to be defined later), $\overline{M}$ is the average molecular weight, and ϕ is a proportionality constant (called the Flory constant) equal to approximately 3×10^{24} mol^{-1}. Substituting $\bar{r}_0^2\alpha^2$ for $\bar{r}^2$, we obtain

$$[\eta] = \frac{\phi(r_0^2\alpha^2)^{3/2}}{\overline{M}}$$

which can be rearranged to

$$[\eta] = \phi(r_0^2\overline{M}^{-1})^{3/2}\overline{M}^{1/2}\alpha^3$$

Since $\bar{r}_0$ and $\overline{M}$ are constants, we can set $K = \phi(\bar{r}_0^2\overline{M}^{-1})^{3/2}$, then

$$[\eta] = K\overline{M}^{1/2}\alpha^3$$

At the theta temperature, $\alpha = 1$ and

$$[\eta] = K\overline{M}^{1/2}$$

For conditions other than the theta temperature, the equation has the form

$$[\eta] = K\overline{M}^a$$

where *a* is a constant that varies with polymer, solvent, and temperature. Known as the *Mark–Houwink–Sakurada equation*, this is an important relationship that, as we shall see in Section 2.5, is used to relate dilute solution viscosity to molecular weight.

Apart from molecular weight determinations, important practical considerations arise from solubility effects. As one moves in the direction of "good" solvent to "poor," and intramolecular forces become more important, the polymer molecules shrink in volume. This increasing compactness leads to reduced "drag" and hence a lower viscosity. Paint chemists are particularly concerned with solvent effects because viscosity is of fundamental importance in spray or brush applications of solvent-based paints or lacquers. Fiber manufacturers can also reduce energy costs involved in spinning fibers from solution (see Chapter 4) by reducing solution viscosity.

2.3 Measurement of Number Average Molecular Weight

2.3.1 End-group Analysis

The number average molecular weight of any linear polymer having end groups that can be measured by chemical or physical means can theoretically be determined if the method of measurement is sensitive enough. It must be remembered, however, that end groups are present in very low concentrations. Presently available techniques[9a,10a] allow a practical upper limit of molecular weight measurement of about 50,000. Some of the current methods of end-group determination include (1) titration, using either indicators or potentiometric techniques; (2) elemental analysis of element-specific end groups; (3) measurement of activity of a radioactive-tagged end group; and (4) ultraviolet spectroscopic determination of an end group with a characterizable chromophore. Infrared and nuclear magnetic resonance (NMR) spectroscopic techniques are of more limited use.

Points to be kept in mind in applying end-group analysis are the following:

1. The method cannot be applied to branched polymers unless the number of branches is known with certainty; thus it is practically limited to linear polymers.

2. In a linear polymer there are twice as many end groups as polymer molecules.

3. If the polymer contains different groups at each end of the chain and only one characterizable end group is being measured, the number of this type is equal to the number of polymer molecules.

4. Measurement of molecular weight by end-group analysis is only meaningful when the mechanisms of initiation and termination are well understood.

As a typical example, let us consider the unsaturated polyesters (Chapter 12). These are linear polymers of relatively low molecular weight prepared from anhydrides or dicarboxylic acids and diols, and they are normally crosslinked in a subsequent fabrication step. (Saturated polyesters of the type used in film and fiber applications are usually of too high molecular weight for end-group analysis.)

To determine the number average molecular weight of the linear polyester before crosslinking, one can titrate the carboxyl and hydroxyl end groups by standard methods. In the case of carboxyl, a weighed sample of polymer is dissolved in an appropriate solvent such as acetone and titrated with standard base to a phenolphthalein end point. For hydroxyl, a sample is acetylated with excess acetic anhydride, and liberated acetic acid, together with carboxyl end groups, is similarly titrated. From the two titrations, one obtains the number of milliequivalents of carboxyl and hydroxyl in the sample. The number average molecular weight (i.e., the number of grams per mole) is then given by:

$$\overline{M}_n = \frac{2 \times 1000 \times \text{sample wt}}{\text{meq COOH} + \text{meq OH}}$$

The 2 in the numerator takes into account that two end groups are being counted per molecule. In polyester processing, the *acid number*, defined as the number of milligrams of base required to neutralize 1 g of polyester, is frequently used to monitor reaction progress.

Other titratable end groups include amino groups in polyamides, acetyl groups in acetyl-terminated polyamides, isocyanate in polyurethanes, and epoxide in epoxy polymers. A number of factors can complicate end-group analysis: lack of solubility, high solution viscosity, severe steric hindrance. But where these difficulties can be surmounted, end-group analysis is preferred for polymers in the 5,000 to 10,000 molecular weight range.

2.3.2 Membrane Osmometry

Of the various methods of number average molecular weight determination based on colligative properties, membrane osmometry is most useful. When pure solvent is separated from a solution by a semipermeable membrane that allows passage of solvent but not solute molecules, solvent will flow through the membrane into the solution. As the liquid level rises in the solution compartment, the hydrostatic pressure increases until it prevents further passage of solvent or, more exactly, until solvent flow is equal in both directions. The pressure at equilibrium is the osmotic pressure. A schematic representation of an osmometer is given in Figure 2.2.

Osmotic pressure may be determined by allowing the system to reach equilibrium and measuring the hydrostatic head that develops. This is referred to as the *static equilibrium method*. Alternatively, one can apply a counterpressure to the measuring tube connected to the solution compartment, which prevents flow of solvent and maintains equal liquid levels in the two measuring tubes. This is called the *dynamic equilibrium method*. Because the former method usually requires long periods of time to attain equilibrium, the dynamic method is preferred.

A variety of dynamic membrane osmometers are manufactured, which usually encompass a horizontal membrane separating solution and solvent cells. The type shown schematically in Figure 2.3 measures osmotic pressure directly via a strain gauge transducer attached to a flexible diaphragm in the solvent cell. Semipermeable membranes are usually constructed from polymeric materials such as cellulose acetate, cellulose nitrate, rubber, and poly(vinyl alcohol).

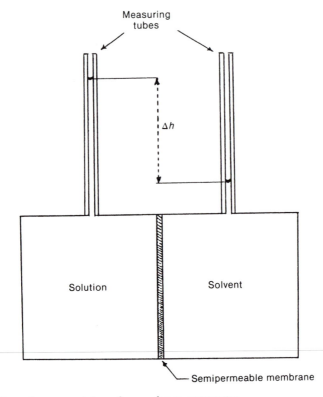

FIGURE 2.2. Schematic representation of a membrane osmometer.

FIGURE 2.3. Automatic membrane osmometer [Courtesy of Wescan Instruments, Inc.]

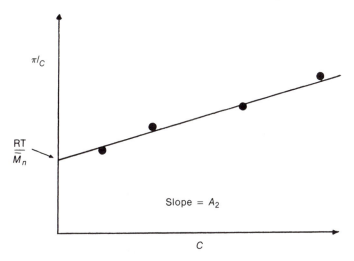

FIGURE 2.4. Plot of reduced osmotic pressure (π/C) versus concentration (C).

Osmotic pressure is related to molecular weight by the van't Hoff equation extrapolated to zero concentration:

$$\left(\frac{\pi}{C}\right)_{C=0} = \frac{RT}{\overline{M}_n} + A_2 C$$

where π, the osmotic pressure, is given by

$$\pi = pg\,\Delta h$$

where R is the gas constant, 0.082 L atm mol^{-1} K^{-1} (CGS) or 8.314 J mol^{-1} K^{-1} (SI); T is the temperature in kelvins; C is the concentration in grams per liter; ρ is the solvent density in grams per cubic centimeter, g is the acceleration due to gravity, 9.81 m/s^2; Δh is the difference in heights of solvent and solution in centimeters; and A_2 is the second virial coefficient (a measure of the interaction between solvent and polymer).

A plot of *reduced osmotic pressure*, π/C, versus concentration (Figure 2.4) is linear with the intercept equal to $RT/\overline{M}_n$ and slope equal to A_2. Units for π/C are dyn L g^{-1} cm^{-1} (CGS) or J kg^{-1} (SI). Because A_2 is a measure of solvent–polymer interaction, the slope is zero at the theta temperature. Thus osmotic pressure measurements may be used to determine theta conditions.

The major source of error in membrane osmometry arises from low-molecular-weight species diffusing through the membrane. It explains why molecular weight values obtained are generally higher than those obtained by other colligative property measurements where *all* species present are counted. Membrane osmometry is generally useful over a molecular weight range of about 50,000 to 2 million, the limiting factors being the permeability of the membrane at the lower end and the smallest measurable osmotic pressure at the upper. Because most polymers of commercial significance fall in this range, membrane osmometry is the most widely used method of determining absolute number average molecular weights.

2.3.3 Cryoscopy and Ebulliometry

The techniques employed in freezing-point depression (cryoscopy) and boiling-point elevation (ebulliometry) are analogous to those for low-molecular-weight compounds, and the
thermodynamic relationships, derived for infinitely dilute solutions, are similar[9c,10c,11a]:

$$\left(\frac{\Delta T_f}{C}\right)_{C=0} = \frac{RT^2}{\rho \Delta H_f \overline{M}_n} + A_2 C \qquad \text{for freezing-point depression}$$

$$\left(\frac{\Delta T_b}{C}\right)_{C=0} = \frac{RT^2}{\rho \Delta H_v \overline{M}_n} + A_2 C \qquad \text{for boiling-point elevation}$$

In these expressions ΔT_f and ΔT_b are the freezing point depression and boiling point elevation, respectively; C is the concentration in grams per cubic centimeter; T is the freezing
point or boiling point of the solvent in kelvins; R is the gas constant; ρ is the solvent density; ΔH_f and ΔH_v are the latent heats of fusion and vaporization, respectively, per gram of
solvent; $\overline{M}_n$ is the number average molecular weight; and A_2 is the second virial coefficient.
Molecular weights are determined from plots of $\Delta T_f/C$ (or $\Delta T_b/C$) versus concentration, analogous to those in osmometry.

The major limitation of cryoscopy and ebulliometry lies in the sensitivity of the meth-
ods of measuring ΔT_f or ΔT_b. As the molecular weight increases, ΔT_f or ΔT_b becomes progressively smaller so that, even with the most sensitive thermistors (some are sensitive to
$1 \times 10^{-4}\,°C$), the upper limit of measurable molecular weight is about 40,000. These methods, together with that of vapor pressure osmometry (discussed below), are, however, preferred when molecular weights are below 20,000.

2.3.4 Vapor Pressure Osmometry

Vapor pressure osmometry[9c,10c] is generally useful for determining molecular weights below about 25,000. While no membrane is involved, the thermodynamic principle involved
is analogous to that of membrane osmometry. A drop of solvent and a drop of solution are
placed by syringe on matched thermistors in an insulated chamber saturated with solvent vapor. Condensation heats the solution thermistor until the vapor pressure is increased to that
of the pure solvent. The temperature change, measured by the change in resistance of the
thermistor, is related to solution molality, and hence molecular weight, by the expression

$$\Delta T = \left(\frac{RT^2}{\lambda 100}\right) m$$

where λ is the heat of vaporization per gram of solvent and m is molality.

2.3.5 Mass Spectrometry

The last two decades of the twentieth century have seen the development of a number of increasingly important mass spectrometric techniques that allow the direct measurement of
molecular weights of macromolecules. Developed initially for characterizing biopolymers,
each of these methods involves removing individual polymer molecules from the condensed

phase to the gaseous phase with concomitant ionization, followed by measurement of the molecular ion mass in the mass spectrometer, usually a time-of-flight (TOF) instrument. Electron bombardment leading to molecular fragmentation, which lies at the heart of conventional mass spectrometry, is not a feature of these "soft ionization" techniques.

The most commonly applied mass spectrometric method for synthetic polymers has the tongue-twisting name of *matrix-assisted laser desorption ionization mass spectrometry*.[12,13] (The abbreviations MALDI-MS or MALDI-TOF are commonly used.) The polymer sample is imbedded in a low-molecular-weight organic compound that absorbs strongly at the wavelength of an ultraviolet laser. Upon irradiation with the laser, desorption (vaporization) of the polymer molecules occurs as the laser energy is transferred from the matrix to the polymer. The presence of metal ions, usually sodium or potassium, in the matrix yields polymer molecules with attached metal ions. For relatively low-molecular-weight polymers, individual cationized molecules are clearly resolved, as shown in Figure 2.5 for the mass spectrum of poly(methyl methacrylate) having an $\overline{M}_n$ of about 6400.[13] The mass difference of 100 between individual molecules is evident in the spectrum. At higher molecular weights the signal-to-noise ratio of the instrument makes resolution increasingly poorer, although peak maxima and distribution can still be measured at molecular weights in the tens of thousands.[13] In the case of monodisperse biopolymers, molecular weights as high as 400,000 can be measured accurately. From the integrated peak areas, reflecting the number of ions (N_i) and the average molecular weight (M_i), both $\overline{M}_n$ and $\overline{M}_w$ can be calculated.

Related mass spectrometric soft ionization methods include *field desorption* (FD-MS),[14] *laser desorption* (LD-MS),[15] and *electrospray ionization* (ESI-MS).[16] Although instrument costs are relatively high, the molecular weight averages as well as the molecular weight dis-

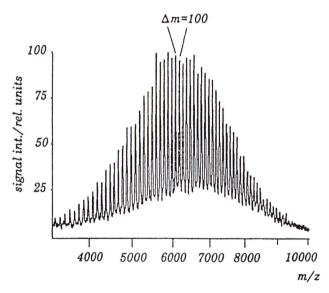

FIGURE 2.5. MALDI mass spectrum of low-molecular-weight poly(methyl methacrylate). [From Bahr, Deppe, Karas, and Hillenkamp,[13] copyright 1992. Reprinted by permission of the American Chemical Society.]

tribution (Section 2.6) are obtained quickly and generally more accurately by mass spectrometry than by the more traditional separation methods.

2.3.6 Refractive Index Measurements

It has been demonstrated that, for certain polymers, a linear relationship exists between index of refraction and inverse number average molecular weight.[17] While the method appears to be most suitable for low-molecular-weight polymers, it does have the advantage of simplicity.

2.4 Measurement of Weight Average Molecular Weight

2.4.1 Light Scattering

Apart from osmometry, light scattering[9d,10d] is the most widely used method of obtaining absolute molecular weights. It relies on the fact that light, when passing through a solvent or solution, loses energy by absorption, conversion to heat, and scattering. The scattering for a pure liquid arises from finite nonhomogeneities in the distribution of molecules within adjacent areas that give rise to differences in density. For solutions, additional scattering is introduced by the solvent molecules. The amplitude or intensity of the scattered light depends on several factors, the most important being concentration, size, and polarizability of the scattering molecules. Refractive index also depends on concentration and amplitude of vibration. To evaluate the turbidity arising from scattering, one combines equations derived from scattering and index of refraction measurements. Turbidity, τ, is related to concentration, c, by the expression

$$\tau = Hc\overline{M}_w$$

where

$$H = \frac{32\pi^3}{3} \frac{n_0^2 (dn/dc)^2}{\lambda^4 N_0}$$

and n_0 is the refractive index of the solvent, λ is the wavelength of the incident light, and N_0 is Avogadro's number. The expression dn/dc, referred to as the *specific refractive increment*, is obtained by measuring the slope of the refractive index as a function of concentration, and it is constant for a given polymer, solvent, and temperature. As molecular size approaches in magnitude the wavelength of the light, corrections must be made for interference between scattered light coming from different parts of the molecules.

To determine molecular weight, the expression for turbidity is rewritten as

$$\frac{Hc}{\tau} = \frac{1}{\overline{M}P(\theta)} + 2A_2 C$$

where $P(\theta)$ is a function of the angle, θ, at which τ is measured, a function that depends on the shape of the molecules in solution. A_2 is the second virial coefficient. Turbidity is then measured at different concentrations as well as at different angles, the latter to compensate for variations in molecular shape. The experimental data are then extrapolated to both zero

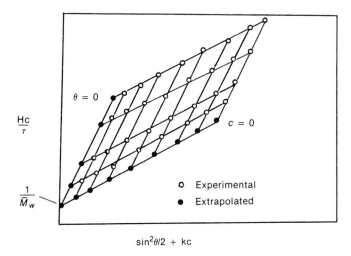

$$\frac{Hc}{\tau}$$

$\theta = 0$

$c = 0$

$\frac{1}{\overline{M}_w}$

o   Experimental

●   Extrapolated

$\sin^2\theta/2 + kc$

FIGURE 2.6. Zimm plot of light-scattering data.

concentration and zero angle, where $P(\theta)$ is equal to 1. Such double extrapolations, shown in Figure 2.6, are called *Zimm plots* (after Bruno Zimm, who originated the method). The factor k on the abscissa is an arbitrary constant. The intercept corresponds to $1/\overline{M}_w$.

A major problem in light scattering is to obtain perfectly clear, dust-free solutions. This is usually accomplished by ultracentrifugation or careful filtration. Despite such difficulties, the light-scattering method is widely used for obtaining weight average molecular weights

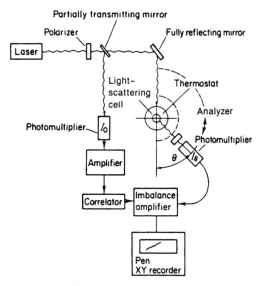

FIGURE 2.7. Schematic of a laser light-scattering photometer. [From Rabek,[9d] copyright 1980. Reprinted by permission of John Wiley & Sons, Ltd.]

between 10,000 and 10,000,000. Older light-scattering photometers employed high-pressure mercury lamps and filters to obtain a monochromatic beam. These have been supplanted with laser light sources. A schematic of a laser light-scattering photometer is given in Figure 2.7.

2.4.2 Ultracentrifugation

The ultracentrifuge is by far the most intricate and expensive instrument for determining molecular weight.[9e,18,19a] It is not as widely used as light scattering or osmometry in determining molecular weights of synthetic polymers, but it has been used extensively with natural polymers, particularly proteins, and it is useful for determining $\overline{M}_z$ of synthetic polymers. The technique is based on the principle that molecules, under the influence of a strong centrifugal field, distribute themselves according to size perpendicularly to the axis of rotation, a process called *sedimentation*, the rate of which is proportional to molecular mass. Centrifugation is accomplished in an evacuated chamber in a cell set in a rotor, both of which are provided with windows so that optical methods, such as index of refraction measurements or interferometry, can be used to observe concentration gradients within the polymer solution. Details of the technique may be found in the references provided.

2.5 Viscometry

Measurements of dilute solution viscosity provide the simplest and most widely used technique for routinely determining molecular weights.[9f,19b] It is not an absolute method; each polymer system must first be calibrated with absolute molecular weight determinations (usually by light scattering) run on fractionated polymer samples. Viscosities are measured at concentrations of about 0.5 g/100 mL of solvent by determining the flow time of a certain volume of solution through a capillary of fixed length. Flow time in seconds is recorded as

A   B

FIGURE 2.8. Capillary viscometers: (A) Ubbelohde, and (B) Cannon–Fenske.

TABLE 2.2. Dilute Solution Viscosity Designations[a]

Common Name	IUPAC Name	Definition
Relative viscosity	Viscosity ratio	$\eta_{rel} = \dfrac{\eta}{\eta_0} = \dfrac{t}{t_0}$
Specific viscosity	—	$\eta_{sp} = \dfrac{\eta - \eta_0}{\eta_0} = \dfrac{t - t_0}{t_0} = \eta_{rel} - 1$
Reduced viscosity	Viscosity number	$\eta_{red} = \dfrac{\eta_{sp}}{C} = \dfrac{\eta_{rel} - 1}{C}$
Inherent viscosity	Logarithmic viscosity number	$\eta_{inh} = \dfrac{\ln \eta_{rel}}{C}$
Intrinsic viscosity	Limiting viscosity number	$[\eta] = \left(\dfrac{\eta_{sp}}{C}\right)_{C=0} = (\eta_{inh})C = 0$

[a] Concentrations (most commonly expressed in grams per 100 mL of solvent) of about 0.5 g/dL.

the time for the meniscus to pass between two designated marks on the viscometer. Viscosities are run at constant temperature, usually $30.0 \pm 0.01°C$.

Two typical viscometers are shown in Figure 2.8. Of the two, the Ubbelohde type is more convenient to use in that it is not necessary to have exact volumes of solution to obtain reproducible results. Furthermore, additional solvent can be added (assuming the reservoir is large enough); thus concentration can be reduced without having to empty and refill the viscometer. Whichever type is used, it is necessary to filter the polymer solutions into the viscometer because dust particles affect flow time. Filtration is conveniently accomplished with micro filter attachments to syringes. Computer-driven viscometer modules are commercially available. They measure capillary flow time photoelectrically and provide automatic dilution and mixing as well as calculation and read-out of viscosity.

Viscosity can be expressed in several ways (Table 2.2). Common names are, at the present time, more widely used than the IUPAC-recommended[20] names. (The latter were adopted by the IUPAC to avoid inconsistencies arising from "viscosity" designations where there were no units of viscosity.)

Relative viscosity (viscosity ratio) (η_{rel}) is the ratio of solution viscosity to solvent viscosity, which is proportional to a first approximation for dilute solutions to the ratio of the corresponding flow times. Viscosity units (commonly expressed as poises) or flow times cancel out in the various viscosity expressions. *Specific viscosity* (η_{sp}) is the fractional increase in viscosity. Both η_{rel} and η_{sp} are dimensionless. As concentration increases, so does viscosity. Hence to eliminate concentration effects, the specific viscosity is divided by concentration and extrapolated to zero concentration to give the *intrinsic viscosity*, $[\eta]$. Not uncommonly viscosities are determined at a single concentration and the *inherent viscosity* (η_{inh}) is used as an approximate indication of molecular weight. Inherent and reduced viscosities extrapolate to the same intrinsic viscosity at zero concentration. Concentration, C, is in grams per 100 mL of solvent or in grams per cubic centimeter, more commonly the former. Thus reduced, inherent, and intrinsic viscosities have units of deciliters per gram or, less commonly, cubic centimeters per gram. Obviously, units of concentration must be specified when viscosity data are reported.

Intrinsic viscosity is the most useful of the various viscosity designations because it can be related to molecular weight by the Mark–Houwink–Sakurada equation:

$$[\eta] = K\overline{M}_v^{\,a}$$

where $\overline{M}_v$, is the *viscosity average molecular weight*, defined as

$$\overline{M}_v = \left(\frac{\sum N_i M_i^{1+a}}{\sum N_i M_i}\right)^{1/a}$$

Log K and a are the intercept and slope, respectively, of a plot of log $[\eta]$ versus log $\overline{M}_w$ or log $\overline{M}_n$ of a series of fractionated polymer samples. Such plots are linear (except at low molecular weights) for linear polymers, thus

$$\log[\eta] = \log K + a \log \overline{M}$$

TABLE 2.3. Representative Viscosity-Molecular Weight Constants[a]

Polymer	Solvent	Temperature, °C	Molecular Weight Range $\times 10^{-4}$	$K^b \times 10^3$	a^b
Polystyrene	Cyclohexane	35[d]	8–42[e]	80	0.50
(atactic)[c]	Cyclohexane	50	4–137[e]	26.9	0.599
	Benzene	25	3–61[f]	9.52	0.74
Polyethylene	Decalin	135	3–100[e]	67.7	0.67
(low pressure)					
Poly(vinyl chloride)	Benzyl alcohol	155.4[d]	4–35[e]	156	0.50
	Cyclohexanone	20	7–13[f]	13.7	1.0
Polybutadiene					
98% *cis*-1,4, 2% 1,2	Toluene	30	5–50[f]	30.5	0.725
97% *trans*-1,4, 3% 1,2	Toluene	30	5–16[f]	29.4	0.753
Polyacrylonitrile	DMF[g]	25	5–27[e]	16.6	0.81
	DMF	25	3–100[f]	39.2	0.75
Poly(methyl methacrylate-*co*-styrene)					
30–70 mol %	1-Chlorobutane	30	5–55[e]	17.6	0.67
71–29 mol %	1-Chlorobutane	30	4.8–81[e]	24.9	0.63
Poly(ethylene terephthalate)	*m*-Cresol	25	0.04–1.2[f]	0.77	0.95
Nylon 66	*m*-Cresol	25	1.4–5[f]	240	0.61

[a] Values taken from Ref. 4e.

[b] See text for explanation of these constants.

[c] Atactic defined in Chapter 3.

[d] θ temperature.

[e] Weight average.

[f] Number average.

[g] N,N-dimethylformamide.

Viscosity average molecular weights lie between those of the corresponding $\overline{M}_w$ and $\overline{M}_n$, but closer to the former. Hence better results are obtained if K and a are determined with fractionated samples of measured $\overline{M}_w$. To evaluate K and a requires considerable manipulation, but fortunately a wide range of values representing a broad spectrum of polymers, solvents, and temperatures has been published.[4e] For most common polymers, values of a vary between 0.5 (for a randomly coiled polymer in a theta solvent) and 0.8; for more rodlike extended-chain polymers where the hydrodynamic volume is relatively large, a may be as high as 1.0, in which case $\overline{M}_v = \overline{M}_w$. K values generally vary between 10^{-3} and 0.5. A few representative values of a and K are given in Table 2.3.

Factors that may complicate the application of the Mark–Houwink–Sakurada relationship are chain branching, too broad a molecular weight distribution in the samples used to determine K and a, solvation of polymer molecules, and the presence of alternating or block sequences in the polymer backbone. Chain entanglement is not usually a problem at such high dilution unless molecular weights are extremely large.

Other types of viscosity measurements based on the principle of mechanical shearing[19b] are also employed, most commonly with concentrated polymer solutions or undiluted polymer; these methods, however, are more applicable to flow properties of polymers (Chapter 3), not molecular weight determinations. Relative methods of determining molecular weight on the basis of polymer fractionation are discussed in the next section.

2.6 Molecular Weight Distribution

Molecular weight distribution is an important characteristic of polymers because it can significantly affect polymer properties. Just as low-molecular-weight polystyrene behaves differently from the high-molecular-weight material, a sample of polystyrene having a narrow molecular weight range will exhibit different properties from one having a broad range, even if the average molecular weights of the two samples are the same.

The relatively new mass spectrometric methods for determining molecular weight were described earlier in Section 2.3.5. As may be seen from Figure 2.5, the molecular weight distribution is also apparent in the mass spectrum. Molecular weight distribution may also be obtained by ultracentrifugation. More widely used techniques for determining molecular weight distribution involve fractionation of the polymer sample and comparison of the fractions thus obtained with samples of known absolute molecular weight by means of some calibration procedure. A variety of fractionation methods have been developed.

2.6.1 Gel Permeation Chromatography (GPC)

Also called *size exclusion chromatography* (SEC), GPC[9g,21,22a] is by far the most widely used method of determining molecular weight distribution. A column chromatography technique, GPC may be used preparatively to obtain narrow molecular weight fractions. Separation is accomplished on a column packed with a highly porous material that separates the polymer molecules according to size, a phenomenon often referred to as *molecular sieving*. Current thinking is that separation is based on the hydrodynamic volume of the molecules rather than on the molecular weight per se.[23] Small molecules are able to diffuse into

Solvent tank

Degasser

High
pressure
pump

Pressure gauge

Surge
tank

Sample
injection

Thermostat

Reference
column

Sample
column

Differential
refractometer

UV/IR
viscosity
detector

Siphon

Fraction
collector

FIGURE 2.9. Schematic representation of a gel permeation chromatograph. [From Rabek,[9g] copyright 1980. Reprinted by permission of John Wiley & Sons, Ltd.]

the pores of the column packing more efficiently, and hence they travel through the column more slowly. Higher molecular weight fractions are thus eluted first.

The essential features of a gel permeation chromatograph are shown in Figure 2.9. Column packing materials come in a variety of forms, but most commonly they consist of fine, semirigid beads of polystyrene crosslinked with divinylbenzene and swollen with solvent, or of rigid porous beads of glass or silica. Detection of polymer fractions in the eluant is most commonly accomplished with refractive index or spectroscopic (ultraviolet and infrared) detectors. Commercially available instruments provide automated sample injection and fraction collection, rapid flow-through by high-pressure pumping, and computer-assisted treatment of data. Closely related to GPC is *gel filtration chromatography* (GFC), a technique used principally to characterize natural polymers in aqueous solution. GFC columns are packed with hydrophilic gels, usually crosslinked dextrans or crosslinked polyacrylamides. As in the case of GPC, separation in GFC columns occurs by molecular sieving, but adsorption, ion exchange, and ion exclusion also play a role.

A typical gel permeation chromatogram (Figure 2.10) plots detector response against the volume of dilute polymer solution that passes through the column (the elution volume, V_r). To obtain molecular weights at a given retention volume, the chromatogram may be compared with a reference chromatogram obtained with fractions of known average molecular weight in the same solvent and at the same temperature. For purposes of comparison, the

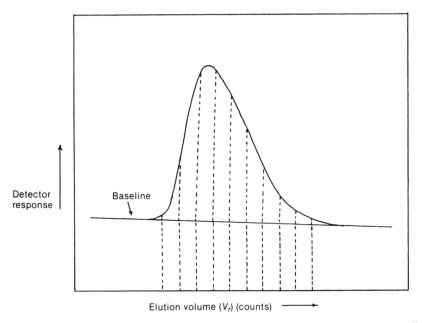

Elution volume (V_r) (counts) ⟶

FIGURE 2.10. Typical gel permeation chromatogram. Dotted lines represent volume "counts."

elution band is divided into "counts" of a specific volume, usually 2.5 or 5.0 mL, as shown, with the height above the baseline being proportional to the amount of polymer (N_1M_1) eluted.

The major problem with calibrating a GPC column for a particular polymer is that few standard samples of narrow molecular weight distribution are available commercially. Polystyrene standards having polydispersity indexes close to unity are available over a wide range of molecular weights (600–2.5 million) and these are often used; if one is dealing with a polymer other than polystyrene, however, the molecular weights thus obtained are at best approximate and may in some instances be seriously in error. To circumvent this difficulty, the *universal calibration* method is employed.

The universal calibration is based on the observation[24] that the product of intrinsic viscosity (limiting viscosity number) and molecular weight is independent of polymer type. This product, $[\eta]M$, is called the *universal calibration parameter*. As shown in Figure 2.11, a plot of log ($[\eta]M$) versus elution volume in tetrahydrofuran (THF) solvent yields a single curve, approximately linear, for a widely disparate group of polymers. Thus log ($[\eta]M$) may be considered a constant for all polymers *for a given column, temperature, and elution volume*. If we assume that the reference polymer (e.g., polystyrene) is polymer 1 and the polymer to be fractionated is polymer 2, it follows that

$$[\eta]_1 M_1 = [\eta]_2 M_2$$

From the Mark–Houwink–Sakurada relationship,

$$[\eta]_1 = K_1 M_1{}^a$$

$$[\eta]_2 = K_2 M_2{}^a$$

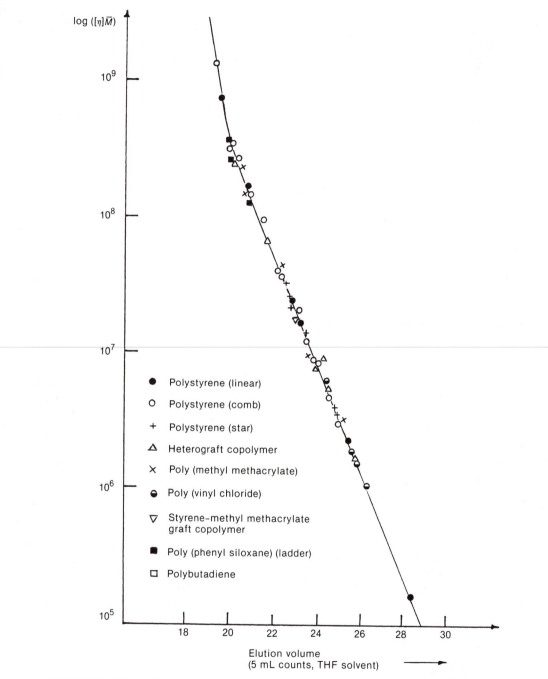

FIGURE 2.11. Universal calibration for gel permeation chromatography. THF, tetrahydrofuran. [From Grubisic, Rempp, and Benoit,[24] copyright 1976. Reprinted by permission of John Wiley & Sons, Inc.]

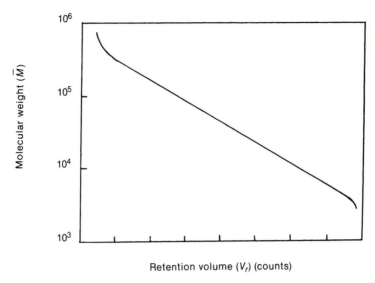

FIGURE 2.12. Typical semilogarithmic calibration plot of molecular weight versus retention volume.

Combining these equations and solving for log M_2, we obtain

$$\log M_2 = \left(\frac{1}{1 + a_2}\right) \log \left(\frac{K_1}{K_2}\right) + \left(\frac{1 + a_1}{1 + a_2}\right) \log M_1$$

To determine the molecular weight (M_2) at a given retention volume, the column must first be calibrated with the standard polystyrene fractions (same solvent, same temperature). This gives a relationship of the type shown in Figure 2.12. Such semilogarithmic calibration plots are generally linear over a broad range of molecular weights, with deviations from linearity occurring at high and low molecular weights, particularly the former. Constants K and a are normally obtainable from a polymer handbook.[4e] Substituting the value of M_1 for a particular retention volume from the calibration plot and the values of K and a in the above equation, M_2 can be readily calculated. GPC thus provides a rapid and convenient method of obtaining the molecular weight distribution once the appropriate calibrations have been worked out. The increasing availability of light scattering detectors for GPC instruments may obviate the need for the universal calibration.

2.6.2 Fractional Solution

In its simplest form, fractional solution[9h,25] might involve extracting a polymer in a Soxhlet-type apparatus to dissolve fractions of increasing molecular weight with time of extraction. More practically it has been adapted as a chromatographic procedure. The chromatographic column is packed with an inert support such as fine sand or glass beads coated with polymer. Nonsolvent is run through the column followed by increasing amounts of solvent mixed with the nonsolvent. Low-molecular-weight polymer dissolves first, followed by increasingly higher-molecular-weight fractions. This is the reverse of GPC in that low-molecular-weight fractions are eluted first. Typical combinations of solvent and nonsolvent are 2-butanone and methanol, respectively, for polystyrene; and xylene and 1-propanol, respectively, for polyethylene. Application of thermal gradients along the column—a method called *solvent-gradient chromatography*—is a variation of fractional solution.

2.6.3 Fractional Precipitation

Fractional precipitation[9h,25] is essentially the reverse of fractional solution. The usual procedure is to add small portions of nonsolvent to a dilute homogeneous solution of polydisperse polymer until high-molecular-weight material precipitates. The precipitate is removed by filtration or decantation and more nonsolvent is added to the solution. The procedure is repeated as many times as necessary to obtain the desired fractionation and all polymer has been recovered.

A variation of fractional precipitation is *turbidimetry*.[22b] As nonsolvent is added to polymer solution, the resultant turbidity, recorded as scattered light in a photocell, is correlated with molecular weight. If temperature changes are used to effect precipitation, the technique is called *thermal gradient turbidimetry*.

2.6.4 Thin-layer Chromatography (TLC)

TLC on alumina- or silica-gel-coated plates has been investigated, not only as a method of determining molecular weight distribution, but also for separating and characterizing polymer types, such as block and graft copolymers.[19c,22c,26–29] There are two principal mechanisms by which polymers are separated on TLC plates: adsorption and precipitation. (Molecular sieving may also occur if a macroporous coating is used.) Adsorption relies on competition between solvent and solute for the adsorbent and is a function primarily of chemical structure. Precipitation involves fractionation by molecular weight; it is the principal mechanism in determining molecular weight distribution.

The obvious advantage of TLC is its low cost and simplicity. As such, it can be used as a routine method for preliminary screening of polymer samples or monitoring polymerization processes. *Paper chromatography* has also been investigated as a method of determining molecular weight distribution,[30] but much less extensively.

References

1. S. G. Weissberg, S. Rothman, and M. Wales, in *Analytical Chemistry of Polymers*, Part 2 (G. M. Kline, Ed.), Wiley-Interscience, New York, 1962, Chap. 1.
2. N. C. Billingham, *Molar Mass Measurements in Polymer Science*, Halsted Press, New York, 1977.
3. H. Morawetz, *Macromolecules in Solution*, 2nd ed., Wiley-Interscience, New York, 1975.
4. J. Brandrup and E. H. Immergut, Eds., *Polymer Handbook*, 3rd ed., Wiley-Interscience, New York, 1989; (a) O. Fuchs, pp. VII/379 ff; (b) J. F. Rudd, p. V/89; (c) E. A. Grulke, pp. VII/519 ff; (d) H.-G. Elias, pp. VII/205 ff; (e) M. Kurata and Y. Tsunashima, pp. VII/1 ff.
5. P. M. Small, *J. Appl. Chem.*, **3**, 71 (1953).
6. K. L. Hoy, *J. Paint Tech.*, **42**, 76 (1970).
7. C. M. Hansen, *J. Paint. Tech.*, **39**, 104, 505, 511 (1967).

8. A. F. M. Barton, *CRC Handbook of Polymer–Liquid Interaction Parameters and Solubility Parameters*, CRC Press, Boca Raton, Florida, 1990.

9. J. F. Rabek, *Experimental Methods in Polymer Chemistry*, Wiley-Interscience, New York, 1980: (a) Chap. 7; (b) Chap. 5; (c) Chap. 6; (d) Chap. 13; (e) Chap. 8; (f) Chap. 9; (g) Chap. 25; (h) Chap. 4.

10. P. E. Slade Jr. (Ed.), *Polymer Molecular Weights*, Part 1, Marcel Dekker, New York, 1975: (a) R. C. Garmon, Chap. 5; (b) R. D. Ulrich, Chap. 2; (c) C. A. Glover, Chap. 4; (d) E. P. Cassassa and G. C. Berry, Chap. 5.

11. L. S. Bark and N. S. Allen (Eds.), *Analysis of Polymer Systems*, Applied Science, London, 1982: (a) G. Davison, Chap. 7; (b) A. R. Cooper, Chap. 8.

12. M. Karas, U. Bahr, A. Deppe, B. Stahl, and F. Hillenkamp, *Makromol. Chem., Macromol. Symp.*, **61**, 397 (1992).

13. U. Bahr, A. Deppe, M. Karas, and F. Hillenkamp, *Anal. Chem.*, **64**, 2866 (1992).

14. L. Prókai, *Field Desorption Mass Spectrometry*, Marcel Dekker, New York, 1989.

15. R. M. O'Malley, M. E. Randazzo, J. E. Weinzierl, J. E. Fernandez, L. N. Nuwaysir, J. A. Castro, and C. L. Wilkins, *Macromolecules*, **27**, 5107 (1994).

16. T. Hohmi and J. B. Fenn, *J. Amer. Chem. Soc.*, **114**, 3241 (1992).

17. R. A. Rhein and D. D. Lawson, *Chemtech*, 122 (Feb. 1971).

18. J. W. Williams, *Ultracentrifugation of Macromolecules*, Academic Press, New York, 1972.

19. P. E. Slade Jr. (Ed.), *Polymer Molecular Weights*, Part 2, Marcel Dekker, New York, 1975: (a) Th. G. Scholte, Chap. 8; (b) D. K. Carpenter and L. Westerman, Chap. 7; (c) A. C. Ouano, E. M. Barrall II, and J. F. Johnson, Chap. 6.

20. International Union of Pure and Applied Chemistry, *J. Polym. Sci.*, **8**, 257 (1952).

21. K. H. Altgelt and L. E. Segal (Eds.), *Gel Permeation Chromatography*, Marcel Dekker, New York, 1971.

22. L. H. Tung (Ed.), *Fractionation of Synthetic Polymers*, Marcel Dekker, New York, 1977: (a) L. H. Tung and J. C. Moore, Chap. 6; (b) H.-G. Elias, Chap. 4; (c) H. Inagaki, Chap. 7.

23. J. C. Moore, in *Liquid Chromatography of Polymers and Related Materials*, Part 3 (J. Cazes, Ed.), Marcel Dekker, New York, 1981, p. 1ff.

24. Z. Grubisic, P. Rempp, and H. Benoit, *J. Polym. Sci.*, **B5**, 753 (1967).

25. R. M. Screaton, in *Newer Methods of Polymer Characterization* (B. Ke, Ed.), Wiley-Interscience, New York, 1964, Chap. 11.

26. E. P. Otocka, in *Polymer Molecular Weight Methods*, Adv. Chem. Ser. 125, American Chemical Society, Washington, D.C., 1973, p. 55.

27. B. G. Benenkii and E. S. Gankina, *J. Chromatog. Rev.*, **141**, 13 (1977).

28. D. W. Armstrong and K. H. Bul, *Anal. Chem.*, **54**, 706 (1981).

29. H. Inagaki, T. Kotaka, and T.-I. Min, *Pure Appl. Chem.*, **46**, 61 (1976).

30. M. I. Siling, V. Ya. Kovner, Yu. Pvyrsky, and O. F. Alkayeva, *J. Chromatog.*, **101**, 83 (1974).

Review Exercises

1. Suppose you wished to make a "model" of a linear polyethylene having a molecular weight of about 170,000 (a reasonable number for a commercial product) using paper clips to represent the repeating unit. How many paper clips would you have to string together?

2. Increasing temperature in general reduces the viscosity of polymer solutions. How might the magnitude of this effect compare in a "poor" solvent and a "good" solvent? (The principle invoked by this exercise is the basis for multiviscosity motor oils.)

3. An experimental method of determining solubility parameters of polymers involves taking lightly crosslinked (network) samples of the polymers in question and immersing them in different solvents whose solubility parameters are known. Explain the rationale behind this method.

4. Determine solubility parameters for amorphous polypropylene and poly(methyl methacrylate). (Densities of the two polymers are 0.905 and 1.18 g/mL, respectively.) Consult a polymer handbook and suggest suitable solvents for the two polymers on the basis of solubility parameters.

5. From the practical standpoint, is it better to use a "good" solvent or a "poor" solvent when measuring polymer molecular weight? Explain.

6. What would be the number average and weight average molecular weight of a sample of propylene oligomer that consists of 5 mol of pentamer and 10 mol of hexamer?

7. Calculate $\overline{M}_n$, $\overline{M}_w$ and the polydispersity index for a hypothetical polymer sample that contains equimolar amounts of polymer having molecular weights of 30,000, 60,000, and 90,000.

8. In exercise 7, change the words "equimolar amounts" to "equal weights" and recalculate.

9. A 0.5000-g sample of an unsaturated polyester resin was reacted with excess acetic anhydride. Titration of the reaction mixture with 0.0102 *M* KOH required 8.17 mL to reach the end point. What is the number average molecular weight of the polyester? Would this method be suitable for determining any polyester? Explain.

10. What is the $\overline{DP}$ of a sample of polyester prepared from 4-hydroxybenzoic acid if the acid number, determined with standard KOH solution, is 11.2?

11. Calculate the freezing point depression, boiling point elevation, and osmotic pressure (in millimeters of water at 25°C) for 1.00% (w/v) aqueous solutions of three samples of a polymer having number average molecular weights of 5000, 50,000, and 100,000. On the basis of your answer, what conclusions might you draw about the relative effectiveness of each method for determining molecular weight?

12. Explain how one might experimentally determine the Mark–Houwink–Sakurada constants *K* and *a* for a given polymer.

13. What would be the viscosity average molecular weight of a hypothetical polymer if the intrinsic viscosity were 2.3 dL/g under theta conditions? Assume a *K* value of 3.6×10^{-3}.

14. Use the Lambert–Beer law to determine the approximate $\overline{M}_n$ and $\overline{DP}$ of a polyimide of structure

if a 0.50-g/L solution in a 1-cm cell exhibits an anthracene end-group absorbance of 0.28 at 360 nm? Assume a molar absorptivity of 7800 for the anthracene unit (M. P. Stevens, *J. Polym. Sci., Polym. Lett. Ed.*, **22**, 467, 1984).

3

CHEMICAL STRUCTURE AND

POLYMER MORPHOLOGY

3.1 Introduction

If asked to list desirable properties of a plastic, one would probably include toughness and durability and possibly, depending on the application, transparency, weather resistance, and heat and flame resistance. For a fiber one would expect high tensile strength, spinnability, and dyeability; for a rubber, resilience; and so on. Because it is properties such as these that make polymers commercially useful, the effect of chemical structure on polymer properties is without doubt the most important aspect of polymer chemistry. It is a subject that recurs throughout the book, and this chapter and the next serve as an introduction to the various factors that contribute to mechanical, physical, or chemical properties in polymeric materials.

We will begin in this chapter with polymer *morphology*—the structure, arrangement, and physical form of polymer molecules—and how morphology relates to chemical structure. In the next chapter we will be concerned more with properties as they relate to commercial application: mechanical strength, thermal and chemical stability, flammability, and so on. The two areas are related. Mechanical and thermal properties of polymers hinge to a large degree on molecular orientation (i.e., morphology) and on chemical constitution.

Generally speaking, two morphologies are characteristic of polymers—*amorphous* and *crystalline*. The former is a physical state characterized by almost complete lack of order among the molecules. The latter refers to the situation where polymer molecules are oriented, or aligned, in a regular array analogous—to a degree—to crystal lattice packing in nonpolymeric solids. Because polymers for all practical purposes never achieve 100% crystallinity, it is more practical to categorize them as amorphous and *semicrystalline*. We will look at characteristics of both amorphous and crystalline (or semicrystalline) states and at the various structural features that give rise to them, including intermolecular and intramolecular forces, stereochemistry, and chemical composition. We will also be concerned with the phenomenon of *liquid crystallinity*, an interesting morphology that walks the tightrope between the purely crystalline solid and the purely amorphous liquid. Finally, we will look at the effect of crosslinking and the increasingly significant science of *polymer blends*, where morphology is of the utmost importance.

3.2 Molecular Weight and Intermolecular Forces

The ultimate properties of any polymer, whether plastic, fiber, or rubber, result from a combination of molecular weight and chemical structure.[1,2] Polymers have to be of a certain molecular weight before they have useful mechanical properties, and the particular molecular weight necessary depends in large measure on chemical structure. These mechanical properties result from attractive forces between molecules: dipole–dipole interaction (including hydrogen bonding), induction forces, dispersion, or London forces between nonpolar molecules. Or they may result from ionic bonding and ion–dipole interactions with polymers containing ionic groups. Some of these intermolecular forces are depicted in Figure 3.1.

Induction (induced dipole) interactions and London forces, particularly the latter, are much weaker than those depicted; nevertheless they become increasingly significant as molecular weight increases. Indeed, with nonpolar polymers such as polyethylene, it is the London forces that give rise to the molecular cohesion and resultant mechanical strength.

As a demonstration of the importance of secondary bonding forces and their relation to molecular weight, polyesters may be compared with polyamides (Figure 3.1). For a polyester and a polyamide having comparable structure—that is, with R being the same in both cases—a higher molecular weight is necessary to achieve good fiber properties in the polyester. The reason lies in the ability of polyamide molecules to associate through hydrogen bonding. Notwithstanding the polarity of the ester group, dipole–dipole interaction is clearly much stronger with polyamides. Similarly, both polyesters and polyamides exhibit strength characteristics at lower molecular weights than does polyethylene. At molecular weights up to about 10,000, polyethylene behaves more like a waxy solid, whereas polyamides are hard, brittle solids at molecular weights as low as 1000 to 2000.

Intermolecular forces drop off very rapidly with distance of separation. It is essential, therefore, that polymer molecules be able to pack together closely to achieve maximum cohesive strength. This phenomenon may be demonstrated very simply by stretching a rubber

FIGURE 3.1. Intermolecular forces in polar polymers: (a) dipole–dipole in a polyester; (b) hydrogen bonding in a polyamide; and (c) ionic in a carboxyl-containing polymer.

band. In the unstretched state, the rubber molecules are randomly distributed and the rubber has a low modulus; it exhibits little resistance to a force applied to it. It is also susceptible to attack by organic solvents. On stretching, however, the molecules become aligned along a common axis, which allows them to pack together more efficiently. At 600% elongation, the rubber band has a modulus about 2000 times greater than in the unstretched state and is considerably more resistant to solvents. Disregarding the structural features that give rise to elastic properties, the unstretched and stretched forms of rubber are, to a rough approximation, analogous to the amorphous and crystalline forms of polymers.

3.3   The Amorphous State—Rheology

The amorphous state[3a] is characteristic of those polymers in the solid state that, for reasons of structure, exhibit no tendency toward crystallinity. It is characteristic of *all* polymers at temperatures above their melting points (except under the special circumstances where liquid crystals may form). If a molten polymer retains its amorphous nature upon cooling to the solid state, the process is called *vitrification* (as opposed to crystallization). In the amorphous state, the polymer resembles a glass. An example of a commercially important vitrified polymer is the poly(ethylene terephthalate) (PET) commonly used in the manufacture of recyclable beverage bottles. In its semicrystalline morphology, on the other hand, poly(ethylene terephthalate) finds use as the common polyester textile fiber.

The amorphous state has been aptly described by one authority[4] as resembling, on a molecular scale, a bowl of cooked spaghetti. The major difference between the solid and liquid amorphous states is that with the former, molecular motion is restricted to very short-range vibrations and rotations, whereas in the molten state there is considerable segmental motion or conformational freedom arising from rotation about chemical bonds. The molten state has been likened[5] on a molecular scale to a can of worms, all intertwined and wriggling about, except that the average "worm" would be extremely long relative to its cross-sectional area. To fully describe the amorphous state, taking into account bond lengths and bond angles, chain conformations, and chain entanglements, requires a statistical treatment[6] that is beyond the scope of this book.

When an amorphous polymer achieves a certain degree of rotational freedom, it can be deformed. If there is sufficient freedom, the polymer flows—the molecules begin to move past one another. The science of deformation and flow is called *rheology*[7–9] and is of fundamental importance in industrial applications. In the traditional sense, rheology is of more interest to the engineer or physicist. It is, after all, the engineer who designs the equipment that converts polymers, usually in the molten state, into useful objects. It is the chemist, however, who designs the molecules the engineer needs; therefore, the two should be on speaking terms. The polymer chemist needs at least a rudimentary understanding of this most important aspect of the amorphous state. A more complete understanding than is provided here requires rigorous mathematical modeling.

To cause a polymer to deform or flow requires the application of a force. If a force is applied, then withdrawn quickly, the polymer molecules tend to revert to their previous, undisturbed configuration, a process called *relaxation.* In other words, the amorphous liquid exhibits a certain *elastic* quality. This elasticity comes about because the molecules were disrupted from what was a thermodynamically favorable arrangement. If the force is applied

PRESSURE

Die swell

Die

Molten polymer

FIGURE 3.2. Schematic representation of polymer flow through a die orifice.

gradually and consistently, the molecules begin to flow irreversibly. Because of chain entanglement and frictional effects, the flowing liquid will be very viscous. This combination of properties—elasticity and viscous flow—is why polymers are referred to as *viscoelastic* materials.[10] A consequence of viscoelasticity is illustrated in Figure 3.2, which depicts molten polymer being forced through a narrow orifice, or die. As the molecules flow rapidly through the die opening, they are compressed. When they emerge, the resultant reduction in pressure causes the molecules to "rebound" to a degree; hence the thickness of the emergent polymer melt will be greater than the width of the die opening. This *die swell*, as it is called, must be taken into account by engineers who design polymer processing machinery. Silly Putty, a siloxane polymer (Chapter 15), is ideal for demonstrating the time dependence of viscoelasticity. If dropped, it bounces, but it can be shaped by the slow application of pressure.

A variety of forces are applicable to polymer deformation, but the one the rheologist is most concerned with is *shear* (or *tangential stress*). Shear is a force applied to one side of a surface in a direction parallel to the surface. If a rectangle is subjected to shear, for example, it becomes a parallelogram. Shear applied to a rectangular is illustrated in Figure 3.3. Shear stress (τ) is defined as the force (F) in dynes (or newtons) per unit of surface area (A) in square centimeters (or square meters); that is,

$$\tau = \frac{F}{A}$$

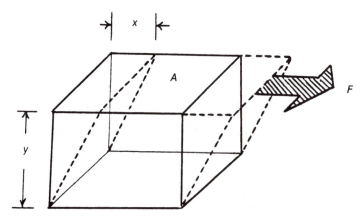

FIGURE 3.3. Representation of shear (tangential stress).

If the solid object is a piece of graphite, the graphite layers, which are held together by secondary bonding forces, slide over each other much as a deck of playing cards moves when shear is applied. It is this property that makes graphite a good lubricant.

When a polymer is in the molten state, shear causes the molecules to flow past one another. Such is the case when a polymer melt is stirred or subjected to pressure, as in Figure 3.2. How readily the molecules flow is a function of temperature (how much kinetic energy they possess); molecular weight (how much the molecules are entangled); and molecular structure (how strong the intermolecular forces are). To shape a plastic or spin a fiber, the polymer molecules have to flow readily—the more readily, the faster the desired object or fiber can be fabricated, and the less costly the process. Viscosity, whether in solution (as is discussed in Chapter 2) or in the melt, is a measure of resistance to flow.

Shear strain, γ, is the amount of deformation of one plane with respect to another. Referring to Figure 3.3,

$$\gamma = \frac{X}{Y}$$

Resistance to shear is the *shear modulus*, G, which is the ratio of shear stress to shear strain,

$$G = \frac{\tau}{\gamma}$$

Shear rate, ($\dot{\gamma}$), also called the *velocity gradient*, is the rate at which the planes (in Figure 3.3) or molecules (in an amorphous liquid) flow relative to one another; that is,

$$\dot{\gamma} = \frac{d\gamma}{dt}$$

If the shear stress increases in proportion to the shear rate, the liquid is said to be *Newtonian* (or ideal) because it follows Newton's law of viscosity, namely

$$\tau = \eta\dot{\gamma}$$

where the proportionality constant, η, is the viscosity. Units of viscosity are poises (dyne s/cm^2) or, in the SI system, Pascal-seconds (Pa s $=$ newton s/m^2). For comparison, a typical gas such as air has a viscosity of the order of 10^{-5}, water about 10^{-3}, glycerine about 1, and molten polymer about 10^2 to 10^6 Pa s. Viscosity is related to temperature by an Arrhenius-type equation:

$$\eta = Ae^{-E_a/RT}$$

where A is a constant related to molecular motion and E_a is the activation energy for viscous flow. E_a is determined mainly by localized segmental motion of the polymer chains, as shown by its relative insensitivity to molecular weight but high dependence on chain structure and branching. The more bulky the chain branch or substituent, the higher the E_a; or the bulkier the group, the more sensitive the polymer viscosity to changes in temperature.

Deviations from ordinary Newtonian behavior are common in polymer flow. One type results when an initial resistance to flow must be overcome, but once flow begins it follows Newtonian behavior. Such a liquid is called a *Bingham Newtonian fluid* (named for Eugene Cook Bingham, who developed much of the modern theory of rheology) and is defined by

$$\tau = \tau_c + \eta\dot{\gamma}$$

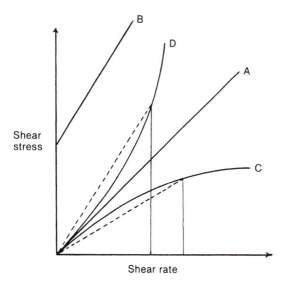

FIGURE 3.4. Types of shear flow: (A) Newtonian; (B) Bingham Newtonian; (C) shear thinning (pseudoplastic); and (D) shear thickening (dilatant).

where τ_c is the critical shear stress, or threshold stress, needed to initiate flow. What causes Bingham-type behavior? Most probably it arises from some type of structured arrangement of the molecules arising from conformational and secondary bonding forces that must be disrupted by the application of an initial stress (τ_c) before the molecules begin to flow.

Other deviations, which are non-Newtonian, occur where shear stress does not increase in direct proportion to shear rate. The deviation may be in the direction of *thinning* (also called pseudoplastic) or *thickening* (dilatant). These various possibilities are illustrated in Figure 3.4. Most commonly, polymers exhibit shear thinning. As might be expected, a Bingham fluid might also exhibit non-Newtonian thinning or thickening. In cases of non-Newtonian flow, it is conventional to define viscosity as the slope of a secant drawn from the origin to any point on the stress–rate curve (shown by the broken lines in Figure 3.4), with the resultant shear rate at that point specified.

Related to non-Newtonian flow with shear thinning is the behavior of *thixotropic* liquids, which have gel-like properties or a high viscosity at low stress but thin out and hence become more workable on being stirred. (Commercial paints are usually thixotropic.) The basic difference between the two is that shear thinning is dependent on shear rate, whereas thixotropic behavior is independent of shear rate but dependent on time at a fixed shear rate.

Flow behavior may also be expressed conveniently in terms of what is referred to as a power law equation:

$$\tau = A\dot{\gamma}^B$$

where A is a constant and B an index of flow. For a Newtonian fluid, $B = 1$ and $A = \eta$. A plot of $\log \tau$ versus $\log \dot{\gamma}$ (Figure 3.5) is linear with slope equal to B and intercept equal to $\log A$.

An obvious question is, *why* do polymers characteristically exhibit shear thinning at high shear rates? On the molecular level this may be explained in terms of molecular entanglement. In the amorphous state there is considerable entanglement of the chains, and while low

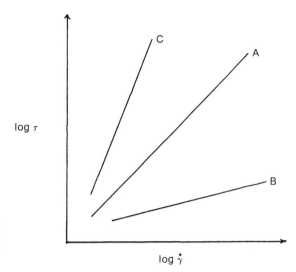

FIGURE 3.5. Power law plots: (A) Newtonian (slope = 1); (B) shear thinning (slope <1); and (C) shear thickening (slope >1).

shear rates may disrupt this to a degree, the mass retains its entangled character. As shear rate increases, disruption may occur faster than the chains can reentangle, and the resultant decreasing entanglement allows the molecules to flow with less resistance, hence the decrease in viscosity. This is shown schematically in Figure 3.6. (Figure 3.2 illustrates the molecules disentangling as they flow rapidly through the die orifice, with subsequent reentanglement in the die swell region.)

If shear rate is increased sufficiently, chain rupture may occur with a resultant drop in molecular weight. (As will be shown later, this may actually be used to advantage to form block copolymers or polymer blends.) The much less common shear thickening may be envisioned as arising from an *increase* in molecular entanglement resulting from some peculiarity of molecular structure as shear rate increases; for example, a reasonably ordered arrangement of molecules might become more disordered, hence more entangled, as the shear rate increases.

Much has been said to this point about molecular entanglement, which is clearly going to increase as the molecular weight increases. Not unexpectedly, therefore, molecular weight is a critical variable in rheology. Interestingly, studies have shown that with flexible chain polymers, there exists a critical molecular weight ($\overline{M}_c$) for entanglement to begin. This is illustrated schematically in Figure 3.7, which plots the log of Newtonian (or zero) viscosity

Increasing shear rate

FIGURE 3.6. Shear thinning arising from disentanglement.

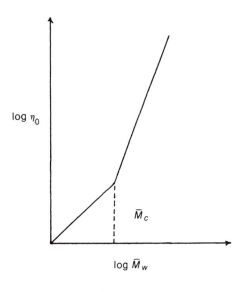

log η_0

$\overline{M}_c$

log $\overline{M}_w$

FIGURE 3.7. Effect of molecular weight on Newtonian viscosity.

(η_0) against the log of weight average molecular weight. Below $\overline{M}_c$ the slope is unity, followed by a sharp increase in η_0. For most common polymers, $\overline{M}_c$ falls in the range 4,000 to 15,000. Although $\overline{M}_c$ varies from one polymer to another, it has been shown, by eliminating mass effects arising from substituents on the chains, that $\overline{M}_c$ is a function of chain length, and chain length is remarkably constant from one polymer to another, corresponding to a $\overline{DP}$ of about 600. In other words, a critical chain length, rather than a critical molecular weight per se, is necessary for entanglement.

Considering the fact that η_0 is increasing *logarithmically* with molecular weight in Figure 3.7, it should be clear why molecular weight control is important in polymer processing. One needs a molecular weight high enough to attain good mechanical properties but not so high that the molten polymer is too viscous to be processed economically. Molecular weight *distribution* also influences flow behavior. In general, the broader the range, the lower the shear rate at which shear thinning develops. However—and this again is important in processing—the narrower the molecular weight distribution, the more pronounced will be the shear thinning once it is manifested, because of the absence of the more highly tangled high-molecular-weight constituents. This is illustrated in Figure 3.8.

Chain branching is another factor that influences flow. The more highly branched a given polymer, the lower will be its hydrodynamic volume and the lower its degree of entanglement at a given molecular weight. We can make the general observation, therefore, that viscosity is higher with linear than with branched polymers at a given shear rate and molecular weight. This does not mean that chain branching is necessarily desirable. As will be seen later, branching results in weaker secondary bonding forces and possibly poorer mechanical properties.

One other factor important in flow is polymer conformation (i.e., shape). In our earlier discussions of the Mark–Houwink–Sakurada equation,

$$[\eta] = KM_v{}^a$$

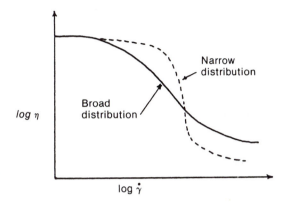

FIGURE 3.8. Effect of molecular weight distribution on shear thinning.

we observed that a varies from 0.5 for a random coil to about 1 for a more rodlike extended shape. While this applies to infinitely dilute solutions, one might expect the same kind of frictional effects in molten polymer, regardless of any reduced entanglement. This is, in effect, realized. More rigid polymers are significantly more viscous, except in the unusual situation where chain conformation results in liquid crystal behavior. (This will be considered later in the chapter.)

One final point needs to be considered: How does one determine viscosity of a polymer melt?[11a] One type of viscometer, the cone–plate rotational viscometer, is shown in Figure 3.9. The molten polymer is contained between the bottom plate and the cone, which is rotated at a constant velocity (Ω). Shear stress (τ) is defined as

$$\tau = \frac{3M}{2\pi R^3}$$

where M is the torque in dynes per centimeter (CGS) or in newtons per meter (SI), and R is the cone radius in centimeters or meters. Shear rate ($\dot{\gamma}$) is given by

$$\dot{\gamma} = \frac{\Omega}{\alpha}$$

where Ω is the angular velocity in degrees per second (CGS) or in radians per second (SI) and α is the cone angle in degrees or radians. Viscosity is then

$$\eta = \frac{\tau}{\dot{\gamma}} = \frac{3\alpha M}{2\pi R^3 \Omega} = \frac{kM}{\Omega}$$

FIGURE 3.9. Cone–plate rotational viscometer.

where

$$k = \frac{3\alpha}{2\pi R^3}$$

a constant defined by viscometer design. Other types of viscometers consist of rotating cylinders immersed in the viscous fluid, or steel capillaries through which the molten polymer is forced at constant pressure or constant flow.

3.4 Glass Transition Temperature

One of the most important characteristics of the amorphous state is the behavior of a polymer during its transition from solid to liquid. If an amorphous glass is heated, the kinetic energy of the molecules increases. Motion is still restricted, however, to short-range vibrations and rotations so long as the polymer retains its glasslike structure. As temperature is increased further, there comes a point where a decided change takes place; the polymer loses its glasslike properties and assumes those more commonly identified with a rubber. The temperature at which this takes place is called the *glass transition temperature* (T_g).[12,13a] (Less commonly it is called the *second-order transition temperature*, to differentiate it from the thermodynamic first-order transition of melting.) If heating is continued, the polymer will eventually lose its elastomeric properties and melt to a flowable liquid.

The importance of the glass transition temperature cannot be overemphasized. It is one of the fundamental characteristics as it relates to polymer properties and processing. What takes place at the molecular level at the glass transition temperature? As one might expect, the transition is accompanied by more long-range molecular motion—greater rotational freedom and consequently more segmental motion of the chains. It is estimated that between 20 and 50 chain atoms are involved in this segmental movement at the T_g. Clearly for this increased motion to take place, the space between the atoms (the *free volume*) must increase, which gives rise to an increase in the specific volume. The temperature at which this change in specific volume takes place, usually observed by dilatometry (volume measurement), may be used as a measure of T_g.

Other changes of a macroscopic nature occur at the glass transition. There is an enthalpy change, which may be measured by calorimetry. The modulus, or stiffness, decreases appreciably, the decrease readily detected by mechanical measurements. Refractive index and thermal conductivity change. We will not be concerned here with details of how T_g is measured (this is covered in Chapter 5), but rather with the nature of the phenomenon and the various factors that influence the temperature at which it occurs.

What determines the glass transition temperature? Clearly chemical structure is most important, and we will examine that in some detail. Other variables, however, make T_g an approximate value at best. Method of measurement is one: A range of several degrees may be observed between a thermal and a mechanical measurement. Difficulties in obtaining reproducible samples is another. The T_g of a polymer may even change with time because of such "aging" factors as oxidation or exposure to ultraviolet light. Finally there is molecular weight. The higher the molecular weight, the fewer the chain ends, which leads to a lower free volume. Polystyrene's T_g, for example, increases from about 40°C at an $\overline{M}_n$ of 3,000 to about 100°C at 300,000.

These various factors combine to make T_g an approximation, and any literature compilation must, therefore, be treated with some caution.[14a] It is not uncommon to see discrepancies of as much as 30°C for a given polymer. With this in mind, let us now turn our attention to the effect of chemical structure. While structural effects are not always predictable, one can begin with some generalizations. Given that T_g is a function of rotational freedom, whatever restricts rotation should increase T_g. Any compound having carbon–carbon single bonds is capable of existing in an infinite number of conformations unless freedom of rotation is impaired. In most instances, energy barriers to rotation are low, but even with ethane there is a torsional barrier of about 3 kcal/mol, which results in a preference for the staggered conformation (**1**). As the hydrogens are replaced with more bulky groups, the energy barrier increases. Butane exists mainly in the *anti* conformation (**2**) which is some 5 kcal/mol more stable than the most energetic conformation in which the methyls are eclipsed. This may be carried to extreme cases where rotation is blocked, as with the *ortho-ortho'* substituted biphenyls (**3**), which are capable of existing as configurational isomers if the substituent groups are large enough.

1   2   3

In the case of polymers, therefore, it follows that the bulkier the substituents on the polymer backbone, the less will be the rotational freedom and the higher the T_g. This is true to a degree; however, as shown with polymers **4** through **23** in Table 3.1, other factors come into play. The first grouping in the table (**4–11**) contains polymers with increasing length of the pendant group. As expected, polypropylene has a higher T_g than polyethylene; but then there is a decrease as the number of methylene groups in R increases, followed by an increase at even longer substituent chain lengths. This behavior is no doubt due to a plasticizing effect of the flexible side chain that is offset at higher chain lengths by entanglement or by side-chain crystallization. (In the next chapter we shall see that plasticizers are very often added to polymers to lower the T_g.) Effect of size is also apparent in comparing polypropylene (**5**) with polystyrene (**21**) and poly(2-vinylnaphthalene) (**22**).

The second grouping in Table 3.1 (**12–16**) contains branched substituent groups. A comparison with polymers in the first grouping, which are structural isomers, shows clearly that branching increases T_g significantly. (Compare **12** with **7**, **13** and **14** with **8**, **15** and **16** with **9**.) A further comparison of **13** with **14** and **15** with **16**, as well as the two vinyl ether polymers (**17** and **18**), confirms this trend.

In addition to size, polarity influences T_g, as shown by poly(vinyl chloride) (**19**) and poly(vinyl alcohol) (**20**). Chloro and hydroxy groups are comparable in size to methyl, but the more polar groups lead to higher T_gs because of increased dipole–dipole interaction. In the case of **20**, one might also envision decreased rotational freedom arising from intramolecular hydrogen bonding. Polarity effects are also evident in comparing polystyrene (**21**) and poly(4-vinylpyridine) (**23**).

Table 3.2 lists T_gs for some representative nonvinyl polymers. A comparison of polyesters and polyamides of similar structure (**25** and **29**) illustrates the importance of hydro-

TABLE 3.1. Glass Transition Temperature (T_g) of Representative Vinyl Polymers[a]

Number	Polymer $\left[CH_2CH \right]_{R}$	R	T_g (°C)
4	Polyethylene	H	−20
5	Polypropylene	CH_3	5
6	Poly(1-butene)	CH_3CH_2	−24
7	Poly(1-pentene)	$n\text{-}C_3H_7$	−40
8	Poly(1-hexene)	$n\text{-}C_4H_9$	−50
9	Poly(1-heptene)	$n\text{-}C_5H_{11}$	−31
10	Poly(1-decene)	$n\text{-}C_8H_{17}$	−41
11	Poly(1-dodecene)	$n\text{-}C_{10}H_{21}$	−6
12	Poly(3-methyl-1-butene)	$CH(CH_3)_2$	50
13	Poly(4-methyl-1-pentene)	$CH_2CH(CH_3)_2$	29
14	Poly(3,3-dimethyl-1-butene)	$C(CH_3)_3$	64
15	Poly(5-methyl-1-hexene)	$CH_2CH_2CH(CH_3)_2$	−14
16	Poly(4,4-dimethyl-1-pentene)	$CH_2C(CH_3)_3$	59
17	Poly(vinyl *n*-butyl ether)	$O\text{-}C_4H_9\text{-}n$	−55
18	Poly(vinyl *t*-butyl ether)	$O\text{-}C_4H_9\text{-}t$	88
19	Poly(vinyl chloride)	Cl	81
20	Poly(vinyl alcohol)	OH	85
21	Polystyrene	C_6H_5	100
22	Poly(2-vinylnaphthalene)		151
23	Poly(4-vinylpyridine)		142

[a] Data from Peyser.[14a]

gen bonding in increasing T_g. Also significant is the chain-stiffening effect of the benzene ring. (Compare polyesters **24** with **26** and **25** with **27**.)

The high T_g of **28**, a special type of polyester called a *polycarbonate*, reflects a further reduction in the flexibility of the polymer backbone. The monomer used to prepare **28**, 2,2-bis(4-hydroxyphenyl)propane, **34** (commonly called bisphenol A), finds use in a variety of polymer formulations where a high T_g is desirable. An extreme case of restricted rotation is shown with the semiladder polybenzimidazole, **33**. These various polymer types are discussed in more detail in later chapters.

34

TABLE 3.2. Glass Transition Temperature (T_g) of Representative Nonvinyl Polymers[a]

Number	Polymer	T_g (°C)
24	$-\overset{O}{\overset{\|}{C}}(CH_2)_4\overset{O}{\overset{\|}{C}}O(CH_2)_2O-$	−63
25	$-\overset{O}{\overset{\|}{C}}(CH_2)_4\overset{O}{\overset{\|}{C}}O(CH_2)_4O-$	−118
26	$-\overset{O}{\overset{\|}{C}}-\bigcirc-\overset{O}{\overset{\|}{C}}O(CH_2)_2O-$	69
27	$-\overset{O}{\overset{\|}{C}}-\bigcirc-\overset{O}{\overset{\|}{C}}O(CH_2)_4O-$	80
28	$-O-\bigcirc-\overset{CH_3}{\underset{CH_3}{\overset{\|}{\underset{\|}{C}}}}-\bigcirc-O\overset{O}{\overset{\|}{C}}-$	149
29	$-\overset{O}{\overset{\|}{C}}(CH_2)_4\overset{O}{\overset{\|}{C}}NH(CH_2)_6NH-$	57
30	$-\overset{O}{\overset{\|}{C}}(CH_2)_4\overset{O}{\overset{\|}{C}}NH(CH_2)_{10}NH-$	50
31	$-NH(CH_2)_5\overset{O}{\overset{\|}{C}}-$	77
32	$-NH(CH_2)_{10}\overset{O}{\overset{\|}{C}}-$	46
33	(benzimidazole structure)	429

[a] Data from Peyser.[14a]

The effect of stereochemistry on T_g is shown in Table 3.3. From the limited data provided it is impossible to make generalizations as to which geometric configuration results in higher glass transition temperatures. Indeed, in the case of 1,4-polyisoprene, it has been demonstrated that there is very little difference between *cis* and *trans* isomers even when there are variations in morphology.[15]

For a polymer to serve as a useful plastic, its glass transition temperature must be appropriately higher than the temperature of its intended work environment. A plastic used in manufacturing coffee cups obviously needs a T_g that is well above the temperature of hot coffee. Glass transition temperatures of elastomers, on the other hand, are *below* room temperature.

TABLE 3.3. Glass Transition Temperature (T_g) of Diene Polymers

	T_g (°C)	
Polymer	*cis*	*trans*
1,4-Polybutadiene[a]	−102	−58
1,4-Polyisoprene[b]	−67	−70
1,4-Polychloroprene[a]	−20	−40

[a] Data from Peyser[14a]

[b] Data from Burfield and Lim.[15]

3.5   Stereochemistry

Before we examine the crystalline morphology, it behooves us to consider one of the most important structural parameters that leads to crystallinity, namely stereoregularity. That Ziegler and Natta shared the 1963 Nobel Prize in Chemistry for their work in this area attests to its significance.

For reasons discussed later, most commercially important vinyl monomers are monosubstituted (CH_2=CHR) or 1,1-disubstituted (CH_2=CR_2, where the R groups may be the same or different). Furthermore, such monomers undergo primarily *head-to-tail polymerization* (reactions 3.1 and 3.2) to yield polymers having the substituent groups on alternate carbon atoms. This is not to say that *head-to-head* or *tail-to-tail* units do not form; they do, but in most instances they contribute very little to the overall polymer structure. Given the fact that each substituent-bearing carbon on the polymer backbone may be a stereogenic (chiral) center, a number of stereochemical variations are possible.[16–18]

$$CH_2{=}CHR \longrightarrow -CH_2-\overset{\displaystyle R}{\underset{\displaystyle |}{CH}}-CH_2-\overset{\displaystyle R}{\underset{\displaystyle |}{CH}}-CH_2-\overset{\displaystyle R}{\underset{\displaystyle |}{CH}}- \qquad (3.1)$$

$$CH_2{=}CR_2 \longrightarrow -CH_2-\overset{\displaystyle R}{\underset{\displaystyle \underset{\displaystyle R}{|}}{\overset{\displaystyle |}{C}}}-CH_2-\overset{\displaystyle R}{\underset{\displaystyle \underset{\displaystyle R}{|}}{\overset{\displaystyle |}{C}}}-CH_2-\overset{\displaystyle R}{\underset{\displaystyle \underset{\displaystyle R}{|}}{\overset{\displaystyle |}{C}}}- \qquad (3.2)$$

For head-to-tail polymers prepared from monomers of type CH_2=CHR, there are two possible stereoregular arrangements—one in which each chiral center has the same configuration, and one in which alternate chiral centers have the same configuration. Natta proposed the terms *isotactic* for the former[19] and *syndiotactic* for the latter.[20] A completely random distribution is called *atactic* or *heterotactic*. The two stereoregular arrangements are represented as idealized extended zig-zag chains in Figure 3.10. As will be seen later, all three types can be synthesized. Where blocks of stereoregular units are formed along the backbone, the term *stereoblock* describes the polymer.

Analogous structures may be drawn for polymers prepared from 1,1-disubstituted monomers having different substituent groups (e.g., methyl methacrylate, **35**), but not from

FIGURE 3.10. Stereoregular polymers derived from monomer CH_2=CHR.

those having identical groups (e.g., vinylidene chloride, **36**), which yield polymers having no chiral centers.

35

36

It is convenient to think of isotactic and syndiotactic in terms of the *meso* and *racemic* structures familiarly used with low-molecular-weight compounds.[21] Let us consider a segment of the polymer chain consisting of two monomeric units. Such is referred to as a *tactic dyad* or *tactic placement.* An isotactic dyad (**37**) has the *meso* configuration (designated *m*), whereas a syndiotactic dyad (**38**) is *racemic* (*r*). Compare, for example, *meso* (**39**) and

37

38

racemic (**40** + **41**) 2,4-dichloropentane. The *r* and *m* designations are particularly useful in classifying heterotactic polymers. Tactic *triads*, for example, have three possible sequences,

meso

39

enantiomer

40

enantiomer

41

mm (**42**), *rr* (**43**), and *mr* (or *rm*) (**44**), corresponding to isotactic, syndiotactic, and heterotactic, respectively. Tetrads have six: *rrr*, *rrm* (or *mrr*), *rmr*, *mrm*, *mmr* (or *rmm*), and *mmm*.

Later we shall see how NMR spectroscopy may be used to identify such sequences in heterotactic polymers.

$$mm$$
42

$$rr$$
43

$$mr$$
44

A variety of other stereochemistries are possible. Monomers of type RCH=CHR (both R groups the same) may also yield isotactic or syndiotactic polymer (Figure 3.11) having adjacent chiral centers. In this case tacticity refers to *every* backbone carbon, not alternate carbons as it is above. Note that when adjacent carbons have the same configuration (isotactic), the eclipsed conformation, **45**, corresponds to **46** in the staggered conformation (the zigzag arrangement shown in Figure 3.11). Similarly the eclipsed conformation of a syndiotactic placement, **47**, corresponds to **48** in the staggered conformation.

45   corresponds to   **46**

47   corresponds to   **48**

Monomers of type RCH=CHR′ (nonidentical R groups) may exhibit *two* isotactic stereochemistries and one syndiotactic, as shown in Figure 3.12. As is the case with low-molecular-weight compounds, the prefixes *threo* and *erythro* refer to similar groups being on the opposite side or on the same side, respectively, when the eclipsed conformations are shown. The corresponding Fischer projections would be

H―R
R′―H

threo

H―R
H―R′

erythro

FIGURE 3.11. Stereoregular polymers derived from mononuclear RCH=CHR.

Only one syndiotactic form exists because the syndiotactic *threo* and *erythro* structures are identical. The fact that they are identical may not be immediately obvious when looking at Figure 3.12, but drawing short-chain segments of *threo* and *erythro* structures or constructing models helps clarify the situation.

Cyclic monomers such as cyclobutene can, in principle, form four stereoregular structures (Figure 3.13). Substituted dienes of type RCH=CH—CH=CHR′ form *tritactic* 1,4-polymers (**49**), the double bond and the two stereogenic centers representing the three centers of tacticity. The R and R′ groups here may be the same or different. Figure 3.14 illustrates two of the many possible stereoregular structures.

$$-CH-CH=CH-CH-CH-CH=CH-CH-$$

with R, R′, R, R′ substituents

49

None of the polymers just described exists in the idealized extended-chain form depicted; rather, steric effects most often lead to a helical conformation. A case in point is the com-

FIGURE 3.12. Stereoregular polymers derived from monomer RCH=CHR′.

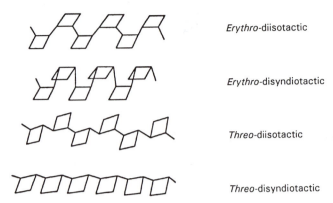

Erythro-diisotactic

Erythro-disyndiotactic

Threo-diisotactic

Threo-disyndiotactic

FIGURE 3.13. Stereoregular polymers derived from cyclobutene.

mercially important isotactic polypropylene in which each methyl group is displaced 120° from the next.

Which of the various stereoregular structures predominates, and the extent of stereoregularity, depend on a variety of factors: whether the monomer double bond is *cis* or *trans*, whether the double bond opens in a *cis* or *trans* manner during polymerization, type of initiator, polymerization temperature, and solvent polarity. All of these are explored later in the book. Some of the structures given are hypothetical; only the *erythro* forms of polycyclobutene, for example, have been identified.

One other aspect of stereoregularity: Polymers have been synthesized in which alternate stereogenic centers are stereoregular, with the remaining centers atactic. The term *hemitactic* has been proposed for such polymers (i.e., hemiisotactic or hemisyndiotactic).[22]

Most important from the standpoint of morphology is that in most instances, regardless of the type of stereochemistry, the tendency for crystalline polymer to form increases with increasing stereoregularity.

Before we leave the subject of stereochemistry, we should reflect a moment on the glass transition temperature. Given that stereoregularity leads to crystallinity, does it also lead to higher T_g's? Prevailing evidence[23] suggests that for a given polymer, syndiotactic usually exhibits a higher T_g than isotactic, as shown by the data in Table 3.4. The atactic T_g is generally

Trans–erythro-diisotactic

Cis–threo-disyndiotactic

FIGURE 3.14. Two stereoregular 1,4-polymers derived from monomer $RCH{=}CH{-}CH{=}CHR'$.

TABLE 3.4. Glass Transition Temperature (T_g) of Polymers of Varying Tacticity[a]

Polymer	T_g (°C)		
	Syndiotactic	*Atactic*	*Isotactic*
Poly(methyl methacrylate)	105	105	38
Poly(ethyl methacrylate)	65	65	12
Poly(*t*-butyl methacrylate)	114	118	7
Polypropylene[b]	−4	−6	−18
Polystyrene	100[c]		99[c]

[a] Data from Peyser[14a] unless otherwise noted.
[b] Data from Burfield and Doi.[23]
[c] Data from Ishihara, Kuramoto, and Uoi, *Macromolecules*, **21**, 3356 (1988).

equal to or close to that of the syndiotactic. Differences are not necessarily pronounced, however, and are subject to variations in the test sample's morphology. It must be remembered that the glass transition is a characteristic of the *amorphous* state, which is still present to varying degrees in semicrystalline polymers. Glass transitions are not always easy to detect in highly crystalline polymers because of the relatively small amount of amorphous material present; nevertheless, it is these amorphous regions that are undergoing the transition.

3.6 Crystallinity

When a polymer has a highly stereoregular structure with little or no chain branching, or when it contains highly polar groups that give rise to very strong dipole–dipole interactions, it may exist in crystalline form.[13b,24–26] Such crystallinity is unlike that of low-molecular-weight compounds, but exists instead in regions of the polymer matrix where polymer molecules order themselves in a thermodynamically favorable alignment. The *fringed micelle* model defines crystallinity in terms of such ordered regions, called *crystallites*, in which any particular polymer chain may extend through a number of crystallites, as shown in Figure 3.15.

Crystallinity is induced in any number of ways—cooling of molten polymer, evaporation of polymer solutions, or heating of a polymer under vacuum or in an inert atmosphere (to prevent oxidation) at a specified temperature, a process called *annealing*. In some instances crystallization may be brought about by stretching a polymer sample at a temperature above its glass transition temperature, a process called *drawing*. One might envision drawing as removing the "kinks" from the polymer chains and extending them along a common axis to promote closer packing. Stretching certain varieties of clear plastic film occasionally allows visual observation of the onset of crystallinity as opaque regions in the film. Drawing is a common commercial process for improving mechanical properties of films and fibers. Each of these methods of inducing crystallization allows the polymer molecules the vibrational and rotational freedom to reorient themselves into a crystalline morphology.

FIGURE 3.15. Schematic representation of a polymer matrix showing crystalline regions (fringed micelle model).

The individual crystalline regions thus formed are interconnected by amorphous regions. Crystalline regions may contain occasional kinks or folds, called *defects*, in the polymer chains. Studies of polymer single crystals[27] using x-ray diffraction techniques indicate a plateletlike structure of about 100 Å thickness composed of polymer chains folded back on themselves to give parallel blocks perpendicular to the face of the crystal. This is called the *folded-chain lamella* model. As depicted in Figure 3.16, the parallel chains may be connected with adjacent reentry folds of uniform or nonuniform thickness or with a more amorphous nonadjacent folding (referred to as the *switchback* model). Which of the lamella models depicted in Figure 3.16 best represents reality, and what the true nature of the amorphous regions between lamellae is, are not known with certainty. Less common are *extended-chain crystals*,[28] characterized by an absence of folding and a crystal size dependent on the extended-chain length of the polymer molecules. Extended-chain crystals, which often take the form of well-developed needles, are usu-

| | | |
| (a) | (b) | (c) |

FIGURE 3.16. Folded-chain lamella model: (a) regular adjacent folds; (b) irregular adjacent folds; and (c) nonadjacent switchback.

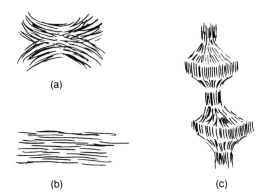

FIGURE 3.17. Some crystalline morphologies: (a) spherulitic; (b) drawn fibrillar; and (c) epitaxial (shish kebab).

(a)

(b)   (c)

ally formed with polymers of relatively low mo-lecular weight by slow crystallization from the melt or under high pressure. Stereoregularity does not automatically lead to crystallinity; a molten stereoregular polymer may, on rapid cooling, solidify in a metastable amorphous state.

Onset of crystallinity is called *nucleation*. Nucleation may occur randomly throughout the matrix as polymer molecules begin to align (homogeneous nucleation) or may occur at the interface of a foreign impurity, perhaps a purposely added nucleating agent such as finely divided silica (heterogeneous nucleation). A considerable variety of crystalline morphologies have been identified by x-ray analysis, three of which are shown in Figure 3.17. *Spherulites* are aggregates of small hairlike strands, called *fibrils*, arranged as clusters in an essentially radial pattern. The spherulitic morphology occurs commonly when nucleation originates in molten polymer or in concentrated polymer solutions in the absence of any applied stress. Drawing forces the spherulitic fibrils into the *drawn fibrillar morphology*. *Epitaxial* crystallinity, which usually arises during crystallization in stirred solutions or melts, is characterized by one crystalline growth on another, as in the so-called shish-kebab morphology that contains lamella growth on long fibrils.

Crystalline polymers are generally tougher, stiffer, more opaque, more resistant to solvents, and of higher density than their amorphous counterparts. (Recall from Chapter 1 the discussion of high- and low-density polyethylene; the linear, high-density form has a high degree of order and is semicrystalline). The higher the degree of crystallinity, the more pronounced the properties. The superior mechanical properties are a reflection of the greater cohesive strength arising from more effective intermolecular secondary forces among the closely packed molecules. Opaqueness arises from scattering of light by the crystallites.

Because crystalline polymers are more resistant to solvents, the fraction of a sample that fails to dissolve may be used as a crude measure of the degree of crystallinity or tacticity. Other methods are based on density measurements, x-ray diffraction, and spectroscopic and thermal measurements.[11b]

Unlike amorphous polymers, which liquify gradually over a broad temperature range above the glass transition temperature, crystalline (or semicrystalline) polymers melt over a relatively narrow range which may be observed most conveniently by thermal analysis (Chapter 5) or by observing the disappearance of birefringence (double refraction) under a polarizing microscope. The value thus obtained is called the *crystalline melting point, T_m*. Sometimes T_m is referred to as the *first-order transition temperature*. Like T_g, the crystalline

TABLE 3.5. Approximate Glass Transition Temperatures (T_g) and Crystalline Melting Temperatures (T_m) of Some Common Polymers[a]

Polymer	T_g		T_m	
	°C	K	°C	K
cis-1,4-Polyisoprene	−73	200	26	299
trans-1,4-Polyisoprene	−58	215	74	347
cis-1,4-Polychloroprene	−20	253	70	343
trans-1,4-Polyisoprene	−40	233	101	374
Poly(methyl vinyl ether)	−31	242	144	417
Poly(vinyl chloride)	81	354	265	538
Polystyrene (isotactic)	100	373	242	515
Poly(ethylene adipate)	−63	210	54	327
Poly(hexamethylene adipamide)	57	330	268	541
Poly(ethylene terephthalate)	69	342	268	541

[a]Data from Peyser[14a] and Miller[14b] (average of reported values).

melting point is not an exact value, but may vary a few degrees from one sample to another because of method of measurement or rate of heating.[14b] Because the same intermolecular forces are responsible for the magnitude of T_g and T_m, it follows that the values should parallel one another to an extent. As a rough approximation, T_gs are generally about one half to two thirds the value of the corresponding T_m, in kelvins. Table 3.5, which lists T_g and T_m values for a few representative polymers, illustrates this relationship.

The importance of stereoregularity in promoting crystallinity may be seen in the free radical polymerization of vinylidene chloride (1,1-dichloroethene) and vinylidene cyanide (1,1-dicyanoethene) to give highly crystalline polymers **50** and **51**, respectively. Under sim-

    50   **51**

ilar conditions, the monosubstituted analogs, vinyl chloride and acrylonitrile, yield atactic polymer with a very low degree of crystallinity. Similarly, polyesters **52** and **54** are crystalline, whereas polyester **53**, which contains a stereogenic carbon, is not.[29] Stereoregularity is not always a prerequisite to crystallinity. Atactic poly(vinyl alcohol) is semicrystalline because the small size of the OH group allows it to fit into the crystalline lattice with the hydrogen atoms.

Crystalline (melting point 220°C)

    52

$$\left[\text{OCH}_2\overset{\overset{\text{CH}_3}{|}}{\text{CH}}\text{CH}_2\text{O}\overset{\overset{\text{O}}{\|}}{\text{C}}-\bigcirc-\overset{\overset{\text{O}}{\|}}{\text{C}}\right] \qquad \text{Noncrystalline}$$

53

$$\left[\text{OCH}_2\underset{\underset{\text{CH}_3}{|}}{\overset{\overset{\text{CH}_3}{|}}{\text{C}}}\text{CH}_2\text{O}\overset{\overset{\text{O}}{\|}}{\text{C}}-\bigcirc-\overset{\overset{\text{O}}{\|}}{\text{C}}\right] \qquad \text{Crystalline (melting point 140°C)}$$

54

From the standpoint of polymer properties, crystallinity is extremely important because of its influence on mechanical properties. This aspect of structure–property relationships is explored in the following chapter.

3.7 Liquid Crystallinity

Liquid crystals[30,31] are neither true liquids nor true solids; hence they have been referred to as a fourth state of matter. Pure liquids are *isotropic*—their molecules lack order. Crystalline materials are *anisotropic*. Liquid crystallinity occurs when molecules become aligned in a crystalline array while still in the liquid state; in other words, the liquid exhibits anisotropic behavior. The ordered regions in the liquid are called *mesophases*. As might be expected intuitively, the types of molecules that exhibit a propensity for forming mesophases are those with a relatively rigid, elongated, or disklike structure. With low-molecular-weight compounds, liquid crystallinity has been recognized for more than a century, although detailed studies and commercial application did not begin until the late 1950s and early 1960s. Liquid crystalline behavior was not recognized in polymers until the 1970s.

Liquid crystals exhibit certain properties not found in liquid or crystalline solid states. For example, their morphology may be influenced by external magnetic or electrical fields, sometimes they change color with temperature, and some exhibit extremely high optical rotation. At the same time they exhibit the fluidity of liquids and the opaqueness of crystalline solids. With low-molecular-weight compounds, a wide variety of ordered liquid crystalline arrangements have been identified; they vary according to temperature, chemical structure, and whether they exist in the melt or in solution. The interested student may pursue this topic further in the references provided. Our concern here is specifically with polymeric liquid crystals.[32–36]

There are two major classifications of liquid crystals: *lyotropic* and *thermotropic*. Lyotropic liquid crystals form under the influence of solvent. Thermotropic liquid crystals form in the melt. These classifications are broken down further (nematic, smectic, cholesteric, etc.) according to how the molecules are oriented in the mesophase. From the standpoint of polymer applications, two properties of liquid crystals are of major interest: the effect of order on viscosity, and the ability of the polymer to retain its ordered configuration in the solid state. Among the first polymers observed to exhibit liquid crystalline behavior were copolyesters prepared from terephthalic acid (**55**), ethylene glycol (**56**), and *p*-hydroxybenzoic acid (**57**).[37] It was found that as the amount of **57** was increased in the

HO$_2$C—⬡—CO$_2$H   HOCH$_2$CH$_2$OH   HO—⬡—CO$_2$H

          55   **56**   **57**

polymer, the melt viscosity initially increased, which was expected because of the decreased flexibility caused by incorporation of the "rigid" *p*-hydroxybenzoate unit. At levels of about 30 mol %, however, the melt viscosity began to *decrease*, reaching a minimum at about 60 to 70 mol %. This is shown in Figure 3.18 at four different shear rates. Furthermore, as the melt viscosity began to decrease, the melt's appearance changed from clear to opaque. The decrease in viscosity and opaqueness arise from the onset of a thermotropic liquid crystalline morphology due to the increased backbone rigidity. The viscosity effect results from the rigid polymeric mesophases becoming aligned in the direction of flow, thus minimizing frictional drag. The implication from the standpoint of polymer processing is obvious: the lower the viscosity, the more readily the polymer can be fabricated into a useful plastic or fiber.

Equally important was the observation that the ordered arrangement of polymer molecules in the melt was retained upon cooling, which was manifested in greatly improved mechanical properties. Thus liquid crystalline behavior is advantageous from the standpoint of both processing and properties. Thermotropic liquid crystalline copolyesters of similar structure are now available commercially.

A commercially available lyotropic liquid crystalline polymer is the aromatic polyamide, **58** (du Pont trade name Kevlar).[38,39] A sulfuric acid solution of the polymer, which exhibits

$$\left[\begin{array}{c} \overset{O}{\underset{\parallel}{C}}-⬡-\overset{O}{\underset{\parallel}{C}}-NH-⬡-NH \end{array}\right]$$

          58

Melt viscosity (275 °C).
poise

Shear rate, s^{-1}

● 15
○ 100
▲ 1600
△ 54,000

0 20 40 60 80 100
p-Hydroxybenzoic
acid, mol–%

FIGURE 3.18. Effect of *p*-hydroxybenzoic acid concentration on melt viscosity of a terephthalic acid/*p*-hydroxybenzoic acid/ethylene glycol copolyester. [From Jackson and Kuhfuss,[37] copyright 1976. Reprinted by permission of John Wiley & Sons, Inc.]

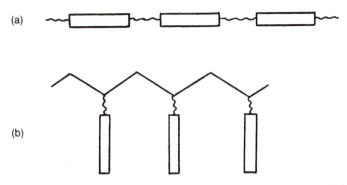

FIGURE 3.19. Representations of mesogenic groups (box) and flexible spacers (∿) in (a) the backbone and (b) the side chain of liquid crystalline polymers.

the liquid crystal phase, is extruded to form a fiber (see Chapter 4), which results in further alignment of the molecules. The product, once the sulfuric acid is removed, is a fiber with a more uniform alignment than could be obtained simply by drawing; as a consequence it has much better mechanical properties. The tensile strength of Kevlar, for example, is considerably higher than that of steel, whereas its density is much lower. Although most Kevlar produced is used in tire cord, the polymer also finds use in specialty clothing. Modern lightweight bulletproof vests contain up to 18 layers of woven Kevlar cloth.

Major drawbacks to the type of rigid polymers that exhibit liquid crystalline behavior are that they have a very high melting point and are difficult, if not impossible, to dissolve in the usual organic solvents. One approach to circumventing these difficulties is to separate the rigid backbone groups (called *mesogens*), which are responsible for the mesophases, with *flexible spacers* such as **59**, **60**, and **61**, or to attach mesogens with flexible spacers to the

$$-[CH_2]_n \qquad -CH_2CH_2O- \qquad \left[-\underset{\underset{CH_3}{|}}{\overset{\overset{CH_3}{|}}{Si}}O-\right]_n$$

59   **60**   **61**

polymer backbone.[40] Both types are illustrated schematically in Figure 3.19. Although the use of flexible spacers does render the polymer more tractable, mechanical properties usually suffer.

3.8 Chemical Crosslinking

The most severe mechanism for decreasing molecular freedom is *chemical crosslinking*—linking the polymer chains together through covalent or ionic bonds to form a network.[41] Occasionally the term *curing* is used to denote crosslinking. There are a number of ways crosslinking can be brought about, but basically they fall into two categories: (1) crosslinking during polymerization by use of polyfunctional instead of difunctional monomers, and (2) crosslinking in a separate processing step after the linear (or branched) polymer is formed. The crosslinks may contain the same structural features as the main chains, which is usually the case with the former, or they may have an entirely different structure, which is more characteristic of the latter. In later chapters we will explore the chemistry of both covalent and ionic crosslinking.

A number of extreme changes accompany crosslinking. If previously soluble, the polymer will no longer dissolve (except in the case of some ionically crosslinked polymers). In the presence of solvent, a crosslinked polymer swells as solvent molecules penetrate the network. The degree of swelling depends on the affinity of solvent and polymer for one another, as well as on the level of crosslinking. It may be recalled (Chapter 2) that a solvent-swollen crosslinked polymer is called a gel. If the gel particles are very small (300–1000 μm), they are called *microgels*. Microgels behave as tightly packed spheres that can be suspended in solvents. Microgels have attracted considerable interest in recent years with the development of solid-phase synthesis and techniques for immobilizing catalysts. Covalently crosslinked polymers also lose their flow properties. They may still undergo deformation, but the deformation will be reversible; that is, the polymer will exhibit *elastic* properties. Ionically crosslinked polymers *will* flow at elevated temperatures, however.

With network polymers it is common to speak of the *crosslink density*, that is, the number of crosslinked monomer units per main chain. Crosslink density (Γ) is defined mathematically as

$$\Gamma = \frac{(\overline{M}_n)_0}{(\overline{M}_n)_c}$$

where $(\overline{M}_n)_0$ is the number average molecular weight of uncrosslinked polymer and $(\overline{M}_n)_c$ is the number average molecular weight between crosslinks. The higher the crosslink density, the more rigid the polymer. Very high crosslink densities lead to embrittlement. Because crosslinking reduces segmental motion, it is frequently employed to increase the glass transition temperature.

Elastomers have a very *low* crosslink density—about one crosslink per 100 monomer units—together with highly flexible main chains to allow large deformations. Elastomeric behavior[13c,42] depends, of course, on polymer structure and crosslink density; but it also depends on morphology and molecular weight. Some of the morphological features that contribute nothing to the ideal elastic behavior of the network (and are thus termed *defects*) are chain ends, loops, and entanglements. One approach to improving elastic properties has been to design telechelic elastomers with reactive chain ends that become incorporated into the network on crosslinking.[43]

3.9 Physical Crosslinking

When polymer chemists use the term *crosslinking*, they usually mean covalent chemical crosslinking. Covalent crosslinking has certain disadvantages, however. Once crosslinked, a polymer cannot be dissolved or molded. Scrap crosslinked polymer cannot be recycled. Polymer chemists have explored ways to circumvent this problem. One approach has been to investigate thermally labile crosslinks, that is, chemical crosslinks that break apart on heating and reform on cooling. Ionic crosslinks fall into this category. Attempts to develop reversible covalent crosslinking are described in Chapter 9. The other approach has been to introduce strong secondary bonding attraction between polymer chains such that the polymer exhibits properties of a thermosetting material while remaining thermoplastic. We call this (for want of a better term) "physical crosslinking." Crystalline polymers fit into this category. Because of the very strong secondary forces arising from close chain packing, many

FIGURE 3.20. Representation of aggregation in an ABA block thermoplastic elastomer (∿ represents end blocks, circle represents microdomains).

of the mechanical and solution properties of crystalline polymers resemble those of crosslinked amorphous polymers. Certain materials intermolecularly associated through hydrogen bonds also behave like crosslinked polymers. Gelatin, an animal-derived protein that exhibits elastomeric properties, is an example.

In recent years the technology of block copolymers has been applied to physical crosslinking.[38,39] The method involves synthesis of block copolymers of the ABA type in which the A and B blocks differ substantially in structure. Consider, for example, a long-chain "flexible" polymer such as polybutadiene, capped at each chain end with short blocks of a "rigid" polymer such as polystyrene. Because polybutadiene and polystyrene are inherently immiscible (incompatible), the polystyrene blocks tend to aggregate and form separate phases (microdomains) within the polymer matrix, as shown in Figure 3.20. The same behavior is observed with homopolymers containing semicrystalline isotactic or syndiotactic end blocks and an amorphous atactic middle block. If the structure of the end blocks is stereoregular, the aggregations may form crystalline microdomains. The aggregations impart a significant degree of elastic behavior, yet the copolymers still exhibit the flow properties of thermoplastics. Appropriately such materials are referred to as *thermoplastic elastomers.*[45,46]

Since the 1960s when they were first introduced, thermoplastic elastomers (or TPEs) are projected to grow to about 7% of the world rubber consumption by the turn of the century. TPEs generally have higher material costs, but these are more than offset by the lower processing costs of thermoplastics. Commercially important TPEs are listed in Table 3.6.

3.10 Polymer Blends

By definition, any physical mixture of two or more different polymers or copolymers that are not linked by covalent bonds is a *polymer blend*, or *polyblend*.[48–51] Some scientists like to make an analogy with metals and call such mixtures *polymer alloys*.[52,53] The concept of blending polymers is not new; the rubber industry has used it for decades. In recent years,

TABLE 3.6. Commercially Important Thermoplastic Elastomers

Type	End Blocks	Middle Block
Styrenic	Polystyrene	Polybutadiene or polyisoprene
Polyolefin[a]	Isotactic polypropylene	Ethylene–propylene copolymer
Polyurethane	Rigid polyurethane	Flexible polyester or polyether
Copolyester	Rigid polyester[b]	Flexible polyester
Polyamide	Rigid polyamide[c]	Flexible polyether

[a] Also manufactured by mechanical blending of isotactic polypropylene and ethylene–propylene–diene copolymer.
[b] Principally poly(ethylene terephthalate).
[c] Nylons 6, 66, 11, 12, and 612.

however, there has been a resurgence of interest arising primarily from the demand for engineering plastics and specialty elastomers and fibers. There are sound economic reasons for this interest. Development of a new polymer to meet a specific need is a costly enterprise. If the desired properties can be realized simply by mixing two or more existing polymers, there is an obvious pecuniary advantage.

TABLE 3.7. Types of Polyblends

Type	Description
Mechanical blends	Polymers are mixed at temperatures above T_g or T_m for amorphous and semicrystalline polymers, respectively
Mechanochemical blends	Polymers are mixed at shear rates high enough to cause degradation. Resultant free radicals combine to form complex mixtures including block and graft components
Solution-cast blends	Polymers are dissolved in common solvent and solvent is removed
Latex blends	Fine dispersions of polymers in water (latexes) are mixed, and the mixed polymers are coagulated
Chemical blends	
Interpenetrating polymer networks (IPN)	Crosslinked polymer is swollen with different monomer, then monomer is polymerized and crosslinked
Semi-interpenetrating polymer networks (semi-IPN)	Polyfunctional monomer is mixed with thermoplastic polymer, then monomer is polymerized to network polymer (also called pseudo-IPN)
Simultaneous interpenetrating polymer networks (SIN)	Different monomers are mixed, then homopolymerized and crosslinked simultaneously, but by noninteracting mechanisms
Interpenetrating elastomeric networks (IEN)	Latex polyblend is crosslinked after coagulation

A number of technologies have been devised to prepare polyblends; these are summarized in Table 3.7. It so happens that most polymers are not compatible. Rather, they separate into discrete phases on being mixed, although an increasing number of completely miscible blends are being developed. Differences between the two types are manifested in appearance—miscible blends are usually clear, immiscible blends are opaque—and in such properties as glass transition temperature—miscible blends exhibit a single T_g intermediate between those of the individual components, whereas immiscible blends exhibit separate T_gs characteristic of each component. Miscibility is by no means prerequisite to commercial utility.

Homogeneous polymer blends are more convenient from the standpoint of being able to predict properties or processing characteristics. If additives are used, for example, there are no problems of migration from one phase to another. Physical or mechanical properties usually reflect, to a degree, the weighted average of the properties of each component. For a binary homogeneous blend, this can be quantified for a particular property (P) using the semiempirical relationship

$$P = P_1\Phi_1 + P_2\Phi_2 + I\Phi_1\Phi_2$$

where Φ is the volume fraction in the mix and I is an interaction term that may be negative, zero, or positive. If the properties are strictly additive, $I = 0$. If I is positive, the property in question is better than the weighted average, and the blend is said to be *synergistic* for that property. If I is negative, the property is worse than the weighted average (*nonsynergistic*). This is illustrated in Figure 3.21 by a plot of property versus composition. What gives rise to synergism or nonsynergism? The former might result from a very favorable dipole–dipole attraction between the polymer components. The latter might arise when a favorable intermolecular interaction is disrupted, for example, the prevention or reduction of crystallinity. Another relationship that works reasonably well for the glass transition temperature of miscible blends is the *simple rule of mixtures*,

$$T_g = w_1 T_{g1} + w_2 T_{g2}$$

where w is the weight fraction and T_g is in kelvins.

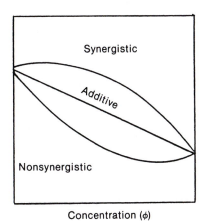

FIGURE 3.21. Hypothetical property–concentration relationships for a binary miscible polyblend.

One example of a commercially important miscible polyblend is the engineering plastic Noryl (General Electric), which is composed of polystyrene, an inexpensive polymer, and poly(oxy-2,6-dimethyl-1,4-phenylene) (62), a relatively expensive polyether known commonly as poly(phenylene oxide) or PPO (also a General Electric trade name). For the most part, the properties of Noryl are additive. For example, Noryl has poorer thermal stability than the polyether alone, but is easier to process. Its single glass transition temperature increases with increasing polyether content. In terms of tensile strength (Chapter 4), however, the polyblend is synergistic.[54]

62

Morphology can have a dramatic effect on properties of polymer blends. A blend of low-density polyethylene (LDPE) with the terpolymer ethylene–propylene–diene monomer rubber (EPDM) exhibits a synergistic effect on tensile strength if the EPDM is partially crystalline, but a nonsynergistic effect if the EPDM is amorphous.[55] The synergism apparently arises from a tendency for crystallites in the LDPE to nucleate crystallization of ethylene segments in the EPDM. Stereochemistry is also important; syndiotactic poly(methyl methacrylate) is miscible with poly(vinyl chloride) at certain concentrations, whereas the isotactic form is immiscible over the entire composition range.[56]

A major problem with developing miscible polyblends is trying to predict miscibility.[57,58] Attempts to predict miscibility using solubility parameters of the type described in Chapter 2 have been largely unsuccessful because strong dipolar interactions are not taken into account. It has been shown, for example, that compatibility may be greatly enhanced by dipole–dipole attraction,[59] hydrogen bonding,[60,61] ion pairing,[62] and charge-transfer complexing,[63] as well as by minimizing dipole moment differences between the individual polymers.[64]

Properties of immiscible polyblends are much less predictable. Attempts to develop additivity relationships are complicated by the effects of varying morphologies that might arise as a result of processing variables. Frequently, one polymer will constitute a continuous phase while the second will be dispersed as a noncontinuous phase in the form of fibrils, spheres, lamellae, and so on. Which polymer is the continuous phase largely determines properties. For example, a 50:50 blend of polystyrene (a hard, glassy polymer) and polybutadiene (an elastomer) will be hard if polystyrene is the continuous phase, but soft if polystyrene is the dispersed phase. Alternatively, an immiscible polyblend may have both components dispersed as continuous phases.

The major problem with immiscible blends is the often poor physical attraction at phase boundaries, which can lead to phase separation under stress with resultant poor mechanical properties. To improve compatibility between immiscible phases, a number of ingenious approaches have been adopted. One is through formation of interpenetrating networks,[65–67] as described in Table 3.7. In such cases the polymers are physically "locked" together by the interdispersed three-dimensional network, a phenomenon referred to as *topological bond-*

FIGURE 3.22. Representation of the use of an AB block copolymer to improve interfacial adhesion in an immiscible polyblend.

ing.[65] Such mixtures still undergo phase separation into microdomains that vary in size according to the degree of immiscibility.

Another approach is to incorporate *compatibilizers* into the blend to improve adhesion between phases.[56a] Let's consider, for example, an incompatible blend of poly(A) and poly(B), the homopolymers being made from hypothetical monomers A and B. If a block copolymer of A and B is added to the blend, the natural affinities of the blocks for their respective homopolymers will serve to localize the copolymer at the phase boundary and help "glue" the phases together, as depicted in Figure 3.22. One drawback is the relatively high cost of block copolymers; even so, at levels of less than 1% by weight, AB block copolymer compatibilizers can significantly increase interfacial adhesion. Not all immiscible blends need an interfacial agent. Commercial varieties of high-impact polypropylene consist of immiscible blends of polypropylene and ethylene–propylene copolymer, which have sufficient interphase bonding arising from the natural affinity of the homopolymer for propylene segments in the copolymer. Adding functional groups to the incompatible polymers, which react upon blending to link the phases with covalent bonds, is another way to improve interfacial adhesion.

The most successful approach to compatibilizing polymer blends from the commercial standpoint is to generate graft copolymers in situ that help bind the immiscible phases together. ABS (acrylonitrile–butadiene–styrene) plastics[69] are an example of this methodology. Typically, an amorphous styrene–butadiene copolymer is dissolved in styrene and acrylonitrile and the monomers are subsequently copolymerized. In the course of polymerization, chain-transfer reactions occur (the mechanisms of which will be described later) to produce grafts of one copolymer to the other. The number of grafts is small, but is sufficient to provide the necessary interfacial adhesion. ABS polymers are widely used as engineering plastics.[70]

Proper control of phase morphology is of much greater importance with immiscible blends. Despite this drawback, they have found much wider acceptance than their less ubiquitous miscible cousins, and include both engineering plastics[71] and fibers.[72]

References

1. R. D. Deanin, *Polymer Structure, Properties and Applications*, Cahners Books, Boston, 1972.
2. M. L. Miller, *The Structure of Polymers*, Rheinhold, New York, 1966.
3. A. Ledwith and A. M. North (Eds.), *Molecular Behavior and the Development of Polymeric Materials*, Wiley-Interscience, New York, 1975: (a) R. N. Hayward, Chap. 12; (b) C. W. Bunn, Chap. 10.
4. A. V. Tobolsky, *Properties and Structure of Polymers*, Wiley-Interscience, New York, 1960.

5. G. L. Wilkes, *J. Chem. Educ.*, **58**, 880 (1981).
6. D. J. Williams, *Polymer Science and Engineering*, Prentice-Hall, Englewood Cliffs, N.J., 1971, Chap. 8.
7. Y. G. Yanovsky, *Polymer Rheology: Theory and Practice*, Chapman and Hall, London, 1993.
8. A. E. Woodward, *Understanding Polymer Morphology*, Hanser–Gardner, Cincinnati, 1995.
9. N. P. Cheremisinoff, *An Introduction to Polymer Rheology and Processing*, CRC Press, Boca Raton, 1993.
10. J. J. Aklonis and W. MacKnight, *Introduction to Polymer Viscoelasticity*, 2nd ed., Wiley-Interscience, New York, 1983.
11. J. F. Rabek, *Experimental Methods in Polymer Chemistry*, Wiley-Interscience, New York, 1980: (a) Chap. 9; (b) Chap. 26.
12. R.-J. Roe, in *Encyclopedia of Polymer Science and Engineering* (H. F. Mark, N. R. Bikales, C. G. Overberger, G. Menges, and J. I. Kroschwitz, Eds.), Vol. 7, Wiley-Interscience, New York, 1987, pp. 531–44.
13. A. D. Jenkins (Ed.), *Polymer Science*, Vol. 1, North-Holland, Amsterdam, 1972: (a) J. A. Brydson, Chap. 3; (b) A. Sharples, Chap. 4; (c) K. J. Smith, Jr., Chap. 5.
14. J. Brandrup and E. H. Immergut (Eds.), *Polymer Handbook*, 3rd ed., Wiley-Interscience, New York, 1975; (a) P. Peyser, pp. VI/209 ff; (b) R. L. Miller, pp. VI/1ff.
15. D. R. Burfield and K. Lim, *Macromolecules*, **16**, 1170 (1983).
16. G. Natta and F. Danusso (Eds.), *Stereoregular Polymers: Stereoregular Polymers and Stereospecific Polymerization*, Pergamon Press, New York, 1967.
17. P. Corradini, in *The Stereochemistry of Macromolecules*, Vol. 3 (A. D. Ketley, Ed.), Marcel Dekker, New York, 1986, Chap. 1.
18. M. L. Huggins, G. Natta, V. Desreux, and H. Mark, *J. Polym. Sci.*, **56**, 152 (1962).
19. G. Natta, *J. Polym. Sci.*, **16**, 143 (1955).
20. G. Natta and P. Corradini, *J. Polym. Sci.*, **20**, 251 (1956).
21. F. A. Bovey and T. K. Kwei, in *Macromolecules: An Introduction to Polymer Science* (F. A. Bovey and F. H. Winslow, Eds.), Academic Press, New York, 1979, Chap. 3.
22. M. Farina, G. DiSilvestro, and P. Sozzani, *Prog. Polym. Sci.*, **16**, 219 (1991).
23. B. R. Burfield and Y. Doi, *Macromolecules*, **16**, 702 (1983).
24. P. H. Geil, E. Baer, and Y. Wada, *The Solid State of Polymers*, Marcel Dekker, New York, 1974.
25. L. Mandelkern, *Crystallization of Polymers*, McGraw-Hill, New York, 1964.
26. H. Tadokoro, *Structure of Crystalline Polymers*, Wiley-Interscience, New York, 1979.
27. P. H. Geil, *Polymer Single Crystals*, Interscience, New York, 1963.
28. B. Wunderlich and L. Melillo, *Makromol. Chem.*, **118**, 250 (1986).
29. R. W. Lenz, *Organic Chemistry of Synthetic High Polymers*, Wiley-Interscience, New York, 1967, Chap. 4.
30. H. Stegemayer (Ed.), *Liquid Crystals*, Springer Verlag, New York, 1994.
31. S. Chandrasekhar, *Liquid Crystals*, Cambridge University Press, London, 1977.
32. M. Gordon (Ed.), *Liquid Crystal Polymers, Adv. Polym. Sci.*, **59**, **60**, **61**, Springer-Verlag, New York, 1984.
33. A. Ciferri, W. R. Krigbaum, and R. B. Meyer (Eds.), *Polymer Liquid Crystals*, Academic Press, New York, 1982.
34. S. K. Varshney, *J. Macromol. Sci., Rev. Macromol. Chem. Phys.*, **C26**, 551 (1986).
35. A. M. Donald and A. H. Windle, *Liquid Crystal Polymers*, Cambridge University Press, Cambridge, 1992.
36. R. Zentel, in *Liquid Crystals* (H. Stegemeyer, Ed.), Springer Verlag, New York, 1994, pp. 103–41.
37. W. J. Jackson, Jr., and H. F. Kuhfuss, *J. Polym. Sci., Polym. Chem. Ed.*, **14**, 2043 (1976).
38. D. Tanner, J. A. Fitzgerald, and B. R. Phillips, *Angew. Chem. Int. Ed. Engl. Adv. Mater.*, **28**, 649 (1989).
39. H. H. Yang, *Kevlar® Aramid Fiber*, Wiley, Chichester, U.K., 1993.
40. C. B. McArdle (Ed.), *Side Chain Liquid Crystal Polymers*, Blackie, London, 1989.
41. A. J. Chompff and S. Newman, *Polymer Networks: Structural and Mechanical Properties*, Plenum Press, New York, 1971.
42. L. R. G. Treolar, *The Physics of Rubber Elasticity*, 3rd. ed., Clarendon, Oxford, 1975.

43. C. A. Uranek, L. H. Hsieh, and O. G. Buck, *J. Polym. Sci.*, **46**, 535 (1960).
44. N. R. Legge, *Chemtech*, **13**, 630 (1983).
45. R. B. Seymour and G. B. Kauffman, *J. Chem. Educ.*, **69**, 967 (1992).
46. N. R. Legge, G. Holden and H. E. Schroeder (Eds.), *Thermoplastic Elastomers: A Comprehensive Review*, Hanser–Gardner, Cincinnati, 1989.
47. M. S. Reich, *Chem. Engr. News*, **74**(32), 10 (Aug. 5, 1996).
48. J. A. Manson and L. H. Sperling, *Polymer Blends and Composites*, Plenum Press, New York, 1976.
49. D. R. Paul and S. Newman (Eds.), *Polymer Blends*, Vols. 1 and 2, Academic Press, New York, 1978.
50. L. H. Sperling, *Macromol. Rev.*, **12**, 141 (1977).
51. H. Rudolph, *Makromol. Chem., Macromol. Symp.*, **16**, 57 (1988).
52. B. J. Schmitt, *Angew. Chem. Int. Ed. English*, **18**, 273 (1979).
53. L. A. Utracki, *Polymer Alloys and Blends: Thermodynamics and Rheology*, Hanser–Gardner, Cincinnati, 1989.
54. A. F. Yee, *Polym. Eng. Sci.*, **17**, 213 (1977).
55. G. H. Lindsay, C. J. Singleton, C. J. Carman, and R. W. Smithe, in *Multiphase Polymers* (S. L. Cooper and G. M. Estes, Eds.), Adv. Chem. Ser. 176, American Chemical Society, Washington, D.C. 1979, p. 367.
56. J. W. Schurer, A. de Boer, and G. Challa, *Polymer*, **16**, 201 (1975).
57. M. M. Coleman, C. J. Serman, D. E. Bhagwager, and P. C. Painter, *Polymer*, **31**, 1187 (1990).
58. O. Olbasi, L. M. Robeson, and M. T. Shaw, *Polymer–Polymer Miscibility*, Academic Press, New York, 1979.
59. R. E. Bernstein, D. R. Paul, and J. W. Barlow, *Polym. Eng. Sci.*, **18**, 683 (1978).
60. M. M. Coleman, J. F. Graf, and P. C. Painter, *Specific Interactions and the Miscibility of Polymer Blends*, Technomic, Lancaster, Pennsylvania, 1991.
61. R. F. M. Lange and E. W. Meijer, *Macromolecules*, **28**, 782 (1995).
62. A. Eisenberg and J. Hara, *J. Polym. Eng. Sci.*, **22**, 307 (1984).
63. V. Percec, H. G. Schild, J. M. Rodriguez-Parada, and C. Pugh, *J. Polym. Sci., Polym. Chem. Ed.*, **26**, 935 (1988).
64. A. Dondos and E. Pierri, *Polym. Bull.*, **16**, 567 (1986).
65. H. I. Frisch, *Brit. Polym. J.*, **17**, 149 (1985).
66. L. H. Sperling, *Interpenetrating Polymer Networks and Related Materials*, Plenum Press, New York, 1981.
67. L. H. Sperling, in *Interpenetrating Polymer Networks* (D. Klempner, L. H. Sperling, and L. A. Utracki, Eds.), Adv. Chem. Ser. 239, American Chemical Society, Washington, D.C., 1994, Ch. 1.
68. S. Datta and D. Lohse, *Polymeric Compatibilizers: Uses and Benefits in Polymer Blends*, Hanser–Gardner, Cincinnati, Ohio, 1996.
69. C. H. Basdekis, *ABS Plastics*, Reinhold, New York, 1964.
70. N. Platzer, *Chemtech*, **7**, 634 (1977).
71. D. R. Paul and J. W. Barlow, in *Multiphase Polymers* (S. L. Cooper and G. M. Estes, Eds.), Adv. Chem. Ser. 176, American Chemical Society, Washington, D.C., 1979, p. 315.
72. S. P. Hersh, in *High Technology Fibers*, Part A (M. Lewin and J. Preston, Eds.), Marcel Dekker, New York, 1985, Chap. 1.

Review Exercises

1. Write concise definitions of the following terms:

   (a) annealing
   (b) atactic
   (c) crosslink density
   (d) *erythro*
   (e) glass transition temperature
   (f) head-to-tail polymer
   (g) hemitactic
   (h) interpenetrating polymer network

   (i) isotactic
   (j) lyotropic liquid crystal
   (k) mesogen
   (l) mesophase
   (m) nucleation
   (n) polyblend
   (o) rheology
   (p) shear stress

(q) shear thinning
(r) spherulite
(s) stereoblock polymer
(t) syndiotactic
(u) telechelic polymer

(v) thermoplastic elastomer
(w) thermotropic liquid crystal
(x) *threo*
(y) viscoelasticity
(z) vitrification

2. Explain the difference between shear thinning and thixotropy.

3. Draw the *trans-threo*-diisotactic and *cis-erythyro*-disyndiotactic forms of the polymer depicted in Figure 3.14.

4. Draw a section of the chain (four or five repeating units) of each of the following head-to-tail polymers:
   (a) syndiotactic poly(vinyl alcohol)
   (b) isotactic poly(α-methylstyrene)
   (c) atactic 1,2-polybutadiene
   (d) syndiotactic poly(1,2-difluoroethylene)
   (e) *threo*-diisotactic poly(1-chloro-2-methoxyethylene)
   (f) *cis*-1,4-polychloroprene
   (g) *threo*-diisotactic polycyclopentene
   (h) *trans*-isotactic-1,4-poly(2-methyl-1,3-pentadiene)

5. What constitutional (i.e., not source-based) name might one assign to the polymer in question 4(h) if the double bond were to be hydrogenated? (*Hint*: Draw a chain segment of the hydrogenated polymer. What is the constitutional repeating unit? What is the stereochemistry?). (M. Farina et al., *Macromolecules*, **15**, 1451, 1982)

6. (a) Draw structures analogous to structure **37** and **38** for the six possible tetrads of poly(methyl methacrylate). (b) How many pentads are possible? Write the *mr* designations (not structures) of each.

7. Which polymer in each of the following pairs would you expect to exhibit the higher glass transition temperature? Explain your choice in each case.

(a) $-\!\!\left[CH_2CH\right]\!\!-$ or $-\!\!\left[CH_2CH\right]\!\!-$

(b) $\left[O-\!\!\bigcirc\!\!-O\overset{O}{\overset{\|}{C}}CH\!\!=\!\!CH\overset{O}{\overset{\|}{C}}\right]$ or $\left[O-\!\!\bigcirc\!\!-O\overset{O}{\overset{\|}{C}}CH_2CH_2\overset{O}{\overset{\|}{C}}\right]$

(c) $-\!\!\left[CH_2CH_2NH\right]\!\!-$ or $-\!\!\left[CH_2CH_2O\right]\!\!-$

(d) $\left[O-\!\!\bigcirc\!\!-\overset{O}{\overset{\|}{C}}\right]$ or $\left[O-\!\!\bigcirc\!\!-\overset{O}{\overset{\|}{C}}\right]$

(e) $\left[\overset{O}{\overset{\|}{C}}-\!\!\bigcirc\!\!-\overset{O}{\overset{\|}{C}}OCH_2CH_2O\right]$ or $\left[\overset{O}{\overset{\|}{C}}-\!\!\bigcirc\!\!-\overset{O}{\overset{\|}{C}}OCH_2CH_2O\right]$

with $CH_2CH_2CH_3$ and $CH(CH_3)_2$ substituents respectively

8. Which of the following "isomeric" polymers would you expect to exhibit the greater crystallinity? Explain.

    (a) $-[CH_2CH_2CH]-$ or $-[CH_2CH]-$
                        $|$ $|$
                      OH OCH_3

    (b) $-[\overset{O}{\overset{||}{C}}(CH_2)_4\overset{O}{\overset{||}{C}}OCH_2\overset{CH_3}{\underset{|}{C}}CH_2O]-$ or $-[\overset{O}{\overset{||}{C}}(CH_2)_4\overset{O}{\overset{||}{C}}OCH_2\underset{\underset{CH_2CH_3}{|}}{C}HCH_2O]-$

    (c) $-[CH_2CF_2]-$ or $-[CHFCHF]-$

9. When poly(tetramethylene adipate) (a polymer of interest for "disappearing" surgical sutures) is subjected to hydrolytic or enzymatic degradation, the polymer's percent crystallinity initially increases. What does this suggest about the degradation process? (A.-C. Albertson and O. Ljunquist, *J. Macromol. Sci.-Chem.*, **A23**, 393, 1986)

10. One method of determining percent crystallinity in a polymer involves inverse gas chromatography; that is, the polymer in question is used as the stationary liquid phase coated on the column packing. Speculate on the rationale underlying this method.

11. When a metal spring, suspended with a weight attached at the bottom, is heated, the spring elongates further because more volume is needed to accommodate the increased atomic vibrations. When a strip of elastomer is similarly suspended, it *contracts* on heating. On the basis of what you have learned in this chapter, suggest a plausible explanation for this difference in behavior.

12. Speculate on why polyester clothing is more wrinkle resistant than cotton. Why is a *hot* iron used to press out the wrinkles?

13. An unusual type of physical crosslinking has been reported for a polyamide containing a macrocyclic 32-crown-10 moiety:

The physical crosslinking is reportedly responsible for the polymer's lack of solubility in any of the common polymer solvents. Polyamides containing the structurally similar but smaller 18-crown-6 unit, however, are soluble. Speculate on what type of physical crosslinking might be involved (Y. Delaviz and H. W. Gibson, *Macromolecules*, **25**, 4859, 1992).

4

CHEMICAL STRUCTURE AND
POLYMER PROPERTIES

4.1 Introduction

In the last chapter we saw how chemical structure influences polymer morphology. Now we will examine the relationship between chemical structure and the properties of major interest in commercial applications. With mechanical properties, in particular, we shall see that the interrelationship between chemical structure and morphology is of overriding importance. Other properties—thermal, chemical, and electrical—also depend to an extent on morphology, but more so on chemical makeup. In many cases the properties under discussion are significant in specialty products that represent only a small fraction of the total synthetic polymer production. Nevertheless, such markets are expanding and, importantly, they illustrate how synthetic polymers can often be tailored for a specific application. We will also look briefly at *additives*—compounds added to polymers to *modify* their properties. Given the fact that virtually all commercial polymer products contain one or more additives, that is clearly an important facet of structure–property relationships.

We will begin with a brief digression from chemistry to see how polymers are converted into the commercial products that benefit from the various properties of interest. While a detailed study of fabrication methods is outside the scope of this book and more relevant to engineering texts, it is instructive to progress from our earlier discussion of polymer flow to recognize how products of the laboratory are transformed into useful articles.

4.2 Fabrication Methods

Polymers are most commonly shaped by one of three basic techniques: molding, extrusion, or casting.[1,2] All three are accomplished at much lower temperatures than those needed for the shaping of steel, aluminum, or glass, and hence energy efficiency is one of the most attractive features of polymer processing. Polymers have one inherent disadvantage, however: They have poor thermal conductivity and melt slowly. To minimize this problem, rotating

FIGURE 4.1. Compression molding

screw devices are used to transport the polymer, thus allowing granular polymer pellets or powder to be heated uniformly by a combination of external heaters and frictional heat generated during transport.

There are two fundamental molding processes: *compression molding* and *injection molding*. Compression molding uses heat and pressure to force molten polymer, introduced between the mating surfaces of a movable mold, into the shape of the mold. In injection molding, molten polymer is compressed into a closed mold cavity. The two methods are shown diagrammatically in Figures 4.1 and 4.2. The former is more commonly used with thermosetting polymers, which cannot be melted once crosslinking has taken place. Crosslinking is therefore accomplished in situ in the mold cavity. In some cases the polymer is partially polymerized so that it will still flow under pressure (this is called *B-staging*); then heat is

FIGURE 4.2. Injection molding. [Reprinted from V. Hopp and I. Hennig, *Handbook of Industrial Chemistry*, copyright 1983, courtesy of McGraw-Hill.]

applied to effect the final crosslinking. Other processes use *preforms* (also called *prepregs*) made of polymeric binder and a reinforcing material such as glass fiber, which are preshaped to fit the mold; then liquid polymer is poured over the preform. Molding pressure distributes the polymer through the preform, and heat activates the crosslinking. The end product is a *fiber-reinforced plastic*.

Injection molding is more commonly applied to thermoplastics. A hydraulic ram or, as illustrated in Figure 4.2, a screw is used to feed the polymer to the mold. Injection molding is generally faster than compression molding.

In recent years, a new type of injection molding called *reaction injection molding* (RIM) has become important for fabricating thermosetting polymers.[3,4] RIM differs from its predecessors in that the finished product is made directly from monomers or low-molecular-weight polymeric precursors, which are rapidly mixed and injected into the mold even as the polymerization reaction is taking place. A schematic of a RIM process is shown in Figure 4.3. For RIM to be successful, the monomers must be fast reacting, and reaction rates must be carefully synchronized with the molding process. There are, however, obvious advantages. Synthesis of polymer prior to molding is eliminated, and the energy requirements for handling of monomers are much lower than those for viscous polymers. Much of the early development of RIM was applied to polyurethanes (Chapter 13), but other polymer types and polymer blends also benefit from this technology. Reinforcing fillers are sometimes injected along with the reactants, a process called *reinforced reaction injection molding* (RRIM).

Another type of molding, particularly useful for manufacturing bottles, is *blow molding*. In this process, polymer tubing (called *parison*) is blown by compressed air or drawn by vacuum into the shape of the mold (Figure 4.4). *Casting*, a much simpler process, involves pouring molten polymer into a mold and allowing the product to cool.

Extrusion involves forcing compacted, molten polymer through a die shaped to give the desired object. An extruder resembles the injection molding apparatus shown in Figure 4.2 except that the mold cavity is replaced with a die, as shown earlier in Figure 3.2. Extrusion is especially useful for making elongated objects such as rods or pipe.

Polymer film is made by casting from solution or by passing polymer under high pres-

FIGURE 4.3. Basic components of a reaction injection molding (RIM) process. [From *Modern Plastics Encyclopedia, 1984–85*. Reprinted with permission of Modern Plastics.]

Extruded Parison–   Mold Closed and   Finished Bottle
Mold Open   Bottle Blown   Removed
                                              from Mold

FIGURE 4.4. Blow molding. [Courtesy of the Society of the Plastics Industry, Inc.]

sure between hot rollers, a process known as *calendering*. Calendering is also used to make thick sheets—for floor tile, for example. Other methods of making film include extrusion through a slit die or through a ring-shaped die. In the latter process, pressure is maintained within the resultant plastic tube to expand it to a cylindrical film that can be cut to flat sheets or processed further as parison in blow molding.

Fibers are made by *spinning*, a process resembling extrusion. Polymer is forced under pressure through a perforated plate called a *spinneret*, as depicted in Figure 4.5. When several threads are combined, the material is called *tow*. Spinning is done either with molten polymer (*melt spinning*) or with polymer solution. In the case of melt spinning, the filaments are cooled rapidly by cold air in the process tank. If polymer solution is used, the solvent is removed in the process tank by evaporation with heat (*dry spinning*) or by leaching with another liquid (*wet spinning*). With some polymers, particularly cellulose (Chapter 17), which does not dissolve in the more common solvents, it may be more advantageous to convert the polymer chemically to a more soluble form and then to regenerate it in the process tank by chemical treatment of the filaments.

Polymeric foams[5,6] are made by incorporating a *blowing agent* into the polymer. There are two types of blowing agents: physical and chemical. Physical blowing agents are gases (air, nitrogen, carbon dioxide) dissolved in the molten polymer under pressure, or low-boiling liquids such as pentane that vaporize on heating or under reduced pressure. Chlorofluorocarbons, once the blowing agents of choice, were discontinued after being linked to stratospheric ozone depletion. Chemical blowing agents are compounds that decompose on heating and give off nitrogen. The commercially important polyurethanes are unusual in that a byproduct of polymerization, carbon dioxide, serves as a blowing agent, usually in combination with a low-boiling liquid that is vaporized by the reaction exotherm. The chemistry of polyurethanes is discussed in Chapter 13.

There are a variety of other fabrication methods, such as *filament winding* to make fiber-reinforced composites, or vapor deposition via monomer pyrolysis. For a more comprehensive discussion, the reader is referred to texts[8,9] that deal with polymer processing in more detail.

FIGURE 4.5. Basic components for spinning.

4.3 Mechanical Properties

Let us consider a typical linear polymer having a uniform repeating unit and discrete end groups. At low molecular weight, the end groups contribute significantly to the overall structure, and this is manifested in such properties as density, refractive index, and spectroscopic absorption, which vary with molecular weight. When a certain molecular weight is attained, say 15,000, the concentration of end groups becomes negligible and these properties become constant, independent of any further increase in molecular weight. This is not the case with mechanical properties, which depend on intermolecular forces. Mechanical properties are much more dependent on molecular weight over a very broad range, although they too level off at the higher end of the molecular weight spectrum. Where they level off depends on structure. For a polyolefin such as polyethylene, where dispersion forces are responsible for mechanical properties, the leveling off occurs at relatively high molecular weight (above 10^5), whereas with very polar polymers such as polyamides, it may occur at a molecular weight as low as about 20,000 to 50,000. This dependence of properties on molecular weight is shown for a hypothetical polymer in Figure 4.6. (The assumption is made here that the "nonmechanical" property increases with molecular weight.) The working range of commercial polymers shown in the figure is at a molecular weight that affords good mechanical properties at a workable viscosity.

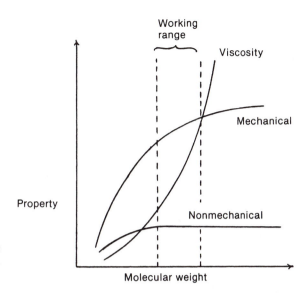

FIGURE 4.6. Dependence of properties on molecular weight (hypothetical polymer).

For the great bulk of commercial polymers, mechanical properties are of fundamental interest.[9–11] Although other properties, such as flame resistance, thermal stability, and chemical resistance, are of concern in more specialized applications, all polymers, regardless of use, must exhibit a specified range of mechanical properties suitable for that application. Among the dozens of interest to polymer manufacturers, *tensile, compressive,* and *flexural strength* (and their corresponding moduli) and *impact resistance* are most important. Related properties include hardness, abrasion resistance, and tear resistance.

To what exactly do these properties refer? In all cases they are a measure of how much stress a sample will withstand before the sample "fails." Tensile strength refers to resistance to stretching. Compressive strength is, in a sense, the opposite of tensile strength; it is the extent to which a sample can be compressed before it fails. Flexural strength is a measure of resistance to breaking, or snapping, when a sample is bent (flexed). Impact strength is a measure of "toughness"—how well a sample will withstand the sudden onset of stress, like a hammer blow. Another mechanical property of interest is *fatigue*—how well a sample will withstand repeated applications of tensile, flexural, or compressive stress. How these various properties are measured is described in the next chapter. Here we are more concerned with the fundamentals of mechanical properties as manifested in stress–strain behavior. Although each mechanical property has its own idiosyncrasies, all such properties result from the same characteristics of structure and morphology and the manner in which the polymeric matrix undergoes molecular reorientation in response to stress. For purposes of our discussion, we will focus on tensile properties.

Tensile strength is determined by stretching a strip of polymer of uniform dimensions. The tensile stress, σ, is the force applied, F, divided by the cross-sectional area, A; that is,

$$\sigma = \frac{F}{A}$$

in units of dynes per square centimeter (CGS) or newtons per square meter (SI) (or pounds per square inch, psi, in the English system). Tensile strain, ε, is the change in sample length, l divided by the original length:

$$\varepsilon = \frac{\Delta l}{l}$$

The ratio of stress to strain is the tensile modulus, E,

$$E = \frac{\sigma}{\varepsilon}$$

which is a measure of the resistance to tensile stress. (Note that these definitions parallel those of shear stress described in Section 3.3.) Because strain is dimensionless, modulus carries the same units as stress.

A distinction among fibers, plastics, and elastomers is often expressed in terms of stress–strain curves, as shown in Figure 4.7. Both plastics and fibers exhibit a steep slope (high modulus), but fibers can sustain greater stress before breaking (end of curve). Elastomers have a low modulus initially, but once in the stretched state the modulus increases sharply. A general stress–strain curve for a plastic such as polyethylene demonstrating various elements of tensile behavior is shown in Figure 4.8. Initially the modulus is high, until a point is reached where the plastic "yields," or deforms. Prior to the yield point, the *elongation* is reversible. At the yield point, enough stress has been applied to cause the molecules to untangle and flow over one another, and further elongation is irreversible. Eventually the sample breaks. The behavior of the polymer—or shape of the curve—beyond the yield point depends to a large degree on the polymer's initial morphology. If the polymer is amorphous or of low crystallinity, application of stress may increase the crystallinity, which in turn increases the modulus. This is the principle behind drawing. Highly crystalline polymers exhibit little change in morphology on drawing and break soon after the yield point, which is the case with the fiber in Figure 4.7.

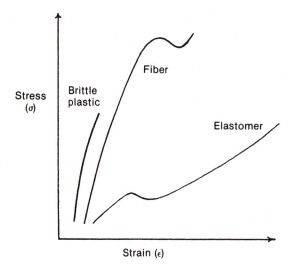

FIGURE 4.7. Characteristics of tensile stress–strain behavior.

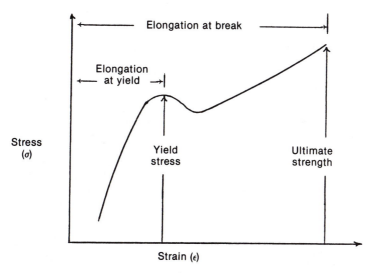

FIGURE 4.8. General tensile stress–strain curve for a typical thermoplastic.

The stress–strain behaviors depicted in Figures 4.7 and 4.8 are at constant temperature. Mechanical properties are, however, very temperature dependent. Let us assume, for example, that we take a strip of plastic similar to the one used to generate the curve shown in Figure 4.8, cool it to a low temperature, and then apply a constant stress. From the elongation ($\Delta l/l$) we can determine the modulus. As the temperature is increased, the modulus will remain high until the glass transition temperature is reached, at which point the modulus drops sharply. This is shown in Figure 4.9. The modulus scale is logarithmic. Very high moduli (of the order of 3×10^9 N/m^2) are typical in the glassy state, but these decline precipitously once the mole-

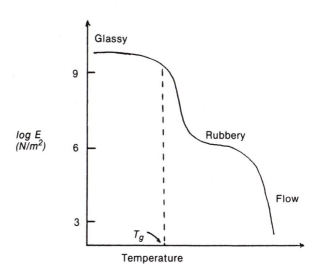

FIGURE 4.9. Effect of temperature on tensile modulus of an amorphous thermoplastic; log E, modulus scale; T_g, glass transition temperature.

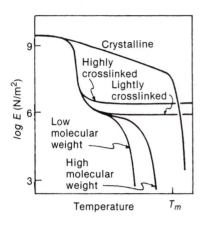

FIGURE 4.10. Effect of temperature on tensile modulus (log *E* scale) of various polymers. T_m, crystalline melting temperature. [Reprinted with permission from J. J. Aklonis, *J. Chem. Educ.*, **58**, 11 (1981).]

cules gain more freedom of movement. The sample remains rubbery until the temperature is sufficiently high to cause flow. As discussed earlier, room temperature would lie somewhere along the glassy plateau for a plastic and along the rubbery plateau for an elastomer.

Figure 4.10 illustrates the modulus–temperature behavior for a variety of polymer types. A thermoplastic amorphous polymer shows an expected molecular weight dependence in the flow region. The higher the molecular weight, the higher the temperature necessary to overcome the increased molecular entanglements. A crosslinked polymer, on the other hand, does not flow. The higher the crosslink density, the greater will be the modulus (the less the elongation) in the rubbery state. Semicrystalline thermoplastic polymers behave much like crosslinked polymers below the melting temperature, T_m, because of the very strong intermolecular forces arising from close chain packing. How sharp a break occurs at the glass transition temperature depends on the degree of crystallinity. Above T_m, of course, the crystalline polymer will flow.

In addition to temperature dependence, there are two important *time*-dependent properties: *creep* (or *cold flow*) and *stress relaxation*. Creep is a measure of the change in strain when a polymer sample is subjected to a constant stress (gravity, for example). Stress relaxation refers to the decrease in stress when a sample is elongated rapidly to constant strain. Both phenomena are characteristic of viscoelastic materials and arise from the slippage of polymer molecules. Plots of stress change versus time bear a distinct resemblance to modulus–temperature plots of the type shown in Figure 4.9. This is not unexpected; increasing temperature, in effect, speeds up the molecular disentanglement. Creep and stress relaxation are of critical concern to plastics manufacturers for the long-term mechanical properties of their materials.

The various characteristics of tensile behavior are readily understandable in terms of the structure–morphology relationships described in Chapter 3, and can be extrapolated to other kinds of mechanical properties. It is difficult to make generalizations when relating structure to properties because so many variables are involved—molecular weight, molecular weight distribution, morphology, method of sample preparation, number and type of additives. One might reasonably assume, however, that the more "flexible" the polymer backbone, the lower will be the tensile, flexural, or compressive strength. Anything that contributes to chain stiffening—bulky side groups, cyclic units, crosslinking, crystallinity—ought to increase these properties at the expense of tensile elongation. Trends in impact resistance are more difficult

TABLE 4.1. Mechanical Properties of Common Homopolymers[a]

| Polymer | Tensile Properties at Break | | | Compressive Strength[b] (MPa) | Flexural Strength[b] (MPa) | Impact Strength[c] (N/cm) |
	Strength[b] (MPa)	Modulus[b] (MPa)	Elongation (%)			
Polyethylene, low density	8.3–31	172–283	100–650	—	—	No break
Polyethylene, high density	22–31	1070–1090	10–1200	20–25	—	0.23–2.3
Polypropylene	31–41	1170–1720	100–600	38–55	41–55	0.23–0.57
Poly(vinyl chloride)	41–52	2410–4140	40–80	55–90	69–110	0.23–1.3
Polystyrene	36–52	2280–3280	1.2–2.5	83–90	69–101	0.20–0.26
Poly(methyl methacrylate)	48–76	2240–3240	2–10	72–124	72–131	0.17–0.34
Polytetra-fluoroethylene	14–34	400–552	200–400	12	—	1.7
Nylon 66	76–83	—	60–300	103	42–117	0.46–1.2
Poly(ethylene terephthalate)	48–72	2760–4140	50–300	76–103	96–124	0.14–0.37
Polycarbonate	66	2380	110	86	93	9.1

[a]Values taken from Agranoff,[12a] converted to SI units, and rounded off.

[b]To convert megapascals to pounds per square inch, multiply by 145.

[c]Izod notched impact test (see Chap. 5). To convert newtons per centimeter to foot pounds per inch, multiply by 1.75.

to predict. Chain stiffening in most instances leads to increased embrittlement and lower impact strength. On the other hand, conformational factors that allow rapid molecular reorientation at the onset of stress lead to improved impact strength. Measures taken to improve impact strength more often than not lead to a diminution of other mechanical properties.

Table 4.1 gives some representative mechanical properties for a few important commercial homopolymers.[12a] The wide range in properties reflects the many variables involved in physical testing. Worth noting is the remarkably high impact strength of polycarbonate. This, together with other strength characteristics, is a key factor in this relatively high-cost material being used widely as an engineering plastic. The significant increase in tensile properties of the mostly linear, crystalline high-density polyethylene, compared with those of the branched low-density polyethylene, arises from close chain packing in the former. Techniques have been developed to refine polyethylene's crystalline morphology further for use in high-strength fiber applications. Called *gel-spinning*, the method involves converting very-high-molecular-weight ($>10^6$) linear polymer to a gel with solvent, followed by spinning and drawing.[13] Gel formation provides the necessary molecular freedom for achieving a high degree of alignment. Similar results have been obtained with specially prepared polyethylene films without resorting to gel formation.[14] The drawable films are deposited directly in a disentangled crystalline state during polymerization onto initiator-treated surfaces.

TABLE 4.2. Fiber Properties[a]

Fiber Type	Tenacity[b] (N/tex)	Specific Gravity
Natural		
Cotton	0.26–0.44	1.50
Wool	0.09–0.15	1.30
Synthetic		
Polyester	0.35–0.53	1.38
Nylon	0.40–0.71	1.14
Aromatic polyamide (aramid)[c]	1.80–2.0	1.44
Polybenzimidazole	0.27	1.43
Polypropylene	0.44–0.79	0.90
Polyethylene (high strength)	2.65[d]	0.95
Inorganic[c]		
Glass	0.53–0.66	2.56
Steel	0.31	7.7

[a] Unless otherwise noted, data taken from L. Rebenfeld, in *Encyclopedia of Polymer Science and Engineering* (H. F. Mark, N. M. Bikales, C. G. Overberger, G. Menges, and J. I. Kroschwitz, Eds.), Vol. 6, Wiley-Interscience, New York, 1986, pp. 647–733.

[b] To convert newtons per tex to grams per denier, multiply by 11.3.

[c] Kevlar (see Chap. 3, structure **58**.)

[d] From *Chem. Eng. News*, **63**(8), 7 (1985).

[e] From V. L. Erlich, in *Encyclopedia of Polymer Science and Technology* (H. F. Mark, N. G. Gaylord, and N. M. Bikales, Eds.), Vol. 9, Wiley-Interscience, New York, 1968, p. 422.

Properties of high-strength polyethylene are compared with properties of other fibers in Table 4.2. Tensile strength is given as *tenacity*, the common tensile parameter used by fiber scientists. Units of tenacity are newtons per tex, where tex is defined as the weight in grams of 1000 meters of the fiber.* Commercial applications for high-strength polyethylene fibers include cut-resistant industrial gloves, protective glove liners for health care workers, bulletproof vests, and crash helmets. The only fiber listed with strength comparable to that of high-strength polyethylene is the aromatic polyamide (aramid) prepared by the liquid crystal method described in Chapter 3.

4.4 Thermal Stability

When organic substances are heated to high temperatures they have a tendency to form aromatic compounds. It follows that aromatic polymers should be resistant to high temperatures. A wide variety of polymers having aromatic repeating units have been developed in recent

*A more traditional unit of tenacity is grams per denier, where denier is defined as the weight in grams of 9000 m of the fiber. The denier evolved from units adopted for numbering raw silk at the Paris International Congress in 1900. The tex is used by the American Society for Testing and Materials (ASTM) as its standard linear density unit, and is also acceptable (along with kilograms per meter) in the SI system.

TABLE 4.3. Representative Thermally Stable Polymers[a]

Type	Structure	Decomposition Temperature (°C)[b]
Poly(p-phenylene)		660
Polybenzimidazole		650
Polyquinoxaline		640
Polyoxazole		620
Polyimide		585[c]
Poly(phenylene oxide)		570
Polythiadiazole		490
Poly(phenylene sulfide)		490

[a] Data from Korshak.[17]

[b] Nitrogen atmosphere unless otherwise indicated.

[c] Helium atmosphere.

years,[15–18] much of the impetus arising from the need of the aerospace industries for high-performance materials that will withstand extremes of temperature. For a polymer to be considered "thermally stable" or "heat resistant," it should not decompose below 400°C and should retain its useful service properties at temperatures near the decomposition temperature. As might be expected, such polymers must have high glass transition or crystalline melting temperatures. Some representative thermally stable polymers, along with their initial decomposition temperatures, are given in Table 4.3. (The chemistry of these polymers is covered in Part III.)

Thermal stability is primarily a function of bond energy. When the temperature increases to the point where vibrational energy causes bond rupture, the polymer degrades. In the case of cyclic repeating units, breaking of one bond in a ring does not lead to a decrease in molecular weight, and the probability of two bonds breaking within one ring is low. Thus, ladder or semi-ladder polymers are expected to have higher thermal stabilities than open-chain polymers.[19] The decomposition temperatures given in Table 4.3 were determined in inert atmospheres. Decomposition in air—a measure of a polymer's *thermooxidative stability*—generally follows a different mechanism. The presence of oxygen, however, in most cases has little effect on the initial decomposition temperature; hence bond rupture is primarily a thermal rather than an oxidative process.

One individual, the late Carl S. Marvel, stands out as the leader in the field of thermally stable polymers; with justification he has been called "the father of high-temperature polymers."[20] Marvel's pioneering work led to commercial development of polybenzimidazole (**1**)[21] (Hoechst Celanese trade name PBI). Polybenzimidazole fiber is used in astronauts' space suits and firefighters' protective clothing, and the plastic finds use in a variety of applications involving high-temperature environments.

1

A wide variety of aromatic and thermally stable organometallic polymers have been developed, many of which will be encountered later in the book. Few have found commercial acceptance because of a combination of high cost and poor processability. Because of their rigid backbone structure, aromatic polymers characteristically exhibit very high glass transition temperatures, high melt viscosities, and low solubility and are therefore more intractable than most other kinds of polymers. It is generally recognized that the upper limits of thermal stability of organic polymers have already been achieved; hence the emphasis in recent years has been to introduce structural variations that allow improved processability. Incorporation of "flexibilizing" groups such as ether or sulfone into the backbone is one strategy. Although these measures often lead to greater solubility and lower viscosity, thermal stability usually suffers.

Another approach has been to introduce cyclic aromatic groups that lie *perpendicular* to the planar aromatic backbone, as in the polybenzimidazole **2**.[22] Such structures, referred to as *cardo polymers* (from the Latin *cardo*, loop),[23,24] usually exhibit improved solubility with no sacrifice of thermal properties.

2

Incorporating reactive groups into the polymer backbone that undergo intramolecular cycloaddition on heating is another way to improve processability. The rationale is that longer

$T_g = 215°C$

$T_g = 265°C$

$$\left(Ar = -\!\!\bigcirc\!\!-O-\!\!\bigcirc\!\! \right)$$

SCHEME 4.1. Increasing T_g of a polyquinoxaline by intramolecular cycloaddition.

flow times are possible because little or no crosslinking takes place; rather, the glass transition temperature increases because of chain stiffening. An example, shown in Scheme 4.1, is the conversion of the aromatic polyquinoxaline **3** into **4** via cycloaddition reactions of the phenylethynyl substituents, resulting in a 50° increase in the glass transition temperature.[25] (Polyquinoxaline synthesis is given in Chapter 15.)

The most productive approach from the standpoint of commercial development has been the synthesis of aromatic oligomers or prepolymers capped with reactive end groups.[26] The end-capped oligomers melt at relatively low temperatures and are soluble in a variety of solvents. On heating they are converted to thermally stable network polymers. End groups that are found in commercially important reactive oligomers are given in Table 4.4. The first four listed have been in commercial use for several years.[27-30] Phenylethynyl-terminated oligomers are being developed for aerospace applications.[31] A variety of other reactive end groups have been investigated.[26,32–39] The chemistry of network polymer formation usually involves cycloaddition or addition polymerization reactions of the end groups. Examples will be encountered in later chapters.

TABLE 4.4. Some Reactive End Groups for
Converting Oligomers to Network Polymers

Type	Structure
Cyanate	—O—C≡N
Ethynyl	—C≡CH
Maleimide	
Nadimide[a]	
Phenylethynyl	

[a] Common name for 5-norbornene-2,3-dicarboximide.

4.5 Flammability and Flame Resistance

Because synthetic polymers are used increasingly in construction and transportation, considerable effort has been expended to develop nonflammable polymers, as well as to understand the mechanism of flame propagation and flame retardation.[40–42] Other concerns include the suppression of smoke and toxic gases formed during combustion, and the development of nonflammable textile fibers.

Some polymers are inherently nonflammable, for example, poly(vinyl chloride) and other polymers having a high halogen content. Others, such as polycarbonates, will burn as long as a source of flame is present, but stop burning when the flame is removed. Such polymers are said to be *self-extinguishing*. Most polymers burn readily. Burning occurs in a series of steps. First, an external heat source increases the polymer temperature to a point where it begins to decompose and release combustible gases. Once the gases ignite, the temperature increases until the release of combustibles is rapid enough for combustion to be self-sustaining so long as sufficient oxygen is available to support the combustion process. With some polymers, such as polystyrene and poly(methyl methacrylate), the combustible gases may be high in monomer because of thermally induced depolymerization of the polymer chains. Monomer breaks down further to lower-molecular-weight combustible products, including hydrogen, as it diffuses toward the flame. Where depolymerization does not occur, surface oxidation plays a role in generation of combustible gases. A simplified picture of polymer burning, similar to that of a candle,[43] is depicted in Figure 4.11 as a closed cycle in which combustion of the diffusing gases generates heat that radiates back to sustain the pyrolysis.

FIGURE 4.11. Representation of polymer combustion. →, gas diffusion; ←, heat flux. [Adapted from Factor.[43]]

Approaches to promoting flame resistance in polymers focus on three strategies: (1) retarding the combustion process in the vapor phase, (2) causing "char" formation in the pyrolysis zone, and (3) adding materials that decompose either to give nonflammable gases or endothermically to cool the pyrolysis zone.

In the vapor phase, combustion occurs by a complex series of free radical propagation and radical transfer reactions. Retardation of combustion may, therefore, be brought about by incorporating into the polymer *radical traps*—compounds that reduce the concentration of radicals in the vapor. Halogenated compounds are particularly effective because the hydrogen halide that is released reacts with free radicals to form less reactive halogen atoms. The commercially available anhydride, **5**, is one example.[44] Polyesters prepared with

5

5 decompose on heating by the retrograde Diels–Alder reaction, and the resultant hexa-

6

chlorocyclopentadiene, **6**, suppresses radical formation in the vapor phase and inhibits free radical depolymerization in the pyrolysis zone. Antimony oxides are often used in combination with halogen compounds because of a synergistic effect believed to arise from formation of antimony halides. A serious disadvantage of halogen compounds is the toxicity of the hydrogen halide formed during burning.

Char formation at the polymer surface reduces flammability by acting as a barrier (principally carbonaceous) to inhibit gaseous products from diffusing to the flame, and to shield the polymer surface from the heat flux. Aromatic polymers have a natural tendency toward char formation, which accounts for their generally low flammability. Crosslinking may also increase char formation; introduction of chloromethyl groups onto polystyrene, for example, is believed to reduce the burning rate of that polymer primarily by promoting crosslinking during pyrolysis.[45] Certain additives, notably phosphorus-containing compounds, also increase char formation.[46,47] Phosphorus compounds are effective, for example, in reducing the flammability of cellulose by promoting dehydration to yield unsaturated compounds that subsequently polymerize to a crosslinked char. Thus cellulosic fibers, especially cotton, are made flame resistant by grafting with phosphorus-containing monomers. Flame retardants that promote formation of a carbonaceous char are referred to as *intumescent* flame retardants.[48]

The third strategy for reducing flammability involves using compounds such as hydrated alumina, $Al_2O_3 \cdot 3H_2O$, that evolve water endothermically to cool the pyrolysis zone, or sodium bicarbonate, which decomposes to form carbon dioxide, which in turn dilutes the combustible gases.

Equally important in considerations of flammability are the generation of toxic decomposition products and smoke. Among the former are hydrogen cyanide evolved from burning nitrogen-containing polymers such as polyamides and polyurethanes (widely used in the manufacture of carpeting, bedding, and furniture upholstery), and hydrogen chloride from poly(vinyl chloride) (used in flooring, wall panels, and siding). Almost all polymers evolve copious amounts of carbon monoxide. Smoke arises from formation of acetylene and benzene, which condense to form soot particles. Ironically, flame retardants that operate in the vapor phase probably promote smoke formation by removing radicals that might otherwise oxidize the soot-forming particles. Although materials such as hydrated alumina are often effective in reducing smoke, toxic byproducts of combustion continue to be a serious impediment to more widespread use of polymers in construction.

4.6 Chemical Resistance

One of the problems that oil companies face with their huge petroleum storage tanks is rusting away of the metal bottom from underneath. Some companies have remedied this by spray-coating the inside floor of the tanks with glass fiber-reinforced unsaturated polyester (Chapter 12). The hard plastic coating significantly lengthens the tank's lifetime and avoids the expense of having to replace the tank bottom with steel, despite the fact that the polymer backbone contains hydrolyzable ester groups.

Not all polyesters have good hydrolytic stability, and people who coat tank bottoms are obviously going to choose a polymer that has proven to be highly resistant to water. With polyesters, two general approaches have been used to increase chemical resistance—one to increase the steric hindrance about the ester groups, and the other to reduce the number of ester groups per unit chain length. Both increase the hydrophobic nature of the polyesters. Two glycols that have found commercial application in chemically resistant polyester formulations are 2,2,4-trimethylpentane-1,3-diol (**7**) and the bisphenol A-propylene oxide derivative **8**.

$$\underset{\textbf{7}}{\underset{\begin{array}{cc}CH_3 & \\ | & OH\end{array}}{HOCH_2C\!-\!-\!CH\!-\!CHCH_3}}$$

$$\underset{\textbf{8}}{HOCHCH_2O\!-\!\!\langle \bigcirc \rangle\!-\!\!\underset{CH_3}{\overset{CH_3}{C}}\!-\!\!\langle \bigcirc \rangle\!-\!OCH_2CHOH}$$

Another approach has been to reduce the hydrophilic nature of the end groups.[49] Phenyl isocyanate, for example, converts hydroxyl groups of polyesters to urethanes,

$$\text{wwwOH} + \phi\text{NCO} \longrightarrow \text{www}\overset{\overset{\displaystyle O}{\|}}{O}\text{CNH}\phi$$

and carboxyl to amide,

$$\text{wwwCO}_2\text{H} + \phi\text{NCO} \longrightarrow \text{www}\overset{\overset{\displaystyle O}{\|}}{C}\text{NH}\phi + \text{CO}_2$$

These examples illustrates how one type of polymer can be made more resistant to corrosive environments by the application of relatively straightforward chemical principles.[50,51] Fluorine has proved to be an element that imparts both water and solvent resistance to a variety of polymers.[52] The inorganic polyphosphazene **9** (Chapter 16), which is very unstable in the presence of moisture, is rendered highly moisture resistant by conversion to **10**.[53] The fluorine, in effect, provides a water-resistant sheath to protect the phosphorus–nitrogen backbone. An ethylene–chlorotrifluoroethylene copolymer is marketed as a chemically resistant coating for underground cables. Because of their chemical inertness, a variety of fluorinated polymers including polytetrafluoroethylene **11**, poly(vinylidene fluoride) **12**, and copolymers such as poly[hexafluoropropylene-*co*-(vinylidene fluoride)] **13** have been developed commercially for use as gaskets, sealants, valves, and so on, where resistance to lubricating fluids is necessary.[54]

$$\underset{\textbf{9}}{\left[\!\!\begin{array}{c} Cl \\ | \\ N\!=\!P \\ | \\ Cl \end{array}\!\! \right]} \qquad \underset{\textbf{10}}{\left[\!\!\begin{array}{c} OCH_2CF_3 \\ | \\ N\!=\!P \\ | \\ OCH_2CF_3 \end{array}\!\! \right]}$$

$$\underset{\textbf{11}}{\left[CF_2CF_2 \right]} \qquad \underset{\textbf{12}}{\left[CH_2CF_2 \right]} \qquad \underset{\textbf{13}}{\underset{\begin{array}{c} | \\ CF_3 \end{array}}{\text{wwCF}_2\text{CF wwwCH}_2\text{CF}_2\text{ww}}}$$

Ozone, formed by the action of ultraviolet light or electrical discharge on oxygen, degrades polymers containing double bonds in the backbone by ozonolysis followed by hydrolysis:

$$
\text{C=C} \xrightarrow{O_3} \overset{O}{\underset{O-O}{\text{C}\diagup\text{C}}} \xrightarrow{H_2O} \text{C=O} + \text{O=C}
$$

This can be a serious problem with many elastomers where double bonds are necessary to effect crosslinking. One way to circumvent this is to replace the more commonly used 1,3-butadiene or isoprene in the elastomer with a cyclic diene such as cyclopentadiene. This results in a cyclopentene unit (**14**) in the backbone by 1,4-addition. Ozonolysis may still occur, but the backbone remains intact.

14

Morphology is also an important variable in chemical resistance. Crystalline polymers are invariably more resistant than their amorphous counterparts because close chain packing reduces permeability. Crosslinking also increases solvent resistance. Crosslinking is particularly important from the standpoint of chemical resistance in the microelectronics industry.[55,56] One step in the manufacture of printed circuits[57,58] involves coating a substrate with a polymer that crosslinks under the influence of light or ionizing radiation. A *mask* carrying the pattern to be transferred to the substrate is placed over the coated surface, then the surface is irradiated. The pattern allows radiation through, and those portions of the polymer thus exposed undergo crosslinking. When the mask is removed, the unexposed parts of the polymer are dissolved with solvent, leaving behind the desired pattern. Such a process, illustrated schematically in Figure 4.12(a), is referred to as a *negative resist*—the exposed portion of the surface becomes *resist*ant to solvent.

Sunlight is another agent that can bring about polymer degradation. Monomers containing ultraviolet-absorbing chromosphores such as 2,4-dihydroxy-4′-vinylbenzophenone (**15**) have been incorporated into vinyl polymers to improve light stability.[52] Photolytic and chemical degradation of polymers is discussed more fully in Chapter 9.

15

4.7 Degradability

Most polymers are very durable; indeed, it is this property that allows them to compete with such other durable materials as glass and metals. The previous section discussed strategies for rendering polymers even more durable. Environmental awareness, on the other hand, teaches us that polymer waste products contribute in ever greater volume to the problems of solid waste

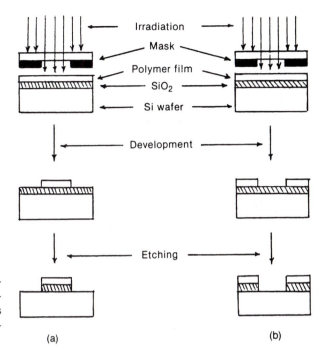

FIGURE 4.12. Schematic of a typical procedure for producing (a) negative resists and (b) positive resists in the manufacture of integrated circuits.

management. As a result, attention has shifted to the opposite end of the durability spectrum—to the synthesis of polymers that are degradable by the environmental effects of sunlight and soil microorganisms.[60,61] Furthermore, recent years have seen new markets develop in agriculture, medicine, and microelectronics for polymers that degrade at a predictable rate.[62–66]

Polymers can be made to degrade photochemically by incorporation of carbonyl groups that absorb ultraviolet (UV) radiation to form excited states energetic enough to undergo bond cleavage. Such processes (referred to as Norrish type II reactions) occur as follows[67]:

Commercially available photodegradable packing materials employ this technology. Microorganisms degrade polymers by catalyzing hydrolysis and oxidation. The lower the molecular weight, the more rapidly the polymer degrades. A combination of light-sensitive and hydrolyzable functional groups are thus more effective in degrading high-molecular-weight polymers in natural environments.[68]

Although the initial impetus for degradable polymers arose from ecological concerns, considerable research is now directed toward resist technology and *controlled release* applications. In the former, degradable polymers are used for *positive resists*, which work in a sense opposite to that of the resists described in the previous section; that is, radiation promotes degradation of the resist exposed by the mask, leaving the unexposed coating intact.[57,58] This is illustrated in Figure 4.12(b).

Controlled release refers to the use of polymers containing agents of agricultural, medicinal, or pharmaceutical activity, which are released into the environment of interest at relatively constant rates over prolonged periods. In the agricultural field, degradable mulches to promote crop growth are composed of combinations of natural polymers (which degrade readily in the presence of soil microorganisms) and synthetic polymers. Examples are starch-*graft*-poly(methyl acrylate)[69] and block copolymers of amylose or cellulose with polyesters.[70] At the completion of the growing season, such mulches may be plowed directly into the soil along with crop residues.

Another application involves binding of agriculture chemicals in polymer formulations for slow release at a rate effective for their intended purpose without the risk of the reagents being washed away by rain or irrigation. The herbicide 2,4-dichlorophenoxyacetic acid (2,4-D), for example, has been incorporated into polymers either as a chelate with iron (**16**)[71] or as a hydrolyzable pendant ester group on a vinyl polymer (**17**).[62a] A different strategy for

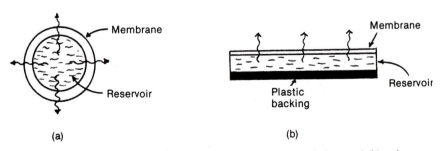

16   **17**

controlled release based on polymer permeability rather than degradation is to encapsulate the active reagent within a polymeric membrane or in a *strip*, as shown in Figure 4.13. Ideally the reagent is contained in the reservoir as a saturated solution with excess in suspension. This allows diffusion through the membrane at a constant rate without loss of activity. Alternatively, the reagent may be dispersed in a polymer matrix and released to the environment by diffusion or extraction. A variety of membrane and matrix devices are commercially available.[62b] Pheromone release strips for insect control[62c] and household fly and cockroach strips for release of insecticide[62d] are also in commercial use.

FIGURE 4.13. Membrane-controlled release devices: (a) microencapsulation, and (b) strip.

Encapsulated pharmaceuticals have been available for many years.[72] Reservoir strips called *transdermal patches* are used to release drugs through the skin, for example, nitroglycerin to treat angina or scopolamine to combat motion sickness. Implanted polymeric matrix devices, including some made with degradable polymers to release the drug through surface erosion, have also been developed. Degradable polyesters are already used as disappearing surgical sutures. The inorganic polyphosphazene (**9**), with amino acids, esters, and steroids in place of the chlorine atoms,[73] are of interest for slow release of the steroid, the other degradation products (amino acid, ester, phosphate, and ammonia) being nontoxic. Another active area of research concerns "smart polymers"—polymers that respond in a predictable way to changes in temperature or polarity. Such a polymer is poly(*N*-isopropylacrylamide), **18**, which shrinks reversibly in response to an increase in temperature.[74] Drugs trapped within an interpenetrating network containing **18**, for example, could be expelled as the network contracts upon encountering body temperature (37°C).

$$\begin{array}{c} +CH_2CH+ \\ | \\ C-NHCH(CH_3)_2 \\ \| \\ O \end{array}$$

18

4.8 Electrical Conductivity

Most polymers are good insulators. Some, for example, poly(N-vinyl-carbazole)[75] (**19**), are *photoconducting*, that is, they conduct electricity to a small degree under the influence of

19

light, and are used in the electrophotography (photocopying) industry.[69] An important discovery in the 1970s was that certain polymers, notably poly(sulfur nitride) (**20**) and polyacetylene (**21**) can be made highly conducting in the presence of certain additives, called *dopants*. This breakthrough initiated a major research effort to elucidate the mechanism of

$$+S{=}N+ \qquad +CH{=}CH+$$

20   **21**

conduction and to apply it to practical light-weight battery technology.[77–83] Although a complete understanding of the conduction mechanism remains elusive, certain structural features are known to influence the level of conductivity. These include:

1. Delocalization. An extended conjugated system is usually necessary for backbone conductivity; however, charge may be transferred in some cases (e.g., **19**) through pendant groups.

Soliton

FIGURE 4.14. Proposed conducting unit of polyacetylene. Soliton may be neutral (radical), positive (carbocation), or negative (carbanion).

2. Doping. Dopants may be electron acceptors such as arsenic pentafluoride or halogen, or electron donors such as alkali metals. Conductivity varies with dopant concentration. Doping may also effect rearrangement of the double bonds of nonconjugated polymers (e.g., polyisoprene) into a conjugated conducting mode.[84]

3. Morphology. Conduction is influenced by configurational and conformational factors, as well as crystallinity. Conductivity of polyacetylene film in the direction of molecular alignment is increased significantly by stretching.

Conductivity (σ) is expressed in units of siemens (S) per centimeter. (A siemen is a reciprocal ohm.) Materials are classified as insulators ($\sigma < 10^{-8}$ S/cm), semiconductors ($\sigma = 10^{-7} - 10^{-1}$ S/cm), and conductors ($\sigma > 10^2$ S/cm). Ordinary poly(sulfur nitride), **20**, has a conductivity of about 100 S/cm, although at very low temperatures (<0.3 K) it exhibits superconductivity. Polyacetylene's intrinsic conductivity is 1.7×10^{-9} S/cm for the *cis* isomer (**22**) and 4.4×10^{-5} S/cm (weakly semiconducting) for the *trans* isomer (**23**).

22   23

Doping increases polyacetylene's conductivity dramatically into the range of metal conductors. Doping apparently converts the red metallic-colored *cis* isomer to blue metallic *trans*. The dopant forms a charge transfer complex with the polymer that is believed to give rise to highly delocalized cation radicals or anion radicals depending on whether the dopant is electron accepting or donating, respectively; for example,

$$2 \cdot \mathrm{[CH{=}CH]} + 3\ I_2 \longrightarrow 2\ \mathrm{[CH{=}CH]}^{\cdot +} + 2\ I_3^{-}$$

$$\mathrm{[CH{=}CH]} + Na \longrightarrow \mathrm{[CH{=}CH]}^{\cdot -} + Na^{+}$$

According to one theory of polyacetylene conductivity, the delocalized regions, called solitons* (Figure 4.14), extend about 15 bond lengths, at which point the energy gain arising from lengthening of double bonds and shortening of single bonds apparently outweighs stabilization arising from delocalization. Conduction is believed to involve movement of electrons intramolecularly and intermolecularly via the positive or negative solitons. Intermolecular conduction is highly dependent on crystallinity, which accounts for the increase in polyacetylene's conductivity (to as much as 1.5×10^5 S/cm) when films of the polymer are properly oriented.

*From *solitary wave solutions*, a mathematical treatment of conductivity.

Although polyacetylene[85] received the most attention initially, its instability to air and moisture have limited commercial exploitation. Instead, attention has shifted to more stable conducting polymers, including polyaniline* (**24**), polypyrrole (**25**), polythiophene (**26**), poly(*p*-phenylene) (**27**), and poly(*p*-phenylenevinylene) (**28**), as well as variously substituted derivatives of each. Doping is again necessary to raise the conductivity of these polymers to practical levels. Typical dopants include AsF_5 and I_2, which are reduced to AsF_6^- and I_3^-, respectively, upon complexation. Polyaniline is unusual in that acid (usually HCl) serves as dopant. Besides conventional chemical synthesis, many conducting polymers are more conveniently synthesized as continuous films from solutions of monomer by anodic polymerization in an electrochemical cell.[86] In such cases, dopant ions are introduced directly from the monomer solution. Details of the synthesis and properties of many of these polymers may be found in later chapters.

24	**25**	**26**	**27**	**28**

Conductivity ranges of doped polymers compared with some metal conductivities are given in Table 4.5. Ranges vary according to method of synthesis, type and concentration of dopant, and polymer morphology. The more stable polymers may not be as conductive as polyacetylene; nevertheless their conductivities are acceptable for commercial development. Also important is that the densities of the polymers (generally around 1 g/cm^3) are much

---

*Polyaniline exists in several oxidation states, only one of which is shown. A more detailed picture is given in Chapter 17.

TABLE 4.5. Conductivities of Metals and Doped Polymers[a]

Material	Conductivity (S/cm)[b]
Copper	5.8×10^5
Gold	4.1×10^5
Polyacetylene	$10^3–10^5$
Poly(sulfur nitride)	$10^3–10^4$
Poly(*p*-phenylene)	10^3
Poly(*p*-phenylenevinylene)	10^3
Polyaniline	$10^2–10^3$
Polypyrrole	$10^2–10^3$
Polythiophene	10^2

[a]Data from J. R. Reynolds, A. D. Child, and M. B. Gieselman, in *Encyclopedia of Chemical Technology*, 4th ed. (J. I. Kroschwitz and M. Howe-Grant, Eds.), Wiley, New York, 1994; and *Chem. Eng. News*, June 22, 1987, p. 20.

[b]I siemen (S) = I ohm^{-1}.

lower than those of the metals (8.92 g/cm^3 for copper, 19.3 g/cm^3 for gold), which makes them attractive for applications where weight is an important consideration. Typical applications include light-weight batteries, thin-film transistors, light-emitting diodes, and anti-static materials. A long-range goal is to fabricate "molecular wires" consisting of conducting oligomers capable of bridging metal surfaces for use in molecular-scale electronic devices.[87]

Another area of conductivity where polymers are of interest is that of solid electrolytes.[88] *Polyelectrolytes* for solid battery applications generally consist of polymers such as poly(ethylene oxide) (**29**) or the polyphosphazene (**30**) in combination with metal salts. A 4:1 ratio of **29** to NaBF$_4$ is typical. To function as an effective polyelectrolyte, the polymer should be highly amorphous and have a low glass transition temperature to allow the freedom of molecular movement necessary for ion transport.

$$\left[CH_2CH_2O\right] \qquad \left[\begin{array}{c} (OCH_2CH_2)_2OCH_3 \\ | \\ P{=}N \\ | \\ (OCH_2CH_2)_2OCH_3 \end{array}\right]$$

29   30

4.9 Nonlinear Optical Properties

Another emerging area in which conjugated systems are destined to play a key role is that of *photonics*: the use of photons instead of electrons for information and image processing. Photonic devices are expected to operate at a much higher rate and to store information much more densely than electronic devices. The materials of choice for photonics devices are polymer films containing chromophores that cause a change in the polymer's refractive index in response to a light-induced electric field. Such materials are said to exhibit *nonlinear optical* (NLO) properties.[89–93]

The chromophores are dipolar moieties containing electron-donating groups at one end and electron-attracting groups at the other, so that a large change in dipole moment is induced upon excitation. For the polymer to exhibit NLO properties, the dipole moments must first be aligned by heating to the glass transition temperature of the polymer, applying a strong electric field to bring about alignment, then cooling. Such a dipole-aligned end product is termed a *poled polymer*. The polymer's glass transition temperature must be high enough to maintain alignment at the operating temperatures of photonic devices, which can be as high as 100°C. If the devices are to be integrated into computer chips, they must also withstand soldering temperatures in excess of 250°C.

The chromophores may be incorporated into the polymer as low-molecular-weight compounds dissolved in the polymer matrix (*host–guest systems*) or, more commonly, they are formulated into the polymer backbone or into a side chain. Crosslinking may also be employed to stabilize the system. A large number of NLO polymers have been reported; examples having NLO chromophores in the backbone and side chain are the polyester **31** and the polyimide **32**, respectively.

31   32

4.10 Additives

Virtually every polymer in commercial use contains additives, usually a combination. Their purpose is twofold: (1) to alter the properties of the polymer and (2) to enhance processability. Property modifiers range from pigments and odorants used for aesthetic reasons to plasticizers for modifying mechanical properties. Processing modifiers vary from lubricants to prevent sticking to fabrication machinery to compounds that alter the chemical structure, such as crosslinking agents. Additives may be completely miscible or, as is the case with reinforcing agents, completely immiscible. Sometimes additives reduce cost. Wood flour, a byproduct of saw mills, is added to phenol–formaldehyde polymers as a cost-reducing *extender*, as well as to reduce brittleness. Additives are a major industry; more than 5×10^9 kg are consumed in the United States alone every year.

Table 4.6 lists commercially important additives according to function. Some appear in more than one category. Plasticizers, for example, increase flexibility, but also reduce the melt viscosity to facilitate molding or extruding. Slip agents minimize adhesion to processing equipment and at the same time prevent film or sheet from sticking together. Some of the additives listed, such as nucleating agents, blowing agents, and flame retardants, have already been mentioned. Others, including crosslinking agents, promoters, emulsifiers and antioxidants, are covered in later chapters. A number of the additives, particularly surface modifiers such as waxes and surfactants, have limited solubility in the polymers, which causes them to migrate to the surface, where their activity is needed.

Plasticizers[98,99] are the most widely used, with di-2-ethylhexyl phthalate (**33**) being the cheapest "general-purpose" plasticizer. Other commonly used plasticizers are listed in Table 4.7. About 80% of all plasticizer consumption goes into a single polymer—poly(vinyl chloride). It is believed that the thermal motion of the low-molecular-weight plasticizers increases the polymer's free volume, allowing more "elbow room" for in-

TABLE 4.6. Polymer Additives

Type	Function
Mechanical property modifiers	
Plasticizers	Increase flexibility
Impact modifiers	Improve impact strength
Reinforcing fillers	Increase strength properties
Nucleating agents	Modify crystalline morphology
Surface property modifiers	
Slip and antiblocking agents	Prevent film and sheet sticking
Lubricants	Prevent sticking to machinery
Antistatic agents	Prevent static charge on surfaces
Coupling agents	Improve bonding between polymer and filler
Wetting agents	Stabilize dispersions of filler
Antifogging agents	Disperse moisture droplets on films
Chemical property modifiers	
Flame retardants	Reduce flammability
Ultraviolet stabilizers	Improve light stability
Antioxidants	Prevent oxidative degradation
Biocides	Prevent mildew
Aesthetic property modifiers	
Dyes and pigments	Impart color
Odorants	Add fragrance
Deodorants	Prevent development of odor
Nucleating agents	Improve light transmission
Processing modifiers	
Plasticizers	Reduce melt viscosity
Slip agents and lubricants	Prevent sticking to processing machinery
Low-profile additives	Prevent shrinkage and warpage
Thickening agents	Increase viscosity of polymer solutions or dispersions
Heat stabilizers	Prevent degradation during processing
Defoaming agents	Reduce foaming
Blowing agents	Manufacture stable foams
Emulsifiers	Stabilize polymer emulsions
Crosslinking (curing) agents	Crosslink polymer
Promoters	Speed up crosslinking (curing)

creased long-range segmental motion of the polymer molecules. For some applications, plasticizer comprises almost 50% of the formulation and, in the case of poly(vinyl chloride), the glass transition temperature may be reduced from about 80°C for unplasticized polymer to below 0°C. For reasons of compatibility, plasticizers are not used with highly crystalline polymers.

TABLE 4.7. Commonly Used Plasticizers

Aromatic
    Di-2-ethylhexyl phthalate
    Di-*n*-octyl phthalate
    Di-*i*-octyl phthalate
    Di-*i*-decyl phthalate
    Di-*n*-undecyl phthalate
    Di-*n*-tridecyl phthalate
    Tri-2-ethylhexyl trimellitate

Aliphatic
    Di-2-ethylhexyl adipate
    Di-2-ethylhexyl sebacate
    Di-2-ethylhexyl azelate

Epoxy
    Epoxidized linseed oil
    Epoxidized soya oil

Polymeric
    Poly(alkylene adipates, sebacates, or azelates)

Fire retardant
    Chlorinated paraffins
    Phosphate esters

$$
\begin{array}{c}
CH_2CH_3 \\
| \\
CO_2CH_2CHCH_2CH_2CH_2CH_3 \\
CO_2CH_2CHCH_2CH_2CH_2CH_3 \\
| \\
CH_2CH_3
\end{array}
$$

33

A number of criteria are considered in choosing a plasticizer, including cost, compatibility, stability, processability, and *permanence*. Permanence refers to how well the plasticizer remains in the polymer: Will it vaporize on heating or be extracted if the plastic comes into contact with solvents or lubricants? An example is the use of phthalate esters in automobile interiors. The "new car smell" enjoyed by many car owners results mainly from plasticizer vaporized in the closed car interior, and actually advertises the deterioration of the vinyl upholstery. If the car is left in the sun and becomes very hot, enough plasticizer may be released to fog the windows with a greasy film. In time the upholstery becomes brittle and cracks. For this reason, higher-molecular-weight phthalates are commonly used for modern car interiors.

Permanence is a phenomenon related to compatibility. Solubility factors, discussed in Chapter 2 for polymer solutions, apply equally as well to plasticizers, except that now the polymer is the major constituent of the solution. Thus a plasticizer having "poor" solubility will reduce the viscosity of a polymer more than one having "good" solubility at equal levels of dilution; however, this reduced compatibility also results in less permanence. Obviously a balance must be struck between the two. In some instances even water functions as a plasticizer, particularly where highly hydrophilic functional groups are present in the polymer. Cotton (cellulose) fibers (Chapter 17) and polyamides (Chapter 13) represent two common polymers in which the effect of water must be taken into account by polymer manufacturers.

Aliphatic esters are usually used for low-temperature flexibility, and epoxy plasticizers are used where good heat and light stability are desirable. Polymeric plasticizers are obviously less volatile, but their relatively high cost limits their use. Polymers plasticized with polymeric plasticizers are, in effect, polyblends[100] (Section 3.10).

The plasticizers just described are *external* plasticizers. *Internal* plasticization is brought about by chemically binding the flexibilizing moiety to the polymer.

Impact modifiers are elastomeric polymers such as polybutadiene. *Impact polystyrene*, for example, is an immiscible polyblend of polystyrene and 5 to 10% polybutadiene. Impact polystyrene is much tougher than polystyrene alone, and accounts for more than half the polystyrene produced.

Reinforcing fillers come in two forms—fibrous and granular (or powder). Carbon black, used to reinforce natural and synthetic rubber, is an example of the latter. Glass fiber is widely used to increase strength properties, particularly in combination with unsaturated polyesters. Such materials are termed *composites*.[101,102] In recent years the development of very-high-strength composites[103] using aromatic polyamide and graphite fibers in combination with engineering plastics has resulted in a wide range of products from tennis rackets and fishing poles to high-performance aerospace components.

The remaining additives listed in Table 4.6 comprise a much smaller percentage of the overall polymer formulation. Details of their composition and function may be obtained from the references provided.

References

1. D. H. Morton-Jones, *Polymer Processing*, Chapman and Hall, New York, 1989.
2. W. Michael, *Plastics Processing: An Introduction*, Hanser, Cincinnati, 1995.
3. W. E. Becker (Ed.), *Reaction Injection Molding*, Van Nostrand Reinhold, New York, 1979.
4. C. Macosko, *Fundamentals of Reaction Injection Molding*, Hanser, Munich, 1988.
5. J. J. Bikerman, *Foams*, Springer-Verlag, New York, 1973.
6. K. C. Frisch and J. H. Saunders, *Plastic Foams*, Marcel Dekker, New York, Part 1, 1972; Part 2, 1973.
7. F. W. Billmeyer, *Textbook of Polymer Science*, 3rd ed., Wiley-Interscience, New York, 1984, Part 6.
8. F. Rodriguez, *Principles of Polymer Systems*, McGraw-Hill, New York, 1970, Chap. 12.
9. I. M. Ward, *Mechanical Properties of Polymers*, Wiley-Interscience, New York, 1971.
10. A. Rudin, *The Elements of Polymer Science and Engineering*, Academic Press, New York, 1982.
11. D. W. Van Krevelen, *Properties of Polymers*, 3rd ed., Elsevier, New York, 1990.
12. J. Agranoff (Ed.), *Modern Plastics Encyclopedia, 1984–1985*, Vol. 61, McGraw-Hill, New York, 1984: (a) pp. 449–704; (b) pp. 106ff.
13. V. A. Marichin, L. P. Mjasnikova, D. Zenke, R. Hirte, and P. Weigel, *Polym. Bull.*, **12**, 287 (1984).
14. P. Smith, H. D. Chanzy, and B. P. Rotzinger, *Polym. Commun.*, **26**, 258 (1985).
15. P. E. Cassidy, *Thermally Stable Polymers*, Marcel Dekker, New York, 1980.
16. J. P. Critchley, G. J. Knight, and W. W. Wright, *Heat-Resistant Polymers*, Plenum, New York, 1983.
17. V. V. Korshak, *Heat-Resistant Polymers*, Halstead, New York, 1972.

18. C. Arnold Jr., *J. Polym. Sci., Macromol. Rev.*, **14**, 265 (1979).
19. C. G. Overberger and J. A. Moore, *Adv. Polym. Sci.*, **7**, 113 (1970).
20. B. C. Anderson and R. D. Lipscomb, *Macromolecules*, **17**, 1641 (1984).
21. G. M. Moelter, R. F. Tetrault, and N. Heffland, *Polym. News*, **9**, 134 (1983).
22. P. R. Srinivasan, V. Mahadevan, and M. Srinivasan, *J. Polym. Sci., Polym. Chem. Ed.*, **20**, 3095 (1982).
23. J. K. Stille, R. M. Harris, and S. M. Padaki, *Macromolecules*, **14**, 486 (1981).
24. V. V. Korshak, S. V. Vinogradova, and Y. S. Vygodskii, *Rev. Macromol. Chem.*, **12**, 45 (1974–1975).
25. F. L. Hedberg and F. E. Arnold, *J. Polym. Sci., Polym. Chem. Ed.*, **14**, 2607 (1976).
26. F. W. Harris and H. J. Spinelli (Eds.), *Reactive Oligomers*, ACS Symp. Ser. 282. American Chemical Society, Washington, D.C., 1985.
27. P. N. Hergenrother, *J. Macromol. Sci., Rev. Macromol. Chem.*, **C19**, 1 (1980).
28. H. D. Stenzenberger, M. Hertzog, W. Römer, R. Scheiblich, and N. J. Reeves, *Br. Polym. J.*, **15**, 1 (1983).
29. W. T. K. Stevenson and I. J. Goldfarb, *J. Appl. Polym. Sci.*, **42**, 665 (1991); W. T. K. Stevenson, I. J. Goldfarb, E. J. Soloski, and M. J. Houtz, *ibid.*, **42**, 679 (1991).
30. S. E. Delos, R. K. Shellenberg, J. E. Smedley, and D. E. Kranbuehl, *J. Appl. Polym. Sci.*, **27**, 4295 (1982).
31. P. M. Hergenrother, R. G. Bryant, B. J. Jensen, and S. J. Havens, *J. Polym. Sci. Part A: Polym. Chem.*, **32**, 3061 (1994).
32. F. W. Harris, A. Pamidimukkala, R. Gupta, S. Das, T. Wu, and G. Mock, *J. Macromol. Sci.-Chem.*, **A21**, 1117 (1984).
33. J. P. Droske and J. K. Stille, *Macromolecules*, **17**, 1 (1984).
34. B. Keszler, V. S. C. Chang, and J. P. Kennedy, *J. Macromol. Sci. Chem.*, **A21**, 307 (1984).
35. R. A. Kirchhoff and K. J. Bruza, *Prog. Polym. Sci.*, **18**, 85 (1993).
36. J. K. Sutter and J. F. Waters, *Polym. Preprints*, **32**(2), 203 (1991).
37. A. L. Rusanov, E. G. Bulycheva, and S. S. Bocharov, *Polym. Preprints*, **33**(1), 916 (1992).
38. I. K. Varma, G. M. Fohlen, and J. A. Parker, *J. Polym. Sci., Polym. Chem. Ed.*, **20**, 283 (1992).
39. A. Sarfaraz, L. D. Kandpal, and I. K. Varma, *J. Appl. Polym. Sci.*, **53**, 1075 (1994).
40. J. W. Lyons, *The Chemistry and Uses of Fire Retardants*, Wiley-Interscience, New York, 1970.
41. M. Lewin, S. M. Atlas, and E. M. Pearce, *Flame Retardant Polymeric Materials*, Plenum Press, New York, 1975.
42. E. M. Pearce, Y. P. Khanna, and D. Raucher, in *Thermal Characterization of Polymeric Materials* (E. A. Turi, Ed.), Academic Press, New York, 1981, Chap. 8.
43. A. Factor, *J. Chem. Educ.*, **51**, 453 (1974).
44. C. T. Vijayakumar and J. K. Fink, *J. Appl. Polym. Sci.*, **27**, 1629 (1982).
45. Y. P. Khanna and E. M. Pearce, *Flame Retard. Polym. Mat.*, **2**, 43 (1978).
46. R. Liepins, J. R. Surles, N. Morosoff, V. Stannett, J. J. Duffy, and F. H. Day, *J. Appl. Polym. Sci.*, **22**, 2403 (1978).
47. J. A. Albright and C. K. Kmiec, *J. Appl. Polym. Sci.*, **22**, 2451 (1978).
48. G. Montaudo, E. Scamporrino, and D. Vitalini, *J. Polym. Sci., Polym. Chem. Ed.*, **21**, 3361 (1983).
49. W. K. Seifert, M. P. Stevens, J. D. Gardner, and D. F. Percival, *J. Appl. Polym. Sci.*, **9**, 1681 (1965).
50. R. B. Seymour and R. H. Steiner, *Plastics for Corrosion Resistant Applications*, Reinhold, New York, 1954.
51. Yu. V. Moiseev and G. E. Zaikov, *Chemical Resistance of Polymers in Aggressive Media*, Consultants Bureau, Plenum Press, New York, 1987.
52. R. E. Uschold, *Polym. J.*, **17**, 253 (1985).
53. H. R. Allcock, *Chem. Rev.*, **72**, 315 (1972).
54. C. A. Sperata, in *Modern Plastics Encyclopedia*, 1984–85, Vol. 61 (J. Agranoff, Ed.), McGraw-Hill, New York, 1984, p. 24.
55. A. Reiser, *Photoreactive Polymers: The Science and Technology of Resists*, Wiley-Interscience, New York, 1989.
56. J. H. Lai (Ed.), *Polymers for Electronics Applications*, CRC Press, Boca Raton, Florida, 1989.
57. H. Steppen, G. Buhr, and H. Vollman, *Angew. Chem. Int. Ed. English*, **21**, 455 (1982).
58. W. Schnabel and H. Sotobayashi, *Prog. Polym. Sci.*, **9**, 297 (1983).

59. D. Bailey, D. Tirrell, C. Pinazzi, and O. Vogl, *Macromolecules*, **11**, 312 (1978).
60. M. Vert, J. Feijen, A. Albertsson, G. Scott, and E. Chiellini, *Biodegradable Polymers and Plastics*, CRC Press, Boca Raton, Florida, 1992.
61. G. S. Kumar, V. Kalpagam, and U. S. Nandi, *J. Macromol. Sci. Rev. Macromol. Chem. Phys.*, **C22**, 225 (1982–1983).
62. D. R. Paul and F. W. Harris (Eds.), *Controlled Release Polymeric Formulations*, ACS Symp. Ser. 33, American Chemical Society, Washington, D.C., 1976: (a) F. W. Harris, A. E. Aulabaugh, R. D. Case, M. K. Dykes, and W. A. Field, pp. 222ff; (b) D. R. Paul, pp. 1ff; (c) B. A. Bierl, E. D. DeVilbis, and J. R. Plimmer, pp. 265ff; (d) A. F. Kydonieus, A. R. Quisumbing, and S. Hyman, pp. 295ff.
63. A.-C. Albertsson and S. Karlsson, *Acta Polym.*, **46**, 114 (1995).
64. W. J. Passl, *Prog. Polym. Sci.*, **14**, 629 (1989).
65. S. Vainionpää, P. Rokkanen, and P. Tormäla, *Prog. Polym. Sci.*, **14**, 679 (1989).
66. X. Zhang, M. F. A. Goosen, U. P. Wyss, and D. Pichora, *J. Macromol. Sci.—Rev. Macromol. Chem. Phys.*, **C33**, 81 (1993).
67. B. Rånby and J. F. Rabek, *Photodegradation, Photooxidation and Photostabilization of Polymers*, Wiley-Interscience, New York, 1975, pp. 48–49.
68. S. J. Huang, C. A. Byrne, and J. A. Pavlisko, in *Modification of Polymers* (C. E. Carraher Jr. and M. Tsuda, Eds.), ACS Symp. Ser. 121, American Chemical Society, Washington, D.C. 1980, p. 299.
69. R. J. Dennenberg, R. J. Bothast, and T. P. Abbott, *J. Appl. Polym. Sci.*, **22**, 459 (1978).
70. K. S. Lee, V. T. Stannett, and R. D. Gilbert, *J. Polym. Sci., Polym. Chem. Ed.*, **20**, 997 (1982).
71. M. L. Beasley and R. L. Collins, *Science*, **169**, 769 (1970).
72. M. Rosoff (Ed.), *Controlled Release of Drugs*, VCH, New York, 1989.
73. H. R. Allock, *J. Polym. Sci., Polym. Symp.*, **70**, 71 (1983).
74. M. J. Snowden, M. J. Murray, and B. Z. Chowdry, *Chem. Ind.*, 551 (July 15, 1996).
75. R. C. Penwell, B. N. Ganguly, and T. W. Smith, *Macromol. Rev.*, **13**, 63 (1978).
76. J. Weigl, *Angew. Chem. Int. Ed. English*, **16**, 374 (1977).
77. T. A. Skotheim (Ed.), *Handbook of Conducting Polymers*, Vols. 1 and 2, Dekker, New York, 1986.
78. C. K. Chiang, M. A. Druy, S. C. Gau, A. J. Heeger, E. J. Louis, A. G. MacDiarmid, Y. W. Park, and H. Shirakawa, *J. Am. Chem. Soc.*, **100**, 1013 (1978).
79. J. Mort, *Science*, **208**, 819 (1980).
80. J. M. Margolis (Ed.), *Conductive Polymers and Plastics*, Chapman and Hall, New York, 1989.
81. G. P. Evans, *Adv. Electrochem. Sci. Engin.*, **1**, 1 (1990).
82. G. Wegner, *Makromol. Chem., Macromol. Symp.*, **1**, 151 (1986).
83. R. H. Baughman, J. L. Bredes, R. R. Chance, R. L. Elsenbaumer, and L. W. Shacklette, *Chem. Rev.*, **82**, 209 (1982).
84. M. Thakur, *Macromolecules*, **21**, 661 (1988).
85. J. C. W. Chien, *Polyacetylene*, Academic Press, New York, 1984.
86. N. Toshima and S. Hara, *Prog. Polym. Sci.*, **20**, 155 (1995).
87. J. M. Tour, *Chem. Rev.*, **96**, 537 (1996).
88. H. Charadame, in *Macromolecules* (H. Benoit and P. Rempp. Eds.), Pergamon Press, New York, 1982, p. 226.
89. P. N. Prasad and D. J. Williams, *Introduction to Nonlinear Optical Effects in Molecules and Polymers*, Wiley, New York, 1991.
90. G. A. Lindsay and K. D. Singer (Eds.), *Polymers for Second-Order Nonlinear Optics*, ACS Symp. Ser. 601, American Chemical Society, Washington, D.C., 1995.
91. H. S. Nalwa and S. Miyata (Eds.), *Nonlinear Optics of Organic Molecules and Polymers*, CRC Press, Boca Raton, Florida, 1996.
92. D. M. Burland, R. D. Miller, and C. A. Walsh, *Chem. Rev.*, **94**, 31 (1994).
93. K. J. Drost, A. K.-Y. Jen, and V. P. Rao, *Chemtech.*, **25**, 16 (Sept. 1995).
94. J. Štěpek and H. Daoust, *Additives for Plastics*, Springer-Verlag, New York, 1983.
95. J. T. Lutz (Ed.), *Thermoplastic Polymer Additives*, Dekker, New York, 1988.
96. E. W. Flick, *Plastics Additives*, Noyes, Park Ridge, New Jersey, 1986.
97. M. P. Stevens, *J. Chem. Educ.*, **70**, 444, 535, 713 (1993).

98. I. Mellan, *Industrial Plasticizers*, Macmillan, New York, 1963.

99. J. K. Sears and J. R. Darby, *The Technology of Plasticizers*, SPE Monograph Series, Wiley-Interscience, New York, 1982.

100. C. F. Hammer, in *Polymer Blends*, Vol. 2 (D. R. Paul and S. Newman, Eds.), Academic Press, New York, 1978, Chap. 17.

101. R. P. Sheldon, *Composite Polymeric Materials*, Applied Science, New York, 1982.

102. R. B. Seymour and R. D. Deanin (Eds.), *History of Polymeric Composites*, VNU Science Press, Utrecht, 1987.

103. P. Beardmore, J. J. Harwood, K. R. Kinsman, and R. E. Robertson, *Science*, **208**, 833 (1980).

Review Exercises

1. Explain the difference between
   (a) tensile strength and flexural strength.
   (b) tensile stress and tensile modulus.
   (c) extrusion and spinning.
   (d) injection molding and compression molding.
   (e) creep and stress relaxation.
   (f) thermal stability and thermooxidative stability.
   (g) positive resist and negative resist.

2. Explain, in terms of structure and morphology, what is happening in the various regions of the stress–strain curve depicted in Figure 4.8.

3. What effect would you expect each of the following to have on creep:
   (a) temperature
   (b) molecular weight
   (c) crosslinking

4. What effect would you expect oxidation to have on stress relaxation over prolonged periods of time?

5. Explain, in terms of structure and morphology, the rationale behind the three strategies described for improving processability of thermally stable polymers: (a) cardo polymers, (b) intramolecular cyclization, (c) reactive oligomers.

6. Poly(vinyl alcohol) is significantly less flammable than the "isomeric" polyoxyethylene. Suggest a reason.

7. The cyanate group (Table 4.4) trimerizes on heating to triazene:

$$3 \ ROCN \ \xrightarrow{\Delta} \ \begin{array}{c} OR \\ N \diagup \diagdown N \\ RO \diagdown N \diagup OR \end{array}$$

   Describe how one might make a semi-IPN (see Chap. 3) employing a cyanate-terminated monomer or oligomer. (Allied Chemical has developed such a product: see *Plast. Technol.*, March, 1985, p. 26; or H. A. Witcoff, *Chemtech*, **17**, 156, 1987.)

8. In elastomer synthesis, 1,4-hexadiene may be used to impart the unsaturation needed for crosslinking. (J. Lal and P. H. Sandstrom, *J. Polym. Sci., Polym. Lett. Ed.*, **13**, 83, 1975). What advantage might this compound offer over butadiene or isoprene?

9. Tygon plastic tubing is commonly used in the laboratory for transporting water or gases. If organic solvents are run through the tubing, it soon becomes stiff. What is the most likely reason?

10. An exercise at the end of Chapter 3 dealt with ironing polyester and cotton fabrics. Why is steam commonly used in ironing? In the same vein, why does paper (almost 100% cellulose) turn "limp" in very humid weather?

11. Suggest a reason why host–guest NLO formulations might be less desirable than main-chain or side-chain NLO polymers.

EVALUATION, CHARACTERIZATION,

AND ANALYSIS OF POLYMERS

5.1 Introduction

Characterizing polymers is much more complex than characterizing low-molecular-weight compounds. Apart from their basic chemical makeup, a variety of configurational and conformational variables relate to their usefulness. If the polymer contains pendant groups, are the groups stereoregular or randomly oriented? Is the polymer linear or branched? If more than one monomeric unit is identified, are the units distributed randomly or in blocks or grafts? What is the average molecular weight, and does the average represent a narrow or a wide (polydisperse) distribution? Is the polymer crosslinked? What is the nature of the end groups? Which additives are present, and at what concentration? If a polymer is being considered for a particular application, how well will it stand up to its use environment? Is it stable enough to withstand the mechanical and thermal stresses of extrusion or molding? Is the polymer tough or brittle? How tough? Will it retain its properties on being heated? To answer this multitude of questions, a broad range of evaluative and analytical procedures must be brought into play.[1–4]

This chapter presents a necessarily brief introduction to many of the analytical and evaluative tools available to the polymer chemist (excluding those covered in Chapter 2 for determining molecular weight). Major focus is placed on spectroscopic and thermal methods because these are used most frequently by the polymer chemist. We shall also explore techniques for surface analysis, a discipline that has assumed increasing importance in the polymer field in recent years. Mechanical and electrical testing are discussed briefly. Because test methods may vary substantially from one laboratory to another, organizations such as the American Society for Testing and Materials (ASTM) and the British Standards Institute (BSI) provide detailed standardized tests that are used throughout the polymer industry.

It is assumed here that readers, in the course of their studies, have received adequate exposure to the fundamentals of chromatography and spectroscopy; hence the discussion in this chapter is concerned only with the application of these techniques to polymer characterization. Fundamentals of the less commonly encountered instrumental methods are provided.

To keep things in perspective, it should be remembered that "wet" chemical analysis lay at the heart of early structural studies of polymers and is still a valuable complement to the instrumental techniques of today.[5,6] We begin, therefore, with a short discussion of chemical analysis.

5.2   Chemical Methods of Analysis

Among the earliest questions of concern to polymer chemists were those of backbone structure, for example, the spacing of substituent groups in vinyl polymers and the location of double bonds in polydienes. That the head-to-tail structure predominates in poly(vinyl alcohol) was demonstrated[7] convincingly by reaction with periodic acid or lead tetraacetate, which causes oxidative cleavage of 1,2-diols:

$$
\underset{\underset{\displaystyle |}{\text{C}}}{\overset{\overset{\displaystyle |}{\text{OH}}}{|}}\!\!-\!\!\underset{\underset{\displaystyle |}{\text{C}}}{\overset{\overset{\displaystyle |}{\text{OH}}}{|}}\!\!- \quad\xrightarrow{\text{HIO}_4 \text{ or Pb(OAc)}_4}\quad -\!\text{C}\!=\!\text{O} + \text{O}\!=\!\text{C}\!-
$$

Neither oxidizing agent reacts to any appreciable extent with poly(vinyl alcohol); however, both cause a reduction in molecular weight as measured by dilute solution viscometry. The results indicate that the polymer has a predominantly head-to-tail structure, but occasional head-to-head units are present, and these undergo chain scission with the subsequent decrease in molecular weight.

Ozonolysis is useful for determining structures of polydienes.[8] Ozonolysis of natural rubber, for example, followed by hydrolysis of the intermediate ozonide under reductive conditions, yields 4-ketopentanal, which establishes the structure as the head-to-tail 1,4-addition polymer of 2-methyl-1,3-butadiene (isoprene):

More recently[9] ozonolysis has been used to establish that polyacetylene, $-\!\!\left(\!\text{CH}=\text{CH}\right)\!\!-$, consists of more than 98% of the linear, conjugated polymer.

Other applications in which chemical methods are useful include the identification of polyesters and polyamides, or other polymers containing functional groups,[10] and the routine testing for polymer additives.[11]

5.3   Spectroscopic Methods of Analysis

Spectroscopic methods commonly employed for studying polymers[12–14] are summarized in Table 5.1. Also listed are non-Raman scattering methods (discussed in the following section), which are not spectroscopic methods per se. Infrared (IR) and nuclear magnetic resonance (NMR) spectroscopy are the most widely used. Spectroscopic and other methods applied to *surface* analysis are covered in Section 5.5. The following discussion follows the order in Table 5.1.

TABLE 5.1. Spectroscopic and Scattering Methods
Commonly Used for Studying Polymers

Vibrational
  Infrared (IR)
  Raman

Spin resonance
  Nuclear magnetic resonance (NMR)
    (proton and carbon-13)
  Electron spin resonance (ESR)

Electronic
  Ultraviolet (UV)–visible
  Fluorescence

Scattering
  X-ray
  Electron
  Neutron

5.3.1 Infrared

The two instrumental variations of IR spectroscopy are the older dispersive method, in which prisms or gratings are used to disperse the IR radiation, and the more recent Fourier transform infrared (FTIR)[15,16], which uses the principle of interferometry. Dispersive spectra of most polymers of commercial import have been recorded,[17,18] and FTIR databases for polymers are available; hence qualitative identification of unknowns can frequently be accomplished by comparison. This includes polymers having varying stereochemistries or monomer sequence distribution, because such differences usually give rise to different spectra. Where comparative spectra are not available, insight into polymer structure may be gained by the usual consideration of functional group absorption bands, or by comparing spectra with those of readily characterizable low-molecular-weight *model compounds* of similar structure. An example of the latter is shown in Figure 5.1.[19] Apart from expected differences in the aromatic C–H bending region (650–900 cm^{-1}) arising from *para*-disubstituted versus monosubstituted benzene rings, the spectra compare favorably.

FTIR brings greater versatility to polymer structural studies. Because spectra can be scanned, recorded, and transformed in a matter of seconds, the technique facilitates the study of such polymer reactions as degradation or crosslinking. The very small sample-size requirements allow coupling of the FTIR instrument with a microscope for analysis of highly localized sections of a polymer sample.[20] And the capability for digital subtraction allows one to generate otherwise "hidden" spectra. An example of spectral subtraction is shown in Figure 5.2 for isotactic polystyrene.[21] The spectral region represented, with the scale greatly expanded, encompasses the carbon–hydrogen bending region of the pendant benzene rings. (Note that these are absorption spectra, rather than the more conventional transmittance spectra shown in Figure 5.1.) Spectrum A is that of semicrystalline polymer obtained by annealing; B is that of the same polymer heated above the T_m and quenched in the amorphous state. Subtraction

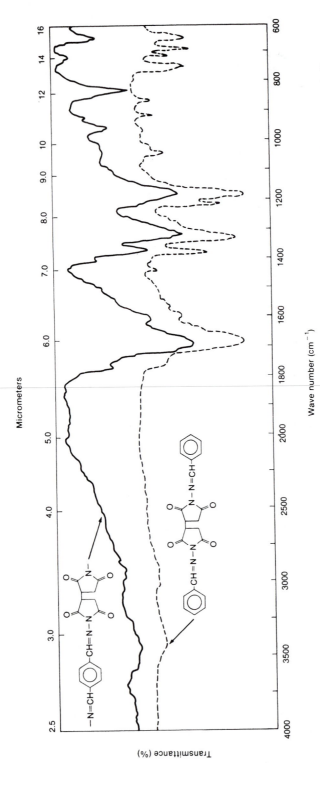

FIGURE 5.1. Infrared spectrum (KBr pellets) of polyimide (——) and model compound (– – –). [Reprinted from Troy, Sobanski, and Stevens,[19] by courtesy of Marcel Dekker, Inc.]

FIGURE 5.2. Fourier transform–infrared spectra of isotactic polystyrene in the 640 to 840 cm^{-1} region: (A) semicrystalline; (B) amorphous; and (C) the difference spectrum obtained by subtracting B from A. [From Painter and Koenig,[21] copyright 1977. Reprinted by permission of John Wiley & Sons, Inc.]

of B from A reveals more well-defined absorbances characteristic of the crystalline regions where the benzene rings are "frozen" into relatively specific conformations.

FTIR is particularly useful in studying polymer blends. Whereas the immiscible blend exhibits an IR spectrum that is the superposition of the two homopolymer spectra, the spectrum of a miscible blend is the superposition of *three* spectra—those of the homopolymers and an *interaction spectrum* arising from chemical or physical interaction between the homopolymers. For the commercial polyblend Noryl, for example, which consists of polystyrene and poly(phenylene oxide) (see Section 3.10), subtraction of the two homopolymer spectra from the spectrum of the blend reveals an interaction spectrum apparently arising from conformational changes brought about by dipole–dipole coupling of the benzene rings of the two homopolymers.[22] The interaction is greatest in blends of 30% polyether to 70% polystyrene, which is also the percentage concentration that offers the best mechanical properties.

5.3.2 Raman

Like IR spectroscopy, Raman[23] derives from vibrational transitions in molecules. When visible light impinges on molecules, the light is scattered; the frequency of the scattered light varies according to the vibrational modes of the scattering molecules. This is referred to as the *Raman effect*. Whereas IR absorption spectra are indicative of unsymmetric bond stretching and bending, the Raman effect responds to the *symmetric* vibrational modes. Thus IR and Raman spectroscopy are complementary. Because the intensity of Raman scattering is extremely low, however, laser light sources are needed for Raman instruments.

Raman is most responsive to symmetrical stretching in carbon–carbon bonds; hence the technique has been applied primarily to conformational studies of polymer chains by comparing spectra with those of long-chain "model" alkanes.[24,25] Other applications include stud-

ies of *cis–trans* isomerism in elastomers,[26] sulfur crosslinks in rubber,[27] and polymer deformations.[28] Because Raman scattering by water is negligible compared with water's intense
IR absorption, Raman is particularly useful in conformational studies of biopolymers in aqueous solution.[29] An added advantage is that Raman lends itself to fiber optics applications and
thus may be used for remote sensing.

5.3.3 Nuclear Magnetic Resonance

No single spectroscopic technique has provided so much valuable information to the polymer chemist as NMR. With the commercial development of NMR in the 1960s, initial applications dealt with proton (^{1}H) spectra of polymer solutions.[30] Thus structural units of polymers are identified from a combination of chemical shift data and spin–spin splitting. With
the very high resolution afforded by modern NMR instruments, polymer scientists may now
gain insight into polymer stereochemistry and monomer sequencing.[31-34] A case in point,
shown in Figure 5.3, is illustrated by the 500-MHz spectra of predominantly (95%) isotactic poly(methyl methacrylate)[35]:

The spectral range shows the methyl resonance as a singlet at about 1.30 ppm and two methylene proton doublets at about 1.62 and 2.33 ppm. The coupling constant for the dissimilar
methylene protons is about 15 Hz. (The methoxyl proton resonance occurs further downfield
at about 3.6 ppm.) The designations *m* and *r* refer to the *meso* and *racemic* dyads, as discussed in Chapter 3:

meso (*m*) dyad   *racemic* (*r*) dyad

Thus the methyl protons are resolved to the pentad (*mmmm*) level and the methylene protons to the hexad (*mmmmm*) level. The *e* and *t* notations refer to the *erythro* and *threo* place-

---

FIGURE 5.3. 500-MHz ^{1}H nuclear magnetic resonance spectrum of isotactic poly(methyl methacrylate) (10% weight/volume solution in chlorobenzene-d_5 at 100°C) at three values of gain. *e, t: erythro*
and *threo* placement of methylene protons with respect to ester groups; *m, r: meso* and *racemic* dyads.
[Courtesy of Dr. Frank Bovey, Bell Laboratories. From Schilling, Bovey, Bruch, and Kozlowski,[35]
copyright 1985. Reprinted by permission of the American Chemical Society.]

ment, respectively, of the methylene protons with respect to the ester groups, the *erythro* resonance being further downfield because of the deshielding influence of the ester group. At the lowest sensitivity (1X) the spectrum resembles that of almost purely isotactic (all *m*) polymer with the exception of the small *mmrm* pentad at about 1.16 ppm. At higher gain (5X and 100X), deviations from purely isotactic are evident. Syndiotactic placement, by contrast, exhibits predominantly *racemic* (*r*) sequences at about 1.05 ppm for methyl and, because both methylene protons lie in identical magnetic environments, one singlet methylene resonance at 1.83 ppm. The methoxyl protons absorb at 3.42 ppm. As might be expected,

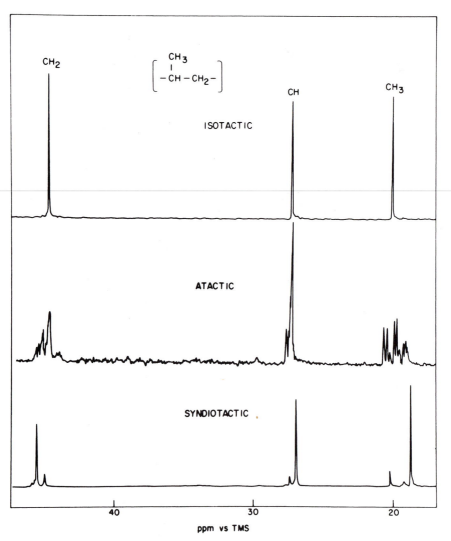

FIGURE 5.4. 24-MHz ^{13}C nuclear magnetic resonance spectrum of polypropylene (3.5% weight/volume in 1,2,4-trichlorobenzene at 135°C.) [Courtesy of Dr. Frank Bovey, Bell Laboratories.]

atactic poly(methyl methacrylate) displays a broad range of sequences in its NMR spectrum, although the largest peaks correspond to those of syndiotactic placement.[35]

Figure 5.4 shows high-resolution proton-decoupled ^{13}C spectra of polypropylene of varying tacticity. Fine structure in the spectra may again be correlated with pentad sequences,[36,37] as discussed previously.

Application of ^{13}C NMR to polymer solids gives rise to very broad resonances (of the order of a kilohertz) arising from dipolar interactions (mainly ^{13}C–H), long relaxation times in the solid state (of the order of minutes), and anisotropy of the chemical shift arising from the molecules being oriented in all possible directions, an effect that is averaged out in polymer solutions. Such difficulties may be resolved by a combination of instrumental refinements: proton decoupling to minimize dipolar broadening; application of radiofrequency pulses to make the ^{13}C and ^1H nuclei spin at the same rate, which effectively speeds up carbon's spin relaxation by transfer of magnetic polarization to the protons, a technique called *cross-polarization*; and rapid spinning of the sample at an angle of 54.7° with respect to the applied magnetic field (*magic angle spinning*), which has the effect of averaging out the spin vectors and minimizing the anisotropy effect. The combination of cross-polarization and magic angle spinning (CP-MAS) along with proton decoupling yields spectra of solids approaching the resolution of those obtained with solutions. As an example, Figure 5.5 shows the solid-state ^{13}C NMR spectrum of polycarbonate,

$$\left[\begin{array}{c} CH_3 \\ | \\ -\!\!\left\langle\!\bigcirc\!\right\rangle\!-\!\overset{}{\underset{|}{C}}\!-\!\left\langle\!\bigcirc\!\right\rangle\!-\!OCO- \\ | \\ CH_3 \end{array} \right]$$

with and without MAS, in the spectral region encompassing the carbonyl and ring carbon resonances.[38] Scientists thus have the capability for spectroscopic studies of crosslinked polymers and bulk polymers containing additives or fillers. CP-MAS also affords chemists the opportunity to study soluble polymers in the physical state in which they are actually used, which leads to a better understanding of solid-state morphology and chain conformation.

The newest application for NMR is the *imaging* of polymers,[39] a technique analogous to magnetic resonance imaging (MRI) for medical diagnosis. Imaging is accomplished by encoding the NMR signals so that the spatial distribution of the nuclei causing the signals

FIGURE 5.5. Solid-state proton-decoupled ^{13}C nuclear magnetic resonance spectrum of polycarbonate (A) with and (B) without cross-polarization magic angle spinning. Peak assignments: (a) carbonyl carbon; (b) substituted ring carbons; and (c) unsubstituted ring carbons. [From Schaefer, Stejskal, and Buchdahl,[38] copyright 1977. Reprinted by permission of the American Chemical Society.]

can be established. NMR imaging is particularly useful in quality control by detecting voids and other defects in fabricated polymer samples. Other applications include characterization of phase boundaries in immiscible polymer blends and monitoring penetration of fluids into polymer samples.

5.3.4 *Electron Spin Resonance*

Also called electron paramagnetic resonance (EPR), ESR works on the same principle as NMR except that microwave rather than radiowave frequencies are employed, and spin transitions of unpaired electrons rather than nuclei are recorded. Unlike NMR spectra, where absorption is recorded directly, ESR spectrometers plot the first derivative of the absorption curve.

Given the fact that the presence of unpaired (paramagnetic) electrons is a prerequisite to obtaining an ESR spectrum, the utility of ESR in polymer chemistry is primarily for studying such free radical processes as polymerization, degradation, and oxidation.[40-44] Figure 5.6, for example, represents the ESR spectrum of the free radical initially formed when poly(vinyl chloride) is irradiated with ultraviolet light[40]:

$$\left[\begin{matrix} CH_2CH \\ | \\ Cl \end{matrix} \right] \xrightarrow{\text{UV}} +CH_2\dot{C}H+ + Cl\cdot$$

The six-line signal arises from interaction of the unpaired electron with the five surrounding protons (4β and 1α), a phenomenon called *hyperfine splitting*. Information on the radical structure can be gained from a consideration of the line shape, intensity, position, and hyperfine splitting.

FIGURE 5.6. Electron spin resonance spectrum of UV-irradiated poly(vinyl chloride) at $-196°C$. [From Yang, Liutkas, and Haubenstock,[40] copyright 1980. Reprinted by permission of the American Chemical Society.]

Another application of ESR is to attach a stable free radical (called a *spin label*) to a polymer by covalent bonding (for example, a transition metal atom or a nitroxide radical), or to mix a compound containing an unpaired electron (called a *spin probe*) with a polymer. Spin labels and spin probes are used primarily for studying molecular motion and relaxation phenomena in polymers.[43,44]

5.3.5 Ultraviolet (UV)–Visible

Ultraviolet–visible absorption spectroscopy is used for the detection and quantitative measurement of chromophores that undergo $n \to \pi^*$ or $\pi \to \pi^*$ transitions. Because of its sensitivity, UV–visible spectroscopy has been particularly useful in identifying and analyzing "foreign" material in polymers—residual monomer, inhibitors, antioxidants, and so on.[45,46] Styrene monomer in polystyrene, for example, may be determined quantitatively using styrene's λ_{max} at 292 nm.[11]

Copolymer composition also lends itself to UV–visible analysis if one of the repeating units has the requisite chromophore, as in the case of styrene copolymers. Beer–Lambert law interpretations of copolymer composition must be treated with some caution, for molar absorptivities vary with tacticity, sequence length, interactions between chromophores and functional groups, and solvent.[47] UV–visible spectroscopy may also be used to identify end groups, and—if one chain end or both ends are known to contain the chromophore—to determine number average molecular weights[43] (see Chapter 2, exercise 14).

Optical rotatory dispersion (ORD), which measures changes in optical rotation as a function of wavelength, is another technique employing UV–visible radiation. Although useful for studies of synthetic polymers containing optically active centers, ORD has been applied primarily to biopolymers.[49]

5.3.6 Fluorescence

When a molecule absorbs light energy to form an excited state, the excited state can lose its acquired energy by a variety of mechanisms. If energy loss occurs with emission of radiation, the process is termed *luminescence*. *Fluorescence* is a luminescence pathway involving energy loss from the lowest excited singlet state of the absorbing molecules. Fluorescence spectrophotometers measure wavelength of the absorbed or emitted light as a function of fluorescence intensity.

Fluorescence spectroscopy has long been studied with biopolymers. Its application to synthetic polymers encompasses two major areas of interest: molecular motions and polymer–polymer compatibility.[50–52] Fluorescent groups are incorporated into polymers either by covalent bonding or by dispersion of low-molecular-weight fluorescent molecules through the polymer matrix. Fluorescence yield—the amount of light emitted relative to that absorbed—varies according to molecular motion in solution or in the melt, and can thus be related to rotational freedom and chain conformation. Alternatively, a fluorescence donor and a fluorescence acceptor can be incorporated into the polymer. In such cases, charge transfer complexes (*exiplexes*) form, which emit at characteristic wavelengths. The ability of donor and acceptor to interact is also a function of molecular motion. In compatibility studies, the fluorescence donor is incorporated into one polymer and the acceptor into the other. Where the two polymers are completely miscible, energy transfer between donor and acceptor is very efficient, but efficiency drops off as miscibility decreases. Such studies have shown that even with incompatible polymers, there is a limiting energy transfer efficiency, which suggests a gradual rather than a sharp boundary separation between the immiscible phases.

5.4 X-ray, Electron, and Neutron Scattering

Scattering methods already discussed are Raman (this chapter) and light scattering for molecular weight determinations (Chapter 2). A vast amount of information on polymer morphology and molecular dimensions can be obtained by scattering of other energy sources, particularly x-rays, electrons, and neutrons. Of the three, x-ray diffraction is most important.[53-55] Indeed, it is possible with x-ray techniques to determine the spatial arrangements of all the atoms in such complex biopolymers as proteins.

X-rays are generated in cathode-ray tubes when high-energy electrons impinge on metal targets. When x-rays are focused on a polymer sample (in pellet or cylinder form), two types of scattering occur. If a sample is crystalline, the x-rays are scattered *coherently*; that is, there is no change in wavelength or phase between the incident and scattered rays. Coherent scattering is commonly referred to as x-ray *diffraction*. If the sample has a nonhomogeneous (semicrystalline) morphology, the scattering is *incoherent*: there is a change in both wavelength and phase. Incoherent scattering (also called *Compton* scattering) is referred to as *diffuse diffraction* or simply as scattering.

Coherent scattering is determined by *wide-angle* measurements and incoherent scattering by *small-angle* measurements, as shown in Figure 5.7. (The acronyms WAXS and SAXS are used to denote wide-angle and small-angle x-ray scattering, respectively.) The wide-angle diffraction pattern consists of a series of concentric cones arising from scattering by the crystal planes; it is recorded as concentric rings on the x-ray plate superimposed on a diffuse background of incoherent scatter. A representative ring pattern is shown in Figure 5.7. As the degree of crystallinity increases, the rings become more sharply defined, and as the crystallites are oriented (as by drawing), the circles give way to a pattern of arcs and spots more nearly resembling diffraction patterns of low-molecular-weight crystalline compounds. Small-angle scatter patterns are very diffuse. For single crystal studies, the crystal is rotated at an angle perpendicular to the incident beam so that diffraction patterns at all possible angles are recorded.

Considerable insight into polymer morphology and structure may be gained from visual examination and mathematical interpretation of the pattern and intensity of scattered radiation, including degree of crystallinity, dimensions of crystalline domains, bond distances and angles, and type of conformation in the crystalline regions.

Electron diffraction is usually accomplished using a transmission electron microscope set in the diffraction rather than the imaging mode. The diffraction pattern is projected on the microscope screen. Polymer samples must be very thin—in the range of several hundred angstrom units. As with x-ray diffraction, information gathered by electron diffraction has to do with morphology—crystal dimensions, degree of crystallinity, and so on. The main advantages of electron diffraction are that (1) both diffraction and transmission measurements are possible with a given sample, (2) sample size requirements are very small, and (3) diffraction intensities and the number of reflections are much higher. A disadvantage is that electrons may cause free radical reactions (e.g., chain scission, crosslinking) in the sample.

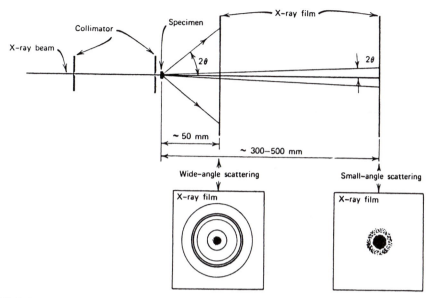

FIGURE 5.7. Wide-angle and small-angle x-ray diffraction techniques. [From E. A. Collins, J. Bares, and F. Billmeyer, *Experiments in Polymer Science*, Wiley-Interscience, copyright 1973. Reprinted by permission of John Wiley & Sons, Inc.]

When the electron microscope is operated in the imaging mode (*transmission electron microscopy*), it is possible to resolve such morphological features as polymer single crystals with a resolution in the range of 2 to 5 Å at magnifications of 200,000 to 500,000.

Neutron scattering[56-58] has the major disadvantages that a nuclear reactor is needed as a neutron source and sophisticated neutron counters are required. There are advantages to the method, however. It is the one scattering technique that provides much useful information from inelastic scattering, especially scattering by hydrogen atoms, where the scattering cross section is significantly higher than it is with other atoms. Thus, where elastic neutron scattering affords the same kind of information as x-ray and electron diffraction, inelastic neutron scattering may be used to study the movement of hydrogen atoms in the polymer. Neutrons are also more sensitive to observations of very-short-chain segments, for example, in studies of chain folding in crystalline lamellae (see Chapter 3, Figure 3.16). A further advantage to neutron scattering is that the sample can be in virtually any shape or form, solid or liquid (including solutions).

5.5 Characterization and Analysis of Polymer Surfaces

Surface analysis has assumed great importance in recent years, particularly in studies of catalysts, semiconductors, and other electronics devices. In the polymer field, coatings, polymer–catalyst and polymer–filler interfaces, surface oxidation, and surface morphology

TABLE 5.2. Methods of Polymer Surface Analyses

Incident Beam	Exit Beam	Technique	Abbreviation
Electrons	Electrons	Scanning electron microscopy	SEM
Photons (IR)	Photons (IR)	Attenuated total reflectance spectroscopy (internal reflectance spectroscopy)	ATR (IRS)
Photons (IR or UV)	Acoustical signal	Photoacoustic spectroscopy	PAS
Photons (x-ray)	Electrons	Electron spectroscopy for chemical analysis (or applications) (x-ray photoelectron spectroscopy)	ESCA (XPS)
Photons (x-ray) or electrons	Electrons	Augér electron spectroscopy	AES
Ions	Ions	Secondary ion mass spectrometry	SIMS
Ions	Ions	Ion-scattering spectroscopy	ISS
—	—	Atomic force microscopy (Scanning force microscopy)	AFM (SFM)

are among the areas of interest. Most methods of surface analysis involve irradiation of the surface with an energy source (photons, electrons, or ions) of sufficient energy to penetrate and cause some type of transition that results in the emission from the surface of an energy beam that can be analyzed. An exception is the relatively new technique of atomic force microscopy, in which the polymer sample is scanned by a fine probe in physical contact with the surface. A great variety of techniques have been developed; those used most commonly for polymers are listed in Table 5.2. Where a method is known by more than one name, the alternative is given in parentheses. Not included is optical microscopy, which is limited to resolving surface characteristics of the order of 2000 Å.

Certain problems are inherent to surface analysis. If ions or electrons are the incident beam, or if electrons or ions are emitted from the surface, an electrical charge develops in the sample which may cause spectral distortion. This is most severe in the case of insulators (the great majority of polymers), in which the charge cannot be dispersed by conduction. Surface charge must be compensated for by calibration methods or, if possible, by charge neutralization. Surface contamination is another problem. Occasionally the contamination itself is the focus of the analysis. When the uncontaminated surface is of interest, careful cleaning and the use of very low pressures (10^{-11}–10^{-7} torr) are often necessary. Low pressures are also necessary to avoid scattering of electrons or ions by air. Finally, the incident beam may cause surface damage. This is not a problem with IR or UV, nor to any great extent with x-rays; but electrons are particularly destructive to organic insulators, which limits the use of electron beam methods in the polymer field. Ion incident beams cause the loss of surface ions—a phenomenon called *sputtering*; measurement of the sputtered ions is, however, the basis of ion beam techniques.

5.5.1 Scanning Electron Microscopy (SEM)

SEM[63] differs from transmission electron microscopy (TEM) in that a very fine electron incident beam is scanned across the sample surface in synchronization with the beam in a cathode-ray tube. Scattered electrons are used to produce a signal that modulates the beam in the cathode-ray tube, producing an image with great depth of field and an almost three-dimensional appearance. SEM is of limited use in studying surface morphology, but it provides useful information on surface topology with a resolution of about 100 Å. Typical applications include studies of pigment dispersions in paints, blistering or cracking of coatings, phase boundaries in immiscible polyblends, cell structure of polymer foams, and adhesive failures. SEM is particularly valuable in evaluating how well polymeric surgical implants react to the host environment.

5.5.2 Attenuated Total Reflectance Spectroscopy (ATR)

Also called *internal reflection spectroscopy* (IRS), ATR[64] is probably the most widely used method of analyzing polymer surfaces because it can be adapted to most IR instruments and is lower in cost. In the ATR method, a polymer sample is attached to a crystal such as ThBr or AgCl that is transparent to IR and has a higher refractive index than that of the polymer. An IR beam is directed into the crystal (sometimes called the *sample plate*) at an angle greater than the critical angle, as shown in Figure 5.8(a). After penetrating into the polymer surface a small distance (a few micrometers), the beam is reflected. Some of the IR frequencies are absorbed by the surface, and the attenuated reflected beam is directed into the spectrometer, where it is plotted as a function of wavelength. Thus the ATR spectrum resembles a normal absorption spectrum. The amount of absorption can be increased by using an internal reflector plate, as shown in Figure 5.8(b), which allows the beam to be reflected a number of times from the surface. This is called *multiple internal reflection* (MIR). How many times the beam is reflected and how deep it penetrates into the polymer surface is a function of the relative refractive indexes, the incident angle, the geometry of the reflectance optics, and the wavelength.

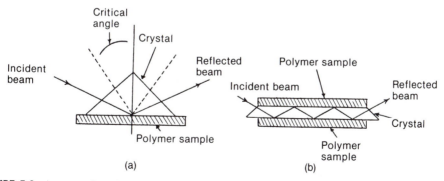

FIGURE 5.8. Attenuated total reflectance (ATR): (a) single reflection, and (b) multiple internal reflection (MIR).

ATR is particularly useful for studying surface coatings, surface oxidation, adhesive–substrate interfaces, and other aspects of polymer surfaces. Microscopy units are also available as adjuncts to FTIR instruments; they allow movement of the sample under study so that different parts of the specimen surface may be analyzed. Because surface penetration is extremely small, ATR spectra should not be interpreted as being necessarily characteristic of the bulk polymer.

5.5.3 Photoacoustic Spectroscopy (PAS)

Although PAS[65,66] encompasses both the IR and the UV–visible range and is applicable to bulk samples, its primary use with polymers is to characterize surfaces using FTIR methodology. A solid polymer sample enclosed in an air-tight chamber is exposed to intensity-modulated IR radiation. The heat energy released following absorption of radiation by the sample causes a modulated pressure change in the chamber atmosphere, which is detected by a sensitive microphone. Figure 5.9 illustrates the process. The spectrum is plotted as frequency versus arbitrary units of the amplified photoacoustic signal, and as such resembles a normal IR spectrum. Because the signal-to-noise ratio of PAS is very low, several thousand scans are needed to obtain a spectrum. The major advantage of PAS is that sample preparation is virtually unnecessary; thus it is particularly useful with very air- or moisture-sensitive polymers such as polyacetylene for which conventional absorption spectra are difficult to obtain. Another feature of PAS is that subsurface layers may be characterized by changing the modulation range, thereby providing a depth profile of the sample.

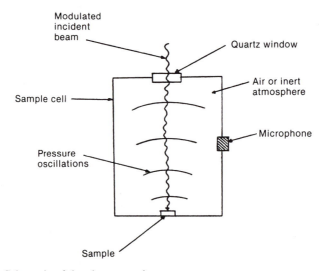

FIGURE 5.9. Schematic of the photoacoustic process.

5.5.4 Electron Spectroscopy for Chemical Analysis (or Applications) (ESCA) and Augér Electron Spectroscopy (AES)

Less commonly called *x-ray photoelectron spectroscopy* (XPS), ESCA[67,68] employs soft x-rays to detach electrons from film or powder surfaces. The difference in energy between the incident x-rays ($h\nu$) and the kinetic energy of the emitted electron (E_{kin}) is equal to the binding energy (E_B) of the electron; that is

$$E_B = h\nu - E_{kin}$$

Two types of electrons are emitted—inner core and the more loosely held valence electrons. When the photon ejects the inner-core electron (photoionization), the effective increase in nuclear charge causes a reorganization of the valence electrons, resulting in excitation of a valence electron from an occupied to an unoccupied level (a process called *shake-up*) and the loss of a valence electron (*shake-off*). The emitted electrons are analyzed by an electrostatic analyzer. A schematic of an ESCA spectrometer is shown in Figure 5.10. The spectrometer is evacuated to 10^{-10} to 10^{-7} torr to preclude electron scattering by air. An ESCA spectrum plots intensity against binding energy of the inner-core electron; thus it provides a surface "map" of the elements present. Peaks corresponding to photoionization are often accompanied by satellite peaks arising from the accompanying shake-up and shake-off, especially where pi bonding is present. Generally, ESCA analyzes the surface to a depth of about 50 Å. Argon ion sputtering may be used to expose greater depths for ESCA analysis.

An example of how ESCA may be used to characterize a film surface is shown in Figure 5.11.[69] Included are spectra of polystyrene and poly(ethylene oxide), and a block copolymer

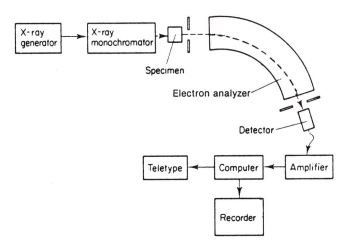

FIGURE 5.10. Schematic diagram of an electron spectroscopy for chemical analysis (ESCA) spectrometer. [From Rabek,[67] copyright 1980. Reprinted by permission of John Wiley & Sons, Ltd.]

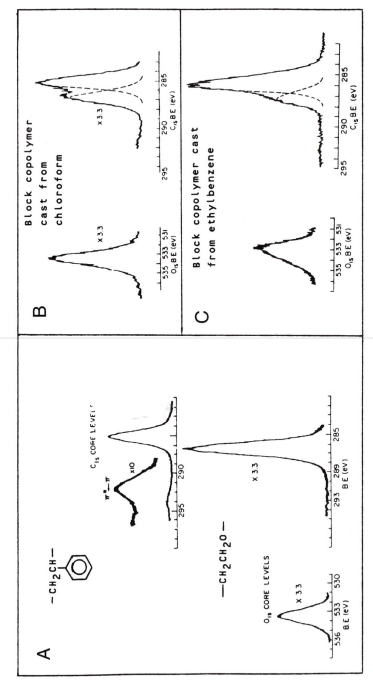

FIGURE 5.11. Electron spectroscopy for chemical analysis (ESCA) spectra of films: (A) Polystyrene and poly(ethylene oxide); and (B), (C) Polystyrene–*block*–poly(ethylene oxide) cast from chloroform and ethylbenzene, respectively. [From Thomas and O'Malley,[69] copyright 1979. Reprinted by permission of the American Chemical Society.]

of the two containing 21 mol % polystyrene cast from two different solvents. Polystyrene exhibits a carbon peak at 285.0 electron volts (eV) and a low-energy $\pi \rightarrow \pi^*$ shake-up peak at 291.6 eV. Poly(ethylene oxide) has peaks at 286.5 eV for carbon (shifted 1.5 eV from that in polystyrene because of the influence of neighboring oxygen atoms) and 533.3 eV for oxygen. Carbon and oxygen peaks both arise from photoionization of core $1s$ electrons. Both film surfaces are seen to be richer in the less abundant polystyrene, especially when ethylbenzene, a better solvent for polystyrene, is used. Interestingly, optical and electron microscopy show variations in surface morphology with the surface concentration of polystyrene.

AES complements ESCA and uses essentially the same instrumentation. When a $1s$ electron is ejected, a $2s$ electron can drop to the $1s$ level with simultaneous loss of a $2p$ electron (also called an Augér electron after the French physicist Pierre Victor Augér, who laid the theoretical groundwork). An AES spectrum plots intensity against binding energy of the Augér electrons. Binding energy, E_B, given by

$$E_B = E_{\text{kin}}(1s) + E_B(2s) + E_B(2p)$$

Unlike ESCA, AES spectra are plotted as derivative peaks to separate the signal from scattered electron background. AES has not been used extensively in polymer studies, but is useful for characterizing polymer–metal interfaces.

5.5.5 Secondary-Ion Mass Spectrometry (SIMS) and Ion-Scattering Spectroscopy (ISS)

SIMS and ISS spectra[70,71] are obtained by irradiating a surface with inert gas ions (called *primary ions*). Two phenomena occur, as shown in Figure 5.12. The primary ions are reflected in an elastic scattering process, with the energy of the scattered ions depending on the atoms causing the scattering. Detection and *kinetic energy* measurement of the scattered ions are the basis of ISS. Additionally, primary ions penetrate the surface and eject atomic and molecular fragments in a process that might be likened to sandblasting. This is the phenomenon of surface sputtering. Detection and mass measurement of the sputtered ions are the basis of SIMS.

The sputtered fragments are primarily neutral; however, enough positive and negative ions are formed to be focused and measured in a medium-resolution mass spectrometer. The

FIGURE 5.12. Representation of surface sputtering and primary ion elastic scattering as the basis of secondary-ion mass spectrometry (SIMS) and ion-scattering spectroscopy (ISS), respectively.

FIGURE 5.13. Secondary-ion mass spectrometry–positive ion spectrum of polytetrafluoroethylene showing surface contamination. (^{20}Ne used as the sputtering gas.) [From Hercules and Hercules,[71] reprinted with permission.]

SIMS spectrum, a plot of mass/charge versus intensity, thus resembles a typical mass spectrum. Because SIMS is very dependent on matrix effects, quantitative measurements are not practical. SIMS is very sensitive, however. Figure 5.13, for example, shows the SIMS–positive ion spectrum of polytetrafluoroethylene,[71] which clearly demonstrates surface contamination by sodium, aluminum, potassium, and saturated hydrocarbons, in addition to the fragmentation pattern of the polymer.

ISS complements SIMS. The great value of ISS lies in the fact that it measures the surface layer composition to the exclusion of subsurface structure. Primary ions that penetrate the monolayer are scattered inelastically and are represented as background in the ISS spectrum. Another advantage is that ISS detects any atom heavier than the primary ion. Thus with helium as the primary ion, elements from lithium up through the periodic table are detectable.

5.5.6   Atomic Force Microscopy (AFM)

AFM [also known as *scanning force microscopy* (SFM)] is a type of scanning probe microscopy similar to scanning tunneling microscopy (STM), allowing three-dimensional topographical imaging of a surface.[72] Common to both AFM and STM is a very fine-tipped probe for scanning the surface. The STM probe is positioned a few angstroms above the surface and measures a tunneling current between the tip of the probe and the surface; thus STM is usually limited to conducting surfaces. The AFM probe, on the other hand, is in physical contact with the surface as it moves over the sample. Because it may be used on any surface, AFM is much more suited to polymer surface analysis.

The essential features of AFM are shown in Figure 5.14. The tip of the probe, which is commonly made of silicon nitride, is attached to a cantilever bearing a reflective surface upon which a laser beam is directed. The sample is mounted on a piezoelectric support, which moves in response to surface variations sensed by the probe. As the tip is scanned (or "rastered") over the surface, topological variations cause deflections in the cantilever that are monitored by recording the path of the reflected laser beam. A computer interprets the deflections as a three-dimensional profile of the polymer surface with resolution in the angstrom range, which is several orders of magnitude better than that obtained by SEM.

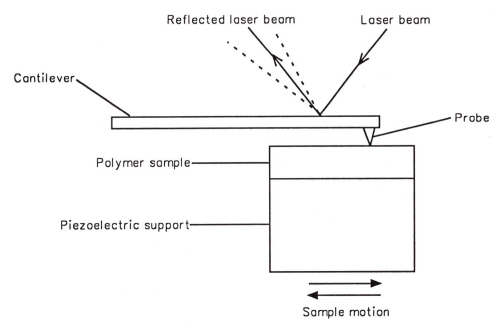

FIGURE 5.14. Schematic diagram of an atomic force microscope.

5.6 Thermal Analysis

A number of thermal properties of polymers are discussed in Chapters 3 and 4, including crystalline melting point, glass transition temperature, flammability, and thermal stability. Flammability and thermal stability are concerned with chemical transformations, whereas melting and glass transition temperatures represent morphological changes.

Glass transition temperatures are most commonly measured by *differential scanning calorimetry* (DSC), *differential thermal analysis* (DTA), or *thermomechanical analysis* (TMA). Thermal degradation is determined by *thermogravimetric analysis* (TGA) and *pyrolysis-gas chromatography* (PGC), the latter frequently in combination with mass spectrometry. For measuring flammability, a variety of standardized procedures are used, from small-scale laboratory to large-scale room tests. These various techniques[73–76] are described below.

5.6.1 Differential Scanning Calorimetry (DSC) and Differential Thermal Analysis (DTA)

DTA is a very old technique, dating to Le Chatelier[77] in the nineteenth century, but it was not applied to polymers until the 1960s.[78] DSC is of more recent vintage, and has become the method of choice for quantitative studies of thermal transitions in polymers.[74] In both DSC and DTA, a polymer sample and an inert reference are heated, usually in a nitrogen atmosphere, and thermal transitions in the sample are detected and measured. The sample holder most commonly used is a very small aluminum cup (gold or graphite is used for analyses above 800°C), and the reference is either an empty cup or a cup containing an inert material, such as anhydrous alumina. Sample sizes vary from about 0.5 to about 10 mg. Although the two methods provide the same type of information, there are significant differences in instrumentation. With DTA, both sample and reference are heated by the same heat source, and the difference in temperature (ΔT) between the two is recorded. When a transition occurs in the sample—for example, the glass transition or a crosslinking reaction—the temperature of the sample will lag behind that of the reference if the transition is endothermic, and will surge ahead if the transition is exothermic. With DSC, sample and reference are provided with individual heaters, and energy is supplied to keep the sample and reference temperatures constant. In this case, the electrical power difference between sample and reference ($d\Delta Q/dt$) is recorded. Schematic representations of DTA and DSC cells are given in Figure 5.15.

Data are plotted as ΔT (for DTA) or $d\Delta Q/dt$ (for DSC) on the ordinate against temperature on the abscissa. Such plots are called *thermograms*. Although ΔT and $d\Delta Q/dt$ are not

FIGURE 5.15. Schematic representations of differential thermal analysis (DTA) and differential scanning calorimetry (DSC) measuring cells. [From Rabek,[76] copyright 1980. Reprinted by permission of John Wiley & Sons, Ltd.]

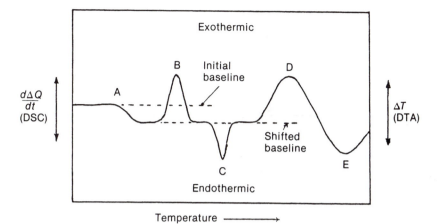

FIGURE 5.16. Idealized differential scanning calorimetry (DSC) or differential thermal analysis (DTA) thermogram: (A) temperature of glass transition, T_g; (B) crystallization; (C) crystalline melting point, T_m; (D) crosslinking; and (E) vaporization. $d\Delta Q/dt$ = electrical power difference between sample and reference; ΔT = difference in temperature between sample and reference.

linearly proportional, they are both related to heat capacity. Thus DSC and DTA thermograms have the same form. The major advantage of DSC is that peak areas of thermograms are related directly to enthalpy changes in the sample, hence may be used for measurements of heat capacities, heats of fusion, enthalpies of reactions, and the like.

An idealized DSC or DTA thermogram for a hypothetical crystallizable polymer is depicted in Figure 5.16, which illustrates the types of transitions of interest to polymer chemists. An actual thermogram[79] of poly(vinyl chloride) is shown for comparison in Figure 5.17. The glass transition causes an endothermic shift in the initial baseline because of the sample's increased heat capacity. In reporting transition temperatures, it is important to indicate whether one is referring to the onset of the transition or to the inflection point or peak maximum, as shown in Figure 5.18. Both conventions are used.

FIGURE 5.17. Differential thermal analysis (DTA) thermogram of poly(vinyl chloride): (A) glass transition, T_g; (B) melting point; (C) oxidative attack; (D) dehydrochlorination; and (E) probable depolymerization. [From Matlack and Metzger,[79] copyright 1966. Reprinted by permission of John Wiley & Sons, Inc.]

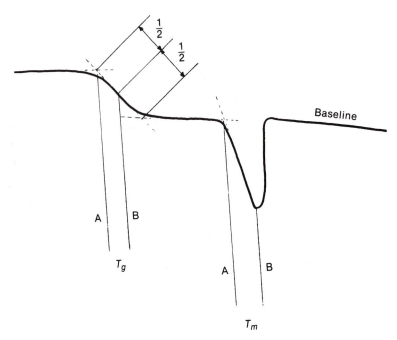

FIGURE 5.18. Methods of reporting transition temperatures: (A) at the onset, and (B) at the inflection point or maximum, T_g = glass transition temperature; T_m = crystalline melting point.

5.6.2 Thermomechanical Analysis (TMA)

TMA[80] employs a sensitive probe in contact with the surface of a polymer sample. As the sample is heated the probe senses thermal transitions such as T_g or T_m, by detecting either a change in volume (dilatometry) or a change in modulus. In the former case a flat probe tip is used, and in the latter a tapered tip, to penetrate the surface. Movement of the probe is detected by means of a variable transformer that records voltage output proportional to the displacement. TMA is generally more sensitive than DSC or DTA for detecting thermal transitions.

Another mechanical method less commonly used is *torsional braid analysis* (TBA). In TBA a glass braid or thread is impregnated with the polymer sample. The sample is then heated as the braid or thread is subjected to torsional oscillations. Variations in oscillation behavior are related to thermomechanical transitions.

5.6.3 Thermogravimetric Analysis (TGA)

TGA is used primarily for determining thermal stability of polymers. Like DTA, TGA is an old technique[81] but has been applied to polymers only since the 1960s.[82] The most widely used TGA method is based on continuous measurement of weight on a sensitive balance (called a *thermobalance*) as sample temperature is increased in air or in an inert atmosphere. This is referred to as *nonisothermal TGA*. Data are recorded as a thermogram of weight ver-

FIGURE 5.19. Thermogravimetric analysis (TGA) thermograms of poly-thiazoles. [From Marvel,[83] reprinted by courtesy of Marcel Dekker, Inc.]

sus temperature. Weight loss may arise from evaporation of residual moisture or solvent, but at higher temperatures it results from polymer decomposition. Besides providing information on thermal stability, TGA may be used to characterize polymers through loss of a known entity, such as HCl from poly(vinyl chloride). Thus weight loss can be correlated with percent vinyl chloride in a copolymer. TGA is also useful for determining volatilities of plasticizers and other additives. Thermal stability studies are the major application of TGA, however. A typical thermogram illustrating the difference in thermal stability between a wholly aromatic polymer and a partially aliphatic polymer of analogous structure is shown in Figure 5.19.[83]

A variation of the method is to record weight loss with time at a constant temperature. Called *isothermal TGA*, this is less commonly used than nonisothermal TGA. Modern TGA instruments allow thermograms to be recorded on microgram quantities of material. Some instruments are designed to record and process DSC and TGA data simultaneously, and may also be adapted for gas chromatographic and/or mass spectrometric analysis of effluent degradation products.

5.6.4 Pyrolysis-gas Chromatography (PGC)

Scientists have been studying polymer pyrolysis almost as long as they have been studying polymers, but it was not until 1954,[84] shortly after the technique of gas chromatography was first reported, that products of polymer pyrolysis were separated by this method. In 1959 the first report of the direct coupling of a pyrolysis device to the column inlet of a gas chromatograph appeared.[85] Since then the combination of pyrolysis and gas chromatography has proved to be a valuable tool for polymer characterization.[86,87]

Pyrolysis is accomplished in one of three ways: *furnace chamber pyrolysis, flash pyrolysis,* or *laser pyrolysis.* In each case the chromatogram of pyrolysis products is called a *pyrogram.* In the furnace chamber method, a polymer sample is either moved into a preheated pyrolysis chamber in a sample "boat" or heated rapidly from ambient to pyrolysis temperature. Gaseous pyrolysis products are swept directly into the gas chromatograph by carrier gas. The flash pyrolysis method employs a high-resistance coil or filament that is coated with polymer sample and attached directly to the sample inlet of the gas chromatograph. The coil is then heated by the application of electric current. Some flash pyrolyzers use ferromagnetic filaments that are heated by high-frequency induction to the Curie point, which brings the sample rapidly to a constant and reproducible temperature. In the laser pyrolysis method, pyrolysis is accomplished with a focused laser beam. Pyrograms thus obtained are usually much simpler than those obtained by the other two methods. Because not all polymers absorb laser light, carbon black is often mixed with the sample to facilitate pyrolysis.

PGC of polymers is useful for both qualitative and quantitative analysis. With the former, polymers are often identifiable from the pattern of peaks (the "fingerprint") in the pyrogram. Even homopolymers of different tacticity, or random, block, and graft copolymers prepared from the same monomers invariably exhibit characteristic pyrograms. An example of quantitative analysis is the determination of vinyl acetate in poly[ethylene-*co*-(vinyl acetate)] by measuring the acetic acid evolved upon pyrolysis[88]:

$$\left[\begin{array}{c} CH_2{-}CH{-}\!\!\!\! \\ \;\;\;|\quad\;\; | \\ \;H\quad O \\ \quad\;\; | \\ O{=}C{-}CH_3 \end{array}\right] \longrightarrow -[CH{=}CH]- \\ \begin{array}{c} O \\ \| \\ +\,HO{-}C{-}CH_3 \end{array}$$

PGC is also useful in elucidating thermal degradation pathways. The pyrogram of polyethylene[89] (Figure 5.20), for example, consists of regularly spaced triplets corresponding to *n*-alkane, 1-alkene, and α,ω-alkadiene (each having the same number of carbons), which indicates that degradation occurs by random scission of the polymer backbone (discussed in Chapter 9). Most pyrograms are much more complex than that in Figure 5.20 because of further degradation of the initially formed pyrolysis products. Some polymers, such as polystyrene and poly(methyl methacrylate), exhibit large monomer peaks in their pyrograms arising from depolymerization reactions. Mass spectrometry (or the combination of GC and mass spectrometry) is also used to identify pyrolysis products.

5.6.5 Flammability Testing

Flammability is a difficult property to measure in a meaningful way because small-scale laboratory tests do not, in general, reflect burning behavior in true fire conditions. A polymer such as polyurethane, for example, which is widely used in carpeting and furniture upholstery, does not burn well if a match is applied to a small test sample; but in a burning room where temperatures are much higher and combustible gases have accumulated, the same sample may burn furiously. Room-size tests have demonstrated that when the accumulated com-

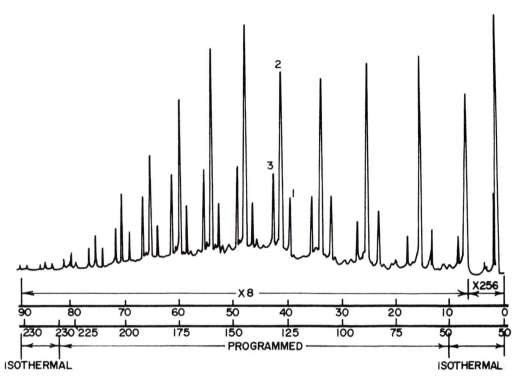

FIGURE 5.20. Pyrogram of high-density polyethylene. Peaks correspond to (1) *n*-alkane, (2) 1-alkene, and (3) α,ω-alkadiene. Conditions: 25-m capillary column coated with poly(phenyl ether); pyrolysis 10 s at 1,000°C. [From Kolb et al.,[89] courtesy of Springer-Verlag.]

bustible gases reach a certain temperature, they spontaneously ignite—a phenomenon called *flashover*—and the fire then spreads almost instantaneously to all other combustible objects in the room. It is not surprising, then, that a large number of tests have been developed.[90–92]

The most versatile small-scale test in widespread laboratory use is the *limiting oxygen index* (LOI) test.[93] The LOI is the minimum percentage of oxygen in an oxygen–nitrogen

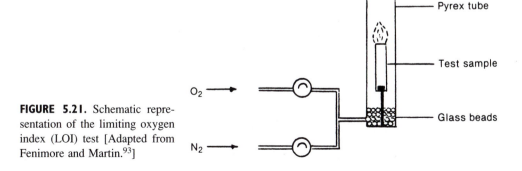

FIGURE 5.21. Schematic representation of the limiting oxygen index (LOI) test [Adapted from Fenimore and Martin.[93]]

TABLE 5.3. Limiting Oxygen Indexes (LOI) of
Some Common Polymers[a]

Polymer	LOI
Polyoxymethylene	15
Poly(ethylene oxide)	15
Poly(methyl methacrylate)	17
Polypropylene	17
Polyethylene	17
Polystyrene	18
Poly(1,3-butadiene)	18
Poly(vinyl alcohol)	22
Polycarbonate	27
Poly(phenylene oxide)	28
Polysiloxane	30
Poly(vinyl chloride)	45
Poly(vinylidene chloride)	60
Polytetrafluoroethylene	95

[a]Values taken from Fenimore and Martin.[93]

mixture that will initiate and support for three minutes the candlelike burning of a polymer sample; that is,

$$LOI = \frac{vol.\ O_2}{vol.\ O_2 + vol.\ N_2} \times 100$$

The procedure involves clamping a test sample of specified dimensions in a Pyrex cylinder, as shown in Figure 5.21. A metered mixture of O_2 and N_2 is fed in at the bottom, and the sample is ignited with a gas flame. Representative LOI values for some common polymers are given in Table 5.3. Noteworthy is the contrast between the "isomeric" poly(ethylene oxide) and poly(vinyl alcohol) (see exercise 6, Chapter 4). Dehydration of the latter results in cooling of the pyrolysis zone with resultant decrease in flammability. The flame-retardant properties of halogen are also apparent.

Another widely used test procedure is to clamp a sample horizontally in air and to apply a flame at one end for 30 seconds. If, after removal of the flame, the sample fails to burn to a mark 4 inches from the point of ignition, it is said to be "self-extinguishing." In the LOI test, self-extinguishing corresponds to an LOI value of about 27.[93]

5.7 Measurement of Mechanical Properties

To measure tensile strength, modulus, and elongation, a test specimen is clamped at each end. One end is fixed, and a gradually increasing load is applied to the other until the sample breaks. Plastics test specimens are usually of the dimensions shown in Figure 5.22(a).

FIGURE 5.22. Typical plastics test specimens for measuring (a) tensile and (b) impact strength.

Fiber and elastomer specimens differ in form, but are tested in essentially the same manner.[94] A typical testing instrument that measures stress and strain automatically with full scale loads from less than a gram to as high as 20,000 pounds is shown in Figure 5.23. Strength and elongation at the yield point, as well as at break, are usually recorded. The same instrument is used to measure compressive and flexural strengths. Repeated flexing of a sample through a given distance often causes the sample to fail at lower stress than it would for a single flexure, a phenomenon known as *fatigue*. Alternating cycles of tensile and compressive stress are also used to evaluate fatigue. In general, fatigue resistance decreases as polymer rigidity or crosslinking increases.

A number of methods are used to measure impact strength; the choice depends on the type of sample. The most widely used test (the Izod test) involves clamping a test specimen at one end and striking the specimen with a weighted pendulum. The distance the pendulum travels after the sample breaks is taken as a measure of impact strength. Notched samples, as shown in Figure 5.22(b), are usually used to give more reproducible results. In such cases the pendulum strikes on the same edge as, and above, the notch, and the strength is calculated on the basis of 1 in. (or 1 m) of thickness. In all mechanical testing, a minimum of ten samples is used because a fairly wide spread of measured values is common with the same polymer batch. Representative mechanical properties are given in Chapter 4 (Table 4.1 and 4.2).

FIGURE 5.23. Instrument for performing tensile, compression, shear, and flexural testing. Shown with tensile testing grips and sample. [Courtesy of Instron Corp.]

Other commonly measured mechanical properties are the following:

Hardness.   This is a measure of a polymer's resistance to surface indentation. The indenting device may be a spring-loaded needle-type indenter (Barcol test) or a weighted steel ball (Rockwell test).

Abrasion resistance. This is normally determined by measuring weight loss of a sample when the sample is subjected to some type of mechanical abrader or finely divided abrasive.

Tear resistance. A weighted pendulum-mounted blade is used to tear apart a test sample, and the distance the blade travels after tearing is correlated with tear resistance.

5.8 Evaluation of Chemical Resistance

Chemical resistance is usually determined by immersing test samples in the reagents of interest, often at elevated temperatures, then measuring some mechanical property such as tensile or flexural strength after a certain length of time.[95] Exposure time may vary from days to months, depending on the polymer and test reagent. Chemical resistance is expressed in terms of the percent retention of properties. Visual effects of failure such as swelling, surface erosion, or *crazing* (the development of stress fractures) are also taken into account. Change in viscosity is a measurement also employed with thermoplastic polymers, particularly where immersion in a corrosive medium is not a consideration, as in photodegradation studies. Resistance to moisture is often measured simply as gain in weight after long-term exposure to water.

Weatherability testing is widely used in the plastics and coatings industries.[96] Test panels set facing south at 45° are commonly used for testing resistance to sunlight. Specially designed weathering chambers equipped with controlled humidity, water sprays, and UV lamps are also employed to duplicate the effects of weather cycles.

5.9 Evaluation of Electrical Properties

Because of the widespread use of polymers in the electronics industry, as well as the traditional use of polymers as electrical insulators, measurement of electrical properties[97–99] is of special significance, more so with the recent discovery of highly conducting polymers. Most important from the standpoint of evaluating polymers as resistors or conductors is the *volume resistivity*, which is determined by attaching electrodes to the polymer specimen, usually a disk as shown in Figure 5.24(a), and applying a potential of about 500 V. The volume resistivity in ohms, ρ_v, which is the inverse of conductivity, is given by

$$\rho_v = \frac{E}{I_v}\frac{A}{t} = \frac{R_v A}{t}$$

where E is the potential in volts, I_v the measured dc current in amps, A the area of the small electrode in square centimeters, t the thickness in centimeters, and R_v the resistance of the sample in ohms. Samples having resistivities of about 10^8 ohm cm or higher are considered good insulators, and those with resistivities below about 10^{-3} ohm cm are good conductors. In between lie semiconductors.

To measure *surface resistivity*, ρ_s, the circuitry shown in Figure 5.24(b) is used. In this case,

$$\rho_s = \frac{E}{I_s}\frac{\pi D_m}{g} = \frac{R_s \pi D_m}{g}$$

Edge view   Top view

FIGURE 5.24. Circuitry for measuring (a) volume resistivity and (b) surface resistivity. Cross-hatched areas are electrodes. [Adapted from Agranoff.[98]]

where I_s is the current (amp), D_m the mean diameter of the gap (cm), g the width of the gap (cm), and R_s the resistance (ohm). *Insulation resistance* refers to the resistance of the sample to direct current both through the sample *and* across the surface.

Other electrical properties of interest are *dielectric strength* (a measure of the maximum voltage a sample can withstand before the sample undergoes physical breakdown); *dielectric constant* (a measure of a sample's contribution to the circuit capacitance); and *arc resistance* (the length of time before a high-voltage discharge arcs between two electrodes attached to the surface).

References

1. E. Schroder, G. Müller, and K.-F. Arndt, *Polymer Characterization*, Hanser, New York, 1989
2. D. Campbell and J. R. White, *Polymer Characterization: Physical Techniques*, Chapman and Hall, New York, 1989.
3. T. R. Crompton, *Analysis of Polymers: An Introduction*, Pergamon Press, New York, 1989.
4. J. Haslam, H. A. Willis, and D. C. M. Squirrell, *Identification and Analysis of Plastics*, 2nd ed., Butterworth, London, 1972, pp. 605ff.
5. D. Braun (Ed.), *Simple Methods for the Identification of Plastics*, 2nd ed., Hanser, Cincinnati, 1987.
6. G. M. Kline (Ed.), *Analytical Chemistry of Polymers*, Wiley-Interscience, New York, Part 1, 1959; Parts 2 and 3, 1962.
7. P. J. Flory and F. S. Leutner, *J. Polym. Sci.*, **3**, 880 (1948); **5**, 267 (1950).
8. C. Marvel and R. E. Light, Jr., *J. Am. Chem. Soc.*, **72**, 3887 (1950).
9. G. C. G. Li, D. A. Kidwell, D. W. Brown, E. E. Hall, and G. E. Wnek, *J. Polym. Sci., Polym. Chem. Ed.*, **21**, 301 (1983).
10. D. Nissen, V. Rossbach, and H. Zahn, *Angew. Chem. Int. Ed. English*, **12**, 602 (1973).
11. T. R. Crompton, *Chemical Analysis of Additives in Plastics*, 2nd ed., Pergamon Press, Oxford, 1971.
12. J. L. Koenig, *Spectroscopy of Polymers*, American Chemical Society, Washington, D.C., 1992.
13. A. H. Fawcett (Ed.), *Polymer Spectroscopy*, Wiley, New York, 1996.
14. W. Klöpffer, *Introduction to Polymer Spectroscopy*, Springer-Verlag, New York, 1984.
15. J. L. Koenig, *Adv. Polym. Sci.*, **54**, 87 (1984).
16. H. Ishida (Ed.), *Fourier Transform Infrared Characterization of Polymers*, Plenum, New York, 1987.
17. D. O. Hummel, *Infrared Spectra of Polymers*, Wiley-Interscience, New York, 1966.
18. R. Zbinden, *Infrared Spectroscopy of High Polymers*, Academic Press, New York, 1967.

19. R. C. Troy, D. B. Sobanski, and M. P. Stevens, *J. Macromol. Sci.-Chem.*, **A23**, 189 (1986).
20. R. G. Messerschmidt and M. A. Harthcock (Eds.), *Infrared Microscopy*, Marcel Dekker, New York, 1988.
21. P. C. Painter and J. L. Koenig, *J. Polym. Sci., Polym. Phys. Ed.*, **15**, 1885 (1977).
22. J. L. Koenig and M. J. M. Tovar Rodriguez, *Appl. Spectrosc.*, **35**, 543 (1981).
23. J. G. Grasselli, M. K. Snavely, and B. J. Bulkin, *Chemical Applications of Raman Spectroscopy*, Wiley, New York, 1981.
24. J. L. Koenig, *Appl. Spectrosc. Rev.*, **4**, 222 (1971).
25. M. W. Urban, in *Structure–Property Relations in Polymers* (M. W. Urban and C. D. Craver, Eds.), Adv. Chem. Ser. 236, American Chemical Society, Washington, D.C., 1993, Chap. 6.
26. S. W. Cornell and J. L. Koenig, *Rubber Chem. Technol.*, **43**, 313, 322 (1970).
27. M. M. Coleman, J. R. Shelton, and J. L. Koenig, *Rubber Chem. Technol.*, **46**, 938 (1973).
28. R. J. Young, in *Spectroscopy of Polymers* (A. H. Fawcett, Ed.), Wiley, New York, 1996, Chap. 8.
29. P. R. Carey, *Biological Applications of Raman and Resonance Raman Spectroscopies*, Academic, New York, 1982.
30. I. Ya Slonim and A. N. Lyubimov, *The NMR of Polymers*, Plenum, New York, 1970.
31. F. A. Bovey and P. Mirau, *NMR of Polymers*, Academic, New York, 1996.
32. L. J. Mathias (Ed.), *Solid State NMR of Polymers*, Plenum, New York, 1991.
33. F. Laupretre, *Prog. Polym. Sci.*, **15**, 425 (1990).
34. V. J. McBrierty and K. J. Packer, *Nuclear Magnetic Resonance in Solid Polymers*, Cambridge University Press, New York, 1993.
35. F. C. Schilling, F. A. Bovey, M. D. Bruch, and S. A. Kozlowski, *Macromolecules*, **18**, 1418 (1985).
36. A. Zambelli, D. E. Dorman, A. I. R. Brewster, and F. A. Bovey, *Macromolecules*, **6**, 925 (1973).
37. F. A. Bovey, *Chain Structure and Conformation of Macromolecules*, Academic Press, New York, 1982, Chap. 3.
38. J. Schaefer, E. O. Stejskal, and R. Buchdahl, *Macromolecules*, **10**, 384 (1977).
39. J. L. Koenig, in *Polymer Spectroscopy* (A. H. Fawcett, Ed.), Wiley, New York, 1996, Chap. 6.
40. N.-L. Yang, J. Liutkas, and H. Haubenstock, in *Polymer Characterization by ESR and NMR* (A. E. Woodward and F. A. Bovey, Eds.), ACS Symp. Ser. 142, American Chemical Society, Washington, D.C., 1980, p. 35.
41. P.-O. Kinell, B. Rånby, and V. Runnstrom-Reio (Eds.), *ESR Applications to Polymer Research*, Wiley-Interscience, New York, 1973.
42. B. Rånby and J. F. Rabek, *ESR Spectroscopy in Polymer Research*, Springer-Verlag, Berlin, 1977.
43. R. F. Boyer and S. E. Keinath (Eds.), *Molecular Motion in Polymers by ESR*, Harwood Academic, New York, 1980.
44. G. G. Cameron and I. G. Davidson, in *Polymer Spectroscopy* (A. H. Fawcett, Ed.), Wiley, New York, 1996, Chap. 9.
45. M. Tyron and E. Horowitz, in *Analytical Chemistry of Polymers* (G. M. Kline, Ed.), Wiley-Interscience, New York, Part 2, 1962, Chap. 7.
46. J. F. Rabek, *Experimental Methods in Polymer Chemistry*, Wiley-Interscience, New York, 1980, Chap. 14.
47. M. Nencioni and S. Russo, *J. Macromol. Sci.-Chem.*, **A17**, 1255 (1982).
48. L. H. Garci Rubio, N. Ro, and R. D. Patel, *Macromolecules*, **17**, 1998 (1984).
49. R. W. Woody, *Macromol. Rev.*, **12**, 181 (1977).
50. H. Itagaki, K. Horie, and I. Mito, *Prog. Polym. Sci.*, **15**, 361 (1990).
51. E. V. Anufrieva and Y. Ya. Gotlib, *Adv. Polym. Sci.*, **40**, 1 (1981).
52. K. P. Ghiggino, A. J. Roberts, and D. Phillips, *Adv. Polym. Sci.*, **40**, 69 (1981).
53. L. E. Alexander, *X-Ray Diffraction Methods in Polymer Science*, Wiley-Interscience, New York, 1969.
54. P. Corradini, in *The Stereochemistry of Macromolecules* (A. D. Ketley, Ed.), Marcel Dekker, New York, 1968, Chap. 1.
55. M. Kakudo and N. Kasai, *X-Ray Diffraction by Polymers*, Elsevier, Amsterdam, 1972.
56. E. W. Fischer, *Polym. J.*, **17**, 307 (1985).
57. R. W. Richards, *Dev. Polym. Charact.*, **5**, 1 (1986).
58. A. R. Rennie, in *Polymer Spectroscopy* (A. H. Fawcett, Ed.), Wiley, New York, 1996, Chap. 13.

59. W. J. Feast and H. S. Munro (Eds.), *Polymer Surfaces and Interfaces*, Wiley-Interscience, New York, 1987.

60. H.-M. Tong and S. P. Kowalczyk (Eds.), *Characterization of Polymers: Surfaces, Interfaces, Thin Films*, Butterworth-Heinemann, Stoneham, Massachusetts, 1993.

61. A. Garton, *Infrared Spectroscopy of Polymer Blends, Composites and Surfaces*, Hanser, Cincinnati, 1992.

62. W. Suetaka and J. T. Yates, *Surface Infrared and Raman Spectroscopy: Methods and Applications*, Plenum, New York, 1995.

63. J. F. Rabek, *Experimental Methods in Polymer Chemistry*, Wiley-Interscience, New York, 1980, Chap. 27.

64. M. W. Urban, *Attenuated Total Reflectance Spectroscopy of Polymers*, American Chemical Society, Washington, D.C., 1996.

65. A. Rosencwaig, *Photoacoustics and Photoacoustic Spectroscopy*, Wiley-Interscience, New York, 1980.

66. J. F. McClelland, *Anal. Chem.*, **55**, 89A (1983).

67. J. F. Rabek, *Experimental Methods in Polymer Chemistry*, Wiley-Interscience, New York, 1980, Chap. 30.

68. D. T. Clark, *Adv. Polym. Sci.*, **24**, 125 (1977).

69. H. R. Thomas and J. J. O'Malley, *Macromolecules*, **12**, 323, (1979).

70. D. Briggs, *Br. Polym. J.*, **21**, 3 (1989).

71. D. M. Hercules and S. H. Hercules, *J. Chem. Educ.*, **61**, 402, 483, 592 (1984).

72. C. J. Roberts, M. C. Davies, K. M. Shakesheff, S. J. B. Tendler, and P. M. Williams, *Trends Polym. Sci.*, **4**, 420 (1996).

73. E. A. Turi (Ed.), *Thermal Characterization of Polymeric Materials*, 2nd ed., Vols. 1 and 2, Academic, New York, 1997.

74. V. B. F. Mathot (Ed.), *Calorimetry and Thermal Analysis of Polymers*, Hanser, Cincinnati, 1994.

75. W. W. Wendlandt and L. W. Collins, *Thermal Analysis*, Wiley-Interscience, New York, 1977.

76. J. F. Rabek, *Experimental Methods in Polymer Chemistry*, Wiley-Interscience, New York, 1980, Chap. 34.

77. H. LeChatelier, *Bull. Soc. Fr. Mineral.*, **10**, 204 (1877).

78. R. C. Mackenzie, *Differential Thermal Analysis*, Vols. 1 and 2, Academic, New York, 1970–72.

79. J. D. Matlack and A. P. Metzger, *J. Polym. Sci.*, **B4**, 875 (1966).

80. W. W. Wendlandt, *Thermal Methods of Analysis*, 2nd ed., Wiley-Interscience, New York, 1974, Chap. 11.

81. K. Honda, *Sci. Repts. Imp. Tohoku Univ.*, **4**, 97 (1915).

82. C. J. Keattch, *An Introduction to Thermogravimetry*, Heyden, London, 1969.

83. C. S. Marvel, in *High Temperature Polymers* (C. L. Segal, Ed.) Marcel Dekker, New York, 1967, p. 9.

84. W. H. T. Davison, S. Slaney, and A. L. Wragg, *Chem. Ind. (London)*, 1356 (1954).

85. R. S. Lehrle and J. C. Robb, *Nature*, **183**, 1671 (1959).

86. S. A. Liebman and E. J. Levy (Eds.), *Pyrolysis and G. C. in Polymer Analysis*, Chromat. Sci. Ser. No. 29, Marcel Dekker, New York, 1985.

87. M. P. Stevens, *Characterization and Analysis of Polymers by Gas Chromatography*, Marcel Dekker, New York, 1969, Chap. 4.

88. E. M. Barrall II, R. S. Porter, and J. F. Johnson, *Anal. Chem.*, **35**, 73 (1963).

89. B. Kolb, G. Kemmner, K. H. Kaiser, E. W. Cieplinski, and L. S. Ettre. *Z. Anal. Chem.*, **209**, 302 (1965).

90. K. Kishore and K. Mohandes, *J. Macromol. Sci.-Chem.*, **A18**, 379 (1982).

91. H. J. Troitzsch, in *Plastics Additives Handbook*, 3rd ed. (R. Gächter and H. Müller, Eds.), Hanser–Oxford, New York, 1990, Chap. 12.

92. R. M. Aseeva and G. E. Zaikov, *Combustion of Polymer Materials*, Hanser, Munich, 1986, Chap. 5.

93. C. P. Fenimore and F. J. Martin, *Modern Plastics*, **44**(11), 141 (1966).

94. N. M. Bikales (Ed.), *The Mechanical Properties of Polymers*, Wiley-Interscience, New York, 1980.

95. R. B. Seymour, in *Treatise on Analytical Chemistry*, Part 3, Vol. 3 (I. M. Kolthoff, P. J. Elving, and F. H. Stross, Eds.), Wiley-Interscience, New York, 1976, pp. 341–92.

96. A. Davis and D. Sims, *Weathering of Polymers*, Applied Science, London, 1983.

97. H. F. Mark, N. M. Bikales, C. G. Overberger, G. Menges, and J. I. Kroschwitz (Eds.), *Encyclopedia of*

Polymer Science and Engineering, 2nd ed. Vol. 5, Wiley-Interscience, New York, 1986: J. R. Perkins Jr., pp. 431–462; K. N. Mathes, pp. 507–587.

98. J. Agranoff (Ed.), *Modern Plastics Encyclopedia, 1984–85*, McGraw-Hill, New York, 1984, pp. 437ff.
99. C. C. Ku and R. Liepins, *Electrical Properties of Polymers*, Hanser, Cincinnati, 1989.

Review Exercises

1. Elemental analysis of a copolymer of propylene and vinyl chloride showed that the polymer contained 25.9% chlorine. What is the molar ratio of vinyl chloride to propylene in the copolymer?

2. Predict the products of ozonolysis (O_3 followed by H_2O/Zn) of the following polymers. Assume head-to-tail polymerization, and disregard products arising from end groups. (If you are unsure of polymer structures, refer to the index.) (a) 1,4-Polybutadiene; (b) 3,4-polyisoprene; (c) 1,2-polyisoprene; (d) 1,4-poly(1,3-cyclohexadiene); (e) polyacetylene.

3. When a random copolymer of 1,3-butadiene (70%) and *o*-chlorostyrene (30%) was subjected to ozonolysis followed by hydrolysis (under oxidative conditions), the major organic products were 3-(*o*-chlorophenyl)hexanedioic acid, 3-carboxyhexanedioic acid, and butanedioic acid. Show the structural units in the copolymer that gave rise to the products (ref. 8).

4. Explain the spectral differences in the 700 to 900-cm^{-1} region between the polymer and model compound shown in Figure 5.1.

5. Suppose you are an employee of a plastics firm and you are assigned the task of identifying a polymer used by a competitor to make certain automotive parts. You prepare a film of the polymer and determine its IR spectrum, as shown below. What information can you deduce from the spectrum? (A table of IR group frequencies might prove helpful.)

6. One stereoregular form of polystyrene (A) has a ^{1}H NMR spectrum containing a triplet centered at about 1.4 ppm and a quintet at about 1.9 ppm. Another stereoregular form (B) has an octet at about 1.6 ppm and a quintet at about 2.3 ppm. (Both spectra also exhibit phenyl proton resonances further downfield.) Interpret the spectra and determine which corresponds to isotactic and which to syndiotactic. (Syndiotactic polystyrene was first reported by N. Ishihara and co-workers in a 1986 paper. You may wish to locate the paper and check your answer.)

7. Figure 5.5 shows a partial ^{13}C (CP-MAS) NMR spectrum of polycarbonate. (a) How many additional absorbances would appear in the full ^{13}C spectrum? (b) Describe or sketch the ^1H NMR spectrum expected for polycarbonate. (Assume high resolution.)

8. Two linear polybenzyls have been reported, the latter by Lim and Choi in 1986. How could you distinguish between the two polymers spectroscopically (IR and NMR)?

9. In addition to giving information on stereochemical sequences, NMR can shed light on copolymer structural sequences. Poly(methyl methacrylate) exhibits a single resonance for the $-OCH_3$ protons at 3.74 ppm downfield from hexamethyldisiloxane (HMDS) in benzonitrile solvent. Copolymers of methyl methacrylate (M_1) and acrylonitrile (M_2) exhibit three methoxyl proton peaks at 3.74, 3.82, and 3.86 ppm. The peaks at 3.82 and 3.86 get progressively larger at the expense of the 3.74 peak as the concentration of M_2 in the copolymer is increased. The peak at 3.86 ppm is the dominant methoxyl resonance at high (>75%) levels of M_2. To what triad sequences of structural units (not tacticity) do these resonances most likely correspond? (See R. Chujo, H. Ubara, and A. Nishioka, *Polym. J.*, **3**, 670, 1972.)

10. How would you expect the x-ray diffraction pattern of a rubber band to change as the band is stretched?

11. In your own words, explain the difference between (a) SEM and TEM, (b) SAXS and WAXS, (c) AES and ESCA, (d) ISS and SIMS, (e) DSC and DTA, (f) isothermal and nonisothermal TGA, (g) self-extinguishing and nonflammable.

12. Suggest a spectroscopic method for determining each of the following and describe the special feature(s) pertaining to each analysis: (a) The amount of styrene in a styrene–butadiene copolymer. (b) The amount of 1,2 polymer in poly(1,3-butadiene). (c) The stereochemistry of double bonds in 1,4-poly(1,3-butadiene). (d) Whether any unreacted polyacrylonitrile (PAN) is present in a sample of carbon fiber prepared by pyrolysis of PAN. (e) Whether a sample of poly(vinyl acetate) has undergone hydrolysis on exposure to moisture. (f) The identity of metallic driers used in a sample of painted wooden fence. (g) How completely a polyester surgical implant has been coated with polytetrafluoroethylene to improve its chemical resistance.

13. From the data in Table 3.5 (Chapter 3), sketch the anticipated DSC thermogram of poly(ethylene adipate) between -100 and $100°C$ (a) if the sample is highly crystalline, and (b) if the sample was initially quenched in the amorphous state. If the sample were to be cycled back to $-100°C$ at the same rate, would you expect the cooling and heating thermograms to appear the same? Explain.

14. A method for determining percent vinyl acetate in poly[ethylene-*co*-(vinyl acetate)] by pyrolysis GC is described in this chapter. PGC may also be used to determine ethyl acrylate ($CH_2 = CHCO_2C_2H_5$) in poly[ethylene-*co*-(ethyl acrylate)]. (ref. 88) What volatile pyrolysis product would be analyzed? What repeating unit would remain following loss of the volatile product? (Hint: Write an analogous mechanism.)

15. A method employing pyrolysis and hydrogenation has been used to determine which products are unsaturated in a pyrogram (ref. 89). How would the chromatogram shown in Figure 5.20 change if this procedure were used? Explain why the largest peak in each triplet is 1-alkene. (Note that chain cleavage occurs homolytically to produce free radicals. You may check the answer in Chapter 9.)

16. An empirical method of determining percent crystallinity in high- and low-density polyethylene using PGC has been reported (D. Deur-Siftar, *J. Gas Chromatog.*, **5**, 72, 1967). If the journal is in your library, read the paper and summarize the method.

17. Given the data in Table 5.3, predict the approximate LOI of polybenzimidazole (structure 1, Chapter 4). Justify your prediction.

P A R T

VINYL POLYMERS

6

FREE RADICAL
POLYMERIZATION

6.1 Introduction

From the commercial standpoint, vinyl polymers are the most important of all polymer types. As discussed in Chapter 1, polyethylene (high and low density), poly(vinyl chloride), polystyrene, and polypropylene lead the market in total volume output and should continue to do so in the foreseeable future. For this reason, a lot more research has been accomplished in the area of vinyl polymerization than with most other polymer types. It is not the purpose of this part of the book to discuss individual vinyl polymers, but rather to emphasize in general terms the differences among the various modes of polymerization and those factors in the polymerization processes responsible for the structure of the final product. For convenience we are classifying all polymers prepared by addition reactions of carbon–carbon double bonds as vinyl polymers. Some polymer scientists prefer to classify olefin, diene, and vinylidene polymers separately.

It will be recalled from Chapter 1 that vinyl polymerization reactions are chain-reaction processes, which is the case with most (but certainly not all) addition polymerizations, and they require an initiator to begin the reaction. Polymerization occurs only at the reactive end of a growing chain; hence high molecular weights are reached rapidly at relatively low percentages of conversion, and monomer is always present in significant quantities during the process. While these generalizations apply to all chain polymerizations, the differences among the various types of initiation and propagation reactions are significant enough to treat each separately. This chapter is concerned only with free radical polymerization.[1,2]

A wide variety of monomer types lend themselves to free radical polymerization, which, as we shall see later, is not the case with ionic polymerization. Commercially important vinyl polymers and copolymers prepared by free radical processes are given in Table 6.1. Not included are *derivatives* of vinyl polymers, such as chlorinated polyethylene or poly(vinyl alcohol); these are discussed in Chapter 9.

TABLE 6.1. Commercially Important Vinyl Polymers Prepared by Free Radical Polymerization[a]

Polymer or Copolymer	Major Uses
Polyethylene, low-density (LDPE)	Packaging film, wire and cable insulation, toys, flexible bottles, housewares, coatings
Poly(vinyl chloride)	Construction, rigid pipe, flooring, wire and cable insulation, film and sheet
Polystyrene[b]	Packaging, foam insulation, appliances, housewares, toys
Polychloroprene (neoprene rubber)	Tires, wire coatings, belting, hoses, shoe heels, coated fabrics
Poly(methyl methacrylate)	Automotive parts, decorative panelling, skylights, glazing
Polyacrylonitrile	Food packaging
Poly(vinyl acetate)	Water-based paints, adhesives, poly(vinyl alcohol) precursor
Poly(vinylidene chloride)[c]	Food packaging
Poly(acrylic acid) and poly(methacrylic acid)	Adhesives, thickening agents, ionomers
Poly(ethyl acrylate) and poly(methyl acrylate)	Inks, varnishes
Polyacrylamide	Thickening agent, flocculant
Polytetrafluoroethylene	Machine parts, gaskets, nonstick cooking utensils
Polytrichlorofluoroethylene	Gaskets, tubing, wire insulation
Poly(vinylidene fluoride)	Protective coatings, gaskets, pipe
Poly(vinyl fluoride)	Protective coatings
Allyl resins[d]	Lenses, electronics parts
Styrene–butadiene rubber (SBR)	Tires, belting, flooring, molded goods, shoe soles, electrical insulation
Acrylonitrile–butadiene copolymer (nitrile rubber)	Fuel tanks, gasoline hoses, adhesives, impregnated paper, leather and textiles
Acrylonitrile–butadiene–styrene copolymer (ABS)	Appliances, business machines, automotive parts, luggage, telephones, electronics components
Styrene–acrylonitrile copolymer (SAN)	Automotive parts, battery cases, kitchenware, appliances, medical supplies, furniture
Styrene–maleic anhydride copolymer (SMA)	Automobile interior parts, trays, tumblers
Ethylene–vinyl acetate copolymer (EVA)	Shrink-wrap food packaging, adhesives, coatings, drum liners, carpet backing, gaskets
Acrylonitrile–vinyl chloride copolymer	Textile fibers

[a] Numerous copolymers are manufactured besides the high-volume ones listed.

[b] Poly(p-methylstyrene) is also available.

[c] Including copolymers, principally with vinyl chloride.

[d] Crosslinked diallyl and triallyl esters.

6.2 Free Radical Initiators

Certain monomers, notably styrene and methyl methacrylate and some strained-ring cycloalkenes, undergo polymerization on heating in the absence of any added free radical initiator. Most monomers, however, require some kind of initiator. A large number of free radical initiators[3,4] are available; they may be classified into four major types: peroxides and hydroperoxides, azo compounds, redox initiators, and certain compounds that form radicals under the influence of light (*photoinitiators*). High-energy radiation (α- and β-particles, γ- and x-rays) can also promote free radical polymerization, although such radiation is used less commonly. *Plasma polymerization*[5–7]—polymerization initiated by a partially ionized gas generated by a radiofrequency glow discharge—is widely used for preparing coatings and insulating layers, especially for microelectronics and medical applications. The polymerization mechanism, however, is complex and appears to involve both free radical and ionic species, and the polymers thus formed are highly crosslinked and bear little resemblance to polymers prepared by more conventional chain polymerization. Indeed, just about *any* organic compound will undergo plasma polymerization.

6.2.1 Peroxides and Hydroperoxides

Of the various types of initiators, peroxides (ROOR) and hydroperoxides (ROOH) are most widely used. They are thermally unstable and decompose into radicals at a temperature and rate dependent on structure. The most commonly used peroxide is benzoyl peroxide (**1**), which undergoes thermal homolysis to form benzoyloxy radicals (reaction 6.1).* The benzoyloxy radicals may undergo a variety of reactions besides adding to monomer, including recombination (reverse of 6.1), decomposition to phenyl radicals and carbon dioxide (6.2),

$$\text{Ph}-\overset{\overset{\displaystyle O}{\|}}{C}-O-O-\overset{\overset{\displaystyle O}{\|}}{C}-\text{Ph} \longrightarrow 2\ \text{Ph}-\overset{\overset{\displaystyle O}{\|}}{C}-O\cdot \qquad (6.1)$$

$$\mathbf{1}$$

$$\text{Ph}-\overset{\overset{\displaystyle O}{\|}}{C}-O\cdot \longrightarrow \text{Ph}\cdot + CO_2 \qquad (6.2)$$

and radical combination (6.3 and 6.4). These secondary reactions occur because of the confining effect of solvent molecules (the *cage effect*), and as a result, the concentration of initiator radicals is depleted. Induced decomposition (6.5) is another "wastage" reaction.

$$2\ \text{Ph}\cdot \longrightarrow \text{Ph}-\text{Ph} \qquad (6.3)$$

$$\text{Ph}-\overset{\overset{\displaystyle O}{\|}}{C}-O\cdot + \text{Ph}\cdot \longrightarrow \text{Ph}-\overset{\overset{\displaystyle O}{\|}}{C}-O-\text{Ph} \qquad (6.4)$$

---

*Ph = phenyl. For radical structures such as the benzoyloxy radical, only the unpaired electron is shown. A more complete representation would include lone pair electrons and contributing resonance structures.

$$\text{Ph}\overset{\overset{\displaystyle O}{\|}}{-\text{C}}-\text{O}-\text{O}-\overset{\overset{\displaystyle O}{\|}}{\text{C}}-\text{Ph} + \text{Ph}\cdot \longrightarrow$$

1

$$\text{Ph}\overset{\overset{\displaystyle O-\text{Ph}}{|}}{\underset{\displaystyle \cdot}{-\text{C}}}-\text{O}-\text{O}-\overset{\overset{\displaystyle O}{\|}}{\text{C}}-\text{Ph} \longrightarrow \text{Ph}-\overset{\overset{\displaystyle O}{\|}}{\text{C}}-\text{O}-\text{Ph} + \cdot\text{O}-\overset{\overset{\displaystyle O}{\|}}{\text{C}}-\text{Ph} \qquad (6.5)$$

Two other common initiators are diacetyl peroxide (**2**) and di-*t*-butyl peroxide (**3** Hydroperoxides such as cumyl hydroperoxide (**4**) decompose to form alkoxy and hydroxy radicals (6.6). Because hydroperoxides contain an active hydrogen atom, induced decomposition occurs readily, for example, by a chain-end radical (6.7). Peroxy radicals may also combine with subsequent formation of oxygen (6.8).

$$\underset{\textbf{2}}{\text{CH}_3\overset{\overset{\displaystyle O}{\|}}{\text{C}}\text{OO}\overset{\overset{\displaystyle O}{\|}}{\text{C}}\text{CH}_3} \qquad \underset{\textbf{3}}{(\text{CH}_3)_3\text{COOC}(\text{CH}_3)_3} \qquad \underset{\textbf{4}}{\text{Ph}-\overset{\overset{\displaystyle OOH}{|}}{\text{C}}(\text{CH}_3)_2}$$

$$\text{ROOH} \longrightarrow \text{RO}\cdot + \cdot\text{OH} \qquad (6.6)$$

$$\text{RO}\left[\text{CH}_2\underset{\underset{\displaystyle R'}{|}}{\text{CH}}\right]_n\text{CH}_2\underset{\underset{\displaystyle R'}{|}}{\text{CH}}\cdot + \text{ROOH} \longrightarrow \text{RO}\left[\text{CH}_2\underset{\underset{\displaystyle R'}{|}}{\text{CH}}\right]_n\text{CH}_2\underset{\underset{\displaystyle R'}{|}}{\text{CH}}_2 + \text{ROO}\cdot \qquad (6.7)$$

$$2\text{ROO}\cdot \longrightarrow [\text{ROO}-\text{OOR}] \longrightarrow 2\text{RO}\cdot + \text{O}_2 \qquad (6.8)$$

The extent to which side reactions occur depends on the structure of the peroxide, the stability of the initially formed radicals, and the reactivity of the monomer. Ideally, a peroxide initiator should be relatively stable at room temperature but should decompose rapidly enough at polymer-processing temperatures to ensure a practical reaction rate. Benzoyl peroxide (half-life 30 minutes at 100°C) has the advantage that benzoyloxy radicals are stable enough that they tend to react with more reactive monomer molecules before eliminating carbon dioxide, thus reducing initiator wastage. Acetoxy radicals formed from **2**, on the other hand, are much less stable, and initiator wastage occurs more readily. More stable than benzoyloxy radicals are *t*-butoxy radicals, which are almost entirely captured by monomer; however, the decomposition temperature is relatively high (half-life 10 hours at 126°C).

Decomposition of peroxides can frequently be induced at lower temperatures by the addition of *promoters*. For example, addition of N,N-dimethylaniline to benzoyl peroxide causes the latter to decompose rapidly at room temperature. Kinetics studies[8,9] indicate that the decomposition involves formation of an unstable ionic intermediate (6.9) that reacts further to give benzoyloxy radical and a radical cation (6.10). The radical cation apparently undergoes reactions other than addition to monomer, since polymers formed by this method contain no nitrogen.

$$\text{Ph}-\overset{\overset{\displaystyle O}{\|}}{\text{C}}-\text{O}-\text{O}-\overset{\overset{\displaystyle O}{\|}}{\text{C}}-\text{Ph} + \text{Ph}-\text{N}(\text{CH}_3)_2 \longrightarrow$$

$$\text{Ph}\overset{\overset{\displaystyle \text{CH}_3}{|}}{-\overset{+}{\underset{\underset{\displaystyle \text{CH}_3}{|}}{\text{N}}}}-\text{O}-\overset{\overset{\displaystyle O}{\|}}{\text{C}}-\text{Ph} + \text{Ph}-\text{CO}_2^- \qquad (6.9)$$

$$Ph\overset{CH_3}{\underset{CH_3}{\overset{|}{\underset{|}{-\overset{+}{N}}}}}\!\!-O\overset{O}{\overset{\|}{-C}}-Ph \longrightarrow Ph\!-\overset{+}{\underset{.}{N}}(CH_3)_2 + Ph\overset{O}{\overset{\|}{-C}}-O\cdot \qquad (6.10)$$

6.2.2 Azo Compounds

The most commonly used azo compounds are those having cyano groups on the carbons attached to the azo linkage—for example, α,α'-azobis(isobutyronitrile) (**5**), which decomposes at relatively low temperatures (half-life 1.3 hours at 80°C). The driving force for decomposition is formation of nitrogen and the resonance-stabilized cyanopropyl radical (6.11). As with peroxide decomposition, the initially formed radicals can combine in the solvent cage to deplete initiator concentration. Combination leads to both tetramethylsuccinonitrile (**6**) (6.12) and the ketenimine (**7**) (6.13).

$$(CH_3)_2\overset{CN}{\overset{|}{C}}\!\!-N\!\!=\!\!N\!\!-\overset{CN}{\overset{|}{C}}(CH_3)_2 \longrightarrow 2(CH_3)_2\overset{CN}{\overset{|}{C}}\cdot + N_2 \qquad (6.11)$$
$$\mathbf{5}$$

$$2(CH_3)_2\overset{CN}{\overset{|}{C}}\cdot \left\{ \begin{array}{l} \longrightarrow (CH_3)_2\overset{CN}{\overset{|}{C}}\!\!-\!\!\overset{CN}{\overset{|}{C}}(CH_3)_2 \qquad (6.12) \\ \qquad\qquad\qquad \mathbf{6} \\ \\ \longrightarrow (CH_3)_2C\!\!=\!\!C\!\!=\!\!N\!\!-\!\!\overset{CN}{\overset{|}{C}}(CH_3)_2 \qquad (6.13) \\ \qquad\qquad\qquad\qquad\quad \mathbf{7} \end{array} \right.$$

6.2.3 Redox Initiators

Production of free radicals by one-electron transfer reactions is particularly useful in initiation of low-temperature polymerization and emulsion polymerization. Some typical examples are given in equations (6.14) to (6.16). Apart from the fact that low temperatures can be employed with redox systems, reaction rates are easy to control by varying the concentration of metal ion or peroxide. The decomposition of cumyl hydroperoxide (6.14) is used commercially for the low-temperature emulsion copolymerization of styrene and butadiene to form styrene–butadiene rubber. Hydrogen peroxide or persulfate systems, (6.15) and (6.16), are used in emulsion polymerization to produce radicals in the aqueous phase. For nonaqueous polymerization, metal ions are generally introduced as the naphthenates. Numerous combinations of possible redox initiator systems are available.

$$Ph\overset{OOH}{\overset{|}{-C}}(CH_3)_2 + Fe^{2+} \longrightarrow Ph\overset{O\cdot}{\overset{|}{-C}}(CH_3)_2 + OH^- + Fe^{3+} \qquad (6.14)$$
$$\mathbf{4}$$

$$HOOH + Fe^{2+} \longrightarrow HO\cdot + OH^- + Fe^{3+} \qquad (6.15)$$

$$^-O_3SOOSO_3^- + S_2O_3^{2-} \longrightarrow SO_4^-\cdot + SO_4^{2-} + \cdot S_2O_3^- \qquad (6.16)$$

6.2.4 Photoinitiators

Peroxides and azo compounds dissociate photolytically as well as thermally. The major advantage of photoinitiation[10] is that the reaction is essentially independent of temperature; thus polymerizations may be conducted even at very low temperatures. Furthermore, better control of the polymerization reaction is generally possible because narrow wavelength bands may be used to initiate decomposition, and the reaction can be stopped simply by removing the light source. A wide variety of photolabile compounds are available, including disulfides (6.17), benzoin (**8**) (6.18), and benzil (**9**) (6.19).

$$RSSR \xrightarrow{hv} 2RS\cdot \qquad (6.17)$$

$$(6.18)$$

8

$$(6.19)$$

9

6.2.5 Thermal Polymerization

Some monomers polymerize slowly on heating in the absence of added initiator.[11] In such cases, free radical initiating species are generated in situ by mechanisms that are not, in most cases, well understood. Of the commercially important vinyl monomers, styrene undergoes thermal polymerization most rapidly. The mechanism[12-14] appears to involve initial formation of a Diels–Alder dimer (**10**) (6.20), which transfers a hydrogen atom to monomer (6.21) to yield an initiator styryl radical (**11**) and a benzylic radical (**12**). This is an example of what is referred to as *molecule-induced homolysis*—the rapid formation of radicals by reaction of nonradical species. Photopolymerization of styrene with no added initiator appears to follow a similar pathway.[15]

$$2CH_2{=}CHPh \longrightarrow \qquad (6.20)$$

10

$$10 + CH_2{=}CHPh \longrightarrow CH_3\dot{C}HPh + \qquad (6.21)$$

11   12

Apart from some limited applications with styrene, thermal polymerization has been used commercially only with certain polymers or oligomers containing thermally labile end groups (see Chapter 4, Table 4.4).

6.2.6 Electrochemical Polymerization

Electrolysis of a solution containing both monomer and electrolyte can be used to initiate polymerization.[16] At the cathode an electrode may be transferred to a monomer molecule to form a radical anion (6.22), and at the anode a monomer molecule may give up an electron to form a radical cation (6.23). The radical ions in turn initiate free radical or ionic polymerization or both, depending on electrolysis conditions. Other species present in solution may also undergo electron transfer to generate species that can initiate polymerization. Electrochemical polymerization is particularly useful for coating metal surfaces with polymer films.

$$RCH{=}CH_2 + e^- \rightarrow R\overset{..}{C}H{-}\overset{.}{C}H_2 \tag{6.22}$$

$$RCH{=}CH_2 \rightarrow R\overset{+}{C}H{-}\overset{.}{C}H_2 + e^- \tag{6.23}$$

6.3 Techniques of Free Radical Polymerization

Free radical polymerization can be accomplished in bulk, suspension, solution, or emulsion.[17] Ionic and other nonradical polymerizations are usually confined to solution techniques. Each of the methods has advantages and disadvantages, as outlined in Table 6.2. In addition, work has also been done on solid- and gas-phase polymerizations of vinyl monomers, but these are of lesser importance. Because polymers are not volatile, the term *gas-phase polymerization* means, in effect, bulk polymerization in which monomer vapors diffuse to the polymerization site.

TABLE 6.2. Free Radical Polymerization Techniques

Method	Advantages	Disadvantages
Bulk	Simple; no contaminants added	Reaction exotherm difficult to control; high viscosity
Suspension	Heat readily dispersed; low viscosity; polymer obtained in granular form and may be used directly	Washing and/or drying required; agglomeration may occur; contamination by stabilizer
Solution	Heat readily dispersed; low viscosity; may be used directly as solution	Added cost of solvent; solvent difficult to remove; possible chain transfer with solvent; possible environmental pollution
Emulsion	Heat readily dispersed; low viscosity; high molecular weight obtainable; may be used directly as emulsion; works on tacky polymers	Contamination by emulsifier and other ingredients; chain transfer agents often needed to control DP; washing and drying necessary for bulk polymer

6.3.1 Bulk

Bulk polymerization is simplest from the standpoint of formulation and equipment, but it is also the most difficult to control, particularly when the polymerization reaction is very exothermic. This, coupled with problems of heat transfer as the monomer–polymer solution increases in viscosity, limits the use of bulk methods in commercial production, although more efficient bulk processes have been developed in recent years.

In cases where polymer is insoluble in monomer, polymer precipitates and the viscosity of the medium does not change appreciably. Problems still arise, however, as a result of free radicals (detectable by ESR) being occluded in the polymer droplet, which can lead to *autoacceleration*, that is, a rapid increase in the polymerization rate.[18] In some instances, particularly with diene monomers, this occlusion effect may lead to formation of insoluble crosslinked polymer nodules, a phenomenon referred to as *popcorn polymerization*.[19] The crosslinked nodules are usually of light weight and occupy considerably more volume than the monomers from which they are derived, which may cause fouling and even fracture of the polymerization apparatus.

The major commercial uses of bulk vinyl polymerization are in casting formulations and low-molecular-weight polymers for use as adhesives, plasticizers, tackifiers, and lubricant additives.

6.3.2 Suspension

Suspension polymerization involves mechanically dispersing monomer in a noncompatible liquid, usually water, and polymerizing the resultant monomer droplets by use of a monomer-soluble initiator. Monomer is kept in suspension by continuous agitation and the use of *stabilizers* such as poly(vinyl alcohol) or methyl cellulose. If the process is carefully controlled, polymer is obtained in the form of granular beads, which are easy to handle and can be isolated by filtration or by spraying into a heated chamber (*spray drying*). A major advantage is that heat transfer is very efficient and the reaction is therefore easily controlled. Suspension polymerization cannot be used for tacky polymers such as elastomers because of the tendency for agglomeration of polymer particles. From the standpoint of kinetics and mechanism, suspension polymerization is identical to bulk polymerization.

Suspension methods are used to prepare a number of granular polymers, including polystyrene, poly(vinyl chloride), and poly(methyl methacrylate).

6.3.3 Solution

Like suspension, solution polymerization allows efficient heat transfer. Solvent must be chosen carefully, otherwise chain transfer reactions (discussed in the next section) may severely limit the molecular weight. Apart from the environmental concerns associated with organic solvents, a major problem in solution polymerization is that it is often difficult to remove solvent completely from the finished polymer. As a result there is considerable interest in using supercritical carbon dioxide* as a polymerization solvent.[20] Advantages are that carbon dioxide is nontoxic, inexpensive, and easily removed from the polymer and recycled.

---

*A supercritical fluid (SCF) is obtained by a combination of heat and pressure to move the gaseous substance beyond its critical point. SCFs exhibit properties intermediate between those of a liquid and a gas, but have the solvating power of liquids.

6.3.4 Emulsion

Developed at Goodyear Tire and Rubber Company in the 1920s, emulsion polymerization resembles suspension polymerization in that water is used as a dispersing medium and heat transfer is very efficient; but there the similarity ends.[21–23] Monomer is dispersed in the aqueous phase by an emulsifying agent such as a soap or detergent. Initiator radicals, usually of the redox type, are generated in the aqueous phase and diffuse into soap micelles swollen with monomer molecules. As monomer is used up in the polymerization reaction, more monomer migrates into the micelles to continue the reaction. Termination of polymerization occurs by radical combination when a new radical diffuses into the micelle. Because only one radical is present in the micelle prior to termination, extremely high molecular weights are obtainable, generally too high to be of practical value unless compounds called *chain transfer agents* are added that control the degree of polymerization. (How chain transfer agents work is discussed later in the chapter.) The overall process is complex, with reaction kinetics differing significantly from that of bulk or solution processes.

Emulsion polymerization is widely used in industry for large-scale preparations, and is particularly useful for manufacturing water-based (latex) paints or adhesives in which the emulsified product is used directly. Emulsion polymerization is also suitable for preparing tacky polymers because the very small particles are stable and resist agglomeration. Two typical emulsion recipes are given in Table 6.3.

A much less commonly used emulsion technique involves dispersing an aqueous solution of monomer in a nonaqueous phase. This is referred to as an *inverse* or *water-in-oil*

TABLE 6.3. Typical Emulsion Polymerization Recipes[a]

Ingredients, Conditions	Styrene-Butadiene Copolymer	Polyacrylate Latex
Ingredients (parts by weight)		
Water	190	133
Butadiene	70	—
Styrene	30	—
Ethyl acrylate	—	93
2-Chloroethyl vinyl ether	—	5
p-Divinylbenzene	—	2
Soap	5	3[b]
Potassium persulfate	0.3	1
1-Dodecanethiol	0.5	—
Sodium pyrophosphate	—	0.7
Conditions		
Time	12 hr	8 hr
Temperature	50°C	60°C
Yield	65%	~100%

[a] Recipes from Cooper.[23]

[b] Sodium lauryl sulfate.

emulsion (as opposed to the more conventional *oil-in-water* type). The mechanism of polymerization is similar for the two techniques, but inverse emulsions tend to be less stable.

6.4 Kinetics and Mechanism of Polymerization

Initiation of free radical chain polymerization involves two reactions: formation of the initiator radical (6.24), and addition of the initiator radical to monomer (6.25). Evidence for the incorporation of initiator radicals arises from spectroscopic and chemical analysis of end groups. Addition of monomer radical to another monomer molecule, followed by successive additions of oligomer and polymer radicals to available monomer (6.26), comprise the propagation reactions.

$$\text{Initiator} \longrightarrow R\cdot \tag{6.24}$$

$$R\cdot + CH_2{=}\underset{\underset{Y}{|}}{CH} \longrightarrow RCH_2\underset{\underset{Y}{|}}{CH}\cdot \tag{6.25}$$

$$RCH_2\underset{\underset{Y}{|}}{CH}\cdot + CH_2{=}\underset{\underset{Y}{|}}{CH} \longrightarrow RCH_2\underset{\underset{Y}{|}}{CH}CH_2\underset{\underset{Y}{|}}{CH}\cdot \xrightarrow{nCH_2{=}CHY}$$

$$RCH_2\underset{\underset{Y}{|}}{CH}{\left[\!{-}CH_2\underset{\underset{Y}{|}}{CH}{-}\!\right]}_{n}CH_2\underset{\underset{Y}{|}}{CH}\cdot \tag{6.26}$$

It was mentioned in Chapter 3 (Section 3.5) that each addition step follows the predominant *head-to-tail* orientation shown in reaction (6.26). This is due to a combination of steric and electronic effects. Steric repulsion favors attack by the radical at the least hindered carbon of the double bond; and resonance stabilization favors formation of the more stable free radical. Head-to-tail polymerization does not occur exclusively. Significant amounts of *head-to-head* structures have been found to occur in certain fluorine-containing polymers,[24] notably poly(vinyl fluoride) (**13**) and poly(vinylidene fluoride) (**14**) (13–17% and 5–6%, respectively), and in poly(allyl acetate) (**15**) (up to 19%, depending on polymerization temperature.[25]) Where the intermediate radical is significantly resonance stabilized, however, head-to-tail structures appear to form almost exclusively.

$$\begin{array}{ccc} {-}{\left[CH_2CHF\right]}{-} & {-}{\left[CH_2CF_2\right]}{-} & {-}{\left[CH_2\underset{\underset{CH_2OCOCH_3}{|}}{CH}\right]}{-} \\[2mm] \textbf{13} & \textbf{14} & \textbf{15} \end{array}$$

Propagation continues until some reaction occurs to terminate it. The two principal ways that termination may occur in free radical polymerization are radical *coupling*, or *combination* (6.27), and *disproportionation* (6.28), which involves transfer of an atom, usually hydrogen, from one chain end to another.[26] Combination leads to initiator fragments at both ends of the polymer chain, whereas disproportionation results in an initiator fragment at one end. Also, combination leads to a head-to-head linkage. Both termination reactions are diffusion controlled and require pairing of electron spins.

$$\text{wwCH-CHww} \quad (6.27)$$
$$\underset{Y \quad Y}{\mid \quad \mid}$$

$$\text{wwCH}_2\text{CH} \cdot + \cdot \text{CHCH}_2\text{ww}$$
$$\underset{Y \qquad Y}{\mid \qquad \mid}$$

$$\text{wwCH=CH} + \text{CH}_2\text{CH}_2\text{ww} \quad (6.28)$$
$$\underset{Y \qquad Y}{\mid \qquad \mid}$$

Whether termination occurs by coupling or by disproportionation depends in large measure on monomer structure or, more exactly, on the structure of the chain-end radical. Polystyryl radicals undergo coupling (6.29) almost exclusively at low temperatures, whereas poly(methyl methacrylate) radicals undergo mainly disproportionation (6.30); but generally both processes occur. What favors one termination reaction over the other? Steric repulsion is one factor. Coupling of poly(methyl methacrylate) radicals would lead to four bulky groups on adjacent carbons. Electrostatic repulsion of polar groups (such as ester) may also raise the activation energy for coupling. Availability of alpha hydrogens for hydrogen transfer is a third factor: poly(methyl methacrylate) radicals have five, compared with two for polystyryl radicals. It is no easy task to predict how termination will occur, however, as shown by the fact that polyacrylonitrile undergoes coupling virtually exclusively at 60°C, whereas poly(vinyl acetate) undergoes disproportionation.

$$2 \text{ wwCH}_2\text{CH} \cdot \qquad \text{wwCH}_2\text{CH---CHCH}_2\text{ww}$$
$$\underset{Ph}{\mid} \quad \longrightarrow \quad \underset{Ph \qquad Ph}{\mid \qquad \mid} \quad (6.29)$$

$$2 \text{ wwCH}_2\overset{CH_3}{\underset{CO_2CH_3}{\overset{\mid}{\underset{\mid}{C}}}} \cdot \quad \longrightarrow \quad \text{wwCH}_2\overset{CH_3}{\underset{CO_2CH_3}{\overset{\mid}{\underset{\mid}{CH}}}} + \text{wwCH}=\overset{CH_3}{\underset{CO_2CH_3}{\overset{\mid}{\underset{\mid}{C}}}}$$

$$\qquad \qquad \qquad (6.30)$$

$$\text{or} \quad \text{wwCH}_2-\overset{CH_2}{\underset{CO_2CH_3}{\overset{\parallel}{\underset{\mid}{C}}}}$$

Another possible termination reaction involves combination of initiator radicals with chain end radicals (6.31). Called *primary radical termination*, this process is significant only at relatively high initiator levels or when very high viscosities limit the diffusion of high-molecular-weight chain end radicals.

$$\text{wwwCH}_2\text{CH} \cdot + \text{R} \cdot \quad \longrightarrow \quad \text{wwwCH}_2\text{CHR} \quad (6.31)$$
$$\underset{Y}{\mid} \qquad \qquad \qquad \underset{Y}{\mid}$$

A simplified picture of the kinetics of free radical polymerization[27,28] assumes that the rates of initiation, propagation, and termination are all different, but that each propagation step is exclusively head-to-tail and goes at the same rate, independent of chain length. This last assumption is basically valid once the chain has grown to about four monomer units, at which point the effect of the end group becomes negligible. Similarly, it is assumed that the rates of termination by combination or disproportionation are independent of chain length. (In cases where there is a significant degree of head-to-head orientation, one must take into account four separate propagation reactions: head-radical-to-head, head-radical-to-tail, tail-radical-to-head, and tail-radical-to-tail.)

Initiation proceeds in two steps: decomposition of initiator to yield initiator radicals, $R\cdot$, followed by addition of $R\cdot$ to monomer, M, to give a new radical, $M_1\cdot$.

Initiation

$$\text{Initiator} \xrightarrow{k_d} R\cdot$$

$$R\cdot + M \xrightarrow{k_i} M_1\cdot$$

Rate constants for the two reactions are k_d and k_i.

In the initial propagation step (rate constant k_p), $M_1\cdot$ adds to another monomer molecule to form a new radical, $M_2\cdot$, which, in turn, adds to M to form $M_3\cdot$, and so on.

Propagation

$$M_1\cdot + M \xrightarrow{k_p} M_2\cdot$$

$$M_2\cdot + M \xrightarrow{k_p} M_3\cdot$$

$$M_x\cdot + M \xrightarrow{k_p} M\cdot_{(x+1)}$$

As was mentioned above, termination occurs principally by radical coupling or disproportionation, where k_{tc} and K_{td} are the respective rate constants.

Termination

$$M_x\cdot + M_y\cdot \xrightarrow{k_{tc}} M_{(x+y)} \qquad \text{Coupling}$$

$$M_x\cdot + M_y\cdot \xrightarrow{k_{td}} M_x + M_y \qquad \text{Disproportionation}$$

Assuming the rate of initiator decomposition is very slow relative to that of addition of initiator radical to monomer (reasonable, given the high reactivity of free radicals), and taking into account the fact that two initiator radicals are formed with each decomposition, the expression for initiation rate, R_i, is

$$R_i = \frac{-d[M\cdot]}{dt} = 2fk_d\,[I]$$

where $[M\cdot]$ is the total concentration of chain radicals, $[I]$ is the molar concentration of initiator, and f is the *initiator efficiency*, that is, the fraction of initiator radicals that actually start a polymer chain and are not consumed in the wastage reactions discussed in Section 6.2. Thus,

$$f = \frac{\text{radicals that initiate a polymer chain}}{\text{radicals formed from initiator}}$$

Efficiency is determined most conveniently by measuring the end groups arising from the addition of initiator radicals (by using isotopically labeled initiator, for example) and comparing with the amount of initiator reacted. For most common free radical polymerization processes, f falls in the range 0.3 to 0.8.

For termination the rate expression is

$$R_t = \frac{-d[M\cdot]}{dt} = 2k_t[M\cdot]^2$$

The factor 2 takes into account that in any termination reaction two radicals are consumed. The rate constant, k_t, represents the summation of k_{tc} and k_{td}.

Because the rate constants for termination are very much greater than those of initiation, we may assume that shortly after the reaction begins, the formation and destruction of radicals occur at the same rates; hence the concentration of radicals, $[M\cdot]$, remains constant. This is referred to as the *steady-state assumption*. Thus, $R_i = R_t$, or

$$2fk_d[I] = 2k_t[M\cdot]^2$$

Solving for $[M\cdot]$ gives

$$[M\cdot] = \sqrt{\frac{fk_d[I]}{k_t}}$$

The rate expression for propagation is

$$R_p = \frac{-d[M]}{dt} = k_p[M][M\cdot]$$

Substituting the expression for $[M\cdot]$ above, we obtain

$$R_p = \frac{-d[M]}{dt} = k_p[M]\sqrt{\frac{fk_d[I]}{k_t}}$$

Because propagation involves large numbers of monomer molecules per chain, whereas initiation consumes only one, the rate of polymerization is, for all practical purposes, equivalent to the rate of propagation. The polymerization rate, therefore, is proportional to the square root of initiator concentration and to the first power of monomer concentration. Thus, doubling the initiator concentration causes the rate to increase by a factor of about 1.4. This relationship has been confirmed experimentally for a variety of free radical polymerizations. Propagation and termination rate constants, as well as the corresponding activation energies, for several commercially important monomers are given in Table 6.4.[29a]

Another important parameter related to polymerization rate is the *average kinetic chain length*, $\bar{\nu}$, which is defined as the average number of monomer units polymerized per chain initiated, which is equal to the rate of polymerization per rate of initiation. Since $R_i = R_t$ under steady-state conditions,

$$\bar{\nu} = \frac{R_p}{R_i} = \frac{R_p}{R_t}$$

Substituting for R_p and R_t from the above expressions:

$$\bar{\nu} = \frac{k_p[M][M\cdot]}{2k_t[M\cdot]^2} = \frac{k_p[M]}{2k_t[M\cdot]}$$

TABLE 6.4. Representative Propagation and Termination Rate Constants, k_p and k_t and Activation Energies, E_p and E_t^a

Monomer	Temperature (°C)	k_p (L/mol s)	E_p^* (kJ/mol)	$k_t \times 10^{-6}$ (L/mol s)	E_t^* (kJ/mol)
Acrylonitrile	60	1960	16.2	782	15.5
Ethylene	83	240	18.4	540	1.3
Methyl acrylate	60	2090	29.7	9.5	22.2
Methyl methacrylate	60	515	26.4	25.5	11.9
Styrene	60	176	26.0	72	8.0
Vinyl acetate	50	2640	30.6	116.8	21.9
Vinyl chloride	50	11,000	16	2100	17.6
Tetrafluoroethylene	40	7400	17.4	7.4×10^{-5}	13.6

^{a} Data from Korus and O'Driscoll.[29a]

Substituting the expression for [$M\cdot$] derived previously,

$$\bar{\nu} = \frac{k_P[M]}{2(fk_tk_d[I])^{1/2}}$$

Kinetic chain length is thus seen to be related to a variety of rate and concentration parameters. It should be noted that $\bar{\nu}$ will decrease as both the initiator concentration and initiator efficiency increase. This is entirely reasonable because increasing the number of growing chains increases the probability of termination. Thus, varying initiator concentration is one way of controlling molecular weight.

In the absence of any side reactions, kinetic chain length is related directly to degree of polymerization (and hence number average molecular weight) according to the mode of termination. If termination occurs exclusively by disproportionation, $\overline{DP} = \bar{\nu}$; if it occurs by coupling, $\overline{DP} = 2\bar{\nu}$.

While these kinetics expressions apply to a great many vinyl polymerization reactions, deviations are common. One type of deviation, variously referred to as the *gel effect*, the *Trommsdorff effect*, or the *Norris–Smith effect*, occurs in bulk or concentrated solution polymerizations when the medium viscosity becomes very high, or in solution polymerizations when the polymer precipitates. In the more viscous medium, chain mobility is reduced, with the result that chain-end radicals have a lower probability of being in position to effect termination. Because the small monomer molecules can still diffuse to the active chain ends even as the termination rate decreases, there is a marked increase in polymerization rate, or autoacceleration. Because the increase in rate is usually accompanied by an increase in reaction exotherm, autoacceleration may cause processing difficulties, particularly in bulk polymerizations. Higher molecular weights also result from the decrease in termination rate.

Deviations in predicted kinetic behavior also arise from *chain transfer reactions*—the transfer of reactivity from the growing polymer chain to another species. Because this process terminates the chain, but at the same time generates a new radical, chain transfer reactions result in lower-molecular-weight product. Since chain transfer reactions invariably arise in most free radical vinyl polymerizations, relatively broad molecular weight distributions are common. Let us examine some of the known chain transfer reactions.

A chain-end radical may abstract a hydrogen atom from a chain, leading to a reactive site for chain branching (6.32). Hydrogen abstraction may also occur intramolecularly, a process re-

$$\text{\tiny www}CH\cdot + \text{\tiny www}CH_2CH\text{\tiny www} \longrightarrow \text{\tiny www}CH_2 + \text{\tiny www}CH_2\dot{C}\text{\tiny www}$$
$$\underset{Y}{|} \qquad \underset{Y}{|} \qquad \underset{Y}{|} \qquad \underset{Y}{|}$$

$$\begin{array}{c} CH_2\dot{C}HY \\ | \\ \text{\tiny www}CH_2C\text{\tiny wwwww} \xleftarrow{\ CH_2=CHY\ } \\ | \\ Y \end{array} \qquad (6.32)$$

ferred to as *backbiting*. Polyethylene, produced by high-pressure free radical polymerization of ethylene, is highly branched, and most of the branches contain a small number of carbon atoms. This has been attributed to backbiting involving five- or six-membered cyclic transition states (6.33). These two types of chain transfer are very important because they lead to branched polymers whose properties may differ markedly from those of the corresponding linear polymers.

$$\begin{array}{ccccc} & CH_2\!-\!CH_2 & & & CH_2\!-\!CH_2 \\ & / \qquad \backslash & & & / \qquad \backslash \\ \text{\tiny www}CH & \quad CH_2 & \longrightarrow & \text{\tiny www}\dot{C}H & \quad CH_2 \\ & \backslash \qquad / & & & \backslash \qquad / \\ & H \rightsquigarrow \cdot CH_2 & & & CH_3 \end{array} \qquad (6.33)$$

Chain transfer may also occur with initiator (6.34) or monomer (6.35), or it may take place with solvent.

$$\text{\tiny wwww}CH_2\dot{C}H\!-\!\!\!\underset{Y}{|} \overbrace{\begin{array}{l} \xrightarrow{\ ROOR\ } \text{\tiny www}CH_2\underset{Y}{\overset{|}{C}}HOR + RO\cdot \qquad\qquad (6.34) \\[2em] \xrightarrow{\ CH_2=CHY\ } \text{\tiny www}CH_2CH_2 + CH_2\!\!=\!\!\underset{Y}{\overset{|}{\dot{C}}}Y \quad (6.35) \end{array}}$$

Polystyrene prepared in carbon tetrachloride, for example, contains chlorine at the chain ends as a result of chlorine transfer (6.36) and initiation by the resultant $\cdot CCl_3$ radicals (6.37). Transfer to monomer is particularly important with monomers containing allylic hydrogen, such as propylene, because the formation of resonance-stabilized allylic radicals (6.38) is highly favorable. Thus, high-molecular-weight polypropylene cannot be prepared under the usual conditions of free radical polymerization. This is not a problem with methyl methacrylate, which also contains allylic

$$\text{\tiny www}CH_2\underset{Ph}{\overset{|}{C}}H\cdot + CCl_4 \longrightarrow \text{\tiny www}CH_2\underset{Ph}{\overset{|}{C}}HCl + \cdot CCl_3 \qquad (6.36)$$

$$CH_2\!\!=\!\!\underset{Ph}{\overset{|}{C}}H + \cdot CCl_3 \longrightarrow Cl_3CCH_2\underset{Ph}{\overset{|}{C}}H\cdot \ etc. \qquad (6.37)$$

$$\underset{}{\overset{CH_3}{|}} \qquad\qquad\qquad \underset{}{\overset{CH_3}{|}}$$
$$\text{\tiny www}CH_2CH\cdot + CH_3CH\!\!=\!\!CH_2 \longrightarrow \text{\tiny www}CH_2CH_2 + [CH_2\!\!\cdots\!\!CH\!\!\cdots\!\!CH_2]\cdot \qquad (6.38)$$

hydrogens, because addition to monomer to form a resonance-stabilized radical is even more favorable.

Most important from the standpoint of molecular weight control is transfer to a modifier, such as a thiol (6.39), a compound with a high affinity for hydrogen transfer. In such

$$\text{wwCH}_2\text{CH} \cdot + \text{RSH} \longrightarrow \text{wwCH}_2\text{CH}_2 + \text{RS} \cdot \qquad (6.39)$$
$$\quad\quad\; | \qquad\qquad\qquad\qquad\qquad | $$
$$\quad\quad Y \qquad\qquad\qquad\qquad\qquad Y$$

cases it is necessary to redefine the kinetic chain length as being the ratio of propagation rate to the combined rates of termination and transfer; that is,

$$\bar{\nu}_{tr} = \frac{R_p}{R_t + R_{tr}}$$

Since transfer reactions are second order, with

$$R_{tr} = k_{tr}[M\cdot][T]$$

where T is the transfer agent, we can rewrite the expression for $\bar{\nu}_{tr}$, taking into account all possible transfer reactions, as

$$\bar{\nu}_{tr} = \frac{k_p[M][M\cdot]}{2k_t[M\cdot]^2 + \Sigma k_{tr}[M\cdot][T]} = \frac{k_p[M]}{2k_t[M\cdot] + \Sigma k_{tr}[T]}$$

Remembering that

$$\bar{\nu} = \frac{k_p[M]}{2k_t[M\cdot]}$$

we can write the reciprocal of the above expression for $\bar{\nu}_{tr}$ as

$$\frac{1}{\bar{\nu}_{tr}} = \frac{1}{\bar{\nu}} + \frac{\Sigma k_{tr}[T]}{k_p[M]}$$

The ratio of the transfer rate constant to that of propagation is commonly defined as the chain transfer constant, C_T, for a particular monomer:

$$\frac{k_{tr}}{k_p} = C_T$$

Substituting, we have

$$\frac{1}{\bar{\nu}_{tr}} = \frac{1}{\bar{\nu}} + \frac{\Sigma C_T[T]}{[M]}$$

Clearly, as the rate of transfer and concentration of the transfer agent increase, the kinetic chain length becomes progressively smaller. Chain transfer constants for a number of compounds and monomers are available in the polymer literature,[29b] several of which are given in Table 6.5. In cases where the highly effective thiols are used to control the kinetic chain length, one can, to a first approximation, ignore other transfer processes to determine the amount of transfer agent needed to obtain a given molecular weight. When the concentra-

TABLE 6.5. Representative Chain Transfer Constants, C_T, for Styrene and Methyl Methacrylate[a]

	$C_T \times 10^4$	
Transfer Agent	*Styrene*	*Methyl Methacrylate*
Benzene	0.023	0.040
Toluene	0.125	0.20
Chloroform	0.5	1.77
Carbon tetrachloride	90	2.40
Carbon tetrabromide	22,000	2700
1-Butanethiol	210,000	6600

[a] Reaction temperature 60°C. Data from Young.[29b]

tion of transfer agent is high and k_{tr} is much greater than k_p, very-low-molecular-weight polymers called *telomers* are obtained. Then the process is called *telomerization.*

Chain transfer reactions can also be used to *prevent* free radical polymerizations. One type of compound, added as a stabilizer to vinyl monomers, is an alkylated phenol (**16**), which can transfer its phenolic hydrogen to form a new radical (6.40) that undergoes coupling reactions (6.41) rather than initiating polymerization. Such compounds, called *inhibitors*, are commonly added to monomers to prevent premature polymerization during shipment or storage. Before the monomer can be used it must be distilled, or the inhibitor must be removed by extraction. Alternatively, sufficient initiator must be added to "use up" the

(6.40)

(6.41)

inhibitor before polymerization can occur. The period during which inhibitor is consumed is known as the *induction period*. Induction periods are common even with purified monomer because of the presence of oxygen, which is, itself, an inhibitor. A variety of compounds are used commercially as inhibitors.

Recently it has been demonstrated that free radical polymerization can actually be made to occur in the *absence* of chain termination or chain transfer. In such cases the radical chain ends remain active when all the monomer is used up. Addition of more monomer results in a further increase in molecular weight or, if the added monomer is different from the starting monomer, in formation of a block copolymer. Polymers possessing still-active chain ends are called *living polymers*. Although recent to free radical polymerization, living polymerization is by no means a new concept; it was first reported in the 1950s by anionic polymerization and, as we shall see in Chapter 7, has been exploited commercially in the manufacture of block copolymers.

One example of living free radical polymerization, termed *atom transfer radical polymerization*,[30] is the polymerization of styrene using 1-chloro-1-phenylethane (**17**, or the bromo analog) as initiator in the presence of a copper(I) bipyridyl* (bpy) complex. Initiation occurs when a halogen atom is transferred from **17** to the complex (6.42). The resultant 1-phenylethyl radical in turn adds to a styrene molecule (6.43). The halogen atom is then reversibly transferred to the styryl radical (6.44), thus preventing radical termination reactions from occurring while allowing the propagation reaction to proceed.

$$
\begin{array}{cc}
CH_3CHCl + Cu(I)(bpy) \rightleftarrows CH_3CH\cdot + Cu(II)(bpy)Cl & (6.42) \\
\quad\ | | & \\
\quad Ph Ph & \\
\quad \mathbf{17} & \\
\end{array}
$$

$$
\begin{array}{cc}
CH_3CH\cdot + CH_2{=}CH \rightarrow CH_3CHCH_2CH\cdot & (6.43) \\
\quad\ | | | | & \\
\quad Ph Ph Ph Ph & \\
\end{array}
$$

$$
\begin{array}{cc}
CH_3CHCH_2CH\cdot + Cu(II)(bpy)Cl \rightleftarrows CH_3CHCH_2CHCl + Cu(I)(bpy) & (6.44) \\
\quad\ | | | | & \\
\quad Ph Ph Ph Ph & \\
\end{array}
$$

Living polymers can also be prepared by adding a stable free radical such as a nitroxide to the polymerization recipe. An example is the polymerization of styrene with benzoyl peroxide in the presence of 2,2,6,6-tetramethylpiperidinyl-1-oxy (TEMPO) (**18**).[31] TEMPO is too stable to initiate polymerization, but it does promote decomposition of benzoyl peroxide to benzoyloxy radicals which initiate chain growth. TEMPO then combines reversibly with the growing polystyrene chain ends (6.45), protecting the chain end radicals from termination. Dissociation in turn frees the chain end radical to add to other styrene molecules.

*Bipyridyl=

$$\text{wwCH}_2\text{CH}\cdot + \cdot\text{O}-\text{N} \rightleftharpoons \text{wwCH}_2\text{CH}-\text{O}-\text{N} \qquad (6.45)$$

18

An important consequence of living polymerization is that the average degree of polymerization is simply equal to the ratio of the initial monomer concentration to the initiator concentration:

$$\overline{\text{DP}} = \frac{[M]_0}{[I]_0}$$

Furthermore, because all the chains are initiated at about the same time and there are no chain termination or transfer reactions, the chains all grow to approximately the same length. Thus low polydispersities are a hallmark of living polymerizations. Polydispersities as low as 1.05 have been reported using atom transfer radical polymerization.[32]

Before leaving the topic of kinetics and mechanism, let us reconsider the case of emulsion polymerization. Unlike the case with free radical polymerization in bulk or solution, increasing rates of emulsion polymerization usually lead to *higher* molecular weights, especially when emulsifier concentration is increased. As was discussed previously under bulk polymerization, increasing the rate of formation of initiator radicals—which increases polymerization rate—also increases the rate of termination. With emulsion polymerization, on the other hand, polymerization rate competes with the rate of diffusion of terminator radicals into the micelles where polymerization is taking place. Once termination occurs, a new radical enters the micelle and polymerization begins again and continues until another terminator radical enters. And so the reaction continues. Assuming a constant rate of diffusion, therefore, each particle contains a growing radical half the time. The overall polymerization rate depends on the polymerization rate within each particle, as well as the number of particles. According to the kinetics treatment originally developed by Smith and Ewart (appropriately called the Smith–Ewart kinetics),[33]

$$R_p = k_p[M]\left(\frac{N}{2}\right)$$

where k_p is the overall polymerization rate constant, $[M]$ is the concentration of monomer in the micelle particles, and N is the number of particles. The factor, $N/2$, which takes into account that only half the particles contain a growing chain at any given time, is equal to the concentration of radicals, $[M\cdot]$. Because the rate of polymerization within each particle is equal to $k_p[M]$, and the rate of termination is equal to the rate of terminator radical entry into the particle, ρ, the average kinetic chain length is given by,

$$\overline{\nu} = \frac{k_p[M]}{\rho}$$

Since ρ is equal to the rate of initiator radical formation divided by the number of particles, the above expression may be rewritten:

$$\bar{\nu} = \frac{k_p[M]}{R_i/N} = \frac{k_p[M]N}{2fk_d[I]}$$

The terminator radical is derived from the initiator; therefore, the molecular weight is not significantly altered by termination, and $\bar{\nu} = \overline{DP}$.

From the above relationship we can see why the molecular weight increases with polymerization rate. If the initiator concentration remains constant, the polymerization rate can be enhanced by adding more emulsifier, which increases the number of micelles, which in turn increases both R_p and $\bar{\nu}$.

6.5   Stereochemistry of Polymerization

Addition of radicals to monomers is not as simple a process as is implied in the previous section, because the stereochemistry of addition can profoundly affect the ultimate properties of the polymer.[34] It will be recalled that for mnosubstituted and 1,1-disubstituted monomers (the two substituents of the latter being different), two stereoregular arrangements are possible in the head-to-tail polymer: *isotactic*, with like substituents all on the same side, and *syndiotactic*, with like substituents of alternate repeating units on the same side. A random arrangement is referred to as *atactic*. (See Figure 3.10, Chapter 3.) There are two factors to consider in describing the stereochemistry of addition of a monomer molecule to a radical chain end: (1) interaction between the terminal chain carbon and the approaching monomer molecule, and (2) the configuration of the penultimate repeating unit in the polymer chain.

Consider first the approach of monomer. The terminal carbon atom presumably has sp^2 hybridization and is planar. For polymerization of monomer $CH_2=CXY$, there are two ways that a monomer molecule can approach the terminal carbon: the mirror approach, with like substituents on the same side (6.46), or the nonmirror approach, with like substituents on opposite sides (6.47). As long as stereochemistry is maintained—that is, free rotation does not occur before the next monomer molecule adds—then the mirror approach (6.46) will always lead to isotactic polymer and the nonmirror approach (6.47) to syndiotactic polymer. If, on the other hand, interactions between the substituents of the penultimate repeating unit and the terminal carbon atom are significant, then conformational factors could also have an influence and the mode of addition could be governed by the conformation that allows the least steric or electrostatic interaction.

$$(6.46)$$

$$(6.47)$$

It was believed for many years that with free radical polymerization these effects were not significant and that only atactic polymer was formed. This was not unreasonable because free radical processes lack the significant steric and electrostatic factors associated with complexes in Ziegler–Natta polymerization or with counterions in ionic polymerization. In 1958, however, it was reported that free radical polymerization of methyl methacrylate at temperatures below 0°C gave crystalline polymer,[35] and this was subsequently shown to be predominantly syndiotactic by the use of high-resolution NMR.[36] Since this initial work, considerable interest has developed in stereoregular free radical polymerization.

One model[37] for the methyl methacrylate polymerization that takes into account penultimate unit interactions and that allows for the least steric hindrance between substituent groups during reaction with monomer is shown in reaction (6.48). (The polymer chain, P, is the bulkiest group.)

$$(6.48)$$

In such an arrangement the radical receives maximum exposure for reaction with monomer. The newly formed radical then assumes the preferred conformation with another monomer molecule, and so the process repeats itself to give the syndiotactic structure. This model assumes an energy barrier to rotation that is considerably higher than that of propagation; otherwise a random distribution would result. This, of course, would be temperature dependent, and as expected, the degree of stereoregularity decreases at higher temperatures.

While the above model satisfactorily explains the formation of syndiotactic poly(methyl methacrylate), it does not take into account possible interaction between chain-end radical and monomer. Using calculations of interaction energies that take into account such factors as distance of separation, angle of approach, and point of radical addition (at the carbon atom or at the π orbital of the double bond), other researchers have concluded that interaction between monomer and terminal unit is the predominant factor resulting in syndiotactic placement as long as the terminal carbon has the sp^2 planar structure.[38] This is the nonmirror arrangement shown in reaction (6.47). If, however, the terminal radical assumes a substantial amount of sp^3 hybridization, which it might do as it enters the transition state, then the resultant change in configuration could cause interactions with the penultimate unit to be of greater significance. Whichever is the case, syndiotactic placement is favored.

Apart from poly(methyl methacrylate), numerous other polymers prepared by free radical initiation have been shown to have varying degrees of syndiotacticity, but no clear-cut trends can be observed.[34] It would seem that increasing the size of substitutent groups would increase the degree of stereoregularity, yet poly(2,4,6-triphenylbenzyl methacrylate) (19) appears to be *less* syndiotactic than poly(methyl methacrylate) prepared under the same conditions. No doubt a combination of steric and polar effects is responsible for the stereochemistry of free radical polymerization, but more work needs to be done before a definitive

19

mechanism can be written. As will be seen in the following chapters, ionic and coordination polymerization can lead to much greater stereochemical control.

6.6 Polymerization of Dienes

6.6.1 Isolated Dienes

Polymerization of isolated dienes leads to crosslinked polymers when the double bonds react independently of one another. In certain cases, cooperative addition reactions of one double bond to the other of the monomer lead to cyclic polymers.[39] This is an example of *cyclopolymerization*—the formation of a cyclic structural unit in a propagation step.[40] Cyclopolymerization of divinylformal (20) (6.49) illustrates this process:

(6.49)

20

The normal (head-to-tail) mode of addition yields the six-membered ring, whereas abnormal (head-to-head) addition (6.50) gives the five-membered ring. The former is preferred, but both are formed in significant amounts.

$$(6.50)$$

Several diallyl monomers are used commercially to prepare highly cross-linked thermosetting *allyl resins* or for crosslinking other polymers.[41] Among the most important are diallyl phthalate (**21**) and diethylene glycol bis(allyl carbonate) (**22**). Monomer **21** is used primarily for manufacturing electrical or electronics items (circuit boards, insulators, television components, etc.) or for preimpregnating glass cloth or fiber for fiber-reinforced plastics. Monomer **22** is used to make polymers for applications requiring good optical clarity, such as eyeware lenses, camera filters, panel covers, and the like.

21   22

6.6.2 Conjugated Dienes

Conjugated dienes such as 1,3-butadiene (**23**) undergo both 1,2- and 1,4-addition. Addition polymerization thus leads by analogy to 1,2- and 1,4-addition polymers via the delocalized radical intermediate **24** (6.51). Thus, 1,2-addition gives polymer (**25**) with pendant vinyl groups, while 1,4-addition leads to polymer with unsaturation in the chain. In the latter case, both *cis* (**26**) and *trans* (**27**) configurations are possible.

$$CH_2{=}CH{-}CH{=}CH_2 \xrightarrow{\text{R·}} [RCH_2{-}CH{=\!=\!=}CH{=\!=\!=}CH_2]\cdot \qquad (6.51)$$

23   24

25   26   27

In general free radical polymerization of 1,3-butadiene results in approximately 20% of the 1,2 structure, while most of the 1,4 polymer has the *trans* structure. As the polymerization temperature is increased, the amount of *cis*-1,4 increases, but the yield of 1,2 polymer remains about the same.

With substituted dienes like isoprene (2-methyl-1,3-butadiene) the situation is more complicated; 1,2 and 3,4 structures (**28** and **29**, respectively) as well as *cis*- and *trans*-1,4 (**30** and **31**), are possible. All are formed in free radical polymerization, but the head-to-tail *trans*-1,4 (**31**) predominates. Natural rubber (discussed in Chapter 17) is a polyterpene consisting of head-to-tail 1,4-polyisoprene units, with the most important type, *Hevea* rubber, having the *cis* structure (**30**).

Table 6.6 lists the percentages of each structural unit at different polymerization temperatures for the three most important diene monomers—butadiene, isoprene, and chloroprene (2-chloro-1,3-butadiene). Several points are worth noting. In the case of isoprene, the amount of *cis*-1,4 structure increases with increasing temperature up to about 100°C, then it decreases. The total amount of 1,2 and 3,4 polymer (approximately 10%) does not change appreciably with temperature; however, the 3,4 structure is favored at higher temperatures. With chloroprene, the *trans*-1,4 structure also decreases with increasing temperature, but the amount of *trans*-1,4 is slightly higher than that of isoprene over the temperature range measured.

TABLE 6.6. Structure of Free Radical–Initiated Diene Polymers[a]

| | | *Percent* | | | |
Monomer	Polymerization Temperature (°C)	cis-1,4	trans-1,4	1,2	3,4
Butadiene	−20	6	77	17	—
	20	22	58	20	—
	100	28	51	21	—
	233	43	39	18	—
Isoprene	−20	1	90	5	4
	−5	7	82	5	5
	50	18	72	5	5
	100	23	66	5	6
	257	12	77	2	9
Chloroprene	−46	5	94	1	0.3
	46	10	81–86	2	1
	100	13	71	2.4	2.4

[a] Data from Cooper[34] p. 275.

Simple diene monomers exist mainly in the *s-trans* conformation. It is believed that stereochemistry is retained during the polymerization process, which accounts for the predominance of *trans* structure.[34] Thus at higher temperatures, as the proportion of *s-cis* conformation increases, the yield of *cis* polymer increases proportionately. Support for this is gained from the fact that 2-(tri-*n*-butyltin)butadiene (**32**), which, because of the large size of the substitutent group, exists almost entirely in the *s-cis* conformation, gives almost 100% *cis*-1,4 polymer at 60°C.

$$CH_2\!\!=\!\!CH$$
$$|$$
$$C-Sn(n\text{-}C_4H_{10})_3$$
$$CH_2\!\!=\!\!$$

32

6.7 Monomer Reactivity

Consideration of free energies of polymerization,[42] that is,

$$\Delta G_p = \Delta H_p - T\Delta S_p$$

suggests that almost any compound containing a double bond is capable of being transformed into polymer. The enthalpies and entropies of polymerization for several common monomers (i.e., the difference in entropy and enthalpy between monomer and polymer) are listed in

TABLE 6.7. Representative Enthalpies, ΔH, and Entropies, ΔS, of Polymerization[a]

Monomer	$-\Delta H$ (kJ/mol)	$-\Delta S$ (J/mol)
Acrylonitrile[b]	77	109
1,3-Butadiene	78	89
Ethylene	109	155
Isoprene	75	101
Methyl methacrylate	65	117[c]
Propylene	84	116
Styrene	70	104
Tetrafluoroethylene	163	—
Vinyl acetate	90	—
Vinyl chloride	71	—

[a] Values selected from Ivin.[29c] Polymerization temperature 25°C unless otherwise indicated.

[b] 74.5°C.

[c] 127°C.

Table 6.7.[29c] Clearly the enthalpy of reaction is very favorable in each case, more than enough to offset the unfavorable entropy effect. A favorable free energy of polymerization, however, does not guarantee a favorable polymerization reaction under any set of conditions. Neither propylene nor isobutylene, for example, undergoes normal free radical polymerization to any extent. Polarity and size of the substituent groups on the double bond as well as the propensity for chain transfer determine the kinetic feasibility of polymerizing a particular monomer.

In free radical polymerizations two factors need to be considered in any discussion of monomer reactivity: the stability of the monomer toward addition of a free radical, and the stability of the monomer radical thus formed. Of the two, the latter is of overriding importance, as shown by the fact that styrene is particularly susceptible to polymerization despite the fact that the vinyl group is conjugated with a benzene ring. Stabilization of the resultant radical (**33**) thus reduces the activation energy for attack by initiator radicals.

33

Nitrile groups are also very effective in stabilizing radicals, as shown by the addition of ethyl radicals to monomers in the gas phase,[43] where the rate order is acrylonitrile > styrene > vinyl acetate. This does not necessarily mean the same order applies in initiation of polymerization, however. Reaction of monomers with ^{14}C-labeled benzoyloxy radicals has been studied[44] by comparing the rates of decarboxylation (6.52) and addition to monomer (6.53).

$$Ph^{14}CO_2\cdot \rightarrow Ph\cdot + {}^{14}CO_2 \qquad (6.52)$$

$$Ph^{14}CO_2\cdot + M \rightarrow Ph^{14}CO_2M\cdot \qquad (6.53)$$

The more reactive the monomer, the greater the amount of initiation by $Ph^{14}CO_2\cdot$ relative to initiation by $Ph\cdot$, hence the more ^{14}C incorporated into the polymer. Such studies indicate an order: styrene > vinyl acetate > acrylonitrile. Clearly polar effects are important in reducing the reactivity of acrylonitrile toward benzoyloxy radicals.

Propagation rate data (Table 6.4) also reflect, to a degree, the inverse relationship between monomer stability and polymerization rate. Vinyl acetate, a very stable monomer, has one of the highest rate constants, the reverse of the situation with styrene. Both steric and polar effects play such an important role in monomer reactivity, however, that it is impossible to make clear-cut generalizations. Acrylonitrile, for example, has a relatively high rate of polymerization despite the stabilizing effect of the nitrile group. The lower rate constant of methyl methacrylate compared with methyl acrylate may be attributed to additional resonance stabilization of the intermediate radical by hyperconjugation or increased steric hindrance, or both.

Steric effects are of particular significance with 1,2-disubstituted monomers,[45] which are difficult to polymerize under ordinary free radical conditions. (Exceptions are fluorine-containing monomers.) This is not because of any increased steric interaction in the polymer. On the contrary, polymers derived from 1,2-disubstituted monomers are generally more thermodynamically stable than those derived from 1,1-disubstituted monomers of analogous structure. Rather, the reduced activity appears to arise from steric interactions between substituent groups of the approaching monomer molecule and the substituent on the carbon adjacent to the chain end radical (**34**).

34

Evidence to support this hypothesis arises from copolymerization studies. Monosubstituted or 1,1-disubstituted monomers, which have only hydrogen atoms attached to the second carbon of the double bond, usually copolymerize readily with 1,2-disubstituted monomers.

One other consideration regarding monomer reactivity involves the reversibility of vinyl polymerization reactions; that is,

$$\text{M}_x\cdot + \text{M} \underset{k_{dp}}{\overset{kp}{\rightleftharpoons}} \text{M}\cdot_{(x+1)}$$

where k_{dp} is the depropagation rate constant. Just as radical stability retards propagation, it enhances depropagation. As the polymerization temperature is raised, the depropagation rate increases until a point is reached where the forward and back reactions are equal. The temperature at which this occurs is called the *ceiling temperature* (T_c), and at that temperature, ΔG of polymerization is zero, or

$$T_c = \frac{\Delta H}{\Delta S}$$

It follows that whatever affects ΔH or ΔS for a given monomer (e.g., monomer concentration, pressure) will also affect T_c. Table 6.8 lists ceiling temperatures for several pure liquid monomers.[46]

TABLE 6.8. Representative Ceiling Temperatures, T_c, of Pure Liquid Monomers[a]

Monomer	T_c (°C)
1,3-Butadiene	585
Ethylene	610
Isobutylene	175
Isoprene	466
Methyl methacrylate	198
α-Methylstyrene	66
Styrene	395
Tetrafluoroethylene	1100

[a]Data from Allcock and Lampe.[46]

Because propagation and depropagation are in equilibrium, there will always be some residual monomer in the polymer. With most common vinyl monomers, however, ceiling temperatures are sufficiently high that the amount of residual monomer is extremely small. A notable exception is α-methylstyrene (**35**), which has a T_c of about 66°C in bulk polymerization.

$$\text{CH}_3$$

a structure showing a benzene ring attached to $\overset{\text{CH}_3}{\underset{}{\text{C}}}{=}\text{CH}_2$

35

Steric effects arising from having bulky methyl and phenyl groups located together on alternate carbon atoms of the polymer, and the relative stability of the tertiary benzylic α-methylstyryl radical, combine to give α-methylstyrene its low T_c. In cases where the T_c is very high (polyethylene or polytetrafluoroethylene, for example), polymers degrade by bond cleavage reactions other than depropagation at temperatures well below the T_c.

6.8 Copolymerization

The mechanism of copolymerization[47] is analogous to that of homopolymerization, but reactivities of the various monomers toward other monomers can vary considerably. Already mentioned in the previous section is the case of 1,2-disubstituted ethylenes, which, in most instances, fail to homopolymerize, yet enter into copolymerization reactions readily with a variety of mono- or 1,1-disubstituted monomers.

Consider the case where two monomers, M_1 and M_2, copolymerize. The monomers can undergo either *self-propagation* ($M_1\cdot$ reacting with M_1, or $M_2\cdot$ reacting with M_2) or *cross-propagation* ($M_1\cdot$ reacting with M_2, or $M_2\cdot$ reacting with M_1). This results in four separate reactions, each with its own rate constant:

$$M_1\cdot + M_1 \xrightarrow{k_{11}} M_1\cdot$$
$$M_1\cdot + M_2 \xrightarrow{k_{12}} M_2\cdot$$
$$M_2\cdot + M_1 \xrightarrow{k_{21}} M_1\cdot$$
$$M_2\cdot + M_2 \xrightarrow{k_{22}} M_2\cdot$$

If we assume that the concentrations of $M_1\cdot$ and $M_2\cdot$ remain constant (steady-state treatment), then the rate of addition of $M_1\cdot$ to M_2 will equal the rate of addition of $M_2\cdot$ to M_1; that is,

$$k_{12}[M_1\cdot][M_2] = k_{21}[M_2\cdot][M_1]$$

The rate of disappearance of M_1 and M_2 may be expressed by the following:

$$\frac{-d[M_1]}{dt} = k_{11}[M_1\cdot][M_1] + k_{21}[M_2\cdot][M_1]$$

$$\frac{-d[M_2]}{dt} = k_{12}[M_1\cdot][M_2] + k_{22}[M_2\cdot][M_2]$$

The ratio of the two rates is then

$$\frac{d[M_1]}{d[M_2]} = \frac{[M_1]}{[M_2]} \left(\frac{k_{11}[M_1\cdot] + k_{21}[M_2\cdot]}{k_{12}[M_1\cdot] + k_{22}[M_2\cdot]} \right)$$

If we combine this equation with the steady-state expression above and define

$$\frac{k_{11}}{k_{12}} = r_1 \quad \text{and} \quad \frac{k_{22}}{k_{21}} = r_2$$

we derive the *copolymer equation* (also called the *copolymer composition equation*).

$$\frac{d[M_1]}{d[M_2]} = \frac{[M_1]}{[M_2]} \left(\frac{r_1[M_1] + [M_2]}{[M_1] + r_2[M_2]} \right)$$

The ratio $d[M_1]/d[M_2]$ represents the molar ratio of the two monomers in the copolymer, while $[M_1]$ and $[M_2]$ are the initial molar concentration of monomers in the reaction mixture (or *feed*, as the mixture is frequently called.) The terms r_1 and r_2, called the *reactivity ratios*, define the relative tendencies of the monomers to self-propagate or cross-propagate. If $r_1 > 1$, then M_1 tends to self-propagate, whereas if $r_1 < 1$, copolymerization is preferred. Thus if we know the reactivity ratios, we can calculate the initial molar ratio of monomers needed to give a desired ratio of monomeric units in the copolymer, at least in the initial stages of copolymerization (up to about 10%). In most cases, one monomer is incorporated in preference to the other; therefore, the monomer composition in the feed changes as the copolymerization reaction proceeds, and the polymer composition changes with increasing conversion. Not infrequently it may be necessary to add one monomer continuously in order to maintain the desired copolymer composition. The validity of the copolymer equation has been confirmed experimentally.

An alternative method of expressing the copolymer equation is in terms of mole fractions. Thus f_1 and f_2 are the mole fractions of M_1 and M_2 in the feed, respectively, and F_1 and F_2 are the mole fractions of M_1 and M_2 in the copolymer. Because

$$f_1 = 1 - f_2 = \frac{[M_1]}{[M_1] + [M_2]}$$

and

$$F_1 = 1 - F_2 = \frac{d[M_1]}{d[M_1] + d[M_2]}$$

the copolymer equation may be rewritten

$$F_1 = \frac{r_1 f_1^2 + f_1 f_2}{r_1 f_1^2 + 2f_1 f_2 + r_2 f_2^2}$$

Determination of reactivity ratios[48–50] is no easy task, requiring analysis of copolymer composition in the early stages of reaction for a variety of monomer feed compositions. More recent methods[51,52] have stressed using NMR triad data. Fortunately, large numbers of reactivity ratios have been compiled[29d,47]; several of them are listed in Table 6.9.

TABLE 6.9. Representative Free Radical Monomer (M) Reactivity Ratios (r)[a]

M_1	M_2	r_1	r_2	Temperature (°C)
Styrene	Methyl methacrylate	0.52	0.46	60
Styrene	Acrylonitrile	0.40	0.04	60
Styrene	Vinyl acetate	55	0.01	60
Styrene	Maleic anhydride	0.041	0.01	60
Styrene	Vinyl chloride	17	0.02	60
Styrene	1,3-Butadiene	0.58	1.35	50
Styrene	Isoprene	0.54	1.92	80
Methyl methacrylate	Vinyl chloride	10	0.1	68
Methyl methacrylate	Vinyl acetate	20	0.015	60
Methyl methacrylate	Acrylonitrile	1.20	0.15	60
Methyl methacrylate	1,3-Butadiene	0.25	0.75	90
Ethylene	Tetrafluoroethylene	0.38	0.1	25
Ethylene	Acrylonitrile	0	7	20
Ethylene	Vinyl acetate	0.97	1.02	130

[a]Data selected from Young[29d] (see this reference for polymerization conditions).

Clearly, a knowledge of reactivity ratios is of immense benefit in formulating feed ratios for copolymers. Let us examine several hypothetical situations involving different values of r_1 and r_2. First we consider the situation where

$$r_1 = r_2 = 1.0$$

In this case the monomers would exhibit no preference for homopolymerization or copolymerization, and a truly random copolymer would result. Under these conditions the copolymer equation reduces to

$$F_1 = f_1$$

and a plot of f_1 versus F_1 (Figure 6.1) would show the same composition in both feed and copolymer. A situation approaching this is the copolymerization of ethylene (M_1) and vinyl acetate (M_2) ($r_1 = 0.97$, $r_2 = 1.02$). This is shown by curve A in Figure 6.1, which is very close to the diagonal.

Next we consider the case where

$$r_1 = r_2 = 0$$

Here, neither monomer would exhibit a tendency to homopolymerize, and a truly *alternating* copolymer would result. Then $F_1 = 0.5$. Approaching this situation is the styrene (M_1)–maleic anhydride (M_2) system where $r_1 = 0.041$ and $r_2 = 0.01$. This is shown by curve B in Figure 6.1.

A far more common situation is where both r_1 and r_2 lie between 0 and 1, such as with the styrene (M_1)–methyl methacrylate (M_2) system ($r_1 = 0.52$, $r_2 = 0.46$), shown by curve C in Figure 6.1. In such cases there is some tendency toward alternation. (If $r_1 = 0.5$, M_1 reacts twice as fast with M_2 as with M_1.) The point at which curve C crosses the diagonal represents a composition where polymerization proceeds to relatively high conversion with no

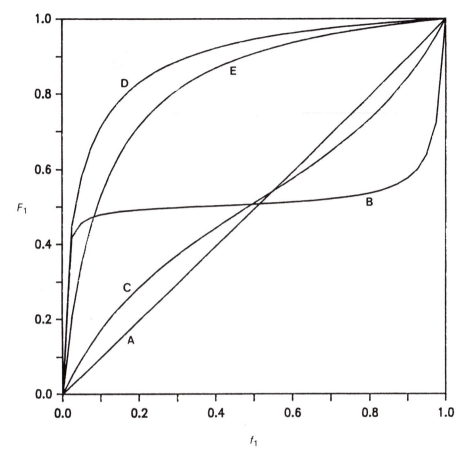

FIGURE 6.1. Variation of F_1, f_1, and r in copolymerization (F_1 = mole fraction of monomer in copolymer; f_1 = mole fraction of monomer in feed; r = reactivity ratio): (A) $r_1 = 0.97$, $r_2 = 1.02$; (B) $r_1 = 0.041$, $r_2 = 0.01$; (C) $r_1 = 0.52$, $r_2 = 0.46$; (D) $r_1 = 55$, $r_2 = 0.01$; (E) $r_1 = 10$, $r_2 = 0.1$.

change in either feed ratio or copolymer composition. Such cases are called *azetropic polymerization*. Since with azeotropic polymerization

$$\frac{d[M_1]}{d[M_2]} = \frac{[M_1]}{[M_2]}$$

the copolymer equation reduces to

$$\frac{[M_1]}{[M_2]} = \frac{1 - r_2}{1 - r_1}$$

under azeotropic conditions. Azeotropic polymerization is also possible, in principle, when both r_1 and r_2 are greater than 1; however, this situation has not been observed in free radical copolymerization.

Now let us consider the case where r_1 is much greater than 1 and r_2 much less than 1; for example, styrene (M_1)–vinyl acetate (M_2) ($r_1 = 55$, $r_2 = 0.01$) or styrene (M_1)–vinyl chloride (M_2) ($r_1 = 17$, $r_2 = 0.02$). Curve D in Figure 6.1 illustrates the former. Such cases lead to incorporation of monomer 1 almost exclusively in the early stages of polymerization. When r_1 is very high and r_2 is close to 0, one obtains essentially a homopolymer of M_1.

It is common among polymer chemists to refer to the situation where $r_1 r_2 = 1$ as *ideal copolymerization* following a parallel with vapor–liquid equilibria in ideal liquid mixtures. One should keep in mind, however, that "ideal" copolymerization in the sense of a random distribution of monomer units in the copolymer occurs only when both r_1 and r_2 are close to 1. The more r_1 and r_2 diverge, the less random will be the distribution. A case in point is the "ideal copolymerization" of methyl methacrylate (M_1) and vinyl chloride (M_2) where $r_1 = 10$ and $r_2 = 0.1$ and $r_1 r_2 = 1$. This is shown with curve E in Figure 6.1.

To simplify the process of determining reactivity ratios, a semiempirical relationship has been developed that expresses the ratios in terms of constants that are characteristic of each monomer but independent of comonomer, thus eliminating the need to determine experimentally the ratios for each pair of monomers in a particular copolymerization. This relationship is called the *Q–e scheme*.[53]

$$r_1 = \left(\frac{Q_1}{Q_2}\right) \exp[-e_1(e_1 - e_2)]$$

$$r_2 = \left(\frac{Q_2}{Q_1}\right) \exp[-e_2(e_2 - e_1)]$$

where Q_1 and Q_2 are measures of the reactivity of monomers M_1 and M_2 and are related to resonance stabilization of monomer. The constants e_1 and e_2 are measures of the polarity of the monomers. Styrene was chosen as the standard and assigned values of $Q = 1.00$ and $e =$

TABLE 6.10. Representative Values for Reactivity (Q) and Polarity (e) of Monomers[a]

Monomer	Q	e
1-Vinylnaphthalene	1.94	−1.12
p-Nitrostyrene	1.63	0.39
p-Methoxystyrene	1.36	−1.11
Styrene	1.00	−0.80
Methyl meth-acrylate	0.74	0.40
Acrylonitrile	0.60	1.20
Methyl acrylate	0.42	0.60
Vinyl chloride	0.044	0.20
Vinyl acetate	0.026	−0.22

[a]Data from Ham.[47]

−0.80. Values of Q increase with increasing resonance stabilization, while e values become less negative as groups attached to the double bond become more electron attracting. A few representative values of Q and e are given in Table 6.10. While the trends in Q and e values are obvious, it should be pointed out that these values are only approximate and provide, at best, a semiquantitative but nevertheless extremely useful means of predicting reactivity ratios where other experimental data are not available. More accurate ways of predicting reactivity ratios have been suggested recently.[54]

Reactivity ratios are the result of a combination of steric, resonance, and polar effects. Steric effects have already been discussed as they apply to 1,2-disubstituted ethylenes. As was mentioned earlier, resonance affects reactivity by stabilization, of the intermediate radical—the more resonance stabilization, the less reactive is the monomer toward propagation. Thus, resonance-stabilized styryl radicals exhibit little tendency to add to vinyl acetate because the resultant radical would not be stabilized. Styrene does, however, copolymerize with methyl methacrylate because the latter, like styrene, forms a delocalized radical (**36**).

36

From the reactivity ratios of the styrene–methyl methacrylate system, it can be seen that each radical has about twice as much tendency to react with its opposite number, a phenomenon attributable to polar effects. The difference in electron-withdrawing or electron-donating effects of the substitutent groups lowers the activation energy for addition of styryl radicals to methyl methacrylate relative to addition to styrene.

Two theories have been proposed to explain those cases where there is a strong tendency toward alternation (r_1 and $r_2 = \sim 0$), one involving polar effects and the other charge-transfer interactions. First, let's consider polar effects as applied to the copolymerization of styrene and maleic anhydride.[47] Addition of maleyl radical to styrene (6.54) leads to a transition state

$$(6.54)$$

that can be stabilized by resonance, including one nonbonded resonance form involving electron transfer. Such structures may be invoked in cases where the monomer contains an abundance of electronegative substituents, as with maleic anhydride. Addition of styryl radical to maleic anhydride (6.55) would give rise to a similarly stabilized transition state. The net effect is that styryl radicals prefer to add to maleic anhydride and maleyl radicals prefer to add to styrene.

$$(6.55)$$

The other interpretation of alternation is based on formation of charge-transfer complexes between comonomers.[55] Such complexes can be detected by observation of characteristic ultraviolet absorption maxima in solutions of monomers. According to this theory, an alternating copolymerization reaction is, in effect, a *homopolymerization of the charge-transfer complex*. Thus the propagating chain end is not a conventional radical, nor is it an ionic species, but rather it is a donor (D)–acceptor (A) complex that reacts preferentially with other charge-transfer complexes (6.56). There is considerable evidence favoring this theory:

$$\text{w(DA)}_n\overset{+}{\text{D}}..\overset{-}{\text{A}} + \overset{+}{\text{D}}..\overset{-}{\text{A}} \longrightarrow \text{w(DA)}_{n+1}\overset{+}{\text{D}}..\overset{-}{\text{A}} \qquad (6.56)$$

1. The rate of polymerization reaches a maximum when the monomer composition ratio is about 1:1, which is where the donor–acceptor complex is at maximum concentration.

2. Alternation is independent of monomer feed ratios, and other reactive monomers included with the feed fail to react while alternating copolymer is forming.

3. The rate is enhanced by addition of Lewis acids, which increase the acceptor properties of one of the monomers.[56,57]

4. Chain transfer agents have little effect on the molecular weight of the copolymer.

Examples of other compounds that undergo free radical-initiated copolymerization with vinyl monomers are carbon monoxide and sulfur dioxide, the former to yield polyketones (6.57) and the latter polysulfones (6.58). These reactions also tend toward alternation, and sulfur dioxide is known to form 1:1 complexes with alkenes.

$$\underset{\text{CH}_2=\text{CH}}{\overset{\text{R}}{|}} + \text{CO} \longrightarrow \left[\text{CH}_2-\underset{|}{\overset{\text{R}}{\text{CH}}}-\overset{\text{O}}{\overset{||}{\text{C}}}\right] \qquad (6.57)$$

$$CH_2{=}CH + SO_2 \longrightarrow \left[CH_2{-}CH{-}SO_2 \right] \quad \quad (6.58)$$

(with R substituent on CH)

Which is responsible for alternation in free radical copolymerization—polar effects in the transition state or charge-transfer complexes, or both—is still open to question[58] despite the volume of literature published on the subject.

References

1. G. E. Ham (Ed.), *Vinyl Polymerization*, Part 1, Marcel Dekker, New York, 1967.
2. G. Moad and D. H. Solomon, *The Chemistry of Free Radical Polymerization*, Elsevier, Tarrytown, New York, 1995.
3. W. A. Pryor, *Free Radicals*, McGraw-Hill, New York, 1966.
4. S. Patai (Ed.), *The Chemistry of Peroxides*, Wiley-Interscience, New York, 1983.
5. N. Inagaki, *Plasma Surface Modification and Plasma Polymerization*, Technomic, Lancaster, Pennsylvania, 1995.
6. M. R. Wertheimer, H. R. Thomas, M. J. Perri, J. E. Klemberg-Sapieha, and L. Martinu, *Pure Appl. Chem.*, **68**, 1047 (1996).
7. M. Yasuda, *Plasma Polymerization*, Academic, San Diego, 1985.
8. C. Walling, *Free Radicals in Solution*, Wiley-Interscience, New York, 1957.
9. W. A. Pryor and W. H. Hendrickson Jr., *Tetrahedron Lett.*, **24**, 1459 (1983).
10. J.-P. Fouassier, *Photoinitiation, Photopolymerization, and Photocuring*, Hanser, Cincinnati, 1995.
11. G. E. Ham, in *Vinyl Polymerization*, Part 1 (G. E. Ham, Ed.), Marcel Dekker, New York, 1967, Chap. 1.
12. W. D. Graham, J. G. Green, and W. A. Pryor, *J. Org. Chem.*, **44**, 907 (1979).
13. A. Husain and A. H. Hamielec, *J. Appl. Polym. Sci.*, **22**, 1207 (1978).
14. O. F. Olaj, H. F. Kauffmann, and J. W. Breitenbach, *Makromol. Chem.*, **177**, 3065 (1976).
15. H. F. Kauffmann, *Makromol. Chem.*, **180**, 2649, 2665, 2681 (1981).
16. B. L. Funt, in *Encyclopedia of Polymer Science and Engineering* (H. F. Mark, N. M. Bikales, C. G. Overberger, G. Menges, and J. I. Kroschwitz, Eds.), Vol. 5, Wiley-Interscience, New York, 1986, pp. 587–601.
17. C. E. Schildknecht, *Polymer Processes*, Wiley-Interscience, New York, 1956.
18. A. D. Jenkins, in *Vinyl Polymerization*, Part 1 (G. E. Ham, Ed.), Marcel Dekker, New York, 1967, Chap. 6.
19. S. Yamamoto and M. Tatsumi, in *Encyclopedia of Polymer Science and Engineering* (H. F. Mark, N. M. Bikales, C. G. Overberger, G. Menges, and J. I. Kroschwitz, Eds.), Vol. 13, Wiley-Interscience, New York, 1988, pp. 453–63.
20. K. A. Shaffer and J. M. DeSimone, *Trends Polym. Sci.*, **3**, 146 (1995).
21. D. C. Blakely, *Emulsion Polymerization—Theory and Practice*, Applied Science, London, 1975.
22. D. R. Bassett and A. E. Hamielec (Eds.), *Emulsion Polymers and Emulsion Polymerization*, ACS Symp. Ser. 165, American Chemical Society, Washington, DC., 1981.
23. W. Cooper, in *Reactivity, Mechanism and Structure in Polymer Chemistry* (A. D. Jenkins and A. Ledwith, Eds.), Wiley-Interscience, New York, 1974, Chap. 5.
24. C. W. Wilson III and E. R. Santee, Jr., *J. Polym. Sci. C*, **8**, 97 (1965).
25. A. Matsumoto, K. Iwanami, and M. Oiwa, *J. Polym. Sci., Polym. Lett. Ed.*, **18**, 211 (1980).
26. A. M. North and D. Postlethwaite, in *Structure and Mechanism in Vinyl Polymerization* (T. Tsuruta and K. F. O'Driscoll, Eds.), Marcel Dekker, New York, 1969, Chap. 4.
27. A. D. Jenkins, in *Reactivity, Mechanism and Structure in Polymer Chemistry* (A. D. Jenkins and A. Ledwith, Eds.), Wiley-Interscience, New York, 1974, Chap. 4.
28. C. H. Bamford, W. G. Barb, A. D. Jenkins, and P. F. Onyan, *The Kinetics of Vinyl Polymerization by Free Radical Mechanisms*, Academic Press, New York (1958).
29. J. Brandrup, E. H. Immergut, and W. McDowell (Eds.), *Polymer Handbook*, 2nd ed., Wiley-Interscience,

New York, 1975: (a) R. Korus and K. F. O'Driscoll, pp. II-45ff and II-451ff; (b) L. J. Young, pp. II-57ff; (c) K. J. Ivin, pp. II-421ff; (d) L. J. Young, pp. II-105ff; (e) L. J. Young pp. II-387ff.

30. J.-S. Wang and K. Matyjaszewski, *J. Am. Chem. Soc.*, **117**, 5614 (1995).
31. M. K. Georges, R. P. N. Veregin, P. K. Kazmaier, and S. K. Hamer, *Macromolecules*, **26**, 2987 (1993).
32. T. E. Patten, J. Xia, T. Abernathy, and K. Matyjaszewski, *Polym. Prepr.*, **37**(1), 575 (1996).
33. W. V. Smith and R. H. Ewart, *J. Chem. Phys.*, **16**, 592 (1948).
34. W. Cooper, in *The Stereochemistry of Macromolecules*, Vol. 2 (A. D. Ketley, Ed.), Dekker, New York, 1967, Chap. 5.
35. T. G. Fox, B. S. Garrett, W. E. Goode, S. Gratch, J. F. Kincaid, A. Spell, and J. D. Stroupe, *J. Am. Chem. Soc.*, **80**, 1768 (1958).
36. F. A. Bovey, *J. Polym. Sci.*, **46**, 59 (1960).
37. D. J. Cram and K. R. Kopecky, *J. Am. Chem. Soc.*, **81**, 2748 (1959).
38. C. E. H. Bawn, W. H. Janes, and A. M. North, *J. Polym. Sci. C*, **4**, 427 (1963).
39. R. J. Cotter and M. Matzner, *Ring-Forming Polymerizations*, Part A, Academic Press, New York, 1969, Chap. 2.
40. G. B. Butler, *Cyclopolymerization and Cyclocopolymerization*, Marcel Dekker, New York, 1992.
41. C. E. Schildknecht, *Allyl Compounds and Their Polymers*, Wiley-Interscience, New York, 1973.
42. K. J. Ivin, in *Reactivity, Mechanism and Structure in Polymer Chemistry* (A. D. Jenkins and A. Ledwith, Eds.), Wiley-Interscience, New York, 1974, Chap. 16.
43. D. G. L. James and T. Ogawa, *Can. J. Chem.*, **43**, 640 (1965).
44. C. A. Barson, J. C. Bevington, and D. E. Eaves, *Trans. Faraday Soc.*, **54**, 1678 (1958).
45. Y. Minouro, in *Structure and Mechanism in Vinyl Polymerization* (T. Tsuruta and K. F. O'Driscoll, Eds.), Marcel Dekker, New York, 1969, Chap. 7.
46. H. R. Allcock and F. W. Lampe, *Contemporary Polymer Chemistry*, 2nd ed. Prentice-Hall, Englewood Cliffs, N. J., 1990, p. 239.
47. G. E. Ham (Ed.), *Copolymerization*, Wiley-Interscience, New York, 1964.
48. D. Braun, H. Cherdron, and W. Kern, *Practical Macromolecular Organic Chemistry*, 3rd ed., Harwood, New York, 1984, pp. 206–221.
49. P. W. Tidwell and G. A. Mortimer, *J. Macromol. Sci.-Rev. Macromol. Chem.*, **C4**, 281 (1970).
50. R. C. McFarlane, P. M. Reilly, and K. F. O'Driscoll, *J. Polym. Sci., Polym. Chem. Ed.*, **18**, 251 (1980).
51. A. Rudin, K. F. O'Driscoll, and M. S. Rumack, *Polymer*, **22**, 740 (1981).
52. J. J. Uebel and F. J. Dinan, *J. Polym. Sci., Polym. Chem. Ed.*, **21**, 917 (1983).
53. T. Alfrey and L. J. Young, in *Copolymerization* (G. E. Ham, Ed.), Wiley-Interscience, New York, 1964, Chap. 2.
54. A. D. Jenkins, *J. Polym. Sci. Part A: Polym. Chem.*, **34**, 3495 (1996).
55. D. Bartlett and K. Nozaki, *J. Am. Chem. Soc.*, **68**, 1495 (1946).
56. H. Hirai, *J. Macromol. Sci.-Rev. Macromol. Chem.*, **11**, 47 (1976).
57. C. H. Bamford, in *Alternating Copolymerization* (J. M. G. Cowie, Ed.), Plenum Press, New York, 1985, Chap. 3.
58. J. R. Ebdon, C. R. Towns, and K. Dodgson, *J. Macromol. Sci., Rev. Macromol. Chem. Phys.*, **C26**, 523 (1986).

Review Exercises

1. Write concise definitions for the following terms:

   (a) autoacceleration
   (b) backbiting
   (c) cage effect
   (d) chain transfer
   (e) cyclopolymerization
   (f) inhibitor
   (g) living polymer
   (h) oil-in-water emulsion

   (i) plasma polymerization
   (j) promoter
   (k) radical combination
   (l) radical disproportionation
   (m) reactivity ratio
   (n) supercritical fluid
   (o) telomerization
   (p) Trommsdorf effect

2. Give reasonable explanations for the following facts:
   (a) Free radical polymerization has received more widespread commercial development than ionic polymerization.
   (b) Almost all substituted ethylene monomers of commercial importance have the structure $CH_2=CHR$ or $CH_2=CRR'$ (where R or R' represent identical or different substituent groups).
   (c) Syndiotactic placement is favored at low temperatures in free radical polymerization.
   (d) In general, increasing polymerization rate decreases molecular weight in bulk polymerization, whereas the reverse is the case in emulsion polymerization.
   (e) Propylene, isobutylene, and allyl acetate cannot be polymerized to high molecular weight under the usual free radical conditions.

3. Discuss the similarities and differences between suspension and emulsion free radical polymerization, including advantages and disadvantages of each.

4. Summarize the causes and ramifications of the gel (Trommsdorff or Norris–Smith) effect.

5. Wastage reaction products of diacetyl peroxide (structure **2**) include carbon dioxide and ethane. Those of di-*t*-butyl peroxide (**3**) include acetone, methane, *t*-butyl alcohol, and isobutylene oxide. Suggest plausible reaction pathways by which these various side products might form.

6. In normal free radical processes with added initiator, the polymerization rate and $\overline{DP}$ are proportional to the first power of monomer concentration. What would be the effect of monomer concentration on $\overline{DP}$ in the thermal polymerization of styrene? Derive a kinetics expression to support your conclusion. (*Reminder*: Initiation involves two monomer molecules.)

7. If equal concentrations of acrylonitrile and methyl methacrylate were each polymerized at 60°C with equal concentrations of the same initiator, which polymer would have the higher $\overline{DP}$ and by how much? Assume polyacrylonitrile undergoes termination only by radical combination and poly(methyl methacrylate) by disproportionation, that no chain transfer occurs, and that initiator efficiencies are the same in both reactions. Which polymer would have the higher $\overline{M}_n$, and by how much?

8. What concentration of benzoyl peroxide ($k_d = 1.45 \times 10^{-6}$ L/mol s at 60°C) would be needed to polymerize a 1.00 M solution of styrene to a molecular weight of 125,000? Assume that termination occurs only by radical combination, that no chain transfer occurs, and that initiator efficiency is 100%.

9. A certain monomer is polymerized by benzoyl peroxide labeled in the benzene ring with carbon-14. (a) Analysis by carbon counting shows that the polymer contains an average of 1.27 labeled phenyl groups per molecule. Assuming no chain transfer occurred, what does this indicate about the mechanism of termination? (b) Vigorous basic hydrolysis removed 60% of the carbon-14. What does this indicate about the mechanism of initiation?

10. Suppose you wanted to repeat the experiment given in question 8 (using the same initiator concentration), but wanted to limit the molecular weight to 15,000. What concentration of *n*-butyl mercaptan (1-butanethiol) would you add?

11. Predict the order of polymer molecular weight obtainable (highest to lowest) by the free radical solution polymerization of styrene in benzene, toluene, ethylbenzene, *i*-propylbenzene, and *t*-butylbenzene. Assume each polymerization is run under identical conditions other than choice of solvent. Justify your answer.

12. Predict how monomer concentration might affect the outcome of cyclopolymerization of an isolated diene.

13. Propylene forms only oligomer under normal free radical conditions, yet the monomer forms high-molecular-weight polymer if it is polymerized while trapped in the channels of a crystalline host substance such as urea or thiourea. Explain [G. Di Silvestro, P. Sozzani, and M. Farina, *Polym. Preprints*, **27**(1), 92, 1986].

14. On the basis of chemical structure, predict the result of free radical copolymerization of the following monomer pairs: (a) acrylamide and acrylonitrile, (b) ethylene and methyl acrylate, (c) 1,3-butadiene and vinyl stearate (vinyl octadecanoate), (d) chlorotrifluoroethylene and ethyl vinyl ether. When you have made your predictions, consult a polymer handbook and check your answers on the basis of reactivity ratios.

15. Predict the ratio of butadiene to styrene repeating units initially produced by free radical copolymerization of an equimolar mixture of the two monomers at 50°C. (Consult Table 6.9.)

16. What weight ratio of the two monomers in the previous problem would be necessary to have twice as many butadiene as styrene units in the copolymer?

17. Compare the reactivity ratios determined for methyl methacrylate and vinyl acetate by the *Q–e* scheme with those determined experimentally (Table 6.9).

18. In the presence of a free radical initiator, furan and maleic anhydride form a 1:1 copolymer at 70°C in benzene solution. In the absence of the initiator, the two reactants form a 1:1 Diels–Alder adduct. The Diels–Alder adduct undergoes polymerization at temperature higher than 70°C to give a polymer identical with the 1:1 copolymer. IR and ^{13}C NMR of the polymer exhibited olefinic unsaturation of the type found in the Diels–Alder adduct. What is the most likely structure of the polymer? What other structures might be feasible in the absence of the spectroscopic evidence? (References pertaining to this problem: N. G. Gaylord et al., *J. Macromol. Sci.-Chem.*, **A6**, 1459, 1972; B. Kamo et al., *Polym. J.*, **6**, 121, 1974; Y. A. Ragab and G. B. Butler, *J. Polym. Sci., Polym. Lett. Ed.*, **14**, 273, 1976; *idem.*, *J. Polym. Sci., Polym. Chem. Ed.*, **19**, 1175, 1981.)

7

IONIC POLYMERIZATION

7.1 Introduction

A fundamental difference between free radical and ionic polymerization reactions is that, with the latter, *counterions* (also called *gegenions*) are present in the reaction medium to preserve electrical neutrality.[1,2] The mechanism of polymerization may be strongly influenced by the counterions, depending on whether they are strongly or weakly associated with the propagating chain. Such association depends in large measure on solvation effects. Ionic polymerizations are therefore considerably more complex than free radical polymerizations, but at the same time they are more versatile in their degree of steric control. They are also more versatile in their scope, having application in ring-opening polymerizations of cyclic ethers (to form polyethers), lactams (to form polyamides), and lactones (to form polyesters), and in the polymerization of aldehydes and ketones (to form polyethers). These applications are covered in later chapters dealing with the appropriate polymer type.

Despite the versatility of ionic polymerization, commercial processes based on ionic vinyl polymerizations are far fewer in number and reflect a much narrower choice of monomers than the corresponding free radical processes. This results from the polar nature of the propagating chain end in ionic polymerization; monomers must contain substituent groups capable of stabilizing carbocations or carbanions. A further limiting factor is the necessity for solution polymerization in most instances, although bulk processes are sometimes used. Examples of vinyl polymers prepared commercially by ionic polymerization are given in Table 7.1.

7.2 Cationic Polymerization

7.2.1 Cationic Initiators

In cationic chain polymerization[3-6] the propagating species is a carbocation.* Initiation is brought about by addition of an electrophile to a monomer molecule (7.1).

$$E^+ + CH_2{=}CR_2 \rightarrow ECH_2\overset{+}{C}R_2 \qquad (7.1)$$

*The terms *carbonium ion* and *carbenium ion* are both used to denote a positively charged carbon species. The IUPAC recommends that the former be reserved for pentavalent carbocations, R_5C^+, and the latter for the more conventional trivalent species, R_3C^+. Because such terminology is not universally accepted, we will simply refer to the propagating species as a carbocation.

TABLE 7.1. Commercially Important Polymers Prepared by Ionic Polymerization

Polymer or Copolymer	*Major Uses*
Cationic[a]	
Polyisobutylene and polybutenes[b] (low and high molecular weight)	Adhesives, sealants, insulating oils, lubricating oil and grease additives, moisture barriers
Isobutylene-isoprene copolymer[c] ("butyl rubber")	Inner tubes, engine mounts and springs, chemical tank linings, protective clothing, hoses, gaskets, electrical insulation
Isobutylene–cyclopentadiene copolymer	Ozone-resistant rubber
Hydrocarbon[d] and polyterpene resins	Inks, varnishes, paints, adhesives, sealants
Coumarone–indene resins[e]	Flooring, coatings, adhesives
Poly(vinyl ether)s	Polymer modifiers, tackifiers, adhesives
Anionic[f]	
cis-1,4-Polybutadiene	Tires
cis-1,4-Polyisoprene	Tires, footware, adhesives, coated fabrics
Styrene-butadiene rubber (SBR)[g]	Tire treads, belting, hose, shoe soles, flooring, coated fabrics
Styrene-butadiene block and star copolymers	Flooring, shoe soles, artificial leather, wire and cable insulation
ABA block copolymers (A = styrene, B = butadiene or isoprene)	Thermoplastic elastomers
Polycyanoacrylate[h]	Adhesives

[a] AlCl$_3$ and BF$_3$ most frequently used coinitiators.

[b] "Polybutenes" are copolymers based on C$_4$ alkenes and lesser amounts of propylene and C$_5$ and higher alkenes from refinery streams.

[c] Terpolymers of isobutylene, isoprene, and divinylbenzene are also used in sealant and adhesive formulations.

[d] Aliphatic and aromatic refinery products.

[e] Coumarone (benzofuran) and indene (benzocyclopentadiene) are products of coal tar.

[f] n-Butyllithium most common initiator.

[g] Contains higher cis content than SBR prepared by free radical polymerization.

[h] Monomer polymerized by adventitious water.

Compounds used most frequently to effect cationic polymerization are mineral acids, particularly H$_2$SO$_4$ and H$_3$PO$_4$, and such Lewis acids as AlCl$_3$, BF$_3$, TiCl$_4$, and SnCl$_4$. Lewis acids are seldom effective alone; rather they require the presence of trace amounts of water or some other proton or cation source, which, on reaction with the Lewis acid, forms the electrophilic species that initiates polymerization. Examples are the reactions of BF$_3$ with water (7.2) and aluminum chloride with an alkyl chloride (7.3). Because it is the proton or carbocation that initiates the polymerization reaction, the compounds that give rise to them are correctly referred to as the *initiators*, and the Lewis acids are *coinitiators*.[5] The com-

$$BF_3 + H_2O \rightleftharpoons HOBF_3^-H^+ \qquad (7.2)$$

$$AlCl_3 + RCl \rightleftharpoons AlCl_4^-R^+ \qquad (7.3)$$

bination of Lewis acid and proton or cation source is the *initiating system*. With certain very active Lewis acids, *autoionization* (7.4) may occur. In such cases the initiator and coinitiator are the same.

$$2AlBr_3 \rightleftharpoons AlBr_4^- AlBr_2^+ \qquad (7.4)$$

A variety of other initiators can bring about cationic polymerization, among them ionizable compounds such as triphenylmethyl halides (7.5) or tropylium halides (7.6), and iodine, which may react by in situ generation of HI (7.7) or by ion-pair formation (7.8). Radical cations formed in electron transfer processes (7.9) in which monomer (M) is the electron donor may also initiate cationic polymerization. The reaction between a nitro compound and N-vinylcarbazole, for example, gives rise to a radical cation (7.10) that initiates polymerization of the vinyl monomer.

$$(C_6H_5)_3CCl \rightleftharpoons (C_6H_5)_3C^+ + Cl^- \qquad (7.5)$$

$$\qquad (7.6)$$

$$\qquad (7.7)$$
$$\qquad (7.8)$$

$$M + A \longrightarrow M\cdot^+ + A\cdot^- \qquad (7.9)$$

$$\qquad (7.10)$$

It should be kept in mind that not all initiating systems are equally effective. Relatively stable carbocations of the triphenylmethyl or tropylium type are only useful with very reactive monomers such as vinyl ethers; and mineral acid initiators seldom lead to very-high-molecular-weight polymer. Furthermore, in most cases of cationic polymerization, a clear understanding of the initiation mechanism is lacking.[7]

7.2.2 Mechanism, Kinetics, and Reactivity in Cationic Polymerization

In carbocationic initiation, addition of the electrophilic species follows Markovnikov's rule, that is, the more stable carbocation intermediate is formed. As would be expected from a consideration of carbocation stability, the rate of addition to aliphatic monomers is in the order

$$(CH_3)_2C=CH_2 > CH_3CH=CH_2 > CH_2=CH_2$$

Of these three monomers, only isobutylene provides the requisite carbocation stability for cationic polymerization. For a series of *para*-substituted styrenes, the reactivity for substituent groups in cationic initiation is in the order of ring activation, namely

$$OCH_3 > CH_3 > H > Cl$$

Because of steric hindrance, *ortho* substituents retard the addition regardless of whether they activate or deactivate the ring. Vinyl ethers are particularly reactive toward cationic initiators (7.11) because the delocalized intermediate carbocation has one contributing resonance structure containing no electron-deficient atoms.

$$CH_2=CH\ddot{O}R \xrightarrow{R'+} R'CH_2-\overset{+}{C}H-\ddot{O}R \longleftrightarrow R'CH_2-CH=\overset{+}{\ddot{O}}R \qquad (7.11)$$

In the propagation step, the reaction is favored by increasing carbocation stability. It may be recalled that in free radical propagation the *less* stable radicals react faster. Possibly cationic propagation involves two steps—a rate-limiting formation of a pi complex between the chain end and an approaching monomer molecule, followed by covalent bond formation (7.12), whereas covalent bond formation is rate limiting in free radical processes.

$$\text{wwww} + + CH_2=CR_2 \xrightarrow{slow} \text{wwww} \overset{CR_2}{\underset{CH_2}{\overset{+\,||}{\cdots}}} \xrightarrow{fast} \text{wwww} CH_2\overset{+}{C}R_2 \qquad (7.12)$$

Solvent polarity influences polymerization rate, but it is extremely difficult to make concrete predictions as to how the effect is manifested. Predictably, increasing solvent polarity favors the initiation step where a charged species is being generated. The opposite is expected in propagation, however, because the charge is dispersed in the transition state. Another complicating factor is the degree of association between the cationic chain end and the anion (A^-). Between the extremes of pure covalent bonds and solvated ions are intimate ion pairs (also called contact ion pairs) and solvent-separated ion pairs (7.13). Which actually

$$\text{wwww}A \rightleftharpoons \text{wwww} + A^- \rightleftharpoons \text{wwww} + \| A^- \rightleftharpoons \text{wwww} + + A^- \qquad (7.13)$$

| Covalent | Intimate ion pair | Solvent-separated ion pair | Solvated ions |

exists in any given solvent depends on both the dielectric strength of the solvent and the solvent's ability to coordinate the cation. (The "solvating power" of an ether solvent, for example, is enhanced by the lone-pair electrons on oxygen.) In general, the more intimate the association, the lower the propagation rate. Thus one might reasonably expect that in poorly solvating solvents, increasing solvent polarity will enhance the propagation rate by shifting the association away from intimate ion pairs. As the solvating power of the solvent increases, however, the shift will be in the opposite direction, and propagation will be retarded. How solvent affects the stereochemistry of propagation is discussed in the next section.

Chain transfer reactions are common in cationic polymerization. Let us consider, for example, the polymerization of styrene with sulfuric acid. Possible transfer reactions include:

1. With monomer:

$$\text{ww}CH_2\overset{+}{C}H\,HSO_4^- + CH_2=CH \longrightarrow \text{ww}CH=CH + CH_3\overset{+}{C}H\,HSO_4^- \qquad (7.14)$$
$$\qquad\quad |\qquad\qquad\quad |\qquad\qquad\qquad\quad |\qquad\qquad\quad |$$
$$\qquad\quad Ph\qquad\qquad\quad Ph\qquad\qquad\qquad Ph\qquad\qquad\quad Ph$$

2. By ring alkylation:

$$\text{\textasciitilde\textasciitilde CH}_2\text{—CH—CH}_2\text{—CH}^+\text{HSO}_4^- + \text{CH}_2\text{=CH} \longrightarrow$$

$$\overset{|}{\text{Ph}} \qquad \overset{|}{\text{Ph}} \qquad \overset{|}{\text{Ph}}$$

$$+ \text{CH}_3\text{—CH}^+\text{HSO}_4^- \qquad (7.15)$$

$$\overset{|}{\text{Ph}}$$

3. By hydride abstraction from the chain to form a more stable ion:

$$\text{\textasciitilde\textasciitilde CH}_2\text{CH}^+\text{HSO}_4^- + \text{\textasciitilde\textasciitilde CH}_2\text{CHCH}_2\text{\textasciitilde\textasciitilde} \longrightarrow \text{\textasciitilde\textasciitilde CH}_2\text{CH}_2 + \text{\textasciitilde\textasciitilde CH}_2\overset{+}{\text{C}}\text{CH}_2\text{\textasciitilde\textasciitilde} \quad (7.16)$$

$$\overset{|}{\text{Ph}} \qquad \qquad \overset{|}{\text{Ph}} \qquad \qquad \overset{|}{\text{Ph}} \qquad \overset{|}{\text{Ph}}$$

with HSO_4^- over the cation.

4. With solvent—for example, benzene—by electrophilic substitution:

$$\text{\textasciitilde\textasciitilde CH}_2\text{CH}^+\text{HSO}_4^- + \bigcirc + \text{CH}_2\text{=CH} \longrightarrow$$

$$\overset{|}{\text{Ph}} \qquad\qquad\qquad \overset{|}{\text{Ph}}$$

$$\text{\textasciitilde\textasciitilde CH}_2\text{CH—}\bigcirc + \text{CH}_3\text{CH}^+\text{HSO}_4^- \qquad (7.17)$$

$$\overset{|}{\text{Ph}} \qquad\qquad\qquad \overset{|}{\text{Ph}}$$

Chain branching occurs via reaction (7.16) or by intermolecular ring alkylation.

Termination reactions are harder to define in cationic processes because they are easy to confuse with chain transfer. Examples of termination are the combination of chain end with counterion (i.e., a change from ionic to covalent bonding) as in the trifluoroacetic acid–initiated polymerization of styrene (7.18),[8] and chain-end chlorination in the $\text{BCl}_3/\text{H}_2\text{O}$-initiated polymerization of isobutylene (7.19).[5]

$$\text{\textasciitilde\textasciitilde\textasciitilde CH}_2\text{CH}^+ \quad {}^-\text{OCCF}_3 \longrightarrow \text{\textasciitilde\textasciitilde\textasciitilde CH}_2\text{CHOCCF}_3 \qquad (7.18)$$

$$\overset{|}{\text{Ph}} \qquad\qquad\qquad\qquad \overset{|}{\text{Ph}}$$

$$\overset{\text{CH}_3}{\underset{|}{\text{\textasciitilde\textasciitilde\textasciitilde CH}_2\overset{+}{\text{C}}}} \quad \text{HOBCl}_3^- \longrightarrow \overset{\text{CH}_3}{\underset{|}{\text{\textasciitilde\textasciitilde\textasciitilde CH}_2\text{CCl}}} + \text{BCl}_2\text{OH} \qquad (7.19)$$

$$\overset{|}{\text{CH}_3} \qquad\qquad\qquad\qquad \overset{|}{\text{CH}_3}$$

Because chain transfer to monomer is so common in cationic polymerization, various strategies have been devised to circumvent it.[5] One involves use of a *proton trap* such as 2,6-di-*t*-butylpyridine, which intercepts the proton (7.20) before it transfers to monomer.[9] The result is a lower overall yield but higher molecular weight and lower polydispersity index. The bulky *t*-butyl groups prevent reaction with electrophiles larger than the proton.

$$\text{\textasciitilde\textasciitilde CH}_2\text{CR}_2^+ + \quad \bigcirc_N \quad \longrightarrow \quad \text{\textasciitilde\textasciitilde CH=CR}_2 + \quad \bigcirc_{\underset{H}{N^+}} \qquad (7.20)$$

Transfer reactions have also been used to advantage to prepare telechelic polymers.[5] A bifunctional compound called an *inifer*, which brings about both *ini*tiation and chain trans*fer*, is used. Steps in the inifer method, using α,α'-dichlorodiisopropylbenzene as the inifer, BCl_3 as cocatalyst, and isobutylene monomer include formation of the initiating cation (7.21), polymerization (7.22), and chain transfer (7.23). Transfer at the other labile chlorine

$$\begin{array}{c}\text{CH}_3 \quad\quad \text{CH}_3 \\ | \quad\quad\quad | \\ \text{Cl}-\text{C}-\bigcirc-\text{C}-\text{Cl} + \text{BCl}_3 \;\rightleftharpoons\; \text{Cl}-\text{C}-\bigcirc-\text{C}^+ \;\; \text{BCl}_4^- \\ | \quad\quad\quad | \\ \text{CH}_3 \quad\quad \text{CH}_3 \end{array} \qquad (7.21)$$

$$\begin{array}{c}\text{CH}_3 \quad \text{CH}_3 \quad\quad \text{CH}_3 \\ \text{Cl}-\text{C}-\bigcirc-\text{C}^+ + \text{CH}_2{=}\text{C} \;\longrightarrow\; \text{Cl}-\text{C}-\bigcirc-\text{C}\text{\textasciitilde\textasciitilde CH}_2\text{C}^+ \\ \text{CH}_3 \quad \text{CH}_3 \quad\quad \text{CH}_3 \end{array} \qquad (7.22)$$

$$\begin{array}{c}\text{CH}_3 \quad \text{CH}_3 \quad \text{CH}_3 \\ \text{\textasciitilde\textasciitilde CH}_2\text{C}^+ + \text{Cl}-\text{C}-\bigcirc-\text{C}-\text{Cl} \;\longrightarrow\; \\ \text{CH}_3 \quad \text{CH}_3 \quad \text{CH}_3 \end{array}$$

$$\text{(7.23)}$$

$$\begin{array}{c}\text{CH}_3 \quad\quad \text{CH}_3 \quad\quad \text{CH}_3 \\ \text{\textasciitilde\textasciitilde CH}_2\text{C}-\text{Cl} + {}^+\text{C}-\bigcirc-\text{C}-\text{Cl} \\ \text{CH}_3 \quad\quad \text{CH}_3 \quad\quad \text{CH}_3 \end{array}$$

leads to polyisobutylene terminated with chloro groups,

$$\begin{array}{c}\text{CH}_3 \; \text{CH}_3 \quad\quad \text{CH}_3 \; \text{CH}_3 \\ \text{Cl}-\text{C}\text{\textasciitilde\textasciitilde C}-\bigcirc-\text{C}\text{\textasciitilde\textasciitilde C}-\text{Cl} \\ \text{CH}_3 \; \text{CH}_3 \quad\quad \text{CH}_3 \; \text{CH}_3 \end{array}$$

that are readily convertible to other functionalities.

When perchloric acid is used to initiate the polymerization of styrene in hydrocarbon solvents, the reaction proceeds at a much slower pace compared with most cationic processes.[10,11] It has been suggested on the basis of spectrophotometric and conductometric studies that the propagating chain end is a covalently bonded perchlorate ester, and that monomer is inserted between carbon–oxygen bonds (7.24). Close ion pairs have also been offered as an explanation.[12] Polymerization of this type have been termed *pseudocationic*.

$$-CH_2CH-OClO_3 + CH_2=CH \longrightarrow -CH_2CHCH_2CH-OClO_3$$

with Ph groups, labeled (7.24)

It was discovered in the 1980s that it is possible to prepare living polymers under cationic conditions.[13,14] (Recall the discussion of living free radical polymers in Chapter 6.) Polymerization of isobutylene with a tertiary ester and BCl_3,[15] for example, involves formation of a tertiary carbocation-initiating species (7.25) and polymerization (7.26) to yield polyisobutylene terminated with what appears to be a very tightly bound—but still active—ion pair. A similar situation occurs when I_2/HI or I_2/ZnI_2 is used as the initiating system.[16–18] In this case, shown for vinyl ether propagation (7.27), the mechanism apparently involves insertion of monomer into an activated carbon–iodine bond. It may be recalled from our discussion in Chapter 6 that low polydispersities are characteristic of living polymers, and that block copolymers are formed upon addition of a second monomer.

$$R_3COCCH_3 + BCl_3 \rightleftharpoons R_3C^+ \quad ^-OCCH_3 \qquad (7.25)$$

$$\xrightarrow{CH_2=C(CH_3)_2} R_3C\left[CH_2\underset{CH_3}{\overset{CH_3}{C}}\right]CH_2C^{\delta+}\cdots{}^{\delta-}OCCH_3 \qquad (7.26)$$

$$(7.27)$$

Because of the complexity of cationic initiation and uncertainty concerning the extent of ion-pair formation at the propagating chain end, any kinetic treatments of cationic polymerization reflect approximations at best. Without specifying the nature of the initiation–coinitiation mechanism, we can express general initiation, propagation, termination, and transfer rates as follows:

$$R_i = k_i[I][M]$$

$$R_p = k_p[M][M^+]$$

$$R_t = k_t[M^+]$$

$$R_{tr} = k_{tr}[M][M^+]$$

where $[I]$, $[M]$, and $[M^+]$ represent molar concentrations of initiator, monomer, and cationic chain end, respectively. As with free radical polymerization (but with less certainty), we can approximate a steady state for the growing chain end; thus $R_i = R_t$, and

$$k_i[I][M] = k_t[M^+]$$

TABLE 7.2. Representative Cationic Propagation Rate Constants, $R_p{}^a$

Monomer	Solvent	Temperature (°C)	Initiator	k_p (L/mol s)
Styrene	None	15	Radiation	3.5×10^6
α-Methylstyrene	None	0	Radiation	4×10^6
i-Butyl vinyl ether	None	30	Radiation	3×10^5
i-Butyl vinyl ether	CH_2Cl_2	0	$C_7H_7{}^+SbCl_6{}^-$	5×10^3
t-Butyl vinyl ether	CH_2Cl_2	0	$C_7H_7{}^+SbCl_6{}^-$	3.5×10^3
Methyl vinyl ether	CH_2Cl_2	0	$C_7H_7{}^+SbCl_6{}^-$	1.4×10^2
2-Chloroethyl vinyl ether	CH_2Cl_2	0	$C_7H_7{}^+SbCl_6{}^-$	2×10^2

^{a}Data from Ledwith and Sherrington.[19]

or

$$[M^+] = \frac{k_i[I][M]}{k_t}$$

Substituting for $[M^+]$ in R_p, one obtains

$$R_p = \frac{k_p k_i [I][M]^2}{k_t}$$

Some representative cationic propagation rate constants are given in Table 7.2.[19]

In the absence of any chain transfer, the kinetic chain length, $\bar{\nu}$, is equal to $\overline{DP}$ and is expressed as

$$\bar{\nu} = \overline{DP} = \frac{R_p}{R_t} = \frac{k_p[M][M^+]}{k_t[M^+]} = \frac{k_p[M]}{k_t}$$

If transfer is the predominant mechanism controlling chain growth,

$$\bar{\nu} = \overline{DP} = \frac{R_p}{R_{tr}} = \frac{k_p[M][M^+]}{k_{tr}[M][M^+]} = \frac{k_p}{k_{tr}}$$

These kinetics expressions illustrate some basic differences between free radical and cationic processes. In the former, the propagation rate is proportional to the square root of initiator concentration, whereas cationic polymerization shows a first-order dependence. Furthermore, the molecular weight in cationic polymerization is independent of initiator concentration, unlike free radical polymerization, where $\overline{DP}$ is inversely proportional to $[I]^{1/2}$ in the absence of chain transfer. The difference arises from radical disproportionation and combination reactions characteristic of free radical termination. Increasing initiator concentration thus increases the probability of radical termination, which is not the case in ionic polymerization.

Diene monomers also undergo cationic polymerization (primarily *trans*-1,4), but their use is limited mainly to copolymer synthesis. Nonconjugated dienes may undergo cationic cyclopolymerization[20,21] by a mechanism analogous to that of free radical cyclopolymerization (Chapter 6, Section 6.6). A typical example is the formation of a six-membered ring-containing polymer from 2,6-diphenyl-1,6-heptadiene (7.28).

$$(7.28)$$

7.2.3 Stereochemistry of Cationic Polymerization

It has been recognized for a long time that cationic polymerization can lead to stereoregular structures. Most of the work in this area has been done on vinyl ethers,[22] although other monomers, notably α-methylstyrene, have also been observed to form stereoregular polymer. Observations resulting from these studies are that (1) greater stereoregularity is achieved at lower temperatures, (2) the degree of stereoregularity can vary with initiator, and (3) the degree and type of stereoregularity (isotactic or syndiotactic) vary with solvent polarity. For example, *t*-butyl vinyl ether forms isotactic polymer in nonpolar solvents and mainly syndiotactic polymer in polar solvents. This behavior can be rationalized in terms of the extent to which the cationic chain end and the counterion are associated. In polar solvents both ions would be strongly solvated and the chain end would exist as a free carbocation surrounded by solvent molecules. The same conformational factors would thus apply here that are responsible for syndiotactic placement in free radical polymerization. In nonpolar solvents, however, association between carbocation chain end and counterion would be strong, and counterion could influence the course of steric control.

How isotactic polymer forms is the subject of considerable debate. Several models have been proposed, the major aspects of which are presented here, but the true course of stereoregular polymerization is still not clear. Two of the models proposed for vinyl ether polymerization[23,24] suggest a six-membered cyclic chain end formed by coordination of the carbocation of the terminal carbon with the oxygen of the alkoxy group attached to the fifth carbon atom (7.29). For this cyclic oxonium ion, the polymer chain, being the bulkiest group, would be equatorial, the alkoxy group on the third carbon would be axial, and the terminal alkoxy group would be equatorial. Attack of monomer at C_1 will be from the back side causing inversion; hence the alkoxy group at what used to be C_1 will have the same configuration as that at the original C_3 (7.30). A new cyclic oxonium ion would then form, and the process would repeat itself to give isotactic polymer. The major difference between the two models proposed lies in the degree of participation by the counterion in complexing the reacting monomer molecule.

$$(7.29)$$

$$\tag{7.30}$$

Another model[25] proposes a strong ion–counterion association at the chain end, with monomer inserting itself in such a way that the alkoxy groups are *anti* to one another (7.31). As long as propagation proceeds faster than rotation about carbon–carbon bonds in the chain, a regular repeating isotactic polymer results. Some coordination of monomer with counterion may also be assumed in this model.

$$\tag{7.31}$$

The major drawback to these mechanisms is that they all assume some tetrahedral character for the carbocation. With vinyl ethers this would diminish the considerable resonance stabilization afforded by delocalization of the unshared p electrons on oxygen, which requires a planar configuration. Furthermore, the first two, involving cyclic oxonium ions, are suitable only for vinyl ether polymerization but are not applicable to such monomers as α-methylstyrene which, in general, exhibit similar steric control in polymerization.

A fourth model has been proposed[26] that takes into account normal sp^2 hybridization of the terminal chain carbon atom and is based on accepted mechanistic theories of displacement reactions. According to this model, in polar solvents the cationic chain end may be considered essentially free, and monomer approaches from the front side in the preferred syndiotactic arrangement analogous to that of syndiotactic free radical polymerization. In nonpolar solvents the chain end exists as an ion pair, the "tightness" of which depends on solvent polarity and counterion. Assuming that considerable energy is required to separate the ion pair, the incoming monomer would then prefer *back-side* attack, or an S_N2 type of reaction leading to isotactic placement. These two possibilities are shown in equations (7.32) and (7.33).

(7.32)

(7.33)

With poly(α-methylstyrene), where two bulky groups are attached to the third carbon of the polymer backbone, back-side attack is hindered and syndiotactic placement is favored in both polar and nonpolar solvents, although the amount of isotacticity does increase as solvent polarity is reduced. Vinyl ethers, being less hindered, exhibit a more pronounced preference for isotactic placement in nonpolar solvents.

The fact that increasing temperature decreases stereoregularity can be interpreted in terms of small differences in activation energy for front- and back-side attack as well as of the effect on conformational mobility of the growing chain. Courterion–monomer interactions are assumed to be negligible, since both may be considered "electron rich."

While the last model (reactions 7.32 and 7.33) satisfactorily explains most of the experimental observations, a point to be gained from this discussion is that, in many polymerization processes, no single theory is necessarily adequate to explain all the facts. One should approach these subjects with an open mind and a realization that considerably more work may be necessary before a mechanism is completely clarified.

7.2.4 Cationic Copolymerization

The copolymerization equation derived for free radical initiation (Chapter 6) may also be applied to cationic copolymerization[27] to determine reactivity ratios, several of which are listed in Table 7.3. The situation is complicated, however, by counterion effects. Unlike free radical copolymerization, reactivity ratios vary with initiator type and solvent polarity. Another significant difference in ionic copolymerization is that there is no apparent tendency for alternating copolymers to form. On the contrary, block copolymers or homopolymer blends are more likely.

Yet another unpredictable effect is that of temperature. In free radical processes, increasing temperature results in greater randomness, which is to be expected, because differences in activation energies between self-propagation and cross-propagation would be re-

TABLE 7.3. Representative Cationic Reactivity Ratios $(r)^a$

Monomer 1	Monomer 2	Coinitiator[b]	Solvent[b]	Temperature (°C)	r_1	r_2
Isobutylene	1,3-Butadiene	$AlEtCl_2$	CH_3Cl	−100	43	0
	1,3-Butadiene	$AlCl_3$	CH_3Cl	−103	115	0
	Isoprene	$AlCl_3$	CH_3Cl	−103	2.5	0.4
	Cyclopentadiene	$BF_3 \cdot OEt_2$	$PhCH_3$	−78	0.60	4.5
	Styrene	$SnCl_4$	EtCl	0	1.60	0.17
	Styrene	$AlCl_3$	CH_3Cl	−92	9.02	1.99
	α-Methylstyrene	$TiCl_4$	$PhCH_3$	−78	1.2	5.5
Styrene	α-Methylstyrene	$SnCl_4$	EtCl	0	0.05	2.90
	p-Methylstyrene	$SnCl_4$	CCl_4	−78	0.33	1.74
	trans-β-Methyl-styrene	$SnCl_4$	CH_2Cl_2	0	1.80	0.10
p-Chlorostyrene	cis-β-Methyl-styrene	$SnCl_4$	$CCl_4/PhNO_2(1:1)$	0	1.0	0.32
	trans-β-Methyl-styrene	$SnCl_4$	$CCl_4/PhNO_2(1:1)$	0	0.74	0.32
Ethyl vinyl ether	i-Butyl vinyl ether	BF_3	CH_2Cl_2	−78	1.30	0.92
2-Chloroethyl vinyl ether	α-Methylstyrene	BF_3	CH_2Cl_2	−23	6.02	0.42

[a] Data from Kennedy and Marechal.[5]
[b] Et = C_2H_5, Ph = phenyl.

duced. Cationic copolymerizations, on the other hand, show greater or lesser tendency toward randomness from one system to another, and no definite trend is observable.

Steric effects are also important in cationic copolymerization, but not always to the extent found in free radical processes. For example, 1,2-disubstituted monomers usually undergo homopolymerization under cationic conditions, which is not the case under free radical conditions. A possible explanation is that the cationic chain end approaches the monomer molecule perpendicularly to form a pi complex (see reaction 7.12) rather than in the sterically unfavorable orientation depicted previously (structure **34**, Chapter 6). (Interestingly, one of the best methods of preparing polymers having substituents on each backbone carbon is by polymerization of diazoalkanes, $RCHN_2$—shown for diazomethane in Chapter 1, reaction 1.15—which also happens to be a cationic process.[28])

The most important of the commercial cationic copolymers is *butyl rubber*, prepared from isobutylene and isoprene. Because of its impermeability, butyl rubber finds wide use in protective clothing and tire inner tubes. It is manufactured by low-temperature (−100°C) copolymerization of about 97% isobutylene and 3% isoprene in chlorocarbon solvents. An ozone-resistant copolymer of isobutylene and cyclopentadiene has been marketed. (Recall the discussion of ozone resistance in Section 4.6.)

7.2.5 Isomerization in Cationic Polymerization

That carbocations undergo 1,2 shifts to form more stable carbocations is well known. This phenomenon is also observed in cationic polymerization.[29] When 3-methyl-1-butene is polymerized using a Lewis acid coinitiator at low temperature (7.34), a hydride shift occurs to yield what is, in effect, a 1,3-addition polymer. The driving force for rearrangement is formation of the more stable tertiary carbocation. Similarly, rearrangement during cationic polymerization of the terpene β-pinene (7.35) relieves strain in the bicyclic system. As is the case with cyclopentadiene, β-pinene also yields an ozone-resistant elastomer upon copolymerization with isobutylene. Numerous examples of isomerization polymerization have been reported.

$$CH_2=CH-\underset{\underset{CH_3}{|}}{\overset{\overset{CH_3}{|}}{CH}} \xrightarrow{X^+} X CH_2-\overset{+}{CH}-\underset{\underset{CH_3}{|}}{\overset{\overset{CH_3}{|}}{CH}} \xrightarrow{H: shift}$$

$$X CH_2-CH_2-\underset{\underset{CH_3}{|}}{\overset{\overset{CH_3}{|}}{\overset{+}{C}}} \xrightarrow{monomer} \left[CH_2-CH_2-\underset{\underset{CH_3}{|}}{\overset{\overset{CH_3}{|}}{C}} \right] \tag{7.34}$$

$$\tag{7.35}$$

7.3 Anionic Polymerization

7.3.1 Anionic Initiators

In anionic vinyl polymerization,[31–34] the propagating chain is a carbanion; hence initiation is brought about by species that undergo nucleophilic addition to monomer (7.36). Monomers having substituent groups capable of stabilizing a carbanion through resonance or induction are most susceptible to anionic polymerization. Examples of such groups are nitro, cyano, carboxyl, vinyl, and phenyl.

$$Nu^- + CH_2=CHR \longrightarrow NuCH_2-\underset{\underset{R}{|}}{CH^-} \tag{7.36}$$

The strength of the base necessary to initiate polymerization depends in large measure on monomer structure. Thus potassium bicarbonate is strong enough to initiate polymerization of 2-nitropropene (7.37), whereas a base as weak as water causes vinylidene cyanide to polymerize violently (7.38). Such high reactivity has been exploited commercially with the development of *cyanoacrylate* adhesives—vinyl monomers substituted on the same carbon with CN and CO_2R groups that polymerize on exposure to atmospheric or surface moisture.

Most applications of anionic polymerization, however, involve initiation by alkali metals or their compounds.

$$CH_2=\underset{\underset{NO_2}{|}}{\overset{\overset{CH_3}{|}}{C}} \xrightarrow{KHCO_3} \left[CH_2-\underset{\underset{NO_2}{|}}{\overset{\overset{CH_3}{|}}{C}} \right] \quad (7.37)$$

$$CH_2=\underset{\underset{CN}{|}}{\overset{\overset{CN}{|}}{C}} \xrightarrow{H_2O} \left[CH_2-\underset{\underset{CN}{|}}{\overset{\overset{CN}{|}}{C}} \right] \quad (7.38)$$

The most typically used anionic initiators may be classified into two basic types—those that react by addition of a negative ion and those that undergo electron transfer. The most common initiators that react by addition of a negative ion are simple organometallic compounds of the alkali metals—for example, butyllithium. Organolithium compounds are generally low melting and soluble in inert organic solvents. Organometallic compounds of the higher alkali metals have more ionic character and are generally insoluble; hence they initiate polymerization by a heterogeneous process. Other initiators of this type include organic compounds of calcium and barium and Grignard reagents, but these are not used as commonly as the alkali metal compounds.

Initiation by charge transfer can be brought about by free alkali metals or by addition complexes of alkali metals and unsaturated or aromatic compounds. Free metals may be employed as solutions in liquid ammonia or certain ether solvents or as suspensions in inert solvents. The latter are prepared by heating the solvent and metal above the melting point of the metal, stirring vigorously to form an emulsion, then cooling to obtain a fine, solid dispersion. The metal may also be used as a free-flowing powder coated on an inert support such as alumina. The polymerization process is heterogeneous if the metal is used as a dispersion and homogeneous if it is in solution. (Not all ammonia solutions of alkali metals react by electron transfer; potassium is believed to form potassium amide, which initiates polymerization by addition of amide ion.) Electron transfer processes involving metal donor, D·, and monomer, M, may be written in general form as follows (7.39):

$$D\cdot + M \rightarrow D^+ + M\cdot^- \quad (7.39)$$

Stable addition complexes of alkali metals that initiate polymerization by electron transfer are formed by reaction of metal and substrate in an inert solvent. Examples are the reactions of sodium with naphthalene (7.40) or stilbene (7.41). A general equation for the

$$Na\cdot + \text{(naphthalene)} \longrightarrow \text{(naphthalene)}^{\overline{\cdot}}\ Na^+ \quad (7.40)$$

$$Na\cdot + Ph\ CH{=}CH\ Ph \rightarrow [Ph\ \overset{\cdot}{C}H{-}\overset{\overline{\ }}{C}H\ Ph]\ Na^+ \quad (7.41)$$

reaction of this type of donor, D·$^-$ with monomer, M, may be written

$$D\cdot^- + M \rightarrow D + M\cdot^- \quad (7.42)$$

Related to the foregoing are addition complexes (called *ketyls*) of alkali metals and nonenolizable ketones, such as benzophenone (7.43). The alkali metal ketyls normally exist in equilibrium with the dianion dimer (7.44), and initiation can be brought about by electron

$$
\text{Na} \cdot + \text{Ph}\overset{\overset{\text{O}}{\|}}{\text{C}}\text{Ph} \longrightarrow \left[\text{Ph}\overset{\overset{\text{O}}{\|}}{\text{C}}\text{Ph} \right]^{\text{-}} \text{Na}^+ \tag{7.43}
$$

transfer or by anion addition. Similar addition complexes are formed with compounds having nitrogen–oxygen, nitrogen–nitrogen, and carbon–nitrogen double bonds.

$$
2 \left[\text{Ph}\overset{\overset{\text{O}}{\|}}{\text{C}}\text{Ph} \right]^{\text{-}} \text{Na}^+ \longrightarrow \overset{\text{NaO}\quad\text{ONa}}{\underset{}{\text{Ph}_2\text{C}-\text{CPh}_2}} \tag{7.44}
$$

7.3.2 Mechanism, Kinetics, and Reactivity in Anionic Polymerization

The mechanism of initiation depends on the type of initiator used. For those that react by ion addition, such as the alkali metal alkyls, essentially two different initiation processes can occur, depending on the degree of association between metal ion and alkyl group. Lithium compounds, for example, may have covalent carbon–metal bonds, whereas the higher alkali metals tend to form more ionic bonds. Between the extremes of pure covalent bonds and solvated ions are ion pairs of varying degrees of association, as is the case with cationic polymerization, with solvent polarity and solvating power playing an equally important role.

In polar solvents where free solvated ions predominate, the mechanism involves simple addition of anion to monomer (7.36). In nonpolar solvents where there is close association between ions, the mechanism is believed to involve initial pi-complex formation, as shown in equation (7.45) for an alkyllithium compound. Which mechanism prevails depends, therefore, on a variety of factors including solvent polarity, type of cation, and temperature.

$$
\text{R}-\text{Li} + \text{CH}_2\!\!=\!\!\underset{\underset{\text{R}'}{|}}{\text{CH}} \rightleftharpoons \text{R}-\text{Li}\text{---}\overset{\text{CHR}'}{\underset{\text{CH}_2}{\|}} \longrightarrow \text{RCH}_2\underset{\underset{\text{R}'}{|}}{\text{CHLi}} \tag{7.45}
$$

The rate of initiation depends on the structure of both initiator and monomer. In some cases initiation occurs much more rapidly than propagation because of the lower reactivity of the propagating carbanion relative to that of initiator.

Initiation by electron transfer is considerably different. Let's consider first initiation by alkali metal. When the metal is dispersed in an inert solvent, the process is heterogeneous. The reaction, shown here for sodium, involves, first, formation of a radical anion (7.46), followed by radical combination to form a dianion (7.47). Evidence that this occurs is based

$$
\text{Na} \cdot + \text{CH}_2\!\!=\!\!\underset{\underset{\text{R}}{|}}{\text{CH}} \longrightarrow \left[\text{CH}_2\!\!=\!\!=\!\!\underset{\underset{\text{R}}{|}}{\text{CH}} \right]^{\text{-}} \text{Na}^+ \tag{7.46}
$$

$$
2\left[\text{CH}_2\!\!=\!\!=\!\!\underset{\underset{\text{R}}{|}}{\text{CH}} \right]^{\text{-}} \text{Na}^+ \longrightarrow \text{Na}^+ \overset{\cdot\cdot}{\underset{\underset{\text{R}}{|}}{\text{C}}}\text{HCH}_2\text{CH}_2\overset{\cdot\cdot}{\underset{\underset{\text{R}}{|}}{\text{C}}}\text{HNa}^+ \tag{7.47}
$$

on a rapidly disappearing ESR signal on mixing initiator and monomer, and the quantitative formation of dicarboxylic acid on addition of carbon dioxide to an equimolar mixture of the two. For alkali metals in solution the process is essentially the same except that, in the case of ammonia, free solvated electrons probably add to monomer rather than being transferred directly from the metal. As was mentioned earlier, potassium in ammonia appears to initiate polymerization by addition of amide ion rather than by electron transfer. Addition complexes such as naphthalenesodium or the alkali metal ketyls also react by electron transfer to form dianions.

The kinetic and mechanistic aspects of anionic polymerization are better understood than those of cationic polymerization. In the case of the potassium amide-initiated polymerization in liquid ammonia, initiation involves dissociation (7.48) followed by addition of amide ion to monomer (7.49).

$$KNH_2 \rightleftharpoons K^+ + NH_2^- \tag{7.48}$$

$$NH_2^- + M \rightarrow H_2N\text{—}M^- \tag{7.49}$$

Because the second step is slow relative to the first,

$$R_i = k_i[NH_2^-][M]$$

Chain termination is known to result primarily by transfer to solvent:

$$H_2N(M)_nM^- + NH_3 \rightarrow H_2N(M)_nMH + NH_2^- \tag{7.50}$$

Rate expressions for propagation and transfer may be written in the conventional way:

$$R_p = k_p[M][M^-]$$

$$R_{tr} = k_{tr}[M^-][NH_3]$$

Assuming a steady state whereby $R_i = R_{tr}$,

$$k_i[NH_2^-][M] = k_{tr}[M^-][NH_3]$$

and

$$[M^-] = \frac{k_i[NH_2^-][M]}{k_{tr}[NH_3]}$$

Substituting in R_p we obtain

$$R_p = \frac{k_p k_i[NH_2^-][M]^2}{k_{tr}[NH_3]}$$

The average kinetic chain length, $\bar{\nu}$, is expressed as

$$\bar{\nu} = \frac{R_p}{R_{tr}} = \frac{k_p[M][M^-]}{k_{tr}[M^-][NH_3]} = \frac{k_p[M]}{k_{tr}[NH_3]}$$

Because $\bar{\nu} = \overline{DP}$ for this process, it is clear that both molecular weight and propagation rate are inversely proportional to the concentration of ammonia.

Other types of transfer reactions are recognized in anionic polymerization. Unsaturated end groups in polyacrylonitrile, for example, are believed to arise from transfer to monomer (7.51) followed by initiation by the resultant vinyl anion (7.52). The predominant chain termination mechanism in methyl methacrylate polymerization is the backbiting nucleophilic displacement of methoxide by the enolate anion chain end to form a cyclohexanone ring (7.53).[35]

$$\text{mCH}_2\text{CH:}^- + \text{CH}_2{=}\text{CH} \longrightarrow \text{mCH}_2\text{CH}_2 + \text{CH}_2{=}\text{C:}^- \qquad (7.51)$$
$$\underset{\text{CN}}{|} \qquad\quad \underset{\text{CN}}{|} \qquad\qquad \underset{\text{CN}}{|} \qquad\quad \underset{\text{CN}}{|}$$

$$\text{CH}_2{=}\text{C:}^- + \text{CH}_2{=}\text{CH} \longrightarrow \text{CH}_2{=}\text{C}{-}\text{CH}_2\text{CH:}^- \qquad (7.52)$$
$$\qquad \underset{\text{CN}}{|} \qquad\quad \underset{\text{CN}}{|} \qquad\qquad \underset{\text{CN}}{|} \quad \underset{\text{CN}}{|}$$

$$\cdots + {}^-\text{OCH}_3 \qquad (7.53)$$

When impurities are rigorously excluded and the polymerization temperature is kept low (about $-78°C$ in most instances), living anionic polymers can be made. This was first demonstrated in the 1950s in a classic experiment by Swarc[36,37] involving the polymerization of styrene with naphthalenesodium initiator. Upon initiation, the green color of the naphthalenesodium changes to the red color of the styryl anion. The red color persists even when the styrene is fully consumed. Addition of more styrene results in further polymerization, while addition of a second monomer leads to a block copolymer. When carbon dioxide is introduced, the anionic chain ends are converted quantitatively to carboxylic acid groups. These early experiments, which established the foundations of living polymers, were followed by commercial development of block copolymers.

In polymerizations such as these, where the rate of initiation is very high relative to that of propagation,

$$\frac{d[M]}{dt} = k_p[I]_0[M]$$

where $[I]_0$ is the initial concentration of initiator. This means, in effect, that all chains begin to grow simultaneously. Integration of the rate expression gives

$$[M] = [M]_0 e^{k[I]t}$$

If no termination or chain transfer reactions occur, the average kinetic chain length is simply equal to the number of monomer molecules reacted divided by the number of chains initiated; that is,

$$\bar{\nu} = \frac{[M]_0 - [M]}{[I]_0}$$

or, as monomer is completely consumed,

$$\bar{\nu} = \frac{[M]_0}{[I]_0}$$

For those cases where initiation is by a simple anionic species (butyllithium, for example), $\bar{\nu} = \mathrm{DP}$. In the case of electron-transfer initiators that combine according to equation (7.47), $\mathrm{DP} = 2\bar{\nu}$ because the polymer chain grows simultaneously from each end of the initiating species.

As was mentioned earlier, an important consequence of having no termination or chain transfer, coupled with the virtually simultaneous initiation of all polymer chains, is that the molecular weight distribution is very narrow. Indeed, anionic polymerization is used to prepare the low-polydispersity standards for calibration in gel permeation chromatography (Chapter 2).

Association between counterion and terminal carbanion is an important factor in propagation rate. In poorly solvating hydrocarbon solvents with Li^+ counterion, for example, association of the type shown in reaction (7.54) impairs reactivity. Larger alkali metal cations,

$$2\text{\tiny{\textbf{ww}}}CH_2CHLi \longrightarrow \text{\tiny{\textbf{ww}}}CH_2CH\begin{subarray}{c}\diagup Li \diagdown \\ \diagdown Li \diagup\end{subarray}CHCH_2\text{\tiny{\textbf{ww}}} \qquad (7.54)$$
$$\quad\quad\;\; | \qquad\qquad\qquad | \quad\quad |$$
$$\quad\quad\;\; R \qquad\qquad\qquad R \quad\;\; R$$

which do not coordinate as strongly as Li^+, generally allow faster propagation under these conditions. In a more polar solvating medium such as an ether, however, Li^+ is the most strongly solvated of the alkali metal cations, and lithium-based initiators afford the fastest rates. The effect of ion-pair association and solvent polarity are evident in the representative propagation rate constants for polystyrene in Table 7.4, where the order of solvating power is seen to be

1,2-dimethoxyethane > tetrahydrofuran > benzene > cyclohexane

TABLE 7.4. Representative Anionic Propagation Rate Constants, k_p, for Polystyrene[a]

Counterion	Solvent	k_p (L/mol s)[b]
Na^+	Tetrahydrofuran	80
Na^+	1,2-Dimethoxyethane	3600
Li^+	Tetrahydrofuran	160
Li^+	Benzene	10^{-3}–10^{-1} [c]
Li^+	Cyclohexane	$(5\text{-}100)\times10^{-5}$ [c]

[a] Data from Morton.[30]

[b] At 25°C unless otherwise noted.

[c] Variable temperature.

Monomer structure is another consideration. For initiation with amide ion, the following order of reactivity has been observed:

$$CH_2{=}CHCN > CH_2{=}\overset{\overset{\displaystyle CH_3}{|}}{C}CO_2CH_3 > CH_2{=}CHC_6H_5 > CH_2{=}CH{-}CH{=}CH_2$$

Methyl substitution on the α-carbon decreases the rate, with reactivity in the order

$$CH_2{=}CHCN > CH_2{=}\overset{\overset{\displaystyle CH_3}{|}}{C}CN > CH_2{=}CHCO_2CH_3 > CH_2{=}\overset{\overset{\displaystyle CH_3}{|}}{C}CO_2CH_3$$

This results from inductive destabilization of the carbanion and steric interference with both chain-end solvation and approach of monomer. Similar effects have been observed in the polymerization of α-methylstyrene and ring-substituted styrenes where alkyl or methoxyl groups decrease the rate.

7.3.3 Stereochemistry of Anionic Polymerization

For nondiene vinyl monomers, stereochemical control in anionic polymerization parallels that of cationic polymerization. With soluble anionic initiators (homogeneous conditions) at low temperatures, polar solvents favor syndiotactic placement, whereas nonpolar solvents favor isotactic placement.[38] Thus, stereochemistry depends in large measure on the degree of association with counterion, as it does in cationic polymerization.

Syndiotactic placement may be rationalized in terms of the preferred stereochemical arrangement between the chain-end carbanion and approaching monomer under conditions where the chain end exists as a solvated ion not closely associated with counterion. This was discussed earlier in this chapter (Section 7.2.3) for cationic polymerization and also in Chapter 6 (Section 6.5) where an analogous situation exists under free radical conditions. The greatest success in obtaining highly syndiotactic or isotactic polymer under anionic conditions has been achieved with acrylic monomers, particularly methyl methacrylate; hence theories pertaining to isotactic placement have centered mainly on this monomer.

Early proposals to explain isotactic placement centered on conformational effects in the polymer backbone[39] or coordination of the anionic chain end with counterion to yield a cyclic unit.[23,24] One example of the latter[23] is shown in reactions (7.55) and (7.56). According to this model, the terminal enolate anion coordinates the carbonyl group of the penultimate ester group via the oxygen to give a six-membered ring that is coordinated strongly with lithium counterion at both oxygens (7.55). Monomer then approaches from below (7.56) because of

$$(7.55)$$

$$(7.56)$$

steric blocking of the top side by the axial methyl group, but also because this is the most favorable orientation to coordinate the counterion. Once addition occurs, the chain end again assumes the preferred cyclic orientation for approach of the next monomer molecule. The level of isotactic placement thus decreases as the solvent polarity is increased, or as lithium is replaced with the less strongly coordinating higher alkali metal ions. Another model, depicted in Scheme 7.1 (and somewhat reminiscent of the front-side/back-side model for cationic polymerization), suggests ion pairing regardless of stereochemistry, with counterion–monomer coordination assisting isotactic placement in poorly solvating solvents, and counterion–solvent interaction blocking isotactic placement in polar solvents.[40]

A number of mechanistic refinements to explain isotactic and syndiotactic placement have appeared over the years,[41] encompassing such monomers as vinyl ketones and vinyl-pyridines as well as acrylic esters. Invariably, coordination of chain-end with counterion and counterion solvation are key ingredients. As is the case with cationic polymerization, a definitive mechanism of stereochemical control is still lacking.

From the commercial standpoint, stereochemical control of diene polymerization[38] is most important, in particular isoprene and 1,3-butadiene (Table 7.1). For both monomers, the level of *cis*-1,4 polymer is increased by using lithium-based initiators in nonpolar solvents. This is especially so with isoprene, where almost entirely *cis*-1,4 polymer is produced using butyllithium initiator in pentane or hexane. Temperature has little effect on stereochemistry. Thus a product virtually identical in structure to the most important form of nat-

SCHEME 7.1. (a) Isotactic approach of methyl methacrylate in a nonpolar solvent. (b) Syndiotactic approach of methyl methacrylate in tetrahydrofuran. (Circles represent backbone or incipient backbone carbons: R = methyl. Backbone hydrogens omitted.)

ural rubber (*Hevea* rubber) is accessible by anionic polymerization. The synthetic product is, in fact, often referred to as "synthetic natural rubber."

Formation of *cis*-polyisoprene has been attributed to lithium's ability either to hold isoprene in its *s-cis* conformation by pi complexation (7.57) or to "lock" the isoprene into a *cis*

$$\text{\sim\sim CH}_2\text{Li} + \text{CH}_2\text{=C}-\text{CH=CH}_2 \longrightarrow \text{\sim\sim CH}_2\text{Li} \qquad (7.57)$$

configuration by forming a six-membered ring transition state:

$$(7.58)$$

Predominantly *cis*-1,4-poly(1,3-butadiene) is also formed under the conditions given above for *cis*-polyisoprene, although at a somewhat lower stereospecificity. This may arise from butadiene's preference for the *s-trans* conformation, whereas the steric effect of the methyl group of isoprene favors *s-cis*:

$$(7.59)$$

7.3.4 Anionic Copolymerization

As is the case with cationic polymerization, relatively few reactivity ratios have been determined for anionic copolymerization processes because of the complicating effects of counterion.[30] Some representative values are given in Table 7.5.

Among the complicating factors are the solvating power of the solvent, the effect of temperature, possible competition between anionic and free radical polymerization when electron transfer initiators are used, and contrasts between homogeneous and heterogeneous polymerization systems. Solvent effects are shown most dramatically with styrene–butadiene copolymerization. A change from the nonpolar hexane to the highly solvating tetrahydrofuran reverses the order of reactivity when the soluble *n*-butyllithium is used as initiator. It should be noted that in homopolymerization, styrene is *more* reactive than butadiene. Clearly the lithium counterion plays a significant role in the preference polystyryl anion exhibits towards butadiene, a role that is effectively nullified in the presence of tetrahydrofuran.

TABLE 7.5. Representative Anionic Reactivity Ratios $(r)^a$

Monomer 1	Monomer 2	Initiator[b]	Solvent[c]	Temperature[d] (°C)	r_1	r_2
Styrene	Methyl methacrylate	Na	NH$_3$		0.12	6.4
		n-BuLi	None		e	e
	Butadiene	n-BuLi	None	25	0.04	11.2
		n-BuLi	Hexane	25	0.03	12.5
		n-BuLi	Hexane	50	0.04	11.8
		n-BuLi	THF	25	4.0	0.3
		n-BuLi	THF	−78	11.0	0.4
		EtNa	Benzene		0.96	1.6
	Isoprene	n-BuLi	Cyclohexane	40	0.046	16.6
	Acrylonitrile	RLi	None		0.12	12.5
	Vinyl acetate	Na	NH$_3$		0.01	0.01
Butadiene	Isoprene	n-BuLi	Hexane	50	3.38	0.47
Methyl methacrylate	Acrylonitrile	NaNH$_2$	NH$_3$		0.25	7.9
		RLi	None		0.34	6.7
	Vinyl acetate	NaNH$_2$	NH$_3$		3.2	0.4

[a] Data from Morton.[30]

[b] Bu = butyl, Et = ethyl, R = alkyl.

[c] THF = tetrahydrofuran.

[d] Temperature not specified in some instances.

[e] No detectable styrene in polymer.

The influence of temperature is also demonstrated by the even greater difference in re-activity ratios at −78°C. One might expect solvation to be felt more strongly at lower temperatures, which is the case in tetrahydrofuran but not to any appreciable extent in hexane.

Differences between homogeneous and heterogeneous processes have been demonstrated in copolymerization of styrene and methyl methacrylate.[42] With soluble n-butyllithium initiator, only poly(methyl methacrylate) is formed. With metallic lithium or methyllithium, both of which are insoluble, block copolymers result, possibly by initial styrene polymerization on the initiator surface followed by detachment, and subsequent initiation of methyl methacrylate by the detached polystyryl anion. Whether or not free radical processes compete with anionic polymerization is difficult to verify experimentally, but it cannot be ruled out when alkali metals or addition complexes are used as initiators.

A variation on the anionic copolymerization process involves nucleophilic displacement of aliphatic dihalides by the dianion formed by electron transfer initiation (7.60).[43] While this reaction clearly does not fit the definition of vinyl polymerization, it is of interest because impurities need not be so rigorously excluded as they have to be in conventional anionic polymerization, and the reaction provides a route to copolymers having a regular unit.

$$Li^+\bar{C}HCH_2CH_2\bar{C}HLi^+ + Cl{-}R{-}Cl \longrightarrow \left[CHCH_2CH_2CH{-}R\right] \quad (7.60)$$
$$\begin{array}{cc} | & | \\ X & X \end{array} \qquad\qquad \begin{array}{cc} | & | \\ X & X \end{array}$$

Most interesting from the standpoint of commercial development is the formation of block copolymers by the living polymer method.[30,44] This is illustrated in reaction (7.61) for polystyrene-*block*-poly(methyl methacrylate). The order of monomer addition is important in block copolymerization.

$$n \ \text{CH}_2{=}\text{CH} \xrightarrow{\text{R:}^-} \left[\text{CH}_2\text{CH} \right]_{n-1} \text{CH}_2\text{CH:}^- \xrightarrow{m \ \text{CH}_2{=}\overset{\text{CH}_3}{\underset{}{\text{C}}}\text{CO}_2\text{CH}_3}$$

(7.61)

$$\left[\text{CH}_2\text{CH} \right]_n \left[\text{CH}_2\overset{\text{CH}_3}{\underset{\text{CO}_2\text{CH}_3}{\text{C}}} \right]_{m-1} \text{CH}_2\overset{\text{CH}_3}{\underset{\text{CO}_2\text{C}}{\text{C:}^-}}$$

In the example given, the copolymer would not form if methyl methacrylate were polymerized first, because living poly(methyl methacrylate) is not basic enough to add to styrene. Nevertheless, large numbers of block copolymers have been synthesized, including some consisting of vinyl blocks coupled with nonvinyl blocks, or consisting of only nonvinyl blocks.

Greatest commercial success has been achieved with ABA triblock polymers, which have proved to be versatile thermoplastic elastomers (Chapter 3, Section 3.9). They can be synthesized by a three-stage sequential formation of each block using a monofunctional initiator such as *n*-butyllithium, or in a two-stage process using a difunctional initiator (i.e., the type that undergoes radical combination) such as naphthalenesodium. The latter is illustrated schematically (7.62) for the commercially important styrene–butadiene–styrene (SBS) triblock polymer. Typically, thermoplastic elastomers of this type have molecular weights of about 50,000 to 70,000 for the polybutadiene blocks and 10,000 to 15,000 for the polystyrene blocks.

$$\text{B} \xrightarrow{\text{initiator}} \text{B:}^- \xrightarrow{\text{combination}} {}^-\text{:BB:}^- \xrightarrow{\text{B}} {}^-\text{:BBBBBBBBBB:}^- \xrightarrow{\text{S}}$$
$${}^-\text{:SSSSSBBBBBBBBBBBSSSSS:}^-$$

(7.62)

Other commercial block copolymers include styrene–butadiene (AB-type) and star-block (also called *radial*) copolymers.[44] The latter are formed by the reaction of living block copolymer with, for example, silicon tetrachloride (7.63). A major advantage of star-block copolymers is that they exhibit much lower melt viscosities, even at very high molecular weights, than their linear counterparts.

$$4 \text{\textasciitilde\textasciitilde:}^- + \text{SiCl}_4 \longrightarrow \text{Si}$$

(7.63)

7.4 Group Transfer Polymerization

In the 1980s a new method for polymerizing acrylic-type monomers, termed group transfer polymerization (GTP), was reported.[45–49] Unlike conventional anionic polymerization, GTP affords low-polydispersity living polymers at room temperature or above. Typically, an organosilicon compound is used to initiate the polymerization in solution in the presence of an anionic

TABLE 7.6. Representative Compounds Used in Group Transfer Polymerization

Monomers[a]	Initiators[a]	Catalysts[a]	Solvents
$CH_2{=}CHCO_2R$	$Me_2C{=}C\begin{smallmatrix}OMe\\OSiMe_3\end{smallmatrix}$	Anionic[b]	Acetonitrile
		HF_2^-	1,2-Dichloroethane[d]
		CN^-	Dichloromethane[d]
		N_3^-	N,N-Dimethylacetamide
$\begin{smallmatrix}Me\\ \mid \\ CH_2{=}CCO_2R\end{smallmatrix}$	$Me_3SiCH_2CO_2Me$	Me_3SiF_2	N,N-Dimethylformamide
		Lewis acid[c]	Ethyl acetate
		ZnX_2	Propylene carbonate
$CH_2{=}CHCONR_2$	Me_3SiCN	R_2AlCl	Tetrahydrofuran
		$(R_2Al)_2O$	Toluene[d]
$CH_2{=}CHCN$	$RSSiMe_3$		
$\begin{smallmatrix}Me\\ \mid \\ CH_2{=}CCN\end{smallmatrix}$	$ArSSiMe_3$		
$\begin{smallmatrix}O\\ \parallel \\ CH_2{=}CHCR\end{smallmatrix}$			

[a]R = alkyl, Ar = aryl, Me = methyl, X = halogen.
[b]0.1 mol % relative to initiator.
[c]10–20 mol % relative to monomer.
[d]Preferred with Lewis acid catalysts.

or Lewis acid catalyst. Examples of each type of compound are given in Table 7.6. The process is illustrated for the polymerization of methyl methacrylate using the methyl trimethylsilyl acetal of dimethylketene* as initiator and bifluoride ion as catalyst (7.64). In each propagation step the SiR_3 group is transferred to the carbonyl oxygen of the incoming monomer, hence the name GTP. If a difunctional initiator is used, the chain propagates from each end[50]:

$$\underset{(R\,=\,CH_3)}{\overset{R}{\underset{R}{>}}C{=}C\overset{OR}{\underset{OSiR_3}{<}}} + CH_2{=}\underset{CO_2CH_3}{\overset{CH_3}{\underset{\mid}{C}}} \xrightarrow{HF_2^-} RO_2C{-}\underset{R}{\overset{R}{\underset{\mid}{C}}}{-}CH_2\overset{CH_3}{\underset{}{C}}{=}C\overset{OSiR_3}{\underset{OCH_3}{<}}$$

$$\xrightarrow[n\ CH_2{=}\overset{CH_3}{\underset{}{C}}{-}CO_2CH_3]{} RO_2C{-}\underset{R}{\overset{R}{\underset{\mid}{C}}}{\Big[}CH_2\underset{CO_2CH_3}{\overset{CH_3}{\underset{\mid}{C}}}{\Big]_n}{-}CH_2\overset{CH_3}{\underset{}{C}}{=}C\overset{OSiR_3}{\underset{OCH_3}{<}}$$

(7.64)

*The initiator is synthesized by converting the appropriate ester to its anion with a strong base, followed by reaction with chlorotrimethylsilane, as follows (R = CH₃):

$$R_2CHCO_2R \xrightarrow{R'_2N^-Li^+} \Big[R_2\ddot{C}CO_2R \longleftrightarrow R_2C{=}C\overset{O^-}{\underset{OR}{<}} \Big] \xrightarrow{R_3SiCl} R_2C{=}C\overset{OSiR_3}{\underset{OR}{<}}$$

$$
\begin{array}{c}
\text{CH}_2\text{SSiMe}_3 \\
| \\
\text{CH}_2\text{SSiMe}_3
\end{array}
+ \text{CH}_2{=}\text{CHCO}_2\text{R} \xrightarrow{\text{ZnI}_2}
\begin{array}{c}
\text{CH}_2\text{S}{-}\!\left[\!\begin{array}{c}\text{CO}_2\text{R}\\ | \\ \text{CH}_2\text{CH} \end{array}\!\right]\!{-}\text{CH}_2\text{CH}{=}\overset{\overset{\displaystyle \text{OSiMe}_3}{|}}{\text{C}}{-}\text{OR} \\[6pt]
\text{CH}_2\text{S}{-}\!\left[\!\begin{array}{c}\text{CH}_2\text{CH}\\ | \\ \text{CO}_2\text{R}\end{array}\!\right]\!{-}\text{CH}_2\text{CH}{=}\underset{\underset{\displaystyle \text{OSiMe}_3}{|}}{\text{C}}{-}\text{OR}
\end{array}
\tag{7.65}
$$

Once the monomer is consumed, a different monomer may be added (to form block copolymer), or the chain can be terminated by removal of catalyst or by protonation (7.66) or alkylation (7.67).

$$
\begin{array}{c}
\text{CH}_3 \\
| \\
\text{wCH}_2\text{C}{=}\text{C}\!\begin{array}{c}\diagup\text{OSiR}_3\\ \diagdown\text{OCH}_3\end{array}
\end{array}
\quad
\begin{cases}
\xrightarrow{\text{CH}_3\text{OH}} \quad \text{wCH}_2\overset{\overset{\displaystyle \text{CH}_3}{|}}{\underset{\underset{\displaystyle \text{CO}_2\text{CH}_3}{|}}{\text{CH}}} & (7.66) \\[18pt]
\xrightarrow{\text{C}_6\text{H}_5\text{CH}_2\text{Br}} \quad \text{wCH}_2\overset{\overset{\displaystyle \text{CH}_3}{|}}{\underset{\underset{\displaystyle \text{CO}_2\text{CH}_3}{|}}{\text{C}}}\text{CH}_2\text{C}_6\text{H}_5 & (7.67)
\end{cases}
$$

It was originally proposed[45] that the propagating chain is completely covalent, and a hypervalent silicon intermediate is formed by activation with the nucleophilic catalyst (Nu⁻). The silyl group is then transferred to the carbonyl group of an incoming monomer molecule via an eight-membered ring transition state (7.68). Where Lewis acid catalysts are used, it was suggested that the catalyst coordinates the carbonyl oxygen of monomer, thereby rendering the monomer more susceptible to nucleophilic attack by the initiator.

$$
\begin{array}{c}
\text{R}\diagdown \quad \diagup\text{OR} \\
\quad\text{C}{=}\text{C} \\
\text{R}\diagup \quad \diagdown\text{OSiR}_3
\end{array}
\xrightarrow{\text{Nu}^-}
\left[
\begin{array}{c}
\text{R}\diagdown \quad \diagup\text{OR} \quad \text{Nu} \\
\quad\text{C}{=}\text{C} \\
\text{R}\diagup \quad \diagdown\text{O}{-}\text{SiR}_3 \\[6pt]
\begin{array}{c}\text{CH}_2 \quad \diagup\text{O}\\ \| \quad \diagdown \\ \text{C}{-}\text{C}\\ | \quad \diagdown\text{OCH}_3 \\ \text{CH}_3\end{array}
\end{array}
\right]^{-}
\tag{7.68}
$$

$$
\begin{array}{c}
\text{R} \\
| \\
\text{R}{-}\text{C}{-}\text{C}\!\begin{array}{c}\diagup\text{OR}\\ \diagdown\text{O}\end{array} \\
| \\
\text{CH}_2{-}\text{C}{=}\text{C}\!\begin{array}{c}\diagup\text{OSiR}_3\\ \diagdown\text{OCH}_3\end{array} \\
| \\
\text{CH}_3
\end{array}
$$

Others, however, have called this mechanism into question[51–55] and have suggested instead that enolate anions are the propagating species, as they are in conventional anionic polymerization of acrylate monomers. Apart from the room temperature conditions appropriate for forming living polymers, GTP and conventional anionic polymerizations are remarkably similar in terms of polymer tacticity and reactivity ratios in copolymerization. In addition, GTP of methyl methacrylate exhibits the same backbiting chain-terminating reaction (7.53)[56]; and the silyl group has been shown to undergo intermolecular exchange,[53] which is inconsistent with the proposed GTP mechanism. An alternative mechanism has been proposed[53] in which enolate anions and silyl ketene acetal chain ends are in rapid equilibrium with a hypervalent silicon complex (7.69). The complex thus provides a low equilibrium concentration of enolate anions for propagation and a mechanism for maintaining living chain ends.

$$
\begin{array}{c}
\text{wwCH}_2 \\
\text{C}=\text{C} \\
\text{CH}_3 \quad \text{OCH}_3
\end{array}
\begin{array}{c}
\text{OSiR}_3 \\
\end{array}
+
\begin{array}{c}
{}^-\text{O} \quad \text{CH}_2\text{ww} \\
\text{C}=\text{C} \\
\text{CH}_3\text{O} \quad \text{CH}_3
\end{array}
\rightleftharpoons
$$

(7.69)

$$
\begin{array}{c}
\text{wwCH}_2 \quad \text{O}-\underline{\text{Si}}(\text{R}_3)-\text{O} \quad \text{CH}_2\text{ww} \\
\text{C}=\text{C} \qquad\qquad\qquad \text{C}=\text{C} \\
\text{CH}_3 \quad \text{OCH}_3 \quad \text{CH}_3\text{O} \quad \text{CH}_3
\end{array}
$$

It remains to be seen whether or not GTP will be used commercially for synthesis of polymers or copolymers now made by conventional anionic polymerization.

References

1. M. Szwarc, *Ionic Polymerization Fundamentals*, Hanser, Cincinnati, 1996.
2. M. Szwarc and M. Van Beylen, *Ionic Polymerization and Living Polymers*, Chapman and Hall, New York, 1993.
3. K. Matyjaszewski (Ed.), *Cationic Polymerizations: Mechanisms, Synthesis and Applications*, Marcel Dekker, New York, 1996.
4. J. P. Kennedy and B. Ivan, *Designed Polymers by Carbocationic Macromolecular Engineering: Theory and Practice*, Hanser, Cincinnati, 1992.
5. J. P. Kennedy and E. Marechal, *Carbocationic Polymerization*, Wiley-Interscience, New York, 1982.
6. J. P. Kennedy, *Cationic Polymerization of Olefins: A Critical Inventory*, Wiley-Interscience, New York, 1975.
7. A. Gandini and H. Cheradame, *Adv. Polym. Sci.*, **34/35**, 1 (1980).
8. J. J. Throssel, S. P. Good, M. Szwarc, and V. Stannett, *J. Am. Chem. Soc.*, **78**, 1122 (1956).
9. J. P. Kennedy, T. Kelen, S. C. Guhaniyogi, and R. T. Chou, *J. Macromol. Sci.-Chem.*, **A18**, 129 (1982), and preceding papers.
10. P. H. Plesch, *Pure Appl. Chem.* **12**, 117 (1966).
11. P. H. Plesch, *Adv. Polym. Sci.*, **8**, 137 (1971).
12. M. Szwarc, *Macromolecules*, **28**, 7309, 7312 (1995).
13. M. Sawamoto, *Prog. Polym. Sci.*, **16**, 111 (1991).
14. B. Ivan and J. P. Kennedy, *Macromolecules*, **23**, 2880 (1990).
15. R. Faust and J. P. Kennedy, *J. Polym. Sci. A. Polym. Chem.*, **25**, 1847 (1987).
16. M. Sawamoto and T. Higashimura, *Makromol. Chem., Makromol. Symp.*, **3**, 83 (1986).
17. T. Higashimura, S. Aoshima, and M. Sawamoto, *Makromol. Chem., Makromol. Symp.*, **3**, 99 (1986).
18. M. Sawamoto, C. Okamoto, and T. Higashimura, *Macromolecules*, **20**, 2693 (1987).
19. A. Ledwith and D. C. Sherrington, in *Reactivity, Mechanism and Structure in Polymer Chemistry* (A. D. Jenkins and A. Ledwith, Eds.), Wiley-Interscience, New York, 1974, Chap. 9.
20. J. P. Kennedy and A. P. Langer Jr., *Adv. Polym. Sci.*, **3**, 508 (1964).
21. R. J. Cotter and M. Matzner, *Ring-Forming Polymerizations*, Part A, Academic Press, New York, 1969, Chap. 2.

22. A. D. Ketley, in *The Stereochemistry of Macromolecules*, Vol. 2 (A. D. Ketley, Ed.), Marcel Dekker, New York, 1967, Chap. 2.
23. D. J. Cram and K. R. Kopecky, *J. Am. Chem. Soc.*, **81**, 2748 (1959).
24. C. E. H. Bawn and A. Ledwith, *Quart. Rev.*, **16**, 361 (1962).
25. T. Higashimura, T. Yonezawa, S. Okamura, and K. Fukui, *J. Polym. Sci.*, **39**, 487 (1959).
26. T. Kunitake and C. Aso, *J. Polym. Sci. A-1*, **8**, 665 (1970).
27. J. P. Kennedy, in *Copolymerization* (G. E. Ham, Ed.), Wiley-Interscience, New York, 1964, Chap. 5.
28. G. W. Cowell and A. Ledwith, *Quart. Rev.*, **24**, 119 (1970).
29. A. D. Ketley and L. P. Fisher, in *Structure and Mechanism in Vinyl Polymerization* (T. Tsuruta and K. F. O'Driscoll, Eds.), Marcel Dekker, New York, 1969, Chap. 12.
30. M. Morton, *Anionic Polymerization: Principles and Practice*, Academic Press, New York, 1983.
31. H. L. Hsieh and R. P. Quirk, *Anionic Polymerization; Principles and Practical Applications*, Marcel Dekker, New York, 1996.
32. P. Rempp and P. Lutz, *Makromol. Chem., Makromol. Symp.*, **67**, 1 (1993).
33. T. E. Hogen-Esch and J. Smid (Eds.), *Recent Advances in Anionic Polymerization*, Elsevier, New York, 1987.
34. M. Szwarc, *Carbanions, Living Polymers, and Electron-Transfer Processes*, Wiley-Interscience, New York, 1968.
35. V. Warzelhan, H. Höcker, and G. V. Schulz, *Makromol. Chem.*, **179**, 2221 (1978).
36. M. Szwarc, *Nature*, **178**, 1168 (1956).
37. M. Szwarc, M. Levy, and R. Milkovich, *J. Am. Chem. Soc.*, **78**, 2656 (1956).
38. D. Braun, in *The Stereochemistry of Macromolecules*, Vol. 2 (A. D. Ketley, Ed.), Marcel Dekker, New York, 1967, Chap. 1.
39. M. Szwarc, *Chem. Ind.* (London), 1589 (1958).
40. W. Fowells, C. Schuerch, F. A. Bovey, and F. P. Hood, *J. Am. Chem. Soc.*, **89**, 1396 (1967).
41. T. E. Hogen-Esch and J. Smid (Eds.), *Recent Advances in Anionic Polymerization*, Elsevier, New York, 1987, pp. 187–272.
42. C. G. Overberger and N. Yamamoto, *J. Polym. Sci. A-1, Polym. Chem.*, **4**, 3101 (1966).
43. D. H. Richards, N. F. Scilly, and F. Williams, *Polymer*, **10**, 603 (1969).
44. A. Noshay and J. E. McGrath, *Block Copolymers: Overview and Critical Survey*, Academic Press, New York, 1977.
45. O. W. Webster, W. R. Hertler, D. Y. Sogah, W. B. Farnham, and T. V. Rajan-Babu, *J. Am. Chem. Soc.*, **105**, 5706 (1983).
46. W. R. Hertler, D. Y. Sogah, O. W. Webster, and B. M. Trost, *Macromolecules*, **17**, 1415 (1984).
47. D. Y. Sogah, W. R. Hertler, O. W. Webster, and G. M. Cohen, *Macromolecules*, **20**, 1473 (1987).
48. O. W. Webster and B.C. Anderson, in *New Methods for Polymer Synthesis* (W. J. Mijs, Ed.), Plenum, New York, 1992, Chap. 1.
49. G. C. Eastmond and O. W. Webster, in *New Methods of Polymer Synthesis* (J. R. Ebdon, Ed.), Blackie, London, 1991, Chap. 2.
50. M. T. Reetz, R. Ostarek, K.-E. Piejko, D. Arlt, and D. Bomer, *Angew. Chem. Int. Ed. Engl.*, **25**, 1108 (1986).
51. A. H. E. Müller, *Makromol. Chem., Macromol. Symp.*, **32**, 87 (1990).
52. S. Bywater, *Makromol. Chem., Macromol. Symp.*, **67**, 339 (1993).
53. R. P. Quirk, J. Ren, and G. Bidinger, *Makromol. Chem., Macromol. Symp.*, **67**, 351 (1993).
54. A. D. Jenkins, *Eur. Polym. J.*, **27**, 649 (1991).
55. A. D. Jenkins, *J. Macromol. Sci.—Pure Appl. Chem.*, **A31**, 1377 (1994).
56. W. J. Brittain and I. R. Dicker, *Macromolecules*, **22**, 1054 (1989).

Review Exercises

1. Give plausible explanations for the following facts:
   a. Polymerization rate and polymer stereochemistry are more sensitive to solvent effects in ionic polymerization than in free radical polymerization.
   b. $\overline{DP} = \bar{\nu}$ in cationic polymerization, but this is not always the case in free radical or anionic polymerization.

c. Ethyl vinyl ether undergoes cationic polymerization faster than β-chloroethyl vinyl ether under the same conditions.

d. Cationic polymerization of methyl α-methylvinyl ether gives predominantly syndiotactic polymer in toluene, whereas methyl vinyl ether yields mainly isotactic polymer.

e. Alkyllithium initiators favor the *cis* stereochemistry in polymerization of conjugated dienes.

f. If methyl methacrylate is polymerized with *n*-butyllithium in toluene at $-30°C$, the resultant polymer is almost 60% isotactic. If 10% of the toluene is replaced with pyridine, polymerization yields polymer that is nearly 60% syndiotactic. (R. L. Miller, *SPE Trans.*, **3**, 1, 1963)

2. Predict the order of reactivity (and justify your prediction): (a) in cationic polymerization: styrene, *p*-methoxystyrene, *p*-chlorostyrene, *p*-methylstyrene; (b) in anionic polymerization: styrene, 2-vinylpyridine, 3-vinylpyridine, 4-vinylpyridine.

3. Write reactions illustrating transfer to monomer and transfer to polymer in the cationic polymerization of propylene and isobutylene. Which type of transfer would you expect to predominate in each case? Suggest a reason why propylene does not form high-molecular-weight polymer under cationic conditions. (*Note*: The order of stability of carbocations is tertiary > allylic > secondary > primary.)

4. If trifluoroacetic acid is added dropwise to styrene, no polymerization occurs. If styrene is added to the acid, however, high-molecular-weight polymer forms rapidly. Suggest an explanation. (See reaction 7.8 and ref. 8.)

5. Rearrangement in cationic polymerization is temperature dependent, with rearranged polymer favored at low temperatures. (a) Explain how NMR might be used to evaluate the degree of rearrangement in the cationic polymerization of 3-methyl-1-butene (reaction 7.34). (b) Predict the principal repeating units for the polymers obtained by low-temperature cationic polymerization of 3,3-dimethyl-1-butene and vinylcyclohexane. (J. P. Kennedy et al., *J. Polym. Sci. A*, **2**, 5029, 1964)

6. Proton and ^{13}NMR of poly(4-methyl-1-pentene) formed under cationic conditions show the presence of five different repeating units. Predict their structure and explain their formation. (G. Ferraris et al., *Macromolecules*, **10**, 188, 1977)

7. What number average molecular weight polystyrene would be formed by polymerization of 2.0 *M* styrene with 1.0×10^{-3} *M* naphthalenesodium in tetrahydrofuran? If the reaction were run at 25°C, how long would it take to reach 90% conversion?

8. Write reactions illustrating the synthesis of poly(ethyl vinyl ether)-*block*-poly(*p*-methoxystyrene).

9. Explain how you could synthesize (a) poly(methyl methacrylate)-*block*-polyacrylonitrile; (b) an ABA block copolymer of the two monomers in part a; (c) carboxyl-terminated telechelic polystyrene.

10. Predict the structure of the soluble polymer formed from 2,6-diphenyl-1,6-heptadiene under anionic conditions. Write a mechanism for its formation.

11. Predict the structure of the polymers formed in the following reactions:

(a) $\xrightarrow[\text{H}_2\text{O}]{\text{AlEtCl}_2}$ (D. Yan and C. Yuan, *Macromolecules*, **23**, 3554, 1990)

(b) $\xrightarrow{\text{SnCl}_4}$ (D. T. Longone and D. T. Glatzhofer, *J. Polym. Sci., Part A: Polym. Chem.*, **24**, 1725, 1986)

12. Predict the structure of the polymer formed by (a) group transfer, (b) cationic, and (c) free radical polymerization of *p*-vinylbenzyl methacrylate, (C. Pugh and V. Percec, *Polym. Bull.*, **14**, 109 (1985)

13. The predominant repeating unit obtained in the following free radical polymerization is as shown:

Would you expect the same result if GTP were used? Explain. [See J. J. Kozakiewicz et al., *Polym. Preprints*, **28**(2), 347, 1987.]

14. Apart from lower viscosity, what other advantage might AB star-block copolymers have over linear AB block copolymers?

15. Silyl vinyl ethers are polymerized in the presence of aldehyde initiators and Lewis acid catalysts to give silylated poly(vinyl alcohol):

$$R'CHO + CH_2{=}CHOSiR_3 \xrightarrow{ZnBr_2} R'CH{-}\left[CH_2CH\right]{-}CH_2CHO$$
$$\qquad\qquad\qquad\qquad\qquad\qquad\quad |\qquad\quad |$$
$$\qquad\qquad\qquad\qquad\qquad\qquad OSiR_3\quad OSiR_3$$

This reaction has been dubbed "aldol-GTP." Propose a mechanism for the polymerization and explain the terminology. (D. Y. Sogah and O. W. Webster, *Macromolecules*, **19**, 1775, 1986)

16. GTP has been referred to as the "mirror image" of pseudocationic polymerization and could be appropriately called "pseudoanionic" (K. Matyjaszewski, *J. Polym. Sci. A., Polym. Chem.*, **25**, 765, 1987). Suggest a mechanism for pseudocationic polymerization consistent with this description.

17. The following sequence of reactions has been used to prepare comb polymers (R. Asami et al., *Polym. Bull.*, **16**, 125, 1986). Give the structures of A through D.

[a]Initiator, [b]Equimolar, [c]Catalyst.

VINYL POLYMERIZATION

WITH COMPLEX

COORDINATION CATALYSTS

8.1 Introduction

In the early 1950s, Karl Ziegler in Germany discovered that certain combinations of transition metal compounds and organometallic compounds polymerize ethylene at low temperatures and pressures to give polyethylene that has an essentially linear structure.[1] Now referred to as high-density polyethylene (HDPE), the product is denser, tougher, and higher melting than the branched low-density polyethylene (LDPE) prepared at high pressures with free radical initiators, because the more regular structure allows closer chain packing and a high degree of crystallinity. Ziegler's remarkable discovery, the culmination of work begun before the Second World War, led to speculation that the high-pressure method of making polyethylene was doomed. In fact, substantial differences in properties between HDPE and LDPE assured ample markets for both, as may be recalled from our discussion of commodity plastics in Chapter 1. A more recent threat to LDPE is what is called *linear low-density polyethylene* (LLDPE), which is actually a copolymer of ethylene and 1-butene (with lesser amounts of 1-hexene and higher 1-alkenes to vary the density) manufactured with Ziegler-type catalysts. First marketed in the late 1970s, LLDPE is attractive because it requires considerably less energy to produce than LDPE.

Following close on the heels of Ziegler's discovery was the recognition by Giulio Natta in Italy that catalysts of the type described by Ziegler were capable of polymerizing 1-alkenes (or *alpha olefins*, as they are called in the chemical industry) to yield stereoregular polymers.[2] Natta and his coworkers, as well as scientists in other laboratories, subsequently extended the range of catalysts, both soluble and insoluble, to yield polymers exhibiting a wide range of stereoregular structures, including those derived from dienes and cycloalkenes.[3,4] Unlike free radical or ionic initiators, the appropriately named *Ziegler–Natta* catalysts are not consumed in the polymerization reaction, although they are not recovered in commercial operations. The catalysts are effective primarily with nonpolar monomers: 1-alkenes, cycloalkenes, dienes, and alkynes.

A variety of commercial products are prepared with Ziegler–Natta catalysts in addition to the HDPE and LLDPE mentioned above. Noteworthy is polypropylene, which cannot be made in sufficiently high molecular weight using free radical or ionic initiators, but which has achieved the status of commodity plastic in its predominately isotactic stereochemistry. Atactic polypropylene, a byproduct of isotactic polypropylene synthesis, is used for modifying asphalt roofing compounds. Other complex catalysts—*reduced metal oxides* and the sodium-based *alfin* catalysts—have been used to manufacture polyethylene and polydienes, respectively. Also certain tungsten- and other transition metal–based catalysts that promote double bond redistribution reactions are used to make commercial polymers from cycloalkenes. Plastics and elastomers manufactured with these various complex coordination catalysts are listed in Table 8.1.

Ziegler–Natta systems are by far the most important of the complex coordination catalysts, with commercial polymer production dating to the 1960s. Over the years many different Ziegler–Natta catalysts have been investigated, both insoluble (heterogeneous) and soluble (homogeneous). The latter, consisting primarily of transition metals sandwiched between cyclopentadiene rings, were not at first considered practical for industrial use because they were not sufficiently active. This changed in the 1980s when it was discovered that catalyst activity could be greatly enhanced by the addition of cocatalysts derived from the reaction of aluminum alkyl with water. As a consequence the "pi sandwich" *metallocenes* have become the object of intense research activity because, in addition to their high catalytic activity, they have been found to offer unprecedented control of polymer stereochemistry. Polymers and copolymers based on metallocene technology, including syndiotactic polystyrene, are already in commercial production, and others, such as syndiotactic polypropylene, are poised to enter the marketplace.

Before we proceed with the fundamentals of complex coordination polymerization, it is worth recalling that Ziegler and Natta shared the Nobel Prize in Chemistry in 1963.

TABLE 8.1. Commercially Available Polymers Synthesized with Complex Coordination Catalysts

Polymer	Principal Stereochemistry	Typical Uses
Plastics		
Polyethylene, high density (HDPE)	—	Bottles, drums, pipe, conduit, sheet, film, wire and cable insulation
Polyethylene, ultrahigh molecular weight (UHMWPE)	—	Surgical prostheses, machine parts, heavy-duty liners
Polypropylene	Isotactic	Automobile and appliance parts, rope, cordage, webbing, carpeting, film
Poly(1-butene)	Isotactic	Film, pipe
Poly(4-methyl-1-pentene)[a]	Isotactic	Packaging, medical supplies, lighting
Polystyrene	Syndiotactic	Specialty plastics

continued

TABLE 8.I. (continued)

Polymer	Principal Stereochemistry	Typical Uses
1,4-Polybutadiene	*trans*	Metal can coatings, potting compounds for transformers
1,4-Polyisoprene	*trans*	Golf ball covers, orthopedic devices
Ethylene-1-alkene[b] copolymer (linear low-density polyethylene, LLDPE)	—	Blending with LDPE, packaging film, bottles
Ethylene-propylene block copolymers (polyallomers)	Isotactic	Food packaging, automotive trim, toys, bottles, film, heat-sterilizable containers
Polydicyclopentadiene[c]	—	Reaction injection molding (RIM) structural plastics
Elastomers		
1,4-Polybutadiene	*cis*	Tires, conveyer belts, wire and cable insulation, footware
1,4-Polyisoprene	*cis*	Tires, footware, adhesives, coated fabrics
Poly(1-octenylene) (polyoctenamer)[c]	*trans*	Blending with other elastomers
Poly(1,3-cyclo-pentenylenevinylene) (norbornene polymer)[c]	*trans*	Molding compounds, engine mounts, car bumper guards
Polypropylene (amorphous)	—	Asphalt blends, sealants, adhesives, cable coatings
Ethylene–propylene copolymer (EPM, EPR)	—	Impact modifier for polypropylene, car bumper guards
Ethylene–propylene–diene copolymer (EPDM)	—	Wire and cable insulation, weather stripping, tire side walls, hose, seals

[a] Usually copolymerized with small amounts of 1-pentene.

[b] 1-Butene, 1-hexene, and 1-octene.

[c] Synthesized by ring-opening metathesis polymerization of the corresponding cycloalkene.

8.2 Heterogenous Ziegler–Natta Polymerization

8.2.1 Heterogeneous Catalysts

The number of compounds and combinations of compounds that fit into the category of Ziegler–Natta catalysts[3,5] are far too numerous to describe here, but a Ziegler–Natta catalyst may be defined as a combination of (1) a transition metal compound of an element from groups IV to VIII, and (2) an organometallic compound of a metal from groups I to III of the periodic table. The transition metal compound is referred to as the catalyst and the organometallic compound as the cocatalyst.

Most commonly the catalyst component consists of halides or oxyhalides of titanium, vanadium, chromium, molybdenum, or zirconium. Compounds of iron and cobalt have, in some instances, also been found to be effective. Some ligands other than halide or oxyhalide that have been investigated include alkoxy, acetylacetonyl, cyclopentadienyl, and phenyl. Cocatalysts are usually hydrides, alkyls, or aryls of metals, such as aluminum, lithium, zinc, tin, cadmium, beryllium, and magnesium. Most important from the commercial standpoint are combinations of titanium trihalides and tetrahalides with trialkylaluminum compounds.

Catalysts are prepared by mixing the components in a dry, inert solvent in the absence of oxygen, usually at a low temperature. They are characterized by having high reactivity toward many nonpolar monomers and are usually capable of affording polymers a high degree of stereoregularity. Catalyst activity usually changes with time, and it is not uncommon for maximum activity to be reached after aging periods of one to two hours. For the $TiCl_4$–AlR_3 (R = alkyl) system it is recognized that initially exchange reactions take place (8.1 to 8.3). Organotitanium compounds are then believed to undergo reduction via homolytic bond cleavage (8.4 and 8.5).

$$AlR_3 + TiCl_4 \rightarrow AlR_2Cl + TiRCl_3 \tag{8.1}$$

$$AlR_2Cl + TiCl_4 \rightarrow AlRCl_2 + TiRCl_3 \tag{8.2}$$

$$AlR_3 + TiRCl_3 \rightarrow AlR_2Cl + TiR_2Cl_2 \tag{8.3}$$

$$TiRCl_3 \rightarrow TiCl_3 + R\cdot \tag{8.4}$$

$$TiR_2Cl_2 \rightarrow TiRCl_2 + R\cdot \tag{8.5}$$

Further reduction may also occur:

$$TiRCl_2 \rightarrow TiCl_2 + R\cdot \tag{8.6}$$

$$TiRCl_3 \rightarrow TiCl_2 + RCl \tag{8.7}$$

In addition, $TiCl_3$ may be formed by the equilibrium

$$TiCl_4 + TiCl_2 \rightleftharpoons 2TiCl_3 \tag{8.8}$$

Radicals formed in these reactions may be removed by combination, disproportionation, or reaction with solvent. While such reactions undoubtedly occur in catalyst formation, it is not known to what extent, and the aging process certainly requires more clarification. Because of the heterogeneous nature of the catalyst, it is certain that reduction of Ti(IV) is not complete.

Although $TiCl_4$ is frequently used in Ziegler–Natta catalysts, better activity is achieved by using $TiCl_3$ directly rather than introducing it by reduction. It is of interest that $TiCl_3$ exists in four different crystalline forms, referred to as α, β, γ, and δ. The α, γ, and δ forms give polymers having a high degree of stereoregularity. All three forms also have similar close-packed layered crystal structures. The β form, however, which has a linear structure, gives much more atactic polymer, indicating that stereoregularity is very much dependent on surface characteristics of the catalyst. The actual amount of stereoregular polymer formed depends on a variety of factors: the nature of the transition metal, the alkyl groups of the cocatalyst, and the presence of additives. Some representative results reported for isotactic polypropylene are shown in Table 8.2.[6,7]

TABLE 8.2. Variation of Polypropylene
Isotacticity with Catalyst[a]

Catalyst[b]	Stereoregularity (%)
$AlEt_3$ + $TiCl_4$	35
$AlEt_3$ + β-$TiCl_3$	45
$AlEt_3$ + α-$TiCl_3$	85
$AlEt_3$ + $ZrCl_4$	55
$AlEt_3$ + VCl_3	73
$AlEt_3$ + $TiCl_4$ + P, As, or Sb compounds	35
$AlEt_2X$ + $TiCl_3$	90–99
$AlEtX_2$ + γ-$TiCl_3$ + amine	99

[a]Data from Jordan[6] and Dawans and Teyssié.[7]

[b]Et = ethyl; X = halogen.

A problem with many of the earlier catalyst systems was their low efficiency. Substantial amounts of catalyst were needed to achieve acceptable yields of polymer, and the spent catalyst had to be removed from the finished product. Efficiency of heterogeneous catalysts can be improved significantly by impregnating the catalyst on a solid support such as $MgCl_2$ or MgO. As an example, a typical $TiCl_3$–AlR_3 catalyst yields about 50 to 200 g of polyethylene per gram of catalyst per hour per atmosphere of ethylene, whereas as much as 7000 g may be produced using a $MgCl_2$-supported catalyst, and without the need for the costly step of removing catalyst from the product. Presumably the support maximizes the number of active sites on the catalyst surface. Such catalyst systems are often referred to as *high-mileage catalysts*.

8.2.2 Mechanism and Reactivity in Heterogeneous Polymerization

Despite the tremendous amount of research that has been conducted in this area,[3,4,8–10] the true mechanism of Ziegler–Natta polymerization is still not entirely clear. It is generally agreed, however, that heterogeneous polymerization occurs at localized active sites on the catalyst surface. The organometallic component is believed to activate the site by alkylation of a transition metal atom at the surface. Of the various mechanisms that have been proposed, the two that are most generally accepted are the so-called *monometallic* and *bimetallic* mechanisms, the former being favored in heterogeneous processes. In both mechanisms, monomer is pictured as being incorporated into the polymer by an insertion reaction between a transition metal atom and a terminal carbon of a coordinated polymer chain.

The monometallic mechanism, with titanium as the transition metal, is shown in Scheme 8.1. According to this mechanism, monomer is complexed at a titanium atom exposed on the catalyst surface by a missing chlorine atom. An insertion reaction takes place, shifting the vacant octahedral position. Then migration of the chain occurs to reestablish the vacant site on the surface.

Either the monometallic or the bimetallic mechanism might occur with soluble catalysts. The bimetallic mechanism, illustrated in Scheme 8.2 for a titanium–aluminum system, in-

SCHEME 8.1. Monometallic mechanism of Ziegler–Natta polymerization.

volves initial pi complexation of monomer to the transition metal (Ti) that is bridged through alkyl groups to an aluminum atom. This is followed by ionization of the transition metal–alkyl bond, formation of a six-membered cyclic transition state, then insertion. In both mechanisms, the polymer chain grows from the catalyst surface by successive insertion reactions of complexed monomer, and the R group originally present in the organometallic cocatalyst ends up as the terminal group of the chain.

As might be expected from a consideration of the metal–alkyl bond at the active site of the catalyst, Ziegler–Natta polymerization bears a superficial resemblance to anionic polymerization. Thus, active hydrogen compounds bring about termination just as they do in anionic processes, and the propagating chains have relatively long lifetimes once monomer is consumed.

A wide variety of nonpolar monomers are polymerizable with Ziegler–Natta catalysts and, predictably, monomer activity decreases with increasing steric hindrance about the double bond. Following is the observed order of reactivity for several 1-alkenes:

$$CH_2{=}CH_2 > CH_2{=}CHCH_3 > CH_2{=}CHCH_2CH_3 >$$
$$CH_2{=}CHCH_2CH(CH_3)_2 > CH_2{=}CHCH(CH_3)_2 >$$
$$CH_2{=}CHCH(CH_2CH_3)_2 > CH_2{=}CHC(CH_3)_3$$

Alkynes are also subject to polymerization with Ziegler–Natta catalysts. Polyacetylene (the conducting polymer discussed in Chapter 3) is an example.[11]

SCHEME 8.2. Bimetallic mechanism of Ziegler–Natta polymerization.

Termination of chain growth may occur a number of ways, leading to both saturated and unsaturated chain ends. As shown for a propagating chain attached to the catalyst surface, these include transfer to monomer (8.9 and 8.10); internal hydride transfer (8.11); transfer to cocatalyst or to an added alkylmetal compound (8.12); and transfer to added hydrogen (8.13). Saturated chain ends are normally more prevalent in Ziegler–Natta polymers.

$$CH_2\text{=}CHR \longrightarrow |\text{--}CH_2CH_2R + CH_2\text{=}\overset{\displaystyle R}{C}\sim\sim\sim \tag{8.9}$$

$$CH_2\text{=}CHR \longrightarrow |\text{--}CH\text{=}CH + CH_3\overset{\displaystyle R}{CH}\sim\sim\sim \tag{8.10}$$

$$|\text{--}CH_2\overset{\displaystyle R}{CH}\sim\sim\sim \longrightarrow |\text{--}H + CH_2\text{=}\overset{\displaystyle R}{C}\sim\sim\sim \tag{8.11}$$

$$AlR'_3 \longrightarrow |\text{--}R' + R'_2Al\text{--}CH_2\overset{\displaystyle R}{CH}\sim\sim\sim \tag{8.12}$$

$$H_2 \longrightarrow |\text{--}H + CH_3\overset{\displaystyle R}{CH}\sim\sim\sim \tag{8.13}$$

Hydrogen is the preferred transfer agent for controlling molecular weight because it reacts cleanly, leaves no residue, and is low in cost. Without the application of such transfer reagents, molecular weight would, in most instances, be too high for commercial use. A true termination reaction is brought about by compounds containing active hydrogen:

$$\underset{\text{│}}{\overset{R}{|}}-CH_2CH\text{\textasciitilde} + HX \longrightarrow \text{│}-X + CH_3\overset{R}{\underset{|}{C}}H\text{\textasciitilde} \tag{8.14}$$

Because of the complexities of Ziegler–Natta polymerization, no comprehensive or unified kinetics scheme has emerged that adequately takes into account surface adsorption, catalyst–cocatalyst interaction, decay of catalyst activity, catalyst morphology, particle size, and so on.[8] Suffice it to say that three types of rate curves have been identified: a constant rate type, a decaying rate type, and one that combines elements of the other two. All three, which are shown schematically in Figure 8.1, exhibit what is referred to as an induction period, that is, a time period leading up to maximum rate. Of the three, the decaying rate type is most common. Rate decay has been attributed to such factors as structural changes that reduce the number or activity of active centers, and encapsulation of active centers by polymer, which prevents approach by monomer.

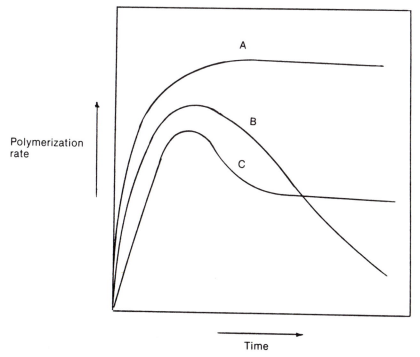

FIGURE 8.1 Types of rate curves observed in Ziegler–Natta polymerization: (A) constant; (B) decaying; and (C) decaying to constant.

Molecular weight distributions are generally broad when insoluble catalysts are used, and much narrower with soluble catalysts.[12] In the former case, the broad distribution may arise from the decay of catalyst activity or from the presence of sites of variable activity.

8.2.3 Stereochemistry of Heterogeneous Polymerization

Without doubt, control of stereochemistry is the most fascinating and useful aspect of Ziegler–Natta polymerization. Isotactic polymers are obtained with insoluble catalysts, with the level of stereoregularity probably depending, to a degree, on how exposed the active site is on the catalyst surface. Presumably, the more exposed the site, the less the stereoregularity. With 1-alkenes, isotactic polymer is formed regardless of the size of the alkyl substituent on the double bond; hence it appears that when monomer is complexed, the smaller CH_2 group is inserted into the catalyst surface with the bulkier end pointing out. When the polymer chain migrates to the carbon bearing the substituent (Schemes 8.1 and 8.2), it thus always approaches from the same side, giving rise to isotactic placement. It is also apparent that the double bond of the monomer undergoes *cis* opening exclusively, that is, *cis* addition to the double bond occurs. This was shown by polymerization of *cis*- and *trans*-1-deuteropropene to give *erythro*-diisotactic and *threo*-diisotactic poly(1-deuteropropylene), respectively (8.15 and 8.16). Similar behavior has been observed with other 1,2-disubstituted olefins. *Trans-i*-butyl propenyl ether, for example, gives *threo*-diisotactic polymer with Ziegler–Natta catalysts. Polymerization of acetylene using an $Al(C_2H_5)_3$–$Ti(OC_4H_9\text{-}n)_4$ catalyst yields *cis* polymer at temperatures below $-78°C$.[11] At 150°C, however, *trans*-polyacetylene is formed, probably by thermal isomerization of the initially formed *cis* polymer. Other Ziegler–Natta catalysts may be used, but this system is the one of choice for achieving good electrical conductivity.

(8.15)

erythro-Diisotactic

(8.16)

threo-Diisotactic

Syndiotactic polypropylene has been prepared under both heterogeneous and homogeneous conditions, but only under homogeneous conditions has a high degree of syndiotactic placement been achieved.

8.2.4 Polymerization of Dienes

The versatility of Ziegler–Natta catalysts is clearly demonstrated in the polymerization of conjugated dienes where several stereochemical arrangements are possible. For 1,3-butadiene, the four possible structures—*cis*-1,4; *trans*-1,4; isotactic 1,2; and syndiotactic 1,2—can all be obtained in relatively pure form with the appropriate choice of catalyst. Some of these catalyst systems are shown in Table 8.3.[2,13,14] Not all catalysts, of course, give such a high degree of stereoregularity; most of the vast number that have been studied give mixtures of stereoisomers. The ones in Table 8.3 were chosen from those reported to give greater than 90% stereospecificity.

Table 8.4 shows the effect of different catalysts on the polymerization of isoprene.[14,15] With this monomer it is possible, in principle, to have *cis*- and *trans*-1,4, 1,2, and 3,4 polymerization. No 1,2 polymer has yet been observed, however, which is not surprising, since the 1,2 bond is more hindered. Butadienes substituted on terminal carbons can be polymerized to give polymers having *cis* or *trans* double bonds and stereogenic carbons in the same structure. *Cis*- or *trans*-1,3-pentadiene, for example, yields isotactic *trans*-1,4 polymer in the presence of AlR_3–$TiCl_3$ or VCl_3 catalyst and mainly isotactic *cis*-1,4 with AlR_3–$Ti(OR)_4$.[14]

It is not possible at the present time to give a definitive mechanism for why any particular stereochemical arrangement is preferred. One theory is based on whether the catalyst coordinates one or both double bonds of the diene. Coordination of one would thus lead to 1,2 polymerization and coordination of both to 1,4. Another theory is that coordination of a π-allylic structure occurs and that the direction of approach of monomer determines the structure. Equations (8.17) and (8.18) illustrate this argument using butadiene as an example. If

TABLE 8.3. Catalysts for the Stereospecific Polymerization of Butadiene

Catalyst[a]	Yield (%)	Polymer structure[b]	Ref. no.
R_3Al + VCl_4	97–98	*trans*-1,4	2
R_3Al + VCl_3	99	*trans*-1,4	2
R_3Al + $VOCl_3$	97–98	*trans*-1,4	2
R_3Al + TiI_4	93–94	*cis*-1,4	2
R_2AlCl + $CoCl_2$	96–97	*cis*-1,4	2
R_3Al + $Ti(OC_6H_9)_4$	90–100	1,2	13
Et_3Al + $Cr(C_6H_5CN)_6$			
Al/Cr = 2	~100	st-1,2	14
Al/Cr = 10	~100	it-1,2	14

[a]Et = ethyl

[b]st = syndiotactic, it = isotactic.

TABLE 8.4. Catalysts for the Stereospecific Polymerization of Isoprene

Catalyst[a]	Yield (%)	Polymer Structure	Ref. No.
$R_3Al + \alpha\text{-}TiCl_3$	91	*trans*-1,4	15
$Et_3Al + VCl_3$	99	*trans*-1,4	14
$Et_3Al + TiCl_4$			
Al/Ti < 1	95	*trans*-1,4	14
Al/Ti > 1	96	*cis*-1,4	14
$Et_3Al + Ti(OR)_4$	95	3,4	14

[a] Et = ethyl

monomer approaches the CH_2–metal (M) bond of the complex, 1,4 polymer forms (8.17). If it approaches the CH–metal bond (8.18), 1,2 polymerization results. This is, of course, a simplified picture that provides no information on the geometric arrangement of the double bond or the tacticity at a stereogenic carbon.

$$(8.17)$$

$$(8.18)$$

Conjugated cyclic dienes such as 1,3-cyclohexadiene (**1**) also undergo Ziegler–Natta polymerization (8.19).

$$(8.19)$$

1

Nonconjugated dienes can be polymerized with coordination catalysts[16,17] to give cyclic repeating units analogous to those already described in free radical and ionic cyclopolymerizations. Typically, 1,5-hexadiene (**2**) is polymerized with $TiCl_4$–$Al(C_2H_5)_3$ to a polymer having repeating cyclopentane rings connected with methylene groups *cis* to each other at positions 1 and 3 (8.20). Soluble catalysts, however, tend to give ordinary 1,2 polymerization.

$$(8.20)$$

2

8.3 Homogeneous Ziegler–Natta Polymerization

8.3.1 Metallocene Catalysts

The earliest metallocene catalysts were bis(cyclopentadienyl)titanium dichloride (**3**, Cp = cyclopentadiene) used in combination with dialkylaluminum chloride compounds (**4**).[18,19] They exhibited low catalytic activity towards ethylene and were generally unreactive towards

$$Cp_2TiCl_2 \qquad\qquad\qquad R_2AlCl$$

3   **4**

propylene. It was later found that addition of water increased the activity substantially. The increase was the result of a reaction between the water and the alkylaluminum cocatalyst to form complex alkylalumoxanes. Subsequently it was shown that especially high activities were realized if methylalumoxanes (MAO) were used in conjunction with metallocene catalysts.[20]

MAO, formed by controlled hydrolysis of trimethylaluminum, has a complex oligomeric structure with molecular weights of 1000 to 1500, most likely consisting of methyl-bridged aluminum atoms alternating with oxygen. Structures **5** and **6** are two proposed structures.[21–23]

5   **6**

MAO is now used with a wide variety of metallocenes having the general structure **7**, in-which the transition metal (M) is usually Zr, Ti, or Hf; X is Cl or alkyl; Z is an optional bridging group, usually $C(CH_3)_2$, $Si(CH_3)_2$, or CH_2CH_2; and R is H or alkyl. Fused ring

7

systems are also found in place of the cyclopentadiene ligands. Examples of catalysts are **8** and **9**, which form isotactic and syndiotactic polypropylene, respectively. For simplification, such catalysts are usually written in condensed form. Compound **8**, for example, which has a dimethylsilyl (Me$_2$Si) bridge and indene (Ind) ligands, is written Me$_2$Si(Ind)$_2$ZrCl$_2$; and **9**, containing an isopropylidene (Me$_2$C) bridge and fluorene (Flu) and cyclopentadiene (Cp) ligands is Me$_2$C(Flu)(Cp)ZrCl$_2$.

$(CH_3)_2Si \quad ZrCl_2$ $(CH_3)_2C \quad ZrCl_2$

8   **9**

8.3.2 Mechanism and Reactivity with Metallocene Catalysts

A major difference between metallocene and conventional heterogeneous Ziegler–Natta catalysts is that the former have well-defined molecular structures, and polymerization occurs at one position in the molecule, the transition metal atom. Polymer chemists, in fact, like to refer to the metallocenes as *single-site catalysts*, in contrast to the multiple active sites characteristic of the more traditional heterogeneous catalysts. The active site is believed to be cationic in nature. A proposed mechanism[21,22] for formation of the active site in a zirconium catalyst, L$_2$ZrCl$_2$ (where L represents the pi ligands), involves initial complexation between MAO and the catalyst, followed by Cl–CH$_3$ exchange to form L$_2$Zr(CH$_3$)$_2$. The methylated zirconocene reacts further with MAO, as shown in Scheme 8.3, to form the active species (**10**).

SCHEME 8.3. Formation of the active site in a zirconocene catalyst.

SCHEME 8.4. Possible polymerization mechanism for ethylene.

The mechanism by which polymerization takes place is the subject of much conjecture,[21,22] but initial pi complexation between monomer and the cationic site is most likely, followed by insertion, as shown in Scheme 8.4 for ethylene polymerization.

Polymers prepared with metallocene catalysts have narrower molecular weight distributions than those prepared with heterogeneous catalysts and consequently better mechanical properties. Polydispersities ($\overline{M}_w/\overline{M}_n$) range from 2 to 2.5 for the former, compared with 5 to 6 for the latter. This is a consequence of the single-site feature of the metallocenes. As mentioned earlier, heterogeneous catalysts exhibit rate decay, and probably contain sites of variable activity, which account for the broader molecular weight distribution. The molecular weight of the metallocene-based polymers decreases with increasing polymerization temperature, increasing catalyst concentration, and addition of hydrogen to the monomer feed.

Activities of metallocene catalysts are from 10 to 100 times higher than those of conventional Ziegler–Natta catalysts. While it is often difficult to correlate structural variables with activity, the following generalizations can be made:

1. For the group 4B metals, the order of activity is Zr > Ti > Hf.
2. Alkyl groups on the cyclopentadiene rings increase catalyst activity if they are not too bulky. Large, bulky alkyl groups and electron-withdrawing groups decrease the activity.
3. Increasing the size of the groups attached to the atom bridging the cyclopentadiene rings (C or Si) reduces the activity.
4. MAO affords much higher catalyst activities than ethyl- or higher alkylalumoxane cocatalysts.

Another way that metallocene catalysts differ from heterogeneous catalysts is that they are capable of polymerizing strained cycloalkenes such as cyclobutene or norbornene (**11**) through the double bond to yield crystalline, high-melting stereoregular polymers (8.21).

Ring-opening polymerization of cycloalkene monomers is more common with heterogeneous Ziegler–Natta catalysts.

(8.21)

11

8.3.3 Stereochemistry of Metallocene-Catalyzed Polymerization

Metallocene catalysts exhibit a remarkable ability to control polymer stereochemistry.[21] Structural variations allow the synthesis of atactic, isotactic, or syndiotactic polypropylene and higher poly(1-alkenes). Achiral catalysts of the type CpZrCl$_2$, bridged or unbridged, yield atactic polymer. Catalysts such as **8** (chiral) and **9** (achiral) yield isotactic and syndiotactic polymer, respectively. The much different sizes of the two pi ligands of **8** is assumed to play a role in the formation of syndiotactic polymer. Remarkably, substitution of a methyl group on the cyclopentadiene ring of **9** results in *hemiisotactic* polypropylene (alternate methyls isotactic, the others atactic). It may be recalled from Chapter 3 (exercise 4h) that hemiisotactic "polypropylene" had been made earlier by hydrogenation of poly(2-methyl-1,3-pentadiene).[24]

Another unusual development concerns catalysts that are capable of producing polypropylene having alternating atactic and isotactic blocks.[25] The zirconium catalyst **12**, for example, can rotate between chiral and achiral geometries (8.22), with the chiral form being responsible for the isotactic blocks and the achiral form the atactic blocks. Because the isotactic blocks form crystalline microdomains in the polymer matrix, the stereoblock

(8.22)

12   **12**

copolymers exhibit elastomeric properties. The degree of isotacticity is sensitive to propylene pressure and reaction temperature. It is now possible, therefore, to make thermoplastic elastomers having a range of properties from a single monomer in a one-pot synthesis!

Exactly why isotactic or syndiotactic chains are favored for a given catalyst is open to various mechanistic interpretations. Most models[21,22] take into account steric interactions involving the pi ligands attached to the metal atom and the last inserted monomer unit of the polymer chain, as the incoming monomer inserts itself into the constrained active site. Whether or not MAO is coordinated closely enough to influence the reaction is open to question. A reaction sequence for isotactic placement involving a catalyst such as **8** is shown in Scheme 8.5. Optically active isotactic polymer would form from a pure enantiomer of **8**.

SCHEME 8.5. A mechanism for isotactic placement with a metallocene catalyst.

In the case of syndiotactic placement at the reactive site of catalyst **9**, the larger fluorenyl group restricts the approach of the propylene molecule such that one side of the coordination site is enantioselective to one face of the monomer, and the other side of the coordination site shows the opposite enantioselectivity. Thus as the coordination site moves from side to side, syndiotactic polymer is formed. A methyl group on the cyclopentadiene ring presumably interferes with the selectivity on one side only. Since only one side of the coordination site is then stereoselective, a hemiisotactic polymer forms.

Depending on structure, metallocenes (like heterogenous catalysts) polymerize conjugated dienes to 1,4-*cis*, 1,4-*trans*, or 1,2 polymers, and isolated dienes to cyclopolymers.

8.4 Ziegler–Natta Copolymerization

Random copolymers of ethylene and 1-alkenes (α-olefins) are obtainable with Ziegler–Natta catalysts,[26,27] the most important being those of ethylene and 1-alkenes (LLDPE) and ethylene with propylene (EPM or EPR and EPDM).* From the representative reactivity ratios presented in Table 8.5, it can be seen that ethylene is much more reactive than higher alkenes with both heterogeneous and homogeneous catalysts. In most instances, $r_1 r_2$ is close to unity. With heterogeneous catalysts, a wide range of compositions is generally obtained, possibly because different active sites may give rise to different reactivity ratios, or because of the previously mentioned encapsulation of active sites leading to decay of activity.

A more homogeneous polymer composition is obtained with soluble catalysts, particularly if monomer composition is carefully controlled to remain relatively constant during

---

*EPM(R) = ethylene–propylene monomer (rubber); EPDM = ethylene–propylene–diene monomer.

TABLE 8.5. Representative Reactivity Ratios in Ziegler–Natta Copolymerization[a]

Monomer 1	Monomer 2	Catalyst[b]	r_1	r_2
Heterogeneous				
Ethylene	Propylene	$TiCl_3/AlR_3$	15.72	0.110
Ethylene	Propylene	VCl_3/AlR_3	5.61	0.145
Ethylene	1-Butene	VCl_3/AlR_3	26.90	0.043
Propylene	1-Butene	VCl_3/AlR_3	4.04	0.252
Homogeneous				
Ethylene	Propylene	Cp_2ZrMe_2	31	0.005
Ethylene	Propylene	$[Z(Ind)_2]ZrCl_2$[c]	6.6	0.06
Ethylene	1-Butene	Cp_2ZrMe_2	55	0.017
Ethylene	1-Hexene	Cp_2ZrMe_2	69	0.02

[a] Data from Boor[3] and Kaminsky.[22]
[b] $R = C_6H_{13}$; Cp = cyclopentadiene; Me = methyl; Z = bridging group; Ind = indene.
[c] $Z = CH_2CH_2$.

polymerization. Such is the case with production of EPM (EPR), which contains about 60 parts ethylene to 40 parts propylene. Crosslinking of such elastomers is accomplished with peroxides.

The other commercially important type of ethylene–propylene copolymer (EPDM) is prepared with small amounts of a nonconjugated diene to facilitate crosslinking.[28] Typical dienes are ethylidenenorbornene (**13**), dicyclopentadiene (**14**), and 1,4-hexadiene (**15**), with **13** being the diene of choice. Both types of elastomers have excellent ozone resistance by virtue of having no unsaturation as an integral part of the backbone. The newer metallocene catalysts produce EPDM copolymers with improved properties.[29]

13   14   15

As was the case with free radical initiation, 1,2-disubstituted monomers, with the exception of cycloalkenes, display little tendency to homopolymerize under Ziegler–Natta conditions. They will copolymerize with ethylene or 1-alkenes[26]; however, it is usually necessary to use a large excess of the 1,2 monomer because of its lower reactivity. It is of interest that there is some tendency toward alternation, with *erythro*-diisotactic stereochemistry (see Chapter 3, Figure 3.13), when sufficient 1,2-disubstituted monomer is used.

A number of block copolymers prepared with Ziegler–Natta catalysts have been reported; in most cases, however, there is serious question as to whether the compositions also include significant amounts of homopolymer.[3] The problem arises with uncertainties over the lifetimes of propagating chains, which may become detached from catalyst by chain transfer, thereby leaving the catalyst still active to initiate homopolymerization of a second monomer. At the present time, therefore, the Ziegler–Natta method is inferior to anionic polymerization for synthesizing carefully tailored block copolymers. Nevertheless, block copolymers of ethylene and propylene (Eastman Kodak's *polyallomers*) have been commercialized. Unlike the elastomeric random copolymers of ethylene and propylene, these are high-impact plastics exhibiting crystallinity characteristics of both isotactic polypropylene and linear polyethylene, and, no doubt, they contain homopolymers in addition to block copolymers. (Ethylene–propylene block copolymers, designated TPR, have also been marketed by Uniroyal.)

Amorphous copolymers of cycloalkenes with ethylene or 1-alkenes (commonly called *cycloolefin copolymers*, or COCs) have been synthesized with metallocene catalysts.[30,31] Copolymers having a wide range of glass transition temperatures can be made simply by varying the type and amount of cycloalkene.

8.5 Supported Metal Oxide Catalysts

Oxides of a variety of metals on finely divided inert support materials initiate polymerization of ethylene and other vinyl monomers by a mechanism that is assumed to be similar to that of heterogeneous Ziegler–Natta polymerization; that is, initiation probably occurs at active sites on the catalyst surface.[3,32] Among the metals that have been investigated are chromium, vanadium, molybdenum, nickel, cobalt, niobium, tantalum, tungsten, and titanium. Typical supports include alumina, silica, and charcoal. Chromium-based catalysts originally developed by the Phillips Petroleum Company are often referred to as *Phillips catalyst*. Other metal oxide catalysts were developed primarily at Standard Oil of Indiana.

Catalysts are prepared by one of two methods. The support material is impregnated with the metal ion, then heated in air at a high temperature to form the metal oxide. Alternatively, when the support material is an oxide such as alumina, the two oxides are coprecipitated and dried in air. In each case the catalyst is activated by treatment with a reducing agent such as hydrogen, metal hydride, or carbon monoxide. Poisoning of the catalyst occurs in the presence of such materials as water, oxygen, or acetylene. The function of the support appears to be more than simply providing a large surface area. Some type of interaction must occur between oxide and support because the oxide alone behaves differently.

Ethylene is the most important monomer used with supported metal oxide catalysts. Much of the high-density polyethylene is, in fact, now manufactured this way. Unlike Ziegler–Natta catalysts, which give rise to polymer having primarily saturated end groups, the metal oxides yield polyethylene with approximately equal amounts of saturated and unsaturated chain ends. Initiation is presumed to involve initial formation of a pi complex which disproportionates to yield both metal–$CH{=}CH_2$ and metal–CH_2CH_3 coordination (8.23, M = metal). Propagation then proceeds via an insertion mechanism.

$$[M] \xrightarrow{C_2H_4} [M] \overset{CH_2}{\underset{CH_2}{\Vert}} \longrightarrow [M]{-}CH{=}CH_2 + [M]{-}CH_2CH_3 \qquad (8.23)$$

The supported metal oxides are not as active as Ziegler–Natta catalysts, and they do not give rise to a high degree of stereoregularity. Propylene forms partially crystalline polymer, but higher 1-alkenes give amorphous product. Styrene is unreactive. Dienes polymerize to yield primarily *trans*-1,4 polymer. Copolymers of ethylene and 1-alkenes have been prepared with chromia and molybdena catalysts.

8.6 Alfin Catalysts

Alfin catalysts[33] are a class of heterogeneous catalysts consisting of an alkenylsodium compound, an alkoxide, and an alkali metal halide. The name is derived from *al*cohol and ole*fin*, which are used to prepare the catalyst. Alfin catalysts are particularly effective in polymerizing butadiene and isoprene to very-high-molecular-weight polymer, although they will polymerize certain other monomers, notably styrene.

The most effective catalyst for diene polymerization consists of allylsodium, sodium isopropoxide, and sodium chloride, a combination typically formed by a reaction sequence such as that shown in reactions (8.24), (8.25), and (8.26).

$$1.5 \ C_5H_{11}Cl + 3Na \rightarrow 1.5 \ C_5H_{11}Na + 1.5 \ NaCl \tag{8.24}$$

$$1.5 \ C_5H_{11}Na + (CH_3)_2CHOH \rightarrow \\ 0.5 \ C_5H_{11}Na + (CH_3)_2CHONa + C_5H_{12} \tag{8.25}$$

$$0.5 \ C_5H_{11}Na + 0.5 \ CH_2{=}CHCH_3 \rightarrow 0.5 \ CH_2{=}CHCH_2Na + 0.5 \ C_5H_{12} \tag{8.26}$$

Alfin catalysts polymerize butadiene within minutes to a polymer having a molecular weight of several million, consisting of about 68% *trans*-1,4 and the remainder divided equally between *cis*-1,4 and 1,2. The process resembles Ziegler–Natta polymerization in that one allyl group is incorporated into each polymer chain. How this comes about, however, is open to speculation; both free radical and anionic mechanisms have been proposed. Butadiene polymerizes about 40 times more rapidly than isoprene with alfin catalyst, and 2,3-dimethylbutadiene is unreactive.

The major drawback to the alfin system is that molecular weights are too high to be practicable; hence it has had little commercial success.

8.7 Metathesis Polymerization

Alkenes undergo a double bond redistribution reaction (8.27) called *olefin metathesis** in the presence of certain complex coordination catalysts.[34] An example is the formation of a 1:2:1 mixture of 2-butene, 2-pentene, and 3-hexene from 2-pentene. Bond redistribution involves cleavage of the double bonds (transalkylidenation). The alternate possibility, scission of the single bonds adjacent to the double bonds (transalkylation) has been shown not to occur by experiments with deuterated alkenes.[35] Tungsten compounds are particularly effective for olefin metathesis. The catalysts are formed by the reaction of tungsten hexachloride (WCl_6) with alcohol and ethylaluminum dichloride ($C_2H_5AlCl_2$). Other transition metals, notably ruthenium, molybdenum, rhenium, and titanium, are also effective.

$$\begin{array}{cc} R_1CH{\dashv}CHR_2 \\ R_1CH{\dashv}CHR_2 \end{array} \rightleftharpoons \begin{array}{cc} R_1CH & CHR_2 \\ \| & \| \\ R_1CH & CHR_2 \end{array} \tag{8.27}$$

*From the Greek *meta* (change) and *tithemi* (place).

There is ample evidence to suggest that this remarkable bond redistribution is brought about by an intermediate metal carbene species (**16**) that reacts with alkene to form a transient metallacyclobutane (**17**) leading subsequently to products (8.28). The metal (M) is bracketed in the equation to indicate that other ligands are present. Metal carbenes have been identified by NMR in metathesis reactions,[36,37] and stable metal carbenes such as $(C_6H_5)_2C=W(CO)_5$ are known to initiate metathesis polymerization.[38]

$$
\begin{array}{ccc}
\underset{[M]}{\overset{C}{\|}} + \underset{C}{\overset{C}{\|}} & \longrightarrow & \underset{[M]-C}{\overset{C-C}{\mid \quad \mid}} & \longrightarrow & \underset{[M]=C}{\overset{C=C}{}} \\
\mathbf{16} & & \mathbf{17} & &
\end{array}
\tag{8.28}
$$

Synthesis of polymers by olefin metathesis assumed prominence in the 1970s,[39,40] and is now recognized as a particularly convenient way to make polymers containing precisely spaced carbon–carbon double bonds in the backbone from either cycloalkenes (8.29) or acylic dienes (8.30).

$$
\text{(structure)} \xrightarrow{\text{CATALYST}} \text{(structure)} \tag{8.29}
$$

$$
\text{(structure)} \xrightarrow{\text{CATALYST}} \text{(structure)} + CH_2=CH_2 \tag{8.30}
$$

8.7.1 Ring-Opening Metathesis Polymerization

Ring-opening metathesis polymerization,[41–44] commonly referred to by its acronym ROMP, yields living polymers with a double bond configuration that is primarily *trans*. One of the propagation steps in the reaction is shown in reaction (8.31). High molecular weights (in excess of 10^5) are usually obtained, and polymerization is accompanied by formation of low-molecular-weight oligomers. The rate of polymerization appears to be a function of metal carbene reactivity rather than of ring strain in the cycloalkene.[45] Even relatively unstrained monomers such as cyclohexene will undergo ring-opening polymerization under certain conditions.[46]

$$
\underset{[M]}{\overset{\sim\sim\sim CH}{\|}} + \underset{}{\overset{}{(CH_2)_n}} \longrightarrow \underset{[M]}{\overset{\sim\sim\sim CH-}{\mid}}(CH_2)_n \longrightarrow \sim\sim\sim CH=CH \underset{[M]=CH}{\overset{(CH_2)_n}{}} \tag{8.31}
$$

One of the most unexpected findings in metathesis polymerization is that certain group VIII metal compounds, for example, $RuCl_3$, catalyze the ring-opening polymerization of monomers such as 7-oxanorbornene derivatives (**18**) to molecular weights in excess of 10^6 *in aqueous solution* (8.32).[47]

$$
\text{(structure)} \xrightarrow{RuCl_3 . H_2O} \text{(structure)} \tag{8.32}
$$

18

This is contrary to the usual polymerization conditions, in which moisture must be rigorously excluded. Furthermore, the catalyst, which appears to owe its activity to the formation of a Ru^{2+}–monomer complex, can be recycled, and the activity actually increases with successive use. Considerably more work in this area may be anticipated in view of the advantages inherent in the use of aqueous polymerization conditions.

Examples of three metathesis polymerizations that have been developed commercially are shown in reactions (8.33) through (8.35). Polyoctenamer (**19**) and norbornene polymer (**20**) are manufactured in Europe, the former by Chemische Werke Huls in Germany

$$\longrightarrow \quad -\!\!\left[CH\!\!=\!\!CH(CH_2)_6\right]\!\!- \qquad\qquad (8.33)$$

19

$$\longrightarrow \quad \left[CH\!\!=\!\!CH-\!\!\bigcirc\right] \qquad\qquad (8.34)$$

20

$$\longrightarrow \qquad\qquad\qquad (8.35)$$

21

(trade name Vestenamer) and the latter by CdF Chemie in France (Norsorex). Polydicyclopentadiene, which is manufactured in the United States by Hercules (Metton), is formed in a two-step reaction, the first involving ROMP of the more strained bridged ring to form a linear polymer (**21**), then polymerization through the remaining double bond by metathesis or alkene addition, or both, to form a crosslinked polymer. Because of the monomer's high reactivity, the process is adaptable to reaction injection molding (see Chapter 4) with monomer plus catalyst constituting one component of the RIM system and monomer plus activator constituting the other.[48]

Metathesis polymerization works equally as well with cyclic polyenes. Polymerization of 1-methyl-1,5-cyclooctadiene (**22**) (8.36) yields an elastomer having the structure of a regularly alternating copolymer of isoprene and butadiene. Similarly, *cis,trans*-cyclodeca-1,5-diene (**23**) forms a polymer having the structure of a regular 2:1 polymer of butadiene and ethylene (8.37). Particularly interesting is the synthesis of polyacetylene (primarily *trans*) by metathesis polymerization of 1,3,5,7-cyclooctatetraene (**24**) (8.38).[49,50]

$$\begin{array}{c} CH_3 \\ \\ \xrightarrow[\ C_2H_5OH\ ]{\ WCl_6, C_2H_5AlCl_2\ } \end{array} \qquad \left[\begin{array}{c} CH_3 \\ \mid \\ CH_2C\!\!=\!\!CHCH_2CH_2CH\!\!=\!\!CHCH_2 \end{array}\right] \qquad (8.36)$$

22

$$\text{23} \xrightarrow[\text{C}_2\text{H}_5\text{ OH}]{\text{WCl}_6\cdot\text{C}_2\text{H}_5\text{AlCl}_2} +(\text{CH}_2\text{CH}_2)(\text{CH}_2\text{CH}=\text{CHCH}_2)_2+ \qquad (8.37)$$

$$\text{24} \xrightarrow[\text{Al(C}_2\text{H}_5)_2\text{Cl}]{\text{W[OCH(CH}_2\text{Cl)}_2]_2 \text{ or } 3} +(\text{CH}=\text{CH})_4+ \qquad (8.38)$$

8.7.2 Acyclic Diene Metathesis Polymerization

Acyclic diene metathesis (commonly called ADMET)[51,52] differs from ROMP in that by-product alkene, usually ethylene, must be removed to obtain high-molecular-weight polymer. Highly purified cation-free tungsten, molybdenum, or ruthenium catalysts are commonly employed. ADMET polymerization of 1,9-decadiene (8.39), for example, yields the same product, polyoctenamer, that is obtained by ROMP of cyclooctene (8.33).

$$\text{CH}_2=\text{CH(CH}_2)_6\text{CH}=\text{CH}_2 \xrightarrow{\text{CATALYST}} +\text{CH}=\text{CH(CH}_2)_6+ + \text{CH}_2=\text{CH}_2 \quad (8.39)$$

ADMET is also useful for the synthesis of functionalized polymers.[53] An example is the synthesis of an unsaturated polymer containing ester functionality (8.40).[54]

$$\text{CH}_2=\text{CH(CH}_2)_8\overset{\overset{\text{O}}{\|}}{\text{C}}\text{OCH}_2\text{CH}_2\text{O}\overset{\overset{\text{O}}{\|}}{\text{C}}(\text{CH}_2)_8\text{CH}=\text{CH}_2 \xrightarrow{\text{CATALYST}}$$

$$(8.40)$$

$$+\text{CH}=\text{CH(CH}_2)_8\overset{\overset{\text{O}}{\|}}{\text{C}}\text{OCH}_2\text{CH}_2\text{O}\overset{\overset{\text{O}}{\|}}{\text{C}}(\text{CH}_2)_8+ + \text{CH}_2=\text{CH}_2$$

Methathesis reactions are equilibrium processes. Given that ethylene is a byproduct of ADMET polymerization, addition of ethylene to an unsaturated polymer can effect *depolymerization*,[55] as shown for polyisoprene (8.41). Such reactions are of interest for degrading scrap rubber into potentially useful products.

$$(8.41)$$

References

1. K. Ziegler, E. Holzkamp, H. Breil, and H. Martin, *Angew. Chem.*, **67**, 541 (1955).
2. G. Natta, *J. Polym. Sci.* **48**, 219 (1960).
3. J. Boor Jr., *Ziegler–Natta Catalysts and Polymerizations*, Academic Press, New York, 1979.
4. G. Natta and F. Danusso (Eds.), *Stereoregular Polymers and Stereospecific Polymerizations*, 2 vols., Pergamon Press, New York, 1967.

5. L. Reich and A. Schindler, *Polymerization by Organometallic Compounds*, Wiley-Interscience, New York, 1966: (a) Chap. 4; (b) Chap. 5.
6. D. O. Jordan, in *The Stereochemistry of Macromolecules*, Vol. 1 (A.D. Ketley, Ed.), Marcel Dekker, New York, 1967, Chap. 1.
7. F. Dawans and P. Teyssié, *Bull. Soc. Chim. Fr.*, 2376 (1963).
8. T. Keii, *Kinetics of Ziegler–Natta Polymerization*, Halsted Press, New York, 1973.
9. V. A. Zakharov, G. D. Bukatov, and Y. I. Yermakov, *Adv. Polym. Sci.*, **51**, 61 (1983).
10. A. Guyot, R. Spitz, C. Journaud, and O. Eisenstein, *Macromol. Symp.*, **89**, 39 (1995).
11. T. Ito, H. Shirakawa, and S. Ikeda, *J. Polym. Sci., Polym. Chem. Ed.*, **12**, 11 (1974).
12. U. Zucchini and G. Cecchin, *Adv. Polym. Sci.*, **51**, 101 (1983).
13. G. Wilke, *J. Polym. Sci.*, **38**, 45 (1959).
14. G. Natta, L. Porri, A. Carbonaro, and G. Stoppa, *J. Makromol. Chem.*, **77**, 114 (1964); G. Natta, L. Porri, and A. Carbonaro, ibid., p. 126.
15. C. E. H. Bawn, A. M. North, and J. S. Walker, *Polymer*, **5**, 419 (1964).
16. W. Marconi, in *The Stereochemistry of Macromolecules*, Vol. 1 (A. D. Ketley, Ed.), Marcel Dekker, New York, 1967, Chap. 5.
17. R. J. Cotter and M. Matzner, *Ring-Forming Polymerizations*, Part A, Academic Press, New York, 1969, pp. 42–47.
18. G. Natta, P. Pino, P. Corradini, F. Danusso, E. Mantica, G. Mazzanti, and G. Moraglio, *J. Am. Chem. Soc.*, **77**, 1708 (1955).
19. D. S. Breslow and N. R. Newburg, *J. Am. Chem. Soc.*, **79**, 5072 (1957).
20. H. Sinn, W. Kaminsky, H. J. Vollmer, and R. Woldt, *Angew. Chem. Int. Ed. Engl.*, **19**, 390 (1980).
21. J. Huang and G. L. Rempel, *Prog. Polym. Sci.*, **20**, 459 (1995).
22. W. Kaminsky, *Macromol. Chem. Phys.*, **197**, 3907 (1996).
23. C. J. Harlan, S. G. Bott, and A. R. Baron, *J. Am. Chem. Soc.*, **117**, 6465 (1995).
24. M. Farina, G. Di Silvestro, and P. Sozzani, *Prog. Polym. Sci.*, **16**, 219 (1991).
25. G. W. Coates and R. M. Waymouth, *Science*, **267**, 217 (1995).
26. G. E. Ham (Ed.), *Copolymerization*, Wiley-Interscience, New York, 1964: C. A. Lukach and H. M. Spurlin, Chap. 4A; G. Crespi, A. Valvassori, and G. Sartori, Chap. 4C.
27. I. Pasquon, A. Valvassori, and G. Sartori, in *The Stereochemistry of Macromolecules*, Vol. 1 (A. D. Ketley, Ed.), Marcel Dekker, New York, 1967, Chap. 4.
28. S. Cesca, *Macromol Rev.*, **10**, 1 (1975).
29. M. Galimberti, E. Martini, F. Piemontesi, F. Sartori, I. Camurati, L. Resconi, and E. Albizzati, *Macromol. Symp.*, **89**, 258 (1995).
30. H. Cherdron, M. J. Brekner, and F. Osan, *Angew. Chem. Int. Ed. Engl.*, **34**, 2273 (1994).
31. W. Kaminsky, A. Bark, and M. Arndt, *Makromol. Chem. Macromol. Symp.*, **47**, 83 (1991).
32. D. R. Witt, in *Reactivity, Mechanism and Structure in Polymer Chemistry*, (A. D. Jenkins and A. Ledwith, Eds.), Wiley-Interscience, New York, 1974, Chap. 13.
33. L. Reich and A. Schindler, *Polymerization by Organometallic Compounds*, Wiley-Interscience, New York, 1966, Chap. 5.
34. K. J. Ivin and I. C. Mol, *Olefin Metathesis and Metathesis Polymerization*, 2nd. ed., Academic, New York, 1996.
35. N. Calderon, E. A. Ofstead, J. P. Ward, W. A. Judy, and K. W. Scott, *J. Am. Chem. Soc.*, **90**, 4133 (1968).
36. J. Kress, M. Wesolek, and J. A. Osborn, *J. Chem. Soc., Chem. Commun.*, 514 (1982).
37. J. Kress, J. A. Osborn, R. M. E. Green, K. J. Ivin, and J. J. Rooney, *J. Am. Chem. Soc.*, **109**, 899 (1987).
38. T. J. Katz, S. J. Lee, and N. Acton. *Tetrahedron Lett.*, 4247 (1976).
39. N. Calderon, E. A. Ofstead, and W. A. Judy, *Angew. Chem. Int. Ed. English*, **15**, 401 (1976).
40. N. Calderon, *J. Macromol. Sci., Rev. Macromol. Chem.*, **C7**, 105 (1972).
41. R. H. Grubbs and W. Tumas, *Science*, **243**, 907 (1989).
42. D. S. Breslow, *Prog. Polym. Sci.*, **18**, 1141 (1993).
43. F. Stelzer, *J. Macromol. Sci.-Pure Appl. Chem.*, **A33**, 941 (1996).
44. A. J. Amass, in *New Methods of Polymer Synthesis* (J. R. Ebdon, Ed.), Blackie, London, 1991.

45. P. A. Patton and T. J. McCarthy, *Macromolecules*, **17**, 2939 (1984).
46. P. A. Patton, C. P. Lillya, and T. J. McCarthy, *Macromolecules*, **19**, 1266 (1986).
47. B. M. Novak and R. H. Grubbs, *J. Am. Chem. Soc.*, **110**, 7542 (1988).
48. A. Bell, in *Catalysis in Polymer Synthesis* (E. J. Vandenberg and J. C. Salamone, Eds.), ACS Symp. Ser. 496, American Chemical Society, Washington, D.C., 1992, Chap. 9.
49. Y. V. Korshak, V. V. Korshak, G. Kanischka, and H. Hocker, *Makromol. Chem., Rapid. Commun.*, **6**, 685 (1985).
50. F. L. Klavetter and R. H. Grubbs, *Polym. Preprints*, **28**(2), 425 (1987).
51. E. A. Ofstead and K. B. Wagener, in *New Methods for Polymer Synthesis* (W. J. Mijs, Ed.), Plenum, New York, 1992, Chap. 8.
52. M. Lindmark-Hamberg and K. B. Wagener, *Macromolecules*, **20**, 2949 (1987).
53. J. E. O'Gara, J. D. Portmess, and K. B. Wagener, *Macromolecules*, **26**, 2837 (1993).
54. C. G. Bauch, K. B. Wagener, and J. M. Boncella, *J. Makromol. Chem., Rapid Commun.*, **12**, 413 (1991).
55. K. B. Wagener and R. D. Puts, *Polym. Prepr.*, **32**(1), 379 (1991).

Review Exercises

1. Set up a table outlining the likelihood of converting the following monomers to high-molecular-weight polymer under free radical, cationic, anionic, group transfer, or coordination conditions: methyl acrylate, 1-pentene, methacrylonitrile, ethylene, propylene, vinyl acetate, isoprene, *i*-propyl vinyl ether, 1,1-dichloroethylene. (List the monomers by structure in the table.)

2. Explain (a) how sites of variable activity in a heterogeneous catalyst might result in a polymer of high polydispersity, (b) why molecular hydrogen is useful for controlling molecular weight in Ziegler–Natta polymerization, and (c) why only coordination catalysts are effective for polymerizing propylene and higher 1-alkenes.

3. Explain how LDPE and LLDPE differ in terms of structure, synthesis, and properties.

4. Explain the advantages of polypropylene elastomers prepared with metallocene catalysts compared with ABA block copolymer elastomers.

5. Give the structure and stereochemical designation of all possible polycyclohexenes that could, in principle, be obtained by stereospecific polymerization (*Pure Appl. Chem.*, **12**, 645, 1966). Which are unlikely to be obtainable in practice?

6. It was mentioned in Section 8.2.4 that 1,3-pentadiene undergoes Ziegler–Natta polymerization to form both *cis*- and *trans*-isotactic-1,4-poly(1,3-pentadiene). There are *nine* other possible stereoregular forms for poly(1,3-pentadiene). What are they?

7. Assuming only head-to-tail polymerization and *cis* double-bond opening occurs, give the structure and stereochemical designation of all stereoregular polymers obtainable in principle by stereospecific polymerization of (a) 2-methyl-1-butene, (b) *trans*-2-pentene, (c) isoprene, and (d) propyne.

8. Draw the structure of (a) *erythro*-diisotactic-*trans*-1,4-poly(2,4-heptadiene) and (b) *erythro*-diisotactic-*cis*-1,4-poly(1,3-cyclohexadiene).

9. Ignoring stereochemistry, predict the repeating units obtained by conventional (not metathesis) Zeigler–Natta polymerization of (a) 2,6-dimethyl-1,6-heptadiene and (b) 1,5-cyclooctadiene.

10. Predict the structure of the linear polymer formed by Ziegler–Natta polymerization of 1,6-heptadiyne.

What type of properties might you expect for such a polymer? (J. K. Stille and D. A. Frey, *J. Am. Chem. Soc.*, **83**, 1697, 1961).

11. Predict the structure of the polymers formed by ROMP of the following monomers (Ref. 39, and T. M. Swager et al., *J. Am. Chem. Soc.*, **110**, 2973, 1988):

$(CH_3)_3Si$ $\quad$ CH_3 $\quad\quad$ C_2H_5

12. Give the repeating unit of the polymers prepared by ADMET polymerization of (a) 1,5-hexadiene and (b) $(CH_3)_2Si(CH_2CH{=}CH_2)_2$. [S. Cummings et al., *Polym. Prepr.*, **37**(2), 192, 1996].

13. When 1,3,5,7-cyclooctatetraene undergoes ring-opening metathesis polymerization (8.38) in dilute solution, benzene is formed irreversibly in 75% yield. Benzene formation is insignificant in bulk metathesis polymerization. Considering the mechanism of metathesis polymerization, suggest a reason for the different behavior. (F. L. Klavetter and R. H. Grubbs, *J. Am. Chem. Soc.*, **110**, 7807, 1988)

CHAPTER 9

REACTIONS OF
VINYL POLYMERS

9.1   Introduction

The subject of polymer reactions is so broad and the literature so extensive that it is not possible to present more than a survey of the field in a book of this type. Chemical modifications encompass a wide range of applications: ion-exchange resins; polymeric reagents and polymer-bound catalysts; polymeric supports for chemical reactions; degradable polymers to address medical, agricultural, or environmental concerns; flame-retardant polymers; and surface, treatments to improve such properties as biocompatibility or adhesion, to name a few. Books and symposium proceedings[1–5] covering reactions of polymers treat individual topics in depth. The purpose of this chapter is to summarize and illustrate chemical modifications of vinyl polymers. For convenience these are grouped into five general categories: (1) reactions that involve the introduction or modification of functional groups, (2) reactions that introduce cyclic units into the polymer backbone, (3) reactions leading to block and graft copolymers, (4) crosslinking reactions, and (5) degradation reactions.

Before we discuss each type, it should be pointed out that while polymers will, in principle, undergo all the reactions common to low-molecular-weight compounds, contrasting behavior may arise from the polymer's molecular size, shape, or morphology. If a polymer is semicrystalline, for example, the crystalline regions, because of their impermeability, are usually inaccessible to chemical reactants, and reaction may be limited to the amorphous regions. Even in a completely amorphous polymer, not all the functional groups of interest are necessarily equally accessible. Conformational or localized steric effects might well influence the rate or extent of the reaction in different parts of the same molecule. At the same time, the proximity of functional groups might enhance a reaction by the *neighboring group effect*. Copolymers of acrylic acid and *p*-nitrophenyl methacrylate, for example, undergo base-catalyzed hydrolysis faster than simple *p*-nitrophenyl esters because of participation by neighboring carboxylate anions (9.1).[6] That neighboring group and conformational effects in polymers are important in influencing chemical reactions is demonstrated in convincing fashion in the area of enzyme chemistry.

$$\text{(9.1)}$$

Another complicating factor in polymer reactions is the possibility that the reaction itself might change the physical form of the polymer. Even at low levels of conversion a reaction could alter the polymer's conformation, thereby influencing the reaction rate at unreacted sites; or it might even cause the polymer to precipitate, which would effectively preclude further reaction in the precipitated regions. The possibility that such variables might play a role should be taken into account in attempting any polymer modification.

9.2 Functional Group Reactions

9.2.1 Introduction of New Functional Groups

Among the oldest commercial processes are the chlorination (9.2) and chlorosulfonation (9.3) of polyethylene.[7,8] The properties of polyethylene are substantially altered by chlorination.

$$\text{(9.2)}$$

$$\text{(9.3)}$$

Flammability is decreased. Solubility is increased or decreased depending on the level of substitution. If the chlorination is run under heterogeneous conditions (polyethylene suspended in an inert medium), the resultant polymer is more crystalline at comparable levels of chlorination than a similar polymer prepared using polyethylene solution. This is not unexpected because the homogeneous process would distribute the chlorine more randomly. Chlorosulfonation provides sites for subsequent crosslinking reactions. Chlorination of poly(vinyl chloride) is also used commercially to increase the glass transition temperature. Free radical substitution reactions of these types cause crosslinking by radical combination reactions; hence the reaction conditions must be carefully controlled. Both chlorinated and chlorosulfonated polyethylene are, however, available as commercial products of long standing,[9] the latter under the well-known Hypalon (du Pont) trademark.

Fluorination may be accomplished by direct reaction between solid polymer and fluorine.[10] Special techniques such as use of mixtures of nitrogen and fluorine are needed to avoid degradation reactions, but polyethylene can be converted almost quantitatively to the fully fluorinated polymer by this technique. Where double bonds or benzene rings are present in the polymer, both substitution and addition occur. Polystyrene, for example, yields the fully fluorinated saturated polymer on reaction with fluorine (9.4). Where the reaction is particularly useful is in surface fluorination to enhance chemical inertness and—in the case of plastic bottles—to improve solvent barrier properties.[11]

$$\left[CH_2CH\atop C_6H_5\right] \xrightarrow[(-HF)]{F_2} \left[CF_2CF\atop C_6F_{11}\right] \tag{9.4}$$

Aromatic substitution reactions (nitration, sulfonation, chlorosulfonation, etc.) occur readily on polystyrene[12,13] and are useful for manufacturing ion-exchange resins or for introducing sites for crosslinking or grafting. Chloromethylation (9.5) of polystyrene is particularly important for introducing new functionalities because the resultant benzylic chloride undergoes nucleophilic displacement so readily. Vinylbenzyl chloride (a mixture of *meta* and *para*) and its polymers are, in fact, available commercially for just such a purpose. The functionalized polymers, crosslinked by including small amounts of divinylbenzene during polymerization, are used for *solid-phase synthesis*, in which the insoluble polymer is used as a "handle" for maintaining a catalyst or reaction product in a solid phase for easy separation from the reaction medium.[14–19] A very important example of this technique for synthesizing proteins is described in Chapter 18.

$$\left[CH_2CH\right] \xrightarrow[AlCl_3]{CH_3OCH_2Cl} \left[CH_2CH\right] + CH_3OH \tag{9.5}$$

Another example that illustrates the value of adding new functionalities is the introduction of ketone groups via the intermediate oxime (9.6) to render polyethylene photodegradable.[20]

$$[CH_2CH_2] \xrightarrow{NOCl} \left[{NOH\atop \|\atop CCH_2}\right] \xrightarrow[H^+]{H_2O} \left[{O\atop \|\atop CCH_2}\right] \tag{9.6}$$

9.2.2 Conversion of Functional Groups

Conversion of one functional group to another is used primarily to obtain polymers difficult or impossible to prepare by direct polymerization. An example is the commercially important synthesis of poly(vinyl alcohol)[21,22] by hydrolysis or alcoholysis of poly(vinyl acetate) (9.7). Poly(vinyl alcohol) cannot be made directly because vinyl alcohol is the unstable enol form of acetaldehyde (see exercise 13, Chapter 1), although the preparation and copolymerization of vinyl alcohol under special conditions has been reported.[23]

$$\left[CH_2CH\right] \xrightarrow[\Delta]{CH_3OH} \left[CH_2CH\right] + CH_3CO_2CH_3 \qquad (9.7)$$
$$\qquad\qquad\overset{|}{\underset{\underset{O}{\|}}{O}CCH_3} \qquad\qquad \overset{|}{OH}$$

Examples of syntheses of other polymers difficult or impossible to obtain directly are the following:

1. Saponification of isotactic or syndiotactic poly(trimethylsilyl methacrylate) (both prepared under anionic conditions) to yield isotactic or syndiotactic poly(methacrylic acid), respectively (9.8).[24]

$$\left[\begin{matrix}CH_3\\|\\-CH_2C-\\|\\C=O\\|\\OSi(CH_3)_3\end{matrix}\right] \xrightarrow[(2)\ H^+]{(1)\ H_2O,\ OH^-} \left[\begin{matrix}CH_3\\|\\-CH_2C-\\|\\CO_2H\end{matrix}\right] \qquad (9.8)$$

2. Hofmann degradation of polyacrylamide to give poly(vinyl amine) (9.9),[25] a reaction complicated by side reactions leading to crosslinking.

$$\left[\begin{matrix}CH_2CH\\|\\CONH_2\end{matrix}\right] \xrightarrow{Br_2,\ OH^-} \left[\begin{matrix}CH_2CH\\|\\NH_2\end{matrix}\right] \qquad (9.9)$$

3. Synthesis of "head-to-head poly(vinyl bromide)" by controlled bromination of 1,4-polybutadiene (9.10).[26]

$$\left[CH_2CH=CHCH_2\right] \xrightarrow{Br_2} \left[\begin{matrix}CH_2CH-CHCH_2\\||\\BrBr\end{matrix}\right] \qquad (9.10)$$

Other types of "classical" functional group conversions include dehydrochlorination of poly(vinyl chloride) (9.11)[27] (an early synthesis of polyacetylene), hydroformylation of polypentenamer (9.12),[28] and hydroboration of 1,4-polyisoprene (9.13).[29]

$$\left[\begin{matrix}CH_2CH\\|\\Cl\end{matrix}\right] \xrightarrow[N,N\text{-dimethylformamide}]{LiCl} \left[CH=CH\right] \qquad (9.11)$$

$$\left[CH=CH(CH_2)_3\right] \xrightarrow[catalyst]{CO,\ H_2} \left[\begin{matrix}CH_2CH(CH_2)_3\\|\\CHO\end{matrix}\right] \qquad (9.12)$$

$$\left[\begin{matrix}CH_3\\|\\CH_2C=CHCH_2\end{matrix}\right] \xrightarrow[(2)\ NaOH,\ H_2O_2]{(1)\ B_2H_6} \left[\begin{matrix}CH_3\\|\\CH_2CHCHCH_2\\|\\OH\end{matrix}\right] \qquad (9.13)$$

A less familiar reaction is the conversion of a fraction of the chloro groups of poly(vinyl chloride) to cyclopentadienyl (9.14) for purposes of crosslinking[30] (discussed in Section 9.4).

$$\left[CH_2CH\right] \xrightarrow{(CH_3)_2Al(C_5H_5)} \left[CH_2CH\right] + (CH_3)_2AlCl \qquad (9.14)$$

Functional group conversions are not limited to pendant groups on the backbone; useful products may also be made by converting the end groups of telechelic polymers. An example is the dehydrochlorination of chlorine-terminated polyisobutylene (9.15) and subsequent epoxidation (9.16). The epoxide-terminated polymer can be cured by reactions analogous to those used with epoxy resins[31] (Chapter 11).

$$\sim\sim\sim CH_2\underset{\underset{CH_3}{|}}{\overset{\overset{CH_3}{|}}{C}}Cl \xrightarrow{t\text{-BuO}^-K^+} \sim\sim\sim CH_2\underset{}{\overset{\overset{CH_3}{|}}{C}}{=}CH_2 \qquad (9.15)$$

$$\sim\sim\sim CH_2\overset{\overset{CH_3}{|}}{C}{=}CH_2 \xrightarrow{ArCO_3H} \sim\sim\sim CH_2\overset{\overset{CH_3}{|}}{\underset{\underset{O}{\diagdown\diagup}}{C}}{-}CH_2 \qquad (9.16)$$

9.3 Ring-Forming Reactions

As might be expected from our earlier consideration of restricted rotation (Chapter 3), the introduction of cyclic units into polymers generally results in greater rigidity, higher glass transition temperatures, and, frequently, improved thermal stability. One of the most important cyclization reactions leading to high thermal stability is the manufacture of *carbon fiber* (also called *graphite fiber*).* It is not a new process; in the 1870s Edison prepared carbon fiber filaments for his first electric light bulbs by pyrolyzing natural cellulose fibers. Most carbon fiber today is made by controlled pyrolysis of polyacrylonitrile fibers.[32–35] This is a complex free radical process involving a series of reactions leading eventually to a highly crosslinked graphitelike polymer (Scheme 9.1) accompanied by loss of HCN and N_2. Quinone-type structures are formed if oxygen is present. If graphitization is applied to fibers under stress, longitudinal order results and the graphite fibers exhibit very high modulus. Such fibers, which may still contain residual nitrogen, are used primarily in high-strength composites and in applications requiring good heat resistance. There is much current interest in *carbon nanotubes*—nanometer-wide, seamless, concentric graphite tubes. Formed at 4000 K in a carbon-arc furnace, the nanotubes are much stronger than conventional graphite fibers, and hold promise of yielding composites with superior properties.[36,37]

*While the terms *carbon fiber* and *graphite fiber* are often used synonymously, many workers in the field prefer to reserve the latter for fibers that have been treated at temperatures in excess of 2000°C, at which point the fibers assume a more truly graphitelike structure.

SCHEME 9.1. Reactions involved in pyrolysis of polyacrylonitrile to form carbon fiber.

Ladder structures are also formed from poly(methyl vinyl ketone) by intramolecular aldol condensation (9.17).[38] Nonladder structures, on the other hand, result from such reactions as dechlorination of poly(vinyl chloride) (9.18).[39] These reactions do not go to completion because cyclization occurs randomly, leaving some functional groups isolated between

(9.17)

(9.18)

ring units. Only when the process is reversible can a cyclization reaction be made to approach its theoretical limits. Such a case is the formation of cyclic acetal groups by the reaction of aldehydes with poly(vinyl alcohol) (9.19) in which more than 90% of the hydroxyl

(9.19)

groups can be converted. This particular reaction is important commercially; the product when R = C_3H_7, commonly called *poly(vinyl butyral)*, is used as a plastic film in laminated safety glass.[22] The commercial product actually contains a relatively high percentage of residual OH groups—close to 40%—for better adhesion to the glass. Another commercially important cyclization is the epoxidation of natural rubber[40] (9.20) to increase oil resistance and decrease gas permeability.

$$\text{wwwCH}_2\overset{\overset{\displaystyle CH_3}{\displaystyle |}}{C}=\text{CHCH}_2\text{www} \xrightarrow[\text{CH}_3\text{CO}_2\text{H}]{\text{H}_2\text{O}_2} \text{wwwCH}_2\overset{\overset{\displaystyle CH_3}{\displaystyle |}}{\underset{\displaystyle O}{C}}\text{—CHCH}_2\text{www} \qquad (9.20)$$

Rubber and other diene polymers undergo cyclization in the presence of acid.[41] A probable pathway is shown for *cis*-1,4-polyisoprene (9.21); however, because of the high probability of transfer reactions occurring, only about one to four rings are fused together in any fused-ring sequence. Quantitative cyclization of 1,2-polybutadiene can be brought about with metathesis catalysts (9.22).[42] As might be expected, flexible polymers lose their elastomeric properties with the introduction of cyclic units.

$$\qquad\qquad\qquad\qquad\qquad\qquad\qquad\qquad\qquad\qquad\qquad\qquad\qquad\qquad\qquad \text{etc.} \quad (9.21)$$

$$\qquad\qquad\qquad\qquad\qquad\qquad\qquad\qquad\qquad\qquad\qquad\qquad\qquad +2\text{CH}_2\text{=}\text{CH}_2 \quad (9.22)$$

9.4 Crosslinking

From the commercial standpoint, crosslinking is the most important reaction of vinyl polymers and is fundamental to the rubber and elastomer industries. It can be brought about by (1) *vulcanization*, using peroxides, sulfur, or sulfur-containing compounds; (2) free radical reactions caused by ionizing radiation; (3) photolysis involving photosensitive functional groups; (4) chemical reactions of labile functional groups; or (5) coulombic interactions of ionic species. The physical and morphological consequences of chemical and physical crosslinking are discussed in Chapter 3. Here we are concerned with the reactions leading to chemical crosslinking.

9.4.1 Vulcanization

Vulcanization is a general term applied to the crosslinking of polymers, particularly elastomers.[42–45] Peroxide-initiated crosslinking of saturated polymers such as polyethylene proceeds by hydrogen abstraction by radicals resulting from homolytic cleavage of peroxide (9.23) followed by radical combination (9.24). With unsaturated polymers, hydrogen abstraction probably occurs at the allylic position (9.25) with subsequent crosslinking again resulting from radical combination (9.26). It is apparent that addition-transfer processes (9.27, 9.28) can also cause crosslinking, because in many instances considerably more crosslinks are formed than would be expected on the basis of only abstraction–combination reactions.

Not all vinyl polymers are crosslinkable with peroxides; polypropylene and poly(vinyl chloride), for example, undergo degradation in preference to crosslinking.

$$RO\cdot + \text{\small ⌇⌇}CH_2CH_2\text{\small ⌇⌇} \longrightarrow \text{\small ⌇⌇}\dot{C}HCH_2\text{\small ⌇⌇} + ROH \tag{9.23}$$

$$\begin{array}{c} \text{\small ⌇⌇}\dot{C}HCH_2\text{\small ⌇⌇} \\ + \\ \text{\small ⌇⌇}\dot{C}HCH_2\text{\small ⌇⌇} \end{array} \longrightarrow \begin{array}{c} \text{\small ⌇⌇}CHCH_2\text{\small ⌇⌇} \\ | \\ \text{\small ⌇⌇}CHCH_2\text{\small ⌇⌇} \end{array} \tag{9.24}$$

$$\text{\small ⌇⌇}CH_2CH{=}CHCH_2\text{\small ⌇⌇} + RO\cdot \longrightarrow \text{\small ⌇⌇}\dot{C}HCH{=}CHCH_2\text{\small ⌇⌇} + ROH \tag{9.25}$$

$$\begin{array}{c} \text{\small ⌇⌇}\dot{C}HCH{=}CHCH_2\text{\small ⌇⌇} \\ + \\ \text{\small ⌇⌇}\dot{C}HCH{=}CHCH_2\text{\small ⌇⌇} \end{array} \longrightarrow \begin{array}{c} \text{\small ⌇⌇}CHCH{=}CHCH_2\text{\small ⌇⌇} \\ | \\ \text{\small ⌇⌇}CHCH{=}CHCH_2\text{\small ⌇⌇} \end{array} \tag{9.26}$$

$$\begin{array}{c} \text{\small ⌇⌇}\dot{C}HCH{=}CHCH_2\text{\small ⌇⌇} \\ + \\ \text{\small ⌇⌇}CH_2CH{=}CHCH_2\text{\small ⌇⌇} \end{array} \longrightarrow \begin{array}{c} \text{\small ⌇⌇}CHCH{=}CHCH_2\text{\small ⌇⌇} \\ \diagdown \\ \text{\small ⌇⌇}CH_2CH{-}\dot{C}HCH_2\text{\small ⌇⌇} \end{array} \tag{9.27}$$

$$\downarrow \text{\small ⌇⌇}CH_2CH{=}CHCH_2\text{\small ⌇⌇}$$

$$\text{\small ⌇⌇}\dot{C}HCH{=}CHCH_2\text{\small ⌇⌇} \quad + \quad \begin{array}{c} \text{\small ⌇⌇}CHCH{=}CHCH_2\text{\small ⌇⌇} \\ \diagdown \\ \text{\small ⌇⌇}CH_2CH{-}CH_2CH_2\text{\small ⌇⌇} \end{array} \tag{9.28}$$

    The oldest method of vulcanization, discovered in 1839 by Charles Goodyear, uses elemental sulfur. The mechanism appears to be ionic in nature, involving addition to a double bond to form an intermediate sulfonium ion (9.29) which then abstracts a hydride ion (9.30) or donates a proton (9.31) to form new cations for propagating the reaction. Termination

$$\text{\small ⌇⌇}CH_2CH{=}CHCH_2\text{\small ⌇⌇} + \text{\small ⌇⌇}\overset{\delta+}{S}{-}\overset{\delta-}{S}\text{\small ⌇⌇} \longrightarrow \text{\small ⌇⌇}CH_2CH{=\!\!=}CHCH_2\text{\small ⌇⌇} \tag{9.29}$$
$$\underset{\underset{+\text{\small ⌇⌇}S^-}{\overset{|}{\overset{\}{S}}}}{\overset{+}{}}$$

$$\begin{array}{c} \text{\small ⌇⌇}CH_2CH{=\!\!=}CHCH_2\text{\small ⌇⌇} \\ \overset{+}{\underset{\overset{|}{S}}{}} \\ + \\ \text{\small ⌇⌇}CH_2CH{=}CHCH_2\text{\small ⌇⌇} \end{array} \left[\begin{array}{l} \longrightarrow \text{\small ⌇⌇}CH_2CH{-}CH_2CH_2\text{\small ⌇⌇} + \text{\small ⌇⌇}\overset{+}{C}HCH{=}CHCH_2\text{\small ⌇⌇} \quad (9.30) \\ \qquad\quad \overset{|}{\underset{\overset{}{S}}{}} \\ \\ \longrightarrow \text{\small ⌇⌇}CH_2CH{-}CH{=}CH\text{\small ⌇⌇} + \text{\small ⌇⌇}CH_2CH_2{-}\overset{+}{C}HCH_2\text{\small ⌇⌇} \quad (9.31) \\ \qquad\quad \overset{|}{\underset{\overset{}{S}}{}} \end{array}\right.$$

occurs by reaction between sulfenyl anions and carbocations. Studies on model compounds indicate that, in addition to simple monosulfide or disulfide linkages, some polysulfide groups and cyclic monosulfide groups resulting from intramolecular addition of pendant sulfur to a double bond are also present.

    The rate of vulcanization with sulfur can be, and normally is, increased by the addition of accelerators such as zinc salts of dithiocarbamic acids (**1**) or organosulfur compounds such as tetramethylthiuram disulfide (**2**). Other compounds, notably zinc oxide and stearic acid, are also added as activators.

$$[(CH_3)_2N\overset{\overset{\displaystyle S}{\|}}{C}S^-]_2Zn^{2+} \qquad\qquad (CH_3)_2N\overset{\overset{\displaystyle S}{\|}}{C}S S\overset{\overset{\displaystyle S}{\|}}{C}N(CH_3)_2$$

$$\mathbf{1} \qquad\qquad\qquad\qquad \mathbf{2}$$

9.4.2 Radiation Crosslinking

When vinyl polymers are subjected to radiation, whether it be photons, electrons, neutrons, or protons, two main types of reaction occur—crosslinking and degradation.[46–50] Generally, both occur simultaneously, although degradation predominates with high doses of radiation. With low doses the polymer structure determines which will be the major reaction. Geminally disubstituted polymers tend to undergo chain scission, with monomer being a major degradation product. Thus, such polymers as poly(α-methylstyrene), poly(methyl methacrylate), or polyisobutylene will decrease in molecular weight on exposure to radiation. Halogen-substituted polymers, such as poly(vinyl chloride), break down with loss of halogen. With most other vinyl polymers, crosslinking predominates. A limitation of radiation crosslinking is that radiation does not penetrate very far into the polymer matrix; hence the method is primarily used with films.

The mechanism of crosslinking is free radical in nature and probably involves initial ejection of a hydrogen atom (9.32), which, in turn, removes another hydrogen atom from an adjacent site on a neighboring chain (9.33). Crosslinking then occurs by radical combina-

$$-CH_2CH_2-\xrightarrow{\text{Radiation}} -\dot{C}HCH_2- + H_2 \tag{9.32}$$

$$-CH_2CH_2- + H\cdot \longrightarrow -\dot{C}HCH_2- + H_2 \tag{9.33}$$
(neighboring chain)

tion. This is a reasonable assumption because hydrogen is a major side product, and random formation of radicals would not give the efficiency of crosslinking that is generally observed.

Chain branches are also ejected. Low-density polyethylene, for example, gives larger amounts of gaseous hydrocarbons on irradiation than does linear high-density polyethylene. Fragmentation reactions of this type, as well as ejection of hydrogen, also lead to double bonds in the polymer chains (9.34) in addition to crosslinking.

$$-CH_2CH- \longrightarrow -\dot{C}HCH- + H\cdot \longrightarrow -CH{=}CH- + RH \tag{9.34}$$
$$\underset{R}{|} \qquad\qquad \underset{R}{|}$$

Radiolysis effects on numerous vinyl polymers have been studied, but polyethylene is most important. Irradiated polyethylene film is used commercially because of its improved tensile and thermal properties. Polystyrene is quite resistant to radiation, a characteristic of aromatic polymers in general, but it can be crosslinked with higher doses.

9.4.3 Photochemical Crosslinking

Ultraviolet or visible light-induced crosslinking[51,52] (*photocrosslinking*) has taken on increasing importance in recent years. Among the numerous applications are printed circuits for electronic equipment; printing inks; coatings for optical fibers; varnishes for paper and carton board; finishes for vinyl flooring, wood, paper, and metal; and curing of dental materials. Photocrosslinking applied to photoresist technology is described in Chapter 4. There are two basic methods for bringing about photocrosslinking: (1) incorporating photosensitizers into the polymer, which absorb light energy and thereby induce formation of free radicals, and (2) incorporating groups that undergo either *photocycloaddition* reactions or light-initiated polymerization.

When triplet sensitizers such as benzophenone are added to polymer, absorption of ultraviolet results in $n \rightarrow \pi^*$ excitation of the sensitizer followed by hydrogen abstraction from the polymer to yield radical sites available for crosslinking by combination reactions. If the chromophore is built into the polymer backbone, as with polymers or copolymers of vinyl ketones, then degradation occurs, either by α-cleavage of the excited polymer (9.35) or by chain cleavage (9.36). The former leads to active sites for crosslinking. (Recall the discussion of Norrish type II reactions in Chapter 4, Section 4.7.) Poly(vinyl ester)s undergo analogous α-cleavage reactions (9.37) with subsequent crosslinking. Copolymers of vinyl esters and fluorinated monomers that can be crosslinked by ultraviolet radiation have been developed for use as weather-resistant wood coatings. In this application, the vinyl ester constitutes about 10% of the copolymer and benzophenone is added as a sensitizer.

$$\text{~CH}_2\text{CHCH}_2\text{CH~} \quad \xrightarrow{h\nu} \quad \begin{array}{l} \text{~CH}_2\text{CHCH}_2\dot{\text{C}}\text{H~} \\ \quad | \\ \quad \text{C=O} \quad + \text{R}\dot{\text{C}}\cdot \\ \quad | \\ \quad \text{R} \end{array} \qquad (9.35)$$

$$\begin{array}{l} \text{~CH}_2\text{CH}_2 + \text{CH}_2\text{=C~} \\ \quad | \qquad\qquad | \\ \quad \text{C=O} \qquad \text{C=O} \\ \quad | \qquad\qquad | \\ \quad \text{R} \qquad\qquad \text{R} \end{array} \qquad (9.36)$$

$$\begin{array}{l} \text{~CH}_2\text{CH~} \\ \quad | \\ \quad \text{C=O} \\ \quad | \\ \quad \text{OR} \end{array} \xrightarrow{h\nu} \begin{array}{l} \text{~CH}_2\dot{\text{C}}\text{H~} \\ \\ \quad \text{O} \\ \quad || \\ + \cdot\text{C—OR} \end{array} \qquad (9.37)$$

As is the case with radiolysis, polymers prepared from 1,1-disubstituted monomers undergo more extensive degradation, particularly to monomer, on photolysis.

A wide variety of functional groups has been used to effect photocycloaddition or light-induced polymerization crosslinking.[51–58] Many of them are listed in Table 9.1. The groups may be present as part of the backbone or, more commonly, as pendant groups. In most cases $2\pi + 2\pi$ cycloaddition occurs to give cyclobutane crosslinks, as shown for cinnamate ester in Scheme

TABLE 9.1. Groups Used to Effect Photocrosslinking[51–58]

Type	Structure
Alkyne	$\text{R—C} \equiv \text{C—R}$
Anthracene	

continued

TABLE 9.1. (continued)

Type	Structure
Benzothiophene dioxide	
Chalcone	ArCH=CHCAr (with C=O)
Cinnamate	ArCH=CHCO$_2$R
Coumarin	
Dibenz[b, f]azepine	
Diphenylcyclopropenecarboxylate	
Episulfide	CH$_2$—CH$_2$ (with S bridge)
Maleimide (R=H, CH$_3$, Cl)	
Stilbazole	R—N$^+$(ring)—CH=CHR, Y$^-$
Stilbene	ArCH=CHAr
Styrene	
1,2,3-Thiadiazole	
Thymine	

SCHEME 9.2. Photocrosslinking (a) by $2\pi + 2\pi$ cycloaddition and (b) by $4\pi + 4\pi$ cycloaddition.

9.2(a). With anthracene, Scheme 9.2(b), cycloaddition is $4\pi + 4\pi$. Free radical vinyl-type poly-merization may accompany cycloaddition, as is the case with maleimide[57]; or vinyl polymer-ization may predominate, as with styrene groups.[53] The reactive groups may be incorporated into the polymer during the polymerization reaction; for example, β-vinyloxyethyl cinnamate (**3**) un-dergoes cationic polymerization through the vinyl ether (9.38) to yield linear polymer contain-ing pendant cinnamate ester.[58] Alternatively, the group may be added to preformed polymer, as in the Friedel–Crafts alkylation of polystyrene with N-chloromethylmaleimide (**4**) (9.39).[57]

$$(9.38)$$

$$(9.39)$$

9.4.4 *Crosslinking Through Labile Functional Groups*

Reaction between appropriate difunctional or polyfunctional reagents with labile groups on the polymer chains can bring about crosslinking, but such reactions have not been exploited nearly as much as the more convenient vulcanization methods.

Polymers containing acid chloride groups, such as the previously described chlorosulfonated polyethylene (9.3), react with diamines (9.40) or diols (9.41) to yield sulfonamide and sulfonate crosslinks, respectively.[59] Dihalogen compounds can be used to crosslink polystyrene by the Friedel–Crafts reaction (9.42).[60]

$$(9.40)$$

$$(9.41)$$

$$(9.42)$$

A particularly interesting example of cycloaddition crosslinking involves the Diels–Alder reaction of cyclopentadiene-substituted polymer (see reaction 9.14). It is well known that cyclopentadiene dimerizes by Diels–Alder cycloaddition, one ring acting as diene, the other as dienophile. At about 180°C the dimer reverts to monomer. Cyclopentadiene substituent groups on polymers undergo an analogous reaction, yielding crosslinked polymer by cycloaddition and linear polymer by retrograde Diels–Alder reaction of the crosslinked polymer at elevated temperatures (9.43).[30] Such reversible systems have potential as thermoplastic elastomers.

$$(9.43)$$

9.4.5 Ionic Crosslinking

Examples of ionic crosslinking are the reaction of chlorosulfonated polyethylene with aqueous lead oxide (9.44) to yield lead sulfonate crosslinks, and the partial conversion of poly[ethylene-*co*-(methacrylic acid)] (**5**) to salts of divalent metals. The latter type, marketed under the du Pont trade name Surlyn, are called *ionomers*.[61,62] (Commercial products contain univalent as well as divalent metal salts.)

$$\text{~~~CH}_2\text{CH~~~} \xrightarrow{\text{PbO, H}_2\text{O}} \text{~~~CH}_2\text{CH~~~} \quad \text{~~~CHCH}_2\text{~~~} \qquad (9.44)$$
$$\text{SO}_2\text{Cl} \qquad\qquad\qquad \text{SO}_2{}^-\text{Pb}^{2+}{}^-\text{O}_2\text{S}$$

$$\text{CH}_3$$
$$\text{~~~CH}_2\text{CH}_2\text{~~~CH}_2\text{C~~~}$$
$$\text{CO}_2\text{H}$$

5

Ionomers have interesting properties compared with the nonionized copolymer. Introduction of ions causes disordering of the semicrystalline structure, which makes the polymer transparent. Crosslinking gives the polymer elastomeric properties, but it can still be molded at elevated temperatures. In fact, molding is generally facilitated by a broader melt range than is the case with the nonionic polymer. Ionic polymers, therefore, also qualify as thermoplastic elastomers. Increased polarity also improves adhesion. Ionomers are used as coatings, as adhesive layers for bonding wood to metal, as blow-molded and injection-molded containers, as golf ball covers, and as blister packaging material. Because of its adhesive qualities, an ionic acrylic copolymer is used as a binder for aluminosilicate dental fillings.

9.5 Block and Graft Copolymer Formation

In Chapters 7 and 8 we encountered the living polymer method of forming block copolymers. In this section we explore some other methods of preparing block copolymers and, for the first time, a few of the numerous procedures developed for preparing graft copolymers.[63–66]

9.5.1 Block Copolymers

Polymers containing functional end groups are generally amenable to forming block copolymers.[67] Styrene polymerized with the H_2O_2–$FeSO_4$ redox system, for example, forms hydroxy-terminated polystyrene. Subsequent reaction with an isocyanate-terminated polymer yields an AB block copolymer via urethane linkages (9.45). Similarly, telechelic polymers can be converted to ABA block copolymers or, if both homopolymers are telechelic, to the $-\!\!+\!\text{AB}\!+\!\!-$type.

$$\text{~~~CHCH}_2\text{OH} + \text{OCN~~~} \longrightarrow \text{~~~ CHCH}_2-\overset{\displaystyle O}{\overset{\|}{\text{OCNH}}}\text{~~~} \qquad (9.45)$$
$$\text{Ph} \qquad\qquad\qquad\qquad\qquad \text{Ph}$$

It is worth noting that polymers containing reactive functional groups are, in a sense, "living," and their conversion to block copolymers might be considered in the same light as the living polymer method of creating block copolymers discussed in the previous three chapters.

Peroxide groups introduced to polymer chain ends can be used to initiate a new chain.[66a] Typically an initiator such as the monohydroperoxide of diisopropylbenzene is used, and the resultant polymer (**6**), which contains an isopropylbenzene end group, can itself be converted to hydroperoxide (9.46).

$$\text{(9.46)}$$

Addition of a second monomer followed by redox or thermal decomposition of the hydroperoxide initiates polymerization at the chain end. Similarly peroxide units can be formed in a polymer backbone by polymerizing a monomer in the presence of small amounts of oxygen (9.47). Thermal cleavage of the peroxide leads to radical-terminated chains capable of initiating a second monomer (9.48).

$$\text{(9.47)}$$

$$\text{(9.48)}$$

Another way to form chain-end radicals is by mechanical degradation of homopolymers using ultrasonic radiation or high-speed stirring.[64] Polyethylene-*block*-polystyrene, for example, has been made by masticating polystyrene in the presence of ethylene. Mastication of mixtures of homopolymers yields similar results. Ionizing radiation may also be used to effect chain cleavage; however, this is more readily applicable to formation of graft copolymers.

9.5.2 Graft Copolymers

There are three general methods of preparing graft copolymers: (1) A monomer is polymerized in the presence of a polymer with branching resulting from chain transfer. (2) A monomer is polymerized in the presence of a polymer having reactive functional groups or positions that are capable of being activated, for example, by irradiation. (3) Two polymers having reactive functional groups are coreacted. Chain transfer grafting will be considered first.

Three components are necessary for grafting by chain transfer: polymer, monomer, and initiator. The initiator may play one of two roles: (1) It polymerizes the monomer to form a polymeric radical (or ion or coordination complex), which, in turn, reacts with the original polymer; or (2) it reacts with the polymer to form a reactive site on the backbone which, in turn, polymerizes the monomer. As with ordinary copolymerization, it is necessary to consider reactivity ratios of monomers to ensure that grafting will occur. It is also necessary to take into account the frequency of transfer to determine the number of grafts. Normally, mixtures of homopolymers result along with graft copolymer.

Grafting generally occurs at sites that are susceptible to transfer reactions, such as at carbons adjacent to double bonds in polydienes or on carbons adjacent to carbonyl groups. An example of the latter is the reaction of poly(vinyl acetate) with ethylene using a free radical initiator.[59] This particular reaction occurs both on the pendant methyl group and on the polymer backbone (9.49). This has been shown to be the case by hydrolysis of the product to give a mixture of poly(vinyl alcohol)-*graft*-polyethylene and long-chain carboxylic acids. Grafting efficiency is improved if a group that undergoes radical transfer readily, such as a mercaptan, is incorporated into the polymer backbone.

$$\text{(9.49)}$$

Cationic chain transfer grafting occurs when styrene is polymerized with BF_3 in the presence of poly(*p*-methoxystyrene) (9.50).[68,69] In this case the activated benzene rings undergo Friedel–Crafts attack by the carbocationic chain end of polystyrene. Isobutylene, which might be expected to behave similarly, fails to form graft copolymer under these conditions.

$$\text{(9.50)}$$

There are numerous examples of grafting by activation by backbone functional groups. An example is the synthesis of poly(*p*-chlorostyrene)-*graft*-polyacrylonitrile by anionic initiation of acrylonitrile using naphthalenesodium in the presence of poly(*p*-chlorostyrene) (9.51).[70,71] A polymeric anion radical is the initiating species.

$$\text{(9.51)}$$

TABLE 9.2. Approximate G Values of Monomers and Polymers[a]

Monomer	G	Polymer	G
Butadiene	Very low	Polybutadiene	2.0
Styrene	0.70	Polystyrene	1.5–3
Ethylene	4.0	Polyethylene	6–8
Acrylonitrile	5.0–5.6	—	—
Methyl methacrylate	5.5–11.5	Poly(methyl methacrylate)	6–12
Methyl acrylate	6.3	Poly(methyl acrylate)	6–12
Vinyl acetate	9.6–12.0	Poly(vinyl acetate)	6–12
Vinyl chloride	10.0	Poly(vinyl chloride)	10–15

[a]Data from Chapiro.[72] G values refer to number of free radicals formed per 100 eV of energy absorbed per gram of material.

Irradiation has been most widely used to provide active sites for graft copolymerization.[64] This is done with ultraviolet or visible radiation, with or without added photosensitizer, or with ionizing radiation, particularly the latter. Free radical reactions are involved in all cases. A major difficulty is that irradiation causes substantial amounts of homopolymerization along with grafting. This has been obviated to some extent by preirradiating the polymer prior to addition of the new monomer. One method is to preirradiate the polymer in the presence of air or oxygen to form hydroperoxide groups on the backbone. Subsequent addition of monomer and heating results in radical polymerization at the peroxide sites accompanied by some homopolymerization initiated by the hydroxy radicals formed on homolysis of hydroperoxide. Preirradiation can also be performed in the absence of air to form free radicals trapped in the viscous polymer matrix. Monomer is then added.

Direct irradiation of monomer and polymer together has been most extensively used. Because homopolymerization can occur, monomer and polymer must be chosen carefully. Generally, the best combination is a polymer that is very sensitive to radiation—that is, one that forms a high concentration of radicals—and a monomer that is not very sensitive. Sensitivity is normally measured in terms of G values, which represent the number of free radicals formed per 100 eV of energy absorbed per gram. Table 9.2 lists some G values for a few common monomers and polymers.[72] According to these data, a good combination would be poly(vinyl chloride) and butadiene. Homopolymerization may be reduced by providing radiation in bursts while monomer is allowed to diffuse through the polymer. Irradiation grafting of polymer emulsions is also an effective way to minimize homopolymerization, since the reaction medium remains fluid even at high conversions.

The third type of grafting process is exemplified by reactions of oxazoline-substituted polystyrene (a commercial product). The oxazoline group forms addition compounds with a variety of other functional groups including carboxylic acid, anhydride, alcohol, amine, epoxide, mercaptan, and phenol. Grafting of an oxazoline-substituted polymer with a carboxyl-terminated polymer is shown in reaction (9.52). Such grafting reactions can be used, for example, for compatibilizing polymer blends, or for improving surface adhesion between polystyrene molded parts and appropriately functionalized surface coatings.

$$\text{(structure)} + HO_2C \text{\wwww} \longrightarrow \text{(structure)} \quad (9.52)$$

9.6 Polymer Degradation

Stability of polymers is of critical interest to manufacturers, and, understandably, polymer degradation reactions have received considerable attention.[73–76] There are three principal methods of degrading polymers: (1) chemical, (2) thermal, and (3) radiative. In addition one can use ultrasonic radiation or mechanochemical techniques (for example, mastication) as described in the previous section.

9.6.1 Chemical Degradation

This discussion is limited to reactions that cause breakdown of the polymer backbone; it does not cover reactions involving pendant groups. Because the backbones of vinyl polymers are made up of carbon chains containing no functional groups other than the double bonds of diene polymers, chemical degradation is essentially limited to oxidation. Most important is oxidation with oxygen because this has a direct bearing on polymer durability.

Saturated polymers are degraded very slowly by oxygen, and the reaction is autocatalytic. It can be speeded up considerably by application of heat or light or by the presence of certain impurities that catalyze the oxidation process. Tertiary carbon atoms are most susceptible to attack, and this is reflected in the following order of resistance to oxidation of three common polymers: polyisobutylene > polyethylene > polypropylene. Reaction products are numerous and include water, carbon dioxide, carbon monoxide, hydrogen, and alcohols. Crosslinking always accompanies degradation. It is believed that decomposition of initially formed hydroperoxide groups is mainly responsible for chain scission (9.53).

$$\underset{R\quad R\quad R}{\text{\wwww}CH_2CHCH_2\overset{OOH}{\underset{|}{C}}CH_2CH\text{\wwww}} \longrightarrow \underset{R}{\text{\wwww}CH_2CHCH_2}\cdot + \underset{R\quad R}{\overset{O}{\underset{||}{C}}CH_2CH\text{\wwww}} + \cdot OH \quad (9.53)$$

Unsaturated polymers undergo oxidative degradation much more rapidly by complex free radical processes involving peroxide and hydroperoxide intermediates. Allylic carbon atoms are most sensitive to attack because resonance-stabilized radicals are formed. Oxidation is inhibited in commercial polymers by addition of antioxidants of the type discussed in Chapter 4 (Section 4.10).

Unsaturated polymers are also very susceptible to attack by ozone. Degradation by ozonolysis is discussed in Chapter 4 (Section 4.6). The discussion will not be repeated here except to remind the reader that an effective strategy for improving ozone resistance is to position the alkene moiety that is necessary for crosslinking such that oxidative bond cleavage causes no reduction in molecular weight. Also discussed in Chapter 4 (Section 4.7) are approaches to *enhancing* the oxidative degradability of vinyl polymers.

9.6.2 Thermal Degradation

Three types of thermal degradation are characteristic of vinyl polymers[75–77]: (1) *nonchain scission*, (2) *random chain scission*, and (3) *depropagation*. Nonchain scission refers to reactions involving pendant groups that do not break the polymer backbone. Typical of such reactions are dehydrochlorination of poly(vinyl chloride) (9.11), elimination of acid from poly(vinyl esters)—for example, poly(vinyl acetate) (9.54)—and elimination of alkene from poly(alkyl acrylate)s (9.55). The first two reactions lead to highly colored residues, indicating that the double bonds formed in the polymer backbone are primarily conjugated. Enough

$$\ce{-[CH2CH]-}\ \underset{}{\overset{\Delta}{\longrightarrow}}\ \ce{-[CH=CH]- + HOCCH3} \tag{9.54}$$

$$\ce{-[CH2CH]-}\ \overset{\Delta}{\longrightarrow}\ \ce{-[CH2CH]- + CH2=CHR} \tag{9.55}$$

side reactions occur, however, that such elimination reactions are not satisfactory for synthesizing polyacetylene. Nonchain scission has, however, been used as one approach to solving the problems of polyacetylene's intractability. The method involves synthesis of a reasonably stable, tractable precursor polymer that can be purified and fabricated, then converted thermally to polyacetylene. An example involves the tricyclic monomer (**7**), which undergoes metathesis polymerization to precursor polymer (**8**) (9.56). Thermal degradation of films of the latter (9.57) yields coherent films of polyacetylene.[78] Developed at the University of Durham, this particular approach to making polyacetylene is referred to as the "Durham route."

$$\tag{9.56}$$

7

8

$$\tag{9.57}$$

A disadvantage of the Durham route is that a relatively large molecule needs to be eliminated. Ring opening of the strained bicyclobutane rings of precursor polymer **9** also yields polyacetylene (9.58) without the necessity of an elimination reaction.[79] (The precurser polymer is prepared by ROMP; see exercise 11 in Chapter 8.)

$$\text{+CH=CH-}\langle\rangle\text{+} \xrightarrow{Hg^{2+}} \text{+CH=CH-CH=CH-CH=CH+} \qquad (9.58)$$

9

Nonchain scission reactions are useful for characterizing copolymers when the amount of a volatile degradation product can be correlated with the concentration of a given repeating unit (see Chapter 5, Section 5.6.4).

Random chain scission results from homolytic bond-cleavage reactions at weak points in the polymer chains. Complex mixtures of degradation products are formed, the origin of which may be explained in terms of radical transfer reactions such as (9.59). It may be recalled (Chapter 5, Figure 5.20) that a gas chromatographic pyrogram of polyethylene exhibits a homologous

$$\text{⁓CH}_2\text{CH}_2\text{CH}_2\text{CH}_2\text{⁓} \rightarrow \text{⁓CH}_2\text{CH}_2\cdot + \cdot\text{CH}_2\text{CH}_2\text{⁓} \rightarrow$$
$$\text{⁓CH=CH}_2 + \text{CH}_3\text{CH}_2\text{⁓} \qquad (9.59)$$

series of triplet peaks corresponding to alkane, 1-alkene, and α,ω-dialkene, with the largest peak in each triplet being that of 1-alkene. This follows from the breakdown pattern shown in Scheme 9.3. Random chain scission occurs with all vinyl polymers to varying degrees, but it occurs less with increasing substitution on the polymer backbone.

Depropagation, or depolymerization (unzipping), to give monomer occurs mainly with polymers prepared from 1,1-disubstituted monomers. Initiation may be at a chain end or at a random site along the backbone. Poly(methyl methacrylate) appears to begin unzipping primarily at the chain ends, whereas poly(α-methylstyrene) does so at random sites along the chain. In both cases tertiary radicals are formed with each depropagation step (9.60). Polymers having single substituents on alternate carbons degrade by both depropagation and random chain scission, with the amount of monomer formed varying with temperature.[82]

$$\text{⁓CH}_2\underset{R}{\overset{R}{C}}\text{CH}_2\underset{R}{\overset{R}{C}}\cdot \longrightarrow \text{⁓CH}_2\underset{R}{\overset{R}{C}}\cdot + \text{CH}_2{=}C{\overset{R}{\underset{R}{\diagdown}}} \qquad (9.60)$$

9.6.3 Degradation by Radiation

It was mentioned earlier that radiation may cause both crosslinking or degradation. Which predominates depends on radiation dosage, polymer structure, and temperature.[42,72] Ultraviolet or visible light causes 1,1-disubstituted polymers to degrade to monomer almost exclusively at elevated temperatures, whereas crosslinking and chain scission reactions predominate at room temperature. Other vinyl polymers undergo crosslinking primarily, regardless of temperature. Rearrangements may also occur as a result of homolysis and recombination reactions, especially with diene polymers.[83,84]

Ionizing radiation leads to much higher yields of monomer from 1,1-disubstituted polymers at room temperature. Degradations of this type initiated with electron beams are put to

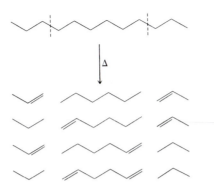

SCHEME 9.3. Random chain scission of polyethylene.

advantage in the manufacture of microcircuits using resist technology. At comparable levels of radiation, polyethylene and monosubstituted polymers undergo mainly crosslinking. All vinyl polymers tend to degrade under very high dosages of radiation.

References

1. E. M. Fettes (Ed.), *Chemical Reactions of Polymers*, Wiley-Interscience, New York, 1964.
2. J. A. Moore (Ed.), *Reactions of Polymers*, Reidel, Boston, 1973.
3. C. E. Carraher, Jr., and M. Tsuda (Eds.), *Modification of Polymers*, ACS Symp. Ser. 121, American Chemical Society, Washington, D.C., 1980.
4. C. E. Carraher Jr., and J. A. Moore (Eds.), *Modification of Polymers*, Plenum Press, New York, 1983.
5. J. L. Benham and J. F. Kinstle (Eds.), *Chemical Reactions on Polymers*, ACS Symp. Ser. 364, American Chemical Society, Washington, D.C., 1988.
6. H. Morawetz and P. E. Zimmering, *J. Phys. Chem.*, **56**, 753 (1954).
7. E. M. Fettes, in *Crystalline Olefin Polymers*, Part 2 (R. A. V. Raff and K. W. Doak, Eds.), Wiley-Interscience, New York, 1965, Chap. 6.
8. M. A. Smook, W. J. Remington, and D. E. Strain, in *Polythene* (A. Renfrew and P. Morgan, Eds.), Wiley-Interscience, New York, 1960, Chap. 14.
9. F. W. Keeley, in *Introduction to Rubber Technology* (M. Morton, Ed.), Reinhold, New York, 1959, Chap. 14.
10. R. J. Lagow and J. L. Margrave, *J. Polym. Sci., Polym. Lett. Ed.*, **12**, 177 (1974).
11. D. M. Buck, P. D. Marsh, and K.J. Kallish, *Polym.-Plast. Technol. Engl.*, **26**, 71 (1987).
12. G. J. Jones, in *Chemical Reactions of Polymers* (E. M. Fettes, Ed.), Wiley-Interscience, New York, 1964, Chap. 3.
13. M. Camps, M. Chatzopoulos, J.-M. Camps, and J.-P. Montheard, *J. Macromol. Sci.-Rev. Macromol. Chem. Phys.*, **C27**, 505 (1987–88).
14. N. K. Mathur, C. K. Narang, and R. E. Williams, *Polymers as Aids in Organic Chemistry*, Academic Press, New York, 1980.
15. W. T. Ford (Ed.), *Polymeric Reagents and Catalysts*, ACS Symp. Ser. 308, American Chemical Society, Washington, D.C., 1986.
16. P. Hodge and D. C. Sherrington (Eds.), *Polymer-Supported Reactions in Organic Synthesis*, Wiley-Interscience, New York, 1980.
17. G. Maneke and W. Storek, *Angew. Chem. Intern. Ed. English*, **17**, 657 (1978).
18. C. G. Overberger and K. N. Sannes, *Angew. Chem. Intern. Ed. English*, **13**, 99 (1974).

19. W. Heitz, *Adv. Polym. Sci.*, **23**, 1 (1977).

20. V. Pozzi, A. E. Silvers, L. Giuffre, and E. Cernia, *J. Appl. Polym. Sci.*, **19**, 923 (1975).

21. C. A. Finch, *Polyvinyl Alcohol: Properties and Applications*, Wiley-Interscience, New York, 1973.

22. J. G. Pritchard, *Poly(vinyl alcohol)*, Gordon and Breach, New York, 1970.

23. A. K. Cederstav and B. M. Novak, *J. Am. Chem. Soc.*, **116**, 4073 (1994).

24. N. N. Aylward, *J. Polym. Sci. A-1*, **8**, 319 (1970).

25. M. Sugiura, M. Ochi, Y. Tani, and Y. Nagai, *Kogyo Kagaku Zasshi*, **72**, 1926 (1969); *Chem. Abstr.*, **72**, 22105b (1970).

26. H. Kawaguchi, P. Loeffler, and O. Vogl, *Polymer*, **26**, 1257 (1985).

27. C. S. Marvel, J. H. Sample, and M. F. Roy, *J. Am. Chem. Soc.*, **61**, 3241 (1939).

28. K. Sanui, W. J. MacKnight, and R. W. Lenz, *Macromolecules*, **7**, 952 (1974).

29. Y. Minoura and H. Ikeda, *J. Appl. Polym. Sci.*, **15**, 2219 (1971).

30. J. P. Kennedy and K. F. Castner, *J. Polym. Sci., Polym. Chem. Ed.*, **17**, 2039, 2055 (1979).

31. J. P. Kennedy, *J. Appl. Polym. Sci., Appl. Polym. Symp.*, **39**, 21 (1984).

32. G. Henrici-Olivé and S. Olivé, *Adv. Polym. Sci.*, **51**, 1 (1983).

33. E. Fitzer, *Angew. Chem. Intern. Ed. English*, **19**, 375 (1980).

34. J.-B. Donnet and R. C. Bansal, *Carbon Fibers*, 2nd ed., Marcel Dekker, New York, 1989.

35. L. H. Peebles, *Carbon Fibers: Formation, Structure, and Properties*, CRC, Boca Raton, Florida, 1995.

36. T. W. Eddesen, *Carbon Nanotubes: Preparation and Properties*, CRC, Boca Raton, Florida, 1996.

37. M. M. J. Treacy, T. W. Ebbesen, and J. M. Gibson, *Nature*, **381**, 678 (1996).

38. T. Ogawa, R. Cedeno, and E. T. Herrera, *Makromol. Chem.*, **180**, 785 (1979).

39. G. J. Smets, in *Chemical Reactions of Polymers* (E. M. Fettes, Ed.), Wiley-Interscience, New York, 1964, Chap. 1D.

40. I. R. Gelling, *Rubber Chem. Technol.*, **58**, 86 (1985).

41. M. A. Golub, in *Polymer Chemistry of Synthetic Elastomers*, Part 2 (J. P. Kennedy and E. G. Tornqvist, Eds.), Wiley-Interscience, New York, 1969, Chap. 10A.

42. G. W. Coates and R. H. Grubbs, *J. Am. Chem. Soc.*, **118**, 229 (1996).

43. D. Craig, in *Chemical Reactions of Polymers* (E. M. Fettes, Ed.), Wiley-Interscience, New York, 1964, Chap. 9C.

44. A. Y. Coran, in *Science and Technology of Rubber* (F. R. Eirich, Ed.), Academic Press, New York, 1978.

45. G. Alliger and I. J. Sjothun (Eds.), *Vulcanization of Elastomers*, Van Nostrand Reinhold, New York, 1964.

46. A. R. Schultz, in *Chemical Reactions of Polymers* (E. M. Fettes, Ed.), Wiley-Interscience, New York, 1964, Chap. 9A.

47. M. Dole (Ed.), *The Radiation Chemistry of Macromolecules*, Vols. 1 and 2, Academic Press, New York, 1973.

48. J. F. Rabek, *Polymer Photodegradation: Mechanisms and Experimental Methods*, Chapman and Hall, London, 1995.

49. R. L. Clough and S. W. Shalaby (Eds.), *Radiation Effects on Polymers*, ACS Symp. Ser. 475, American Chemical Society, Washington, D.C., 1991.

50. D. R. Randell (Ed.), *Radiation Curing of Polymers*, CRC Press, Boca Raton, Fla., 1987.

51. G. E. Green, B. P. Stark, and S. A. Zahir, *J. Macromol. Sci., Revs. Macromol. Chem.*, **C21**, 187 (1981–82).

52. G. Oster and N. Yang, *Chem. Rev.*, **68**, 125 (1968).

53. M. J. Farrall, M. Alexis, and M. Trecarton, *Polymer*, **24**, 114 (1983).

54. A. O. Patil, D. D. Deshpande, and S. S. Talwar, *Polymer*, **22**, 434 (1981).

55. O. Zimmer and H. Meier, *J. Chem. Soc., Chem. Commun.*, 481 (1982).

56. M. Tsunooka, T. Ueda, and M. Tanaka, *J. Polym. Sci., Polym. Lett. Ed.*, **19**, 201 (1981).

57. M. P. Stevens and A. D. Jenkins, *J. Polym. Sci., Polym. Chem. Ed.*, **17**, 3675 (1979).

58. M. Kata, M. Hasegawa, and T. Ichijo, *J. Polym. Sci. B*, **8**, 263 (1970).

59. N. J. Gaylord and F. S. Ang, in *Chemical Reactions of Polymers* (E. M. Fettes, Ed.). Wiley-Interscience, New York, 1964, Chap. 10B.

60. N. Grassie and J. Gilks, *J. Polym. Sci., Polym. Chem. Ed.*, **11**, 1531 (1973).

61. R. Longwirth, in *Ionic Polymers* (L. Holliday, Ed.), Applied Science, London, 1975, Chap. 2.

62. R. H. Kinsey, *J. Appl. Polym. Sci., Appl. Polym. Symp.*, **11**, 77 (1969).

63. A. Noshay and J. E. McGrath, *Block Copolymers: Overview and Critical Survey.* Academic Press, New York, 1977.
64. H. A. J. Battaerd and G. W. Tregear, *Graft Copolymers*, Wiley-Interscience, New York, 1967, Chap. 3.
65. R. J. Ceresa (Ed.), *Block and Graft Copolymerization*, Wiley-Interscience, New York, 1973.
66. G. E. Ham (Ed.), *Copolymerization*, Wiley-Interscience, New York, 1964: (a) A. S. Hoffman and R. Bacskai, Chap. 6; (b) G. Bier and G. Lehman, Chap. 4B.
67. R. Jerome, R. Fayt, and T. Ouhadi, *Progr. Polym. Sci.*, **10**, 87 (1984).
68. H. C. Hass, P. M. Kamath, and N. W. Schuler, *J. Polym. Sci.*, **24**, 85 (1957).
69. H. C. Hass and P. M. Kamath, *J. Polym. Sci.*, **24**, 143 (1957).
70. G. Greber, J. Tölle, and W. Burchard, *Makromol. Chem.*, **71**, 47 (1964).
71. G. Greber and J. Tölle, *Makromol. Chem.*, **53**, 208 (1962).
72. A. Chapiro, *Radiation Chemistry of Polymeric Systems*, Wiley-Interscience, New York, 1962.
73. O. Cicchetti, *Adv. Polym. Sci.*, **7**, 70 (1970).
74. L. Reich and S. S. Stivala, *Elements of Polymer Degradation*, McGraw-Hill, New York, 1971.
75. S. L. Madorsky, *Thermal Degradation of Organic Polymers*, Wiley-Interscience, New York, 1964.
76. A. L. Bhuiyan, *Adv. Polym. Sci.*, **47**, 43 (1982).
77. J. H. Flynn and R. E. Florin, in *Pyrolysis and GC in Polymer Analysis* (S. A. Liebman and E. J. Levy, Eds.). Marcel Dekker, New York, 1985, Chap. 4.
78. J. H. Edwards, W. J. Feast, and D. C. Bott, *Polymer*, **25**, 395 (1984).
79. T. M. Swager, D. A. Dougherty, and R. H. Grubbs, *J. Am. Chem. Soc.*, **110**, 2973 (1988).
80. L. A. Wall, in *Analytical Chemistry of Polymers* (G. M. Kline, Ed.), Wiley-Interscience, New York, 1964, Part 2, Chap. 5.
81. M. P. Stevens, *Characterization and Analysis of Polymers by Gas Chromatography.* Marcel Dekker, New York, 1969, Chap. 4.
82. F. A. Lehmann and G. M. Brauer, *Anal. Chem.*, **33**, 673 (1961).
83. M. A. Golub and M. L. Rosenberg, *J. Polym. Sci., Polym. Chem. Ed.*, **18**, 2543 (1980).
84. S. W. Shalaby, *Macromol. Rev.*, **14**, 419 (1979).

Review Exercises

1. Write a concise definition of the following terms: (a) depropagation; (b) ionomer; (c) nonchain scission; (d) random chain scission; (e) vulcanization.

2. Give the product expected from the reaction of polystyrene with (a) HNO_3/H_2SO_4; (b) $Br_2/FeBr_3$; (c) H_2SO_4/SO_3; (d) N-bromosuccinimide (NBS); (e) $CH_3COCl/SnCl_4$.

3. Give the main product(s) expected from the reaction of 1,4-polyisoprene with the following reagents. (You may wish to consult your organic chemistry text.) For those reactions that do not result in degradation, indicate how the polymer's properties might be altered. (a) Br_2; (b) F_2; (c) H_2O, H^+; (d) H_2, Ni; (e) $CHCl_3$, base; (f) *m*-chloroperbenzoic acid; (g) $KMnO_4$, base; (h) $KMnO_4$, H^+; (i) 1. $CH_3OCH_2OCH_2CH_2OH$, $Hg(OAc)_2$; 2. $NaBH_4$. (The product of this two-step reaction sequence on 1,3-butadiene has been suggested as a polyelectrolyte by S. E. Bell and D. J. Bannister, *J. Polym. Sci. C., Polym. Lett.*, **24**, 165, 1985.)

4. Commercial styrene–diene AB and ABA block copolymers are often hydrogenated. What advantage does this procedure offer?

5. When colorless poly(1,3-butadiene) is treated with potassium *t*-butoxide, the polymer turns blue-black. Speculate on what chemical change might have occurred. Propose a mechanism for the change. (A. J. Dias and T. J. McCarthy, *Macromolecules*, **18**, 869, 1985)

6. Suggest a way to synthesize (a) head-to-head polypropylene [O. Vogl, *Polym. Preprints*, **20**(1), 154, 1979]; (b) head-to-head poly(vinyl alcohol); (c) the "alternating copolymer" of ethylene and vinylidene chloride (1,1-dichloroethene).

7. Using designations of the type used in exercise 6, how might you describe (a) the polymer prepared by oxidative cleavage of 1,4-poly(1,3-cyclohexadiene) (see reaction 8.19); (b) the polymer formed by hydration (H^+, H_2O) of 1,4-polyisoprene; (c) the product of hydroboration of 1,4-polyisoprene (reaction 9.13)?

8. Show how polyethylene modified according to reaction (9.6) undergoes photodegradation. Do the same for poly(methyl vinyl ketone). (Cf. Section 4.)

9. Cyclization of 1,4-polyisoprene under the influence of acid is shown in reaction (9.21). Under acidic conditions, 3,4-polyisoprene also undergoes cyclization. Give the likely structure of the product and show how it is formed.

10. In reaction 9.38, explain why linear rather than crosslinked polymer is formed.

11. The chloro group of chloromethylated polystyrene (see reaction 9.5) undergoes facile displacement with SH^-. Show how this could be used to advantage to prepare polystyrene-*graft*-poly(methyl methacrylate).

12. Show how chloromethylated polystyrene could be converted to polystyrene-*graft*-polyisobutylene.

13. What advantages does the macromonomer route (Chapter 1) have over conventional methods of preparing graft copolymers?

14. The ease of air oxidation of vinyl polymers is in the order polybutadiene > polypropylene > polyethylene > polyisobutylene. Explain why.

15. List at least two compounds that can be used to formulate more ozone-resistant elastomers, and write the structural units for the compounds as they are incorporated into the polymer.

16. Give the products of reductive ozonolysis of the following polymers discussed in Chapter 8: (a) polyoctenamer (reaction 8.33); (b) norbornene polymer (reaction 8.34); (c) the polymer formed by ROMP of 1-methyl-1,5-cyclooctadiene (reaction 8.36).

17. How might one conveniently determine whether thermal depropagation of a vinyl polymer is initiated at the chain ends or at random sites along the backbone?

18. On exposure to uv light, natural rubber (1,4-polyisoprene) undergoes extensive rearrangement leading to (among other structures) some 1,2- and 3,4-polyisoprene linkages. Show how these structures are likely to arise.

19. An ionomer (see Section 9.4.5) behaves as a thermoplastic elastomer even if the counterion is monovalent. Explain. (V. K. Daltye and R. L. Taylor, *Macromolecules*, **17**, 414, 1984)

P A R T

NONVINYL POLYMERS

10

STEP-REACTION AND

RING-OPENING

POLYMERIZATION

10.1 Introduction

The preceding four chapters dealt with carbon-chain polymers prepared from alkene and diene monomers. With the exception of ring-opening metathesis, such polymers are almost always prepared by chain-reaction addition polymerization. The remainder of this book is devoted almost exclusively to polymers having functional groups in the backbone that, in most instances, are prepared by step-reaction or ring-opening polymerization. The purpose of this chapter is to explore briefly the characteristics of these two polymerization processes. More detailed mechanistic treatments of individual polymer types are provided in the appropriate chapters. We shall also look briefly at methods for synthesizing the increasingly important dendritic polymers.

10.2 Step-Reaction Polymerization—Kinetics

At this point it behooves the student to review the material in Chapter 1 (Sections 1.4 and 1.5) that compares step-reaction and chain-reaction polymerizations. Among the distinctive features[1-5] of the former are:

1. Linear polymers are synthesized either from difunctional monomers of the AB type (where A and B represent coreactive functional groups) or from a combination of AA and BB difunctional monomers.

2. Network polymers are formed from monomers having a functionality greater than two.

3. Polymers retain their functionality as end groups at the completion of polymerization.

4. A single reaction (or reaction sequence) is responsible for all steps contributing to polymer formation (in contrast to initiation, propagation, and termination in chain-reaction polymerization).

5. Molecular weight increases slowly even at high levels of conversion. This precept is embodied in the Carothers equation relating $\overline{DP}$ to monomer conversion (p):

$$\overline{DP} = \frac{1}{1 - p}$$

6. High-yield reactions and an exact stoichiometric balance are necessary to obtain high-molecular-weight linear polymer. While an exact stoichiometric balance is readily achieved by using highly purified AB-type monomer, it is much more difficult to achieve when using AA and BB monomers. And in both cases, side reactions (e.g., dehydration of alcohol in polyesterification) will upset the balance.

From the practical standpoint, it should be realized that the very high molecular weights one can obtain routinely by chain-reaction polymerization are very difficult to achieve by step polymerization of difunctional monomers. Furthermore, many step polymerization reactions are equilibrium condensation processes; it is necessary in these cases to provide an efficient means of removing byproducts to drive the reaction to high conversion. A further complication arises from the tendency of certain monomers to form cyclic compounds, particularly where five- or six-membered rings are involved. Thus ethylene glycol, in the presence of acid, forms 1,4-dioxane (10.1) in preference to polyether (10.2), and 4-aminobutanoic acid forms cyclic amide (lactam) (10.3) in addition to polyamide (10.4).

$$\text{HOCH}_2\text{CH}_2\text{OH} \xrightarrow[\;(-2\text{H}_2\text{O})\;]{\text{H}^+}$$

(10.1)

$$-\!\!\left[\text{OCH}_2\text{CH}_2\right]\!\!-$$

(10.2)

(10.3)

$$\text{H}_2\text{NCH}_2\text{CH}_2\text{CH}_2\text{CO}_2\text{H} \xrightarrow[\;(-\text{H}_2\text{O})\;]{\Delta}$$

$$\left[\text{NHCH}_2\text{CH}_2\text{CH}_2\overset{\overset{\text{O}}{\|}}{\text{C}}\right]$$

(10.4)

Table 10.1 lists most of the commercially important polymers prepared by step-reaction polymerization in the order in which they appear in subsequent chapters. A wealth of step-reaction polymers have been investigated besides those listed. From the standpoint of mechanism and kinetics, polyesters and polyamides have been most thoroughly studied. Such studies allow us to make certain simplifying assumptions in order to relate $\overline{DP}$ to polymerization rate in a manageable way. One of the most important is that the rate constant is independent of chain length. Indeed, this has been shown to be the case for simple esterification of a homologous series of linear carboxylic acids in which the rate becomes virtually constant once the chain length reaches four carbons. Clearly the situation is more complicated where vis-

TABLE 10.1. Commercially Important Polymers Prepared by Step-Reaction Polymerization

Polymer Type	Repeating Functional Unit	Chapter
Polyether	—Ar—O—	11
Polyether [poly(phenylene oxide)]		11
Polyether (epoxy)[a]	OH | —CH$_2$CHCH$_2$OAr—	11
Polysulfide	—ArS—	11
Poly(alkylene polysulfide)	—RS$_x$—	11
Polysulfone	—ArSO$_2$—	11
Polyester	O || —RCO—	12
Polycarbonate	O || —ROCO—	12
Polyamide	O || —RCNH—	13
Polyurea	O || —RNHCNH—	13
Polyurethane	O || —ROCNH—	13
Phenol–formaldehyde[b]		14
Urea–formaldehyde[c]	O || >NCNHCH$_2$—	14
Melamine–formaldehyde[b]		14
Polyimide		15

continued

TABLE 10.1. (continued)

Polymer Type	Repeating Functional Unit	Chapter
Polybenzimidazole		15
Polyphenylene		17
Poly(*p*-xylylene)	$-CH_2-\bigcirc-CH_2-$	17

[a] Formed from epoxide-terminated oligomer.
[b] Complex network structure.
[c] Network or linear; see Chap. 14 for discussion.

cosity effects might hinder the rate at which carboxyl and hydroxyl groups diffuse together. Because most step polymerization reactions have relatively high activation energies, the rate of diffusion is usually much higher than the rate of the functional group reaction. Where diffusion rate might be more critical, however, it is assumed that once the groups are close enough to react, they are more favorably disposed to do so because the diffusion of groups *away* from one another is equally slowed by viscosity effects. Another simplifying assumption, which is more intuitively reasonable, is that the mechanism of the reaction remains constant throughout.

If the polymerization reaction is first order with respect to each functional group reactant, A and B, then the rate expression is

$$-\frac{d[A]}{dt} = k[A][B]$$

Given the fact that an exact stoichiometric balance is usually a necessary condition for achieving high molecular weight, we can equate [A] and [B], and the rate expression then becomes

$$-\frac{d[A]}{dt} = k[A]^2$$

or, by integration,

$$\frac{1}{[A]} - \frac{1}{[A_0]} = kt$$

At any particular time, t, in the polymerization process, $\overline{DP}$ is equal to the ratio of monomer molecules present initially to the total number at that time; that is,

$$\overline{DP} = \frac{[A_0]}{[A]}$$

Combining this expression with the Carothers equation and solving for $[A]$, we have

$$[A] = [A_0](1 - p)$$

By substitution in the second-order rate expression, we obtain

$$\frac{1}{[A_0](1 - p)} - \frac{1}{[A_0]} = kt$$

Since $\overline{DP} = 1/(1 - p)$.

$$\frac{\overline{DP}}{[A_0]} - \frac{1}{[A_0]} = kt$$

or

$$\overline{DP} = [A_0]kt + 1$$

Thus if one knows the rate constant, one can calculate the time needed to reach any number average molecular weight for a given concentration of monomer.

While the foregoing relationships are valid for a great many step-reaction polymerizations, they do not apply to the very important uncatalyzed polyesterification reactions in which carboxylic acid is monomer. Simple esterification is normally acid catalyzed. In preparing polyesters, however, where high temperatures are usually necessary to reduce viscosity and facilitate removal of water, an acid catalyst might well have an adverse effect, causing discoloration and even decomposition of monomer or polymer. In the absence of added acid the carboxylic acid assumes the role of catalyst, and the rate expression then becomes second order in acid, or third order overall:

$$-\frac{d[A]}{dt} = k[A]^2[B]$$

Again, assuming $[A]$ and $[B]$ are equal,

$$-\frac{d[A]}{dt} = k[A]^3$$

Integration then gives

$$\frac{1}{[A]^2} - \frac{1}{[A_0]^2} = 2kt$$

Substituting for $[A]$ as above and rearranging,

$$\frac{1}{[A_0]^2(1 - p)^2} - \frac{1}{[A_0]^2} = 2kt$$

or

$$\frac{1}{(1 - p)^2} = 2kt[A_0]^2 + 1$$

and

$$(\overline{DP})^2 = 2kt[A_0]^2 + 1$$

Thus, in the case of uncatalyzed polyesterification, the molecular weight increases more gradually than when acid catalyst is added.

These relationships, which have been confirmed experimentally, do not apply over the full polymerization reaction. Plots of $\overline{DP}$ versus t deviate from linearity at both low and high levels of functional group conversion. The reason cited most often for deviations in the early stages of polyesterification has to do with changes in polarity of the reaction medium as the highly polar alcohol and acid are converted to the much lower polarity ester. The more polar the medium, the more association through hydrogen bonding inhibits reactivity. In the uncatalyzed polyesterification of adipic acid and diethylene glycol (10.5), for example, 80% of the functional groups are consumed before the third-order relationship given assumes linearity. (As may be seen by application of the Carothers equation, this still represents a very low molecular weight.) Deviations at very high levels of conversion arise partly from difficulties in removing water from the highly viscous reaction medium and, in the case of acid-catalyzed reactions, from the reaction of catalyst with hydroxyl end groups.

$$HO_2C(CH_2)_4CO_2H + HOCH_2CH_2OH \longrightarrow \left[\begin{matrix} O & & O \\ \| & & \| \\ C(CH_2)_4COCH_2CH_2O \end{matrix} \right] + H_2O \quad (10.5)$$

10.3 Stoichiometric Imbalance

Despite the fact that an exact stoichiometric balance is necessary to achieve high molecular weights in most instances of step-reaction polymerization, in some important applications it is necessary to limit the molecular weight. Formation of epoxy resins (Chapter 11) is one. As footnoted in Table 10.1, these are polyethers that are processed to an oligomer stage and are subsequently converted to network polymer by appropriate reactions of terminal epoxide groups. Similarly, unsaturated polyesters (Chapter 12) are processed to moderately low molecular weights before being crosslinked. With polyamides for fiber applications, molecular weight must often be limited because too high a viscosity is detrimental to extrusion of filaments through the fine holes of a spinneret.

There are three ways to limit molecular weight in step-reaction polymerization. One is to quench the polymerization reaction by rapid cooling when the desired molecular weight (or viscosity) is attained. This is routinely done in the manufacture of unsaturated polyesters. Another is to use an excess of one monomer when two difunctional monomers (AA and BB) are polymerized. This is the method of choice with epoxy resins to ensure that epoxide end groups are present in the oligomer. The third method is to use small amounts of monofunctional reactant. Addition of acetic acid, for example, is one way to limit the molecular weight of nylon 66 (prepared from adipic acid and 1,6-hexanediamine) for fiber manufacture.

When a nonstoichiometric amount of functional groups is used, the relationship between $\overline{DP}$ and reaction conversion can be quantified by a modification of the Carothers equation. At this point we introduce a factor, r, representing the *stoichiometric imbalance*. If the number of A and B functional groups present at the beginning of an AA + BB polymerization is N_A^0 and N_B^0, respectively,

$$r = \frac{N_A^0}{N_B^0}$$

By convention, r is always less than unity (except, of course, when $N_A^0 = N_B^0$). As before, p is the reaction conversion, which in this case represents the fraction of A groups that have reacted. Because the reaction of each A group consumes one B group, the fraction of B reacted at conversion p is equal to pN_A^0, or prN_B^0. The number of *unreacted* groups, N_A and N_B, is then given by

$$N_A = (1 - p)N_A^0 \quad \text{and} \quad N_B = (1 - pr)N_B^0 = (1 - pr)\frac{N_A^0}{r}$$

At this point, the number of A and B end groups is equal to $N_A + N_B$, and, because there are two end groups per molecule, the number of molecular chains, N, is given by

$$N = \frac{1}{2}(N_A + N_B)$$

that is,

$$N = \frac{1}{2}\left[(1 - p)N_A^0 + (1 - pr)\frac{N_A^0}{r}\right]$$

which reduces to

$$N = \frac{N_A^0}{2}\left(1 + \frac{1}{r} - 2p\right)$$

We know that one repeating unit is formed every time an A and a B functional group react; therefore, the total number of repeating units, N_r, is given by

$$N_r = \frac{1}{2}(N_A^0 + N_B^0)$$

Since $r = N_A^0/N_B^0$,

$$N_r = \frac{1}{2}\left(N_A^0 + \frac{N_A^0}{r}\right) = \frac{N_A^0}{2}\left(\frac{r + 1}{r}\right)$$

The average degree of polymerization is equal to the number of monomeric units divided by the number of chains; that is

$$\overline{\text{DP}} = \frac{N_r}{N} = \frac{(N_A^0/2)[(r + 1)/r]}{N_A^0/2[1 + (1/r) - 2p]}$$

which reduces to

$$\overline{\text{DP}} = \frac{1 + r}{r + 1 - 2rp}$$

Given the stoichiometric imbalance, therefore, we can calculate the extent of reaction necessary to achieve a given degree of polymerization. If $r = 1$, the relationship reduces to

$$\overline{\text{DP}} = \frac{1}{1 - p}$$

which is, of course, the Carothers equation. When monomer A is completely used up in the polymerization (that is, when $p = 1$), the equation becomes

$$\overline{\text{DP}} = \frac{1 + r}{1 - r}$$

Now, what about the situation where monofunctional reagent is added to control molecular weight? According to our definition of r, these must necessarily be B groups. The imbalance factor is then redefined as

$$r' = \frac{N_A^0}{N_B^0 + 2N_{B'}^0}$$

where $N_{B'}^0$ is the number of monofunctional B groups. The factor 2 takes into account the fact that each monofunctional B' molecule is equally as effective as one excess BB monomer in limiting the molecular weight. The factor r' is substituted for r in the previous equations.

Up to this point we have only considered reactions involving AA and BB. For the AB type, $r = 1$ and molecular weight is controlled by addition of monofunctional reagent. The same relationships apply, except that in the definition of r', N_A^0 is necessarily always equal to N_B^0.

10.4   Molecular Weight Distribution

Using statistical methods derived by Flory,[1,2,6] it is possible to relate the molecular weight distribution in step polymerization to the reaction conversion. In any polymerization involving difunctional monomer (AB or AA + BB), each step reaction links two molecules together; thus the number of monomer units in the polymer is always one more than the number of functional group reactions. If a polymer contains x monomer units ($\overline{\text{DP}} = x$), the number of A or B groups that have reacted is $x - 1$. The remaining unreacted A or B groups are at the ends of the chain. It is assumed that each reaction occurs in random fashion throughout the polymerization medium. The problem is to determine the probability of finding a chain containing x monomer units and a single unreacted A or B group at time t.

The probability that $x - 1$ A or B groups has reacted is p^{x-1}, where p is the reaction conversion, defined previously as

$$p = \frac{N_0 - N}{N_0}$$

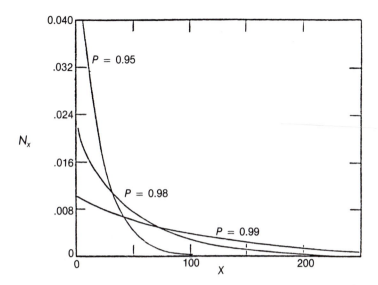

FIGURE 10.1 Mole fraction distribution in linear step-reaction polymers. [From Flory,[2] copyright 1946. Reprinted by permission of the American Chemical Society.]

The probability of finding an *unreacted* group is $1 - p$. The probability of finding a molecule containing x units and an unreacted A or B group is then $p^{x-1}(1 - p)$. If the total number of molecules present at time t is N, then the fraction that contains x units, N_x, is given by

$$N_x = Np^{x-1}(1 - p)$$

Knowing that $N/N_0 = 1 - p$ (Carothers equation), we can rewrite the expression for N_x as

$$N_x = N_0(1 - p)^2 p^{x-1}$$

where N_0 is the number of monomer units present initially. This relationship defines the number or mole fraction distribution at time t. Plots of N_x versus x at three different reaction conversions are shown in Figure 10.1.[2] Interestingly, even at 99% conversion, monomer still represents the most abundant species present. This is misleading, however, because monomer contributes a small percentage to the overall mass. A more reasonable picture is obtained by expressing the distribution in terms of the weight fraction. The weight fraction, w_x, of molecules containing x units is equal to the ratio of the mass of those molecules divided by the total mass; that is

$$w_x = \frac{xN_xM_0}{N_0M_0} = \frac{xN_x}{N_0}$$

where M_0 is the mass of the repeating unit. Substituting in the previous expression for N_x, we obtain

$$w_x = x(1 - p)^2 p^{x-1}$$

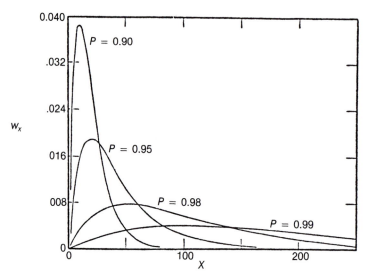

FIGURE 10.2 Weight fraction distribution in linear step-reaction polymers. [From Flory,[2] copyright 1946. Reprinted by permission of the American Chemical society.]

which is the weight fraction distribution. A plot of w_x versus x at four levels of conversion are shown in Figure 10.2.[2] Both figures confirm what we already know from the Carothers equation, that very high conversions are necessary to obtain high degrees of polymerization.

To determine the polydispersity index ($\overline{M}_w/\overline{M}_n$) at a given conversion, we need to define $\overline{M}_w$ and $\overline{M}_n$ in terms of p. Given that $\overline{M}_n$ is the product of $\overline{DP}$ and M_0, and $\overline{DP} = 1/(1 - p)$, we can write

$$\overline{M}_n = \frac{M_0}{1 - p}$$

To obtain a similar expression for $\overline{M}_w$, we take the expression

$$\overline{M}_w = \Sigma\, w_x M_x$$

and rewrite it for x units as

$$\overline{M}_w = \Sigma\, w_x x M_0$$

Substituting the expression for w_x above, we obtain

$$\overline{M}_w = M_0(1 - p)^2 \Sigma x^2 p^{x-1}$$

The series $\Sigma x^2 p^{x-1}$ reduces to $(1 + p)/(1 - p)^3$; therefore

$$\overline{M}_w = \frac{M_0(1 + p)}{1 - p}$$

Dividing $\overline{M}_w$ by $\overline{M}_n$, we arrive at the polydispersity index

$$\frac{\overline{M}_w}{\overline{M}_n} = 1 + p$$

which tells us that at very high levels of conversion (as p approaches 1), the most probable polydispersity approaches 2. This has been confirmed experimentally.

It should be noted that in the case of condensation polymerization, each polymer molecule contains the mass of the byproduct in its end groups. This usually represents a minor contribution to the overall mass, however, and is ignored in the above equations.

10.5 Network Step Polymerization

If monomers containing a functionality greater than two are used in step polymerization, chain branching results. An important application of this type of polymerization is the manufacture of alkyd-type polyester resins (Chapter 12) in which polyfunctional alcohols such as glycerol are used. Pheno–, urea–, and melamine–formaldehyde polymers (Chapter 14) are other important examples. If the reaction is carried to a high enough conversion, gelation occurs. The onset of gelation, or *gel point*, is accompanied by a sudden increase in viscosity such that the polymer undergoes an almost instantaneous change from a liquid to a gel. Bubbles no longer rise in the mixture, and stirring becomes impossible. If the polymerization is being done in solution, the gel is swollen with solvent but is otherwise insoluble. At this point the polymer is a true network polymer, and the crosslink density will continue to rise with a concomitant increase in rigidity if the reaction is allowed to continue. Gelation during polymer synthesis is normally something to be avoided because it can have disastrous consequences if it occurs in a large-scale reactor. Clearly it is desirable to be able to predict the gel point.

One way to predict the onset of gelation is to use a modified form of the Carothers equation that takes into account the *average functionality* of the monomers involved.[7] Suppose, for example, one wanted to prepare a polyester using an equimolar mixture of phthalic anhydride (**1**) (which reacts as a diacid), trimellitic acid (**2**), ethylene glycol (**3**), and glycerol (**4**).

| **1** | **2** |

$$\text{HOCH}_2\text{CH}_2\text{OH} \qquad \text{HOCH}_2\overset{\displaystyle\text{OH}}{\underset{\displaystyle|}{\text{CH}}}\text{CH}_2\text{OH}$$

| **3** | **4** |

The average functionality, f_{av}, is $(2 + 3 + 2 + 3)/4 = 2.5$. Given that

$$p = \frac{N_0 - N}{N_0}$$

where N_0 and N are the number of monomer molecules initially and at conversion p, respectively, then the number of functional groups that have reacted is $2(N_0 - N)$. The number of functional groups initially is $N_0 f_{av}$. Thus,

$$p = \frac{2(N_0 - N)}{N_0 f_{av}}$$

Since $\overline{DP} = N_0/N$, the above expression may be rewritten as

$$p = \frac{2}{f_{av}} - \frac{2}{\overline{DP} f_{av}}$$

It is assumed that gelation occurs when $\overline{DP}$ becomes infinite, at which point the second term becomes zero. Then

$$p_c = \frac{2}{f_{av}}$$

where p_c denotes the *critical reaction conversion* at which gelation occurs. For the system given above, this relationship predicts that gelation would occur at a reaction conversion of 80%. This may be contrasted with a system consisting solely of difunctional monomers in which an 80% conversion leads to a $\overline{DP}$ of only 5.

Similarly, a mixture of 3 mol of **1** and 2 mol of **4**, which has an average functionality of

$$\frac{(3 \times 2) + (2 \times 3)}{5} = 2.4$$

is predicted to gel at 83% conversion. In practice, such a mixture gels at about 77% conversion. The discrepancy arises primarily from the greater contribution of higher-molecular-weight fractions.

Both examples given have *equivalent* amounts of acid and alcohol functional groups. Where nonequivalent amounts are involved, the determination of f_{av} takes a different form.[8] Let's consider a mixture consisting of three monomers A, B, and C, in which A and C have the same functional groups but different functionalities (e.g., a mixture of two acids such as **1** and **2**), B contains a different functional group (e.g., an alcohol such as **3** or **4**), and the B groups are in excess. Then the average functionality is given by

$$f_{av} = \frac{2r f_A f_B f_C}{f_A f_C + r\rho f_A f_B + r(1 - \rho) f_B f_C}$$

where the constants r and p are given by

$$r = \frac{N_A f_A + N_C f_C}{N_B f_B}$$

and

$$\rho = \frac{N_C f_C}{N_A f_A + N_C f_C}$$

The critical conversion, p_c, then refers to the extent of reaction of the A groups only.

Statistical methods have also been developed that predict gelation at a lower level of conversion than that predicted by the Carothers equation.[1,9] For this case, with f_A and f_B each equivalent to 2, and $f_C > 2$.

$$p_c = \frac{1}{[r + r\rho(f - 2)]^{1/2}}$$

where f is the functionality of C. Experimental values of p_c fall between the values calculated by the statistical and nonstatistical methods. Deviations from the former have been attributed to intramolecular branching reactions leading to what are called *wasted loops*, and to differing reactivities of functional groups, for example, the lower reactivity of the secondary alcohol group of glycerol relative to that of the two primary groups.

10.6 Step-Reaction Copolymerization

A copolymer is defined in step polymerization as one having more than one repeating unit. Thus a polyester (**5**) prepared from terephthalic acid and ethylene glycol is a homopolymer, but a polyester (**6**) made with a 1:1:2 mixture of terephthalic acid, isophthalic acid, and ethylene glycol is a copolymer. A variety of copolymer compositions is employed in industrial step polymerization. An example that appeared earlier in the book (Chapter 3) was a liquid crystalline copolyester prepared from terephthalic acid, *p*-hydroxybenzoic acid, and ethylene glycol.

5

6

In synthesizing copolymers such as **6**, the distribution of monomer units is random because the two dicarboxylic acids have virtually equal reactivity. It is possible in many instances to synthesize alternating step copolymers. Consider the hypothetical case of two different monomers, AA and BB, both of which react with monomer CC. A 1:1:2 mixture of AA, BB, and CC yields a random copolymer. If, however, AA is first reacted with the two equivalents of CC and then in a separate step the product is reacted with BB, an alternating copolymer is formed. This is shown schematically in reaction (10.6). Such reactions are used to synthesize some important aromatic polyamides, as described in Chapter 13. The method

$$AA + 2CC \rightarrow CC—AA—CC \xrightarrow{BB} +CC—AA—CC—BB+ \qquad (10.6)$$

does not *always* lead to alternating copolymer. Polyesters prepared with dibasic acids and diols may undergo transesterification in the second step and scramble the monomer distribution, but where such complications can be avoided, alternating polymers are formed without difficulty.

Because step polymers are true telechelic polymers, it is a relatively straightforward matter to prepare block copolymers by linking homopolymers together through coreactive functional groups. Typical examples are the reaction of a hydroxyl-terminated polyether and an isocyanate-terminated polyurethane (10.7), and the reaction of an acid chloride-terminated polyester with an amine-terminated polyamide (10.8). Alternatively, one could react an isocyanate-terminated polyurethane with two equivalents of hydroxy-terminated polyester to form an ABA block copolymer (10.9).

HO-polyether-OH + OCN-polyurethane-NCO $\longrightarrow$

$$\left[\begin{array}{c} \underset{\parallel}{O} \qquad\qquad\qquad \underset{\parallel}{O} \\ +O\text{-polyether-OCNH-polyurethane-NHC}+ \end{array} \right] \qquad (10.7)$$

$$\underset{\parallel}{O} \qquad \underset{\parallel}{O}$$
ClC-polyester-CCl + H$_2$N-polyamide-NH$_2$ $\longrightarrow$

$$\left[\begin{array}{c} \underset{\parallel}{O} \qquad\qquad \underset{\parallel}{O} \\ +C\text{-polyester-CNH-polyamide-NH}+ \end{array} \right] \qquad (10.8)$$

2HO-polyester-OH + OCN-polyurethane-NCO $\longrightarrow$

$$\underset{\parallel}{O} \qquad\qquad\qquad\qquad \underset{\parallel}{O}$$
HO-polyester-OCNH-polyurethane-NHCO-polyester-OH $\qquad (10.9)$

Irreversible reactions of the types depicted here are necessary to control placement of the blocks. Numerous examples of AB diblock, +AB+ multiblock, and ABA triblock copolymers have been reported.[10,11] A number of these have commercial application.

10.7 Step Polymerization Techniques

A significant difference between vinyl and nonvinyl polymerization is that with the former, a large enthalpy factor contributes to the change from double to single bonding, resulting in generally low activation energies and significant reaction exotherms. Dissipation of the exotherm is often a critical factor in large-scale bulk polymerization of vinyl monomers. Nonvinyl polymerization, on the other hand, is more often than not characterized by high activation energies and low reaction exotherms. As a result, a good many step polymerizations are run at elevated temperatures.

There are basically four processing techniques used to make step-reaction polymers, two homogeneous (bulk and solution), and two heterogeneous (interfacial and phase-transfer catalyzed). Dispersion or emulsion methods have been used rarely. Of the four principal techniques, bulk and solution are by far the most important from the standpoint of commercial production.

Bulk polymerization has the obvious advantage of providing a product free of contaminants other than residual byproducts or impurities arising from side reactions. The major disadvantage is that high viscosities necessitate the use of elevated temperatures for effective stirring and, in the case of condensation polymerization, removal of byproduct. As a rule, bulk polymerizations are done under an inert atmosphere. A high vacuum is usually employed to facilitate removal of byproducts. In the case of unsaturated polyesters and alkyd resins (Chapter 12), which are processed to much lower molecular weights, byproducts are removed by blowing large volumes of inert gas, most often nitrogen, through the reaction mixture in the later stages of polymerization, a process called *sparging*.

Solvent polymerization minimizes the problems arising from high viscosity and can assist in removal of byproduct by azeotropic distillation. As will be seen in later chapters, methods have been developed to remove water in situ by use of effective dehydrating agents, which often allow the polycondensation to proceed at ambient temperatures. The major disadvantage of solvent processing is the necessity of removing solvent once the reaction is finished. The problem is obviated when solvent-based coatings or solution-extruded fibers are the final product.

Interfacial polymerization[12,13] involves solutions of the two monomers in separate, immiscible solvents, one usually being water. When the two solutions are brought into contact, polymer is formed at the interface. A typical example is the Schotten–Baumann synthesis of polyamide from a diacid chloride dissolved in an organic solvent, and a diamine dissolved in aqueous base (10.10). The base is needed to neutralize the byproduct HCl, which would otherwise react with the diamine to form amine hydrochloride. If the reaction is done in an

$$
\text{Cl}\!-\!\overset{\overset{\text{O}}{\|}}{\text{C}}\!-\!\text{R}\!-\!\overset{\overset{\text{O}}{\|}}{\text{C}}\!-\!\text{Cl} + \text{H}_2\text{N}\!-\!\text{R}'\!-\!\text{NH}_2 \xrightarrow[\text{H}_2\text{O. CH}_2\text{Cl}_2]{\text{NaOH}}
$$

$$
\left[\!+\!\overset{\overset{\text{O}}{\|}}{\text{C}}\!-\!\text{R}\!-\!\overset{\overset{\text{O}}{\|}}{\text{C}}\!-\!\text{NH}\!-\!\text{R}'\!-\!\text{NH}\!+\!\right] \tag{10.10}
$$

open beaker, the polymer film at the interface may be grasped with tweezers and pulled out as a continuous "filament" of collapsed film until monomer is used up. This very effective demonstration (Figure 10.3) is aptly called the "nylon rope trick."[14] Rapid stirring to maximize the interfacial area increases the yield of polymer. (A kitchen blender is often used for interfacial laboratory preparations.) The method may also be used to synthesize polyesters from diacid chlorides and bisphenols; however, dialcohols are not sufficiently reactive, and hydrolysis of the diacid chloride takes precedence. Apart from some miscellaneous aromatic polyamides and polyesters, commercial exploitation of interfacial polymerization has been primarily limited to polycarbonates (Chapter 12).[15]

Interfacial polymerization differs significantly from bulk or solution polymerization:

1. The reaction goes rapidly at low temperatures.
2. Because the reaction is so rapid, diffusion of monomer to the interface is rate determining.

Polyamide

Diamine in water

Diacid chloride in
halogenated solvent

FIGURE 10.3 Interfacial synthesis of polyamide.

3. The polymerization takes on some of the characteristics of chain polymerization as monomer reacts with the growing chains at the interface more rapidly than it diffuses through the polymer film to initiate new chains; hence molecular weights tend to be significantly higher.

4. As a consequence of the last point, an exact stoichiometric balance is not necessary.

With so many apparent advantages, the obvious question is, why is interfacial polymerization not used more widely in industry? The answer has to do with the economics of the process. The high cost of acid chlorides and the large volumes of solvent required make the interfacial method prohibitively expensive for many polymers relative to bulk or solution processes.

Phase-transfer catalysis (PTC), although commonly employed in organic synthesis, has had more limited application in polymerization reactions.[16–18] In its most common form, the PTC method involves an aqueous phase and an organic phase, each containing one of the monomers. Hence it is also an interfacial technique. The catalyst, often a quaternary ammonium salt, functions by transporting a nucleophilic monomer from the aqueous phase to the organic phase, where its nucleophilicity is greatly enhanced because of reduced solvation effects. A case in point is the synthesis of polymer (**9**) by the reaction of α,α'-dichloro-*p*-xylene (**7**) and *t*-butyl cyanoacetate (**8**) using benzyltriethylammonium chloride catalyst (10.11).[17] In this case, the anion derived from **8** by reaction with NaOH is transported to the benzene solution as the soluble benzyltriethylammonium salt, where it reacts rapidly with **7** by nucleophilic displacement. Polymer is formed because **8** contains two active hydrogens. Other examples of PTC polymerization are given in later chapters.

$$ClCH_2 - \langle \bigcirc \rangle - CH_2Cl + NCCH_2CO_2C(CH_3)_3$$

$$\textbf{7} \qquad\qquad\qquad \textbf{8}$$

$$\xrightarrow[\text{NaOH, H}_2\text{O/benzene}]{\phi CH_2\overset{+}{N}(C_2H_5)_3Cl^-} \left[CH_2 - \langle \bigcirc \rangle - CH_2\underset{\underset{CO_2C(CH_3)_3}{|}}{\overset{\overset{CN}{|}}{C}} \right] \qquad (10.11)$$

$$\textbf{9}$$

10.8   Dendritic Polymers

One of the more intriguing aspects of step-growth polymerization is the synthesis of dendritic polymers or dendrimers[19-24] (cf. Chapter 1). Since the first Starburst dendrimers were reported in the 1980s,[25] these aesthetically pleasing macromolecules have now reached the point of commercial development. Drug delivery systems, controlled release of agricultural chemicals, molecular sensors, and rheology modifiers all represent potential applications. Some of the characteristic features of dendrimers are:

1. They are defined by their three component parts: a central core, an interior dendritic structure, and an exterior surface.

2. Their macromolecular dimensions are easily controlled because each level of synthesis is normally by a repetitive sequence of steps.

3. They tend to be more soluble than linear polymers because of their high surface functionality.

4. Because they lack the chain entanglement of linear polymers, their solutions usually have relatively low viscosities.

5. Supramolecular assemblies may be constructed by incorporating guest molecules among the interior branches of the dendrimer.[26]

How are dendrimers constructed? There are two general approaches: *divergent*, in which branch points are constructed in a stepwise fashion from a central core; and *convergent*, in which branch segments are constructed separately and then joined to a multifunctional core. A general strategy for divergent synthesis is exemplified by the polyamidoamine (PAMAM) dendrimers.[24] Using an ammonia core, the first step involves exhaustive Michael addition to methyl acrylate (10.12). The resultant triester (**10**) is then reacted with excess 1,2-ethanediamine to form amide–amine **11** (10.13). These two steps represent the "first generation." Each new generation results from a repetition of the two-step reaction sequence, as shown

$$NH_3 + 3\ CH_2{=}CHCO_2CH_3 \rightarrow N(CH_2CH_2CO_2CH_3)_3 \qquad (10.12)$$

$$\textbf{10}$$

$$\textbf{10} + H_2NCH_2CH_2NH_2 \rightarrow N(CH_2CH_2CONHCH_2CH_2NH_2)_3 \qquad (10.13)$$

$$\textbf{11}$$

SCHEME 10.1. Formation of second and third generations of PAMAM dendrimer. Reaction sequence: (1) excess $CH_2=CHCO_2CH_3$; (2) excess $H_2NCH_2CH_2NH_2$.

schematically for generations 2 and 3 in Scheme 10.1. With each new generation, the number of surface functional groups doubles, reaching 3072 by the tenth generation. "Starburst" is thus an apt metaphor for these polymers.

An example of convergent synthesis of a polyether dendrimer[22] is shown in Schemes 10.2 and 10.3. The branching unit, 3,5-dihydroxybenzyl alcohol (**13**), is first reacted with

SCHEME 10.2. Formation of a polyether dendritic segment.

SCHEME 10.3. Convergent synthesis of a dendrimer from dendritic segments.

the benzylic bromide **12** in the presence of base to yield **14**. The alcohol group of **14** is then converted to bromide, and reacted again with **12**. The steps are repeated until the desired segment size is reached. The polyether dendrimer is then assembled by condensing a benzylic bromide-terminated segment with a polyfunctional phenol (Scheme 10.2).

There is a limit to the size of dendrimers or dendrimer segments because of steric congestion; nevertheless, dendrimers with molecular weights in the hundreds of thousands have been achieved. Such materials are sometimes referred to as *nanostructures* because their molecular dimensions lie in the nanometer range, defined as 1–100 nm. There is much current interest in nanostructures[27] because of their potential for mimicking the behavior of nanoscopic biomolecules and for their possible use in molecular devices.

Numerous types of dendrimers have been reported.[19] The related *hyperbranched polymers* have more random architectures and do not emanate from a central core. They are synthesized from monomers having A$_x$B functionality, where *x* is greater than 1. Polymerization occurs via dendritic branching with no possibility of crosslinking (other than by side reactions). An example, used to make hyperbranched polyethers comparable to the polyether dendrimer shown above, is 3,5-dihydroxybenzyl bromide (**15**). Examples of hyperbranched polyesters are given in Chapter 12.

15

10.9   Ring-Opening Polymerization

Ring opening is the third general method of preparing linear polymers.[28–30] Examples of commercially important types are given in Table 10.2 in the order in which they appear in later chapters. Of those listed, only the metathesis-derived polymers (discussed in Chapter 8) are composed solely of carbon chains. Those that have enjoyed the longest history of commercial exploitation are polyethers prepared from three-membered ring cyclic ethers (epoxides), polyamides from cyclic amides (lactams), and polysiloxanes from cyclic siloxanes. It

TABLE 10.2. Commercially Important Polymers Prepared by Ring-Opening Polymerization

Polymer Type	Polymer Repeating Group	Monomer Structure	Monomer Type	Chapter
Polyalkene	$+CH{=}CH(CH_2)_x+$	$(CH_2)_x$ with $CH{=}CH$	Cycloalkene	8
Polyether	$+CH_2O+$	Trioxane ring	Trioxane	11
Polyether	$+(CH_2)_xO+$	$(CH_2)_xO$	Cyclic ether[a]	11
Polyester[b]	$+(CH_2)_x\overset{O}{\overset{\|}{C}}O+$	$(CH_2)_x$ lactone ring	Lactone	12
Polyamide	$+(CH_2)_x\overset{O}{\overset{\|}{C}}NH+$	$(CH_2)_x$ lactam ring (NH)	Lactam	13
Polysiloxane	$+\overset{CH_3}{\underset{CH_3}{Si}}O+$	$[Si(CH_3)_2]_x$ ring	Cyclic siloxane	16
Polyphosphazene	$+\overset{Cl}{\underset{Cl}{P}}{=}N+$	Hexachlorocyclotriphosphazene ring ($Cl_2P{=}N$ trimer)	Hexachloro-cyclotriphos-phazene[c]	16
Polyamine	$+CH_2CH_2NH+$	$\overset{NH}{CH_2{-}CH_2}$	Aziridene[d]	17

[a] Epoxide ($x = 2$); oxetane ($x = 3$).

[b] Ring opening of cyclic oligomers has also been developed.

[c] Phosphonitrilic chloride trimer.

[d] Also called alkyleneimine.

should be remembered that although these reactions (other than metathesis) lead to condensation-type polymers, they are not polycondensation reactions because no byproduct is formed. (Exceptions to this general rule are found in Chapters 12 and 13.)

Mechanisms of ring-opening polymerization vary according to monomer type and initiator, but in most instances they fit one of two general forms:

1. Monomer is attacked by some ionic or coordination species, designated X*, at the functional group (G) that causes ring opening. This is followed by attack of the ring-opened monomer on another cyclic unit, and so on (10.14).

$$\bigcirc\!\!G \xrightarrow{\ X^* \ } X\,\text{ᴡᴡᴡ}\,G^* \xrightarrow{\ \bigcirc\!G\ } X\,\text{ᴡᴡᴡ}\,G\,\text{ᴡᴡᴡ}\,G^* \xrightarrow{\ \bigcirc\!G\ } \text{etc.} \qquad (10.14)$$

2. Monomer is attacked by X* to form a coordination species (most frequently a cation) that undergoes reaction with a second monomer molecule to open the ring, and so on (10.15).

$$\bigcirc\!\!G \xrightarrow{\ X^* \ } \bigcirc\!\!G^*\!\!-X \xrightarrow{\ \bigcirc\!G\ } \bigcirc\!\!G^*\,\text{ᴡᴡᴡ}\,G\!\!-\!X \xrightarrow{\ \bigcirc\!G\ }$$

$$\bigcirc\!\!G^*\,\text{ᴡᴡᴡ}\,G\,\text{ᴡᴡᴡ}\,G\!\!-\!X \quad \text{etc.} \qquad (10.15)$$

Although these mechanisms suggest a chain-growth process, this is not necessarily the case. Epoxide polymerization under anionic conditions, for example, exhibits characteristics of both step and chain growth depending on choice of initiator. Lactams may polymerize by a combination of step and chain growth with a single initiator. Thus it is impossible to generalize in terms of an overall kinetics scheme for ring-opening polymerization. As indicated above, ionic or coordination compounds are used most commonly to initiate ring-opening polymerization, although molecular compounds (water, alcohols, amines, etc.) are also used. In such cases, the initiator (XY) serves to open the ring, then polymerization proceeds by step growth (10.16). Free radical ring-opening reactions are far less common, although an important example leading to polyester formation is given in Chapter 12.[31]

$$\bigcirc\!\!G \xrightarrow{\ XY \ } X\,\text{ᴡᴡᴡ}\,G\!\!-\!Y \xrightarrow{\ \bigcirc\!G\ } X\,\text{ᴡᴡᴡ}\,G\,\text{ᴡᴡᴡ}\,G\!\!-\!Y \xrightarrow{\ \bigcirc\!G\ } \text{etc.} \qquad (10.16)$$

As might be expected from considerations of ring strain, the most reactive monomers are usually those containing three- and four-membered rings.[29,32] With seven-membered and larger rings, transannular interactions provide the thermodynamic favorability for ring opening. Five- and six-membered rings are virtually free of angle strain; however, torsional strain arising from conformational eclipsing of C—H bonds is the major factor responsible for the lower thermodynamic stability of cyclopentane relative to cyclohexane. Incorporation of heteroatoms into the ring influences ring stability in ways that are not always predictable. Six-membered cyclic esters (lactones) form polyester quite readily, for example, whereas six-membered lactams are reluctant to form polyamide. The reverse is the case with five-membered rings. Cyclic ethers, which exhibit ring strain similar to that encountered in cycloalkanes, exhibit the following order of reactivity in terms of ring size:

$$3 > 4 > 8 > 7 > 5 > 6$$

Tetrahydropyran, the six-membered ring, resists polymerization. Ring strain does not, of itself, guarantee that ring-opening polymerization will occur. Cycloalkanes, which have no functional groups to provide a kinetic pathway, do not polymerize apart from some highly strained cyclic or polycyclic compounds. Ceiling temperatures are often quite low in ring-opening polymerizations compared with vinyl polymerization, particularly where five- or six-membered rings are involved.

Copolymers may be prepared by ring-opening polymerizations, including some with different functional groups. Large differences in reactivity, however, often result in significant block formation. Deliberately formed block copolymers represent the most interesting types. A wide variety of AB, $+$AB$+$, and ABA block copolymers have been prepared by the living polymer method, including some containing both vinyl and nonvinyl blocks.[11]

Ring-opening polymerizations are generally restricted to solution or bulk processes.

References

1. P. J. Flory, *Principles of Polymer Chemistry*, Cornell University Press, Ithaca, N.Y., 1953, Chaps. 3, 8, and 9.
2. P. J. Flory, *Chem. Rev.*, **39**, 137 (1946).
3. D. H. Solomon (Ed.), *Step-Growth Polymerization*, Marcel Dekker, New York, 1972.
4. R. W. Lenz, *Organic Chemistry of Synthetic High Polymers*, Wiley-Interscience, New York, 1967, Chap. 3.
5. P. E. M. Allan and C. R. Patrick, *Kinetics and Mechanisms of Polymerization Reactions*, Wiley-Interscience, New York, 1974, Chap. 5.
6. L. H. Peebles, Jr., *Molecular Weight Distributions in Polymers*, Wiley-Interscience, New York, 1971.
7. W. H. Carothers, *Trans. Faraday Soc.*, **32**, 39 (1936).
8. S. H. Pinner, *J. Polym. Sci.*, **21**, 153 (1956).
9. W. H. Stockmayer, *J. Polym. Sci.*, **9**, 69 (1952); **11**, 424 (1953).
10. R. Jerome, R. Fayt, and T. Ouhadi, *Progr. Polym. Sci.*, **10**, 87 (1984).
11. A. Noshay and J. E. McGrath, *Block Copolymers: Overview and Critical Survey*, Academic Press, New York, 1977.
12. F. Millich and C. E. Carraher Jr. (Eds.), *Interfacial Synthesis*, Vols. 1 and 2, Marcel Dekker, New York, 1975 and 1977.
13. P. W. Morgan, *J. Macromol. Sci.-Chem.*, **A15**, 683 (1981).
14. P. W. Morgan and S. L. Kwolek, *J. Chem. Educ.*, **36**, 182 (1959); G. C. East and S. Hassell, *ibid.* **60**, 69 (1983).
15. E. D. Oliver and Y.-C. Chen, in *Interfacial Synthesis*, Vol. 2 (F. Millich and C. E. Carraher Jr., Eds.), Marcel Dekker, New York, 1977, Chap. 11.
16. L. J. Mathias and C. E. Carraher Jr. (Eds.), *Crown Ethers and Phase Transfer Catalysis in Polymer Science*, Plenum Press, New York, 1984.
17. Y. Imai, *J. Macromol. Sci.-Chem.*, **A15**, 833 (1981).
18. L. J. Mathias, *J. Macromol. Sci.-Chem.*, **A15**, 853 (1981).
19. B. I. Voit, *Acta Polymer.*, **46**, 87 (1995).
20. D. A. Tomalia and H. D. Durst, in *Supramolecular Chemistry I: Directed Synthesis and Molecular Recognition* (E. Weber, Ed.), Topics Curr. Chem. 165, Springer-Verlag, New York, 1993.

21. D. A. Tomalia, *Adv. Mater.*, **6**, 529 (1994).
22. J. M. J. Fréchet, *Science*, **263**, 1710 (1994).
23. G. R. Newkome (Ed.), *Advances in Dendritic Macromolecules*, JAI, Greenwich, Connecticut, 1993.
24. D. A. Tomalia, A. Naylor, and W. Goddard III, *Angew. Chem., Int. Ed. Engl.*, **29**, 138 (1990).
25. D. A. Tomalia, H. Baker, J. Dewald, M. Hall, G. Kallos, S. Martin, J. Roek, J. Ryder, and P. Smith, *Polym. J.*, **17**, 117 (1985).
26. J. F. G. A. Jansen, E. M. M. de Brabender–van den Berg, and E. W. Meijer, *Science*, **266**, 1226 (1994).
27. G. W. Whitesides, J. P. Mathias, and C. T. Seto, *Science*, **254**, 1312 (1991).
28. D. J. Brunelle (Ed.), *Ring-Opening Polymerization: Mechanism, Catalysis, Structure, Utility*, Hanser, Cincinnati, 1992.
29. K. C. Frisch and S. L. Reegan (Eds.), *Ring-Opening Polymerization*, Marcel Dekker, New York, 1969.
30. O. Vogl and J. Furukawa (Eds.), *Polymerization of Heterocycles*, Marcel Dekker, New York, 1973.
31. W. J. Bailey, in *Ring-Opening Polymerization* (J. E. McGrath, Ed.), ACS Symp. Ser. 286, American Chemical Society, Washington, D.C., 1985, pp. 46–65.
32. J. E. McGrath, in *Ring-Opening Polymerization* (J. E. McGrath, Ed.), ACS Symp. Ser. 286, American Chemical Society, Washington, D.C., 1985, pp. 1–22.

Review Exercises

1. Considering the reaction conditions for bulk polycondensations, suggest reasons other than those given in the chapter for deviations in $\overline{DP}$ versus t plots for polyesters.

2. Plots of $1/(1-p)^2$ versus t for the uncatalyzed polyesterification of adipic acid and ethylene glycol are linear between 80 and 93% conversion of monomer. To what number average molecular weights do these limits correspond?

3. To what percent conversion should a batch of 11-aminoundecanoic acid be processed to yield polyamide having a number average molecular weight of 2.50×10^4?

4. Rate constants for polyesterification of adipid acid and 1,10-decanediol at 161°C are 0.0012 (eq/kg)^2/min (uncatalyzed) and 0.079 (eq/kg)/min (catalyzed) (S. D. Hamann et al., *J. Macromol. Sci.-Chem.*, **A2**, 153 1968). How long would it take to prepare a polyester of number average molecular weight 15,000 (bulk processing) under each set of conditions? (Assume equimolar amounts of each reactant and 2 equivalents per mole, and that bulk mass remains constant.) What would the polydispersity index be for the polymer?

5. If 5 mol % excess diol were to be used in the previous problem, what number average molecular weight would be obtained if the reaction were carried to the same conversion?

6. Using the Carothers relationship, calculate the percent conversion at the gel point for (a) a 3:1:3 and (b) a 1:1:2 molar ratio of phthalic anhydride, trimellitic acid, and glycerol.

7. If one were to react 50.0 g of terephthalic acid (TPA) (1,4-benzenedicarboxylic acid) with 20.0 g of ethylene glycol, (a) how many grams of TPA would have to be consumed in order to obtain a $\overline{DP}$ of 20? (b) If the reaction were to be continued until the TPA was completely consumed, what would be the number average molecular weight of the polyester?

8. Explain why aliphatic polyamides but not aliphatic polyesters can be prepared by interfacial polymerization.

9. Write equations illustrating the synthesis of (a) an alternating copolyamide from isophthaloyl chloride, terephthaloyl chloride, and *m*-diaminobenzene, and (b) an ABA block copolymer of poly(11-undecanoamide) (A) with the polyamide (B) formed in (a).

10. Which reactions would be expected to yield higher-molecular-weight polyester: the polycondensation of 5-hydroxypentanoic acid or 6-hydroxyhexanoic acid? Explain.

11. Give structures of cyclic monomers that might be used to prepare:

(a) $+\text{NHCH}_2\text{CH}_2\text{CH}_2\overset{\displaystyle \overset{O}{\|}}{\text{C}}+$

(b) $+\overset{\displaystyle \overset{\text{CH}_3}{|}}{\underset{\displaystyle \underset{\text{CH}_3}{|}}{\text{C}}}\text{CH}_2\overset{\displaystyle \overset{O}{\|}}{\text{C}}\text{O}+$

(c)

11

POLYETHERS, POLYSULFIDES,

AND RELATED POLYMERS

11.1 Introduction

Polyethers and related sulfur-containing polymers are the most diverse of the nonvinyl polymers in the structural variety that have achieved commercial prominence (Table 11.1). They include polyacetals synthesized by chain-reaction polymerization of aldehydes (a reaction akin to chain polymerization of vinyl monomers), polyethers prepared by ring-opening and step-reaction polymerization, and the structurally related polysulfides and other sulfur-containing polymers. Many of these polymers are engineering plastics, and their use is expanding accordingly.

This chapter is divided into three sections. Section 11.1 includes polyethers made by chain polymerization of carbonyl compounds and ring-opening polymerization of cyclic ethers. Section 11.2 is concerned with step-polymerization methods that have proven most fruitful in development of epoxy polymers and a number of aromatic polyethers for engineering applications. Certain types of polyacetals and polyketals are also covered in this section. Section 11.3 deals with sulfur-containing polymers: polysulfides (sometimes called *polythioethers*), poly(alkylene polysulfides), and polysulfones. It should be noted that there are no oxygen-containing counterparts to these last two polymers, although unstable copolymers of oxygen and vinyl monomers containing peroxide groups in the backbone have been used to synthesize block copolymers (see Chapter 9) and poly(styrene peroxide)[1] has been suggested as a solid propellant because of its autopyrolytic properties.[2]

11.2 Preparation of Polyethers by Chain-Reaction and Ring-Opening Polymerization

11.2.1 Polymerization of Carbonyl Compounds

Aldehydes have a tendency to form cyclic trimers or tetramers, and polymerization conditions are chosen to circumvent this reaction.[3,4] Formaldehyde is an exception in that while its cyclic trimer, *trioxane* (1), is well known, it is not so easy to prepare, although it can be

TABLE 11.1. Commercially Available Polyethers and Sulfur-Containing Polymers

Type	*Structure*	*Typical Uses*
Polyacetal[a,b]	$-\!\!\left[CH_2O\right]\!\!-$	Engine, appliance, and plumbing parts; electronics components; zippers; buckles
Poly(ethylene oxide)[c]	$-\!\!\left[CH_2CH_2O\right]\!\!-$	Surfactants, hydraulic fluids, lubricants, water-soluble packaging film, textile size
Poly(propylene oxide)[c]	$-\!\!\left[CH_2CHO\right]\!\!-$, with CH_3	Polyurethane intermediates, lubricants, surfactants
Poly(hexafluoropropylene oxide)	$-\!\!\left[CF_2CFO\right]\!\!-$, with CF_3	Lubricating oils, greases
Poly[3,3-(dichloromethyl)trimethylene oxide]	$\left[CH_2\overset{\overset{\displaystyle CH_2Cl}{\mid}}{\underset{\underset{\displaystyle CH_2Cl}{\mid}}{C}}CH_2O\right]$	Corrosion-resistant tanks and tank liners, gaskets, machine parts
Polytetrahydrofuran	$-\!\!\left[(CH_2)_4O\right]\!\!-$	Thermoplastic elastomers, artificial leather
Polyetherimide[a,d] (PEI)		Automotive and electronics parts, composites, wire and cable insulation
Polyetheretherketone[a,e] (PEEK)		Wire and cable insulation, radiation-resistant parts for nuclear plants, machine parts
Poly(phenylene oxide)[a] (PPO)		Telecommunications and computer equipment, automotive parts, appliance housing, electrical equipment
Epoxy[d]	$CH_2CHCH_2\!\!-\!\!\left[OR\right]\!\!-\!\!OCH_2CHCH_2$ (with epoxide rings)	Coatings, laminated circuit boards, adhesives, composites, bridge road surfaces, highway "rumble strips"
Polysulfide[a]		Electronics, electrical, automotive, and machine parts; small appliances; chemical processing equipment
Poly(alkylene polysulfide)[d]	$-\!\!\left[RS_x\right]\!\!-$	Sealants, gaskets, hose, solid rocket fuel binder

continued

TABLE 11.1. (continued)

Type	Structure	Typical Uses
Polysulfone[a,d]	$-\!\!\left[ArSO_2\right]\!\!-$	Medical and food-processing equipment, electrical and electronics components, camera cases, piping

[a] Engineering plastics.
[b] Homopolymer and copolymer.
[c] Block copolymers of ethylene oxide and propylene oxide are also available as surfactants.
[d] See text for details of structural units.
[e] Related commercial ether–ketene polymers are PEK, PEKK, and PEEKK.

made by condensing formaldehyde vapors at low temperatures. Acetaldehyde, on the other hand, forms cyclic trimer (*paraldehyde*, **2**) and tetramer (*metaldehyde*, **3**) under the influence of acids.

        1   **2**   **3**

Formaldehyde is by far the most important aldehyde monomer because it forms polymers having the combination of physical and mechanical properties and economy of processing to make it attractive for commercial use. *Polyoxymethylene* (**4**), alternatively referred to as *polyformaldehyde*, *polyformal*, or *acetal* polymer,[5,6] is a highly crystalline (75–80%), high-strength polymer that replaces metals in machine parts, springs, zippers, and a host of other applications. Commercialized in 1960, it is marketed under the du Pont trade name Delrin.[7] Polyoxymethylene has a relatively low ceiling temperature (about 120°C) and undergoes depolymerization initiated at the hydroxyl end groups; hence the commercial product is stabilized by end-capping with acetic anhydride:

$$HO\!-\!\!\left[CH_2O\right]\!\!-\!CH_2OH \xrightarrow{(CH_3CO)_2O} CH_3\overset{\displaystyle O}{\overset{\displaystyle \|}{C}}O\!-\!\!\left[CH_2O\right]\!\!-\!CH_2O\overset{\displaystyle O}{\overset{\displaystyle \|}{C}}CH_3 \qquad (11.1)$$
$$\mathbf{4}$$

A number of methods have been employed to prepare polyoxymethylene. One is to condense gaseous formaldehyde on a cold surface in the presence of such initiators as metal alkoxides and metal alkyls, Lewis acids, and HCl. Free radical processes do not appear to be significant in polymerization of carbonyl compounds. The commercial product is prepared under anionic or cationic conditions in a hydrocarbon solvent from which the polymer precipitates. Interestingly, formaldehyde polymerizes spontaneously in aqueous solution (11.2) by successive condensation reactions of the hydrate, **5**. Evaporation of the solvent yields solid polymer. Known in this form as *paraformaldehyde*, the polymer serves as a convenient source of anhydrous, gaseous formaldehyde because it depolymerizes at 180 to 200°C.

$$nCH_2(OH)_2 \rightarrow HO[CH_2O]_nH + (n-1)H_2O \tag{11.2}$$
$$\mathbf{5}$$

Cationic polymerization of aldehydes[4] illustrated in reaction (11.3) for the BF_3–H_2O initiating system, is analogous to cationic vinyl polymerization. The carbon–oxygen double bond is highly polar, and addition of the cationic species is regiospecific at the electronegative oxygen. Chain termination occurs primarily by transfer reactions such as (11.4).

$$F_3\bar{B}:\overset{+}{O}H_2 + RCHO \longrightarrow \left[HO-\overset{+}{\underset{R}{C}}H \longleftrightarrow H\overset{+}{O}=\underset{R}{C}H \right] F_3\bar{B}:OH \xrightarrow{nRCHO}$$

$$HOCH\!\!\left[\!\!\underset{R}{OCH}\!\!\right]_{n-1}\!\!\!\overset{+}{\underset{R}{O}CH} \; F_3\bar{B}:OH \tag{11.3}$$

$$\underset{R}{\wedge\wedge\wedge\overset{+}{O}CH} \; F_3\bar{B}:OH + H_2O \longrightarrow \underset{R}{\wedge\wedge\wedge OCHOH} + F_3\bar{B}:\overset{+}{O}H_2 \tag{11.4}$$

Alkoxides and metal alkyls are the best anionic initiators for aldehyde polymerization (11.5); however, the aldol condensation is a serious side reaction with aldehydes that contain alpha hydrogen. Spectroscopic detection of ester end groups suggests that chain transfer occurs by hydride transfer (11.6), a reaction analogous to the crossed Cannizaro reaction. As is the case with anionic vinyl polymerization, there is no inherent termination reaction. Living polymers that lend themselves to block copolymerization may be formed.

$$RCHO + R'M \longrightarrow R'\underset{R}{CHO^-}M^+ \xrightarrow{nRCHO} R'\!\!\left[\!\!\underset{R}{CHO}\!\!\right]_{n-1}\!\!\!\underset{R}{CHO^-}M^+ \tag{11.5}$$

$$\underset{R}{\wedge\wedge\wedge OCHO^-}M^+ \xrightarrow{RCHO} \wedge\wedge\wedge O\underset{R \;\; O-M}{C} \longrightarrow \wedge\wedge\wedge O\underset{R}{C}=O + RCH_2O^-M^+ \tag{11.6}$$

Polymers of acetaldehyde and higher aldehydes have considerably lower ceiling temperatures than polyoxymethylene (usually below 0°C), although stability can be improved by endcapping. Polychloral (polytrichloroacetaldehyde, **6**), of interest because of its flame resistance and good mechanical properties, has been formed into objects by an interesting monomer casting technique that takes advantage of the low ceiling temperature of 58°C.[8] Monomer and initiator are mixed above 58°C, poured into a mold, then cooled to room temperature or below to effect polymerization. Halogen substituents, particularly fluorine,[9] facilitate anionic polymerization of aldehydes by inductive stabilization of the propagating alkoxide ion.

$$\left[\begin{array}{c} CH-O \\ | \\ CCl_3 \end{array}\right]$$

6

Unsaturated aldehydes, especially acrolein (**7**), have also been studied; however, the resultant polymers, prepared under free radical or ionic conditions, are insoluble, infusible, and complex in structure.[10] Polyacrolein is of interest because of its ability to bind antibodies and enzymes with little loss of biological activity.[11]

$$CH_2=CH-CH=O$$

7

Dialdehydes also form polymers containing cyclic ether units.[12] Gluturaldehyde, for example, polymerizes under cationic conditions (11.7) by a mechanism analogous to that of polymerization of isolated dienes.

$$R^+ + \quad \longrightarrow \quad R \quad \longrightarrow \quad R \quad \longrightarrow \text{etc.} \qquad (11.7)$$

Ketones are polymerizable at low temperatures, but the polymers are unstable. Polyacetone (**8**) prepared at liquid nitrogen temperatures, for example, decomposes rapidly at room temperature.[4]

$$\left[\begin{array}{c} CH_3 \\ | \\ C-O \\ | \\ CH_3 \end{array}\right]$$

8

11.2.2 Stereochemistry of Aldehyde Polymerization

Polymerization of acetaldehyde and the higher aldehydes leads to polymers having stereogenic carbon atoms and the resultant potential for stereoregularity. Not as much work has been done in this area as with stereoregular vinyl polymerization, but it is apparent that isotactic sequences are favored by anionic initiators and by bulky monomers.[12]

With acetaldehyde, cationic initiators give amorphous atactic polymer, or, at best, very small amounts of crystalline material. Crystalline polymers are formed, however, with higher aldehydes under cationic conditions. This may not necessarily be a function of the cationic mechanism but may reflect conformational effects resulting from the size of the monomer, because some higher aldehydes such as isobutyraldehyde and heptanal form crystalline polymer under high pressure in the absence of catalyst.

Anionic polymerization using metal alkyls results in isotactic polyaldehyde with the degree of stereoregularity increasing with increasing size of the aldehyde. Because the reaction is believed to occur via alkoxide addition reactions (11.5), a mechanism of stereoregular polymerization has been proposed in which stereochemistry is controlled by coordination of the metal atom of the alkoxide chain end and the carbonyl oxygen of monomer. The resultant complex (**9**) can then follow the pathway of hydride transfer leading to the well-known Meerwein–Ponndorf–Verley reduction (11.8), or it can polymerize through alkoxide addition (11.9). At low temperatures polymerization is favored, but chain transfer can occur by the hydride transfer mechanism (see reaction 11.6).

$$
\begin{array}{ccc}
\underset{\substack{\\ \textbf{9}}}{\overset{\displaystyle \;\; }{\begin{array}{c} O-C\!\!\stackrel{R}{\diagup} \\ M \qquad H \\ O=CHR \end{array}}}
& \xrightarrow{\text{hydride transfer}} &
\begin{array}{c} O=C\!\!\stackrel{R}{\diagup} \\ M \qquad H \\ O-CHR \end{array}
\end{array}
\qquad (11.8)
$$

$$
\begin{array}{c} M \qquad O\dot CHR \\ O-CHR \end{array}
\;\xrightarrow{\text{RCHO}}\;
\begin{array}{c} O\dot CHR \\ \;\;\;\;\;R \\ O-C \\ M \qquad H \\ O=CHR \end{array}
\qquad (11.9)
$$

(polymerization)

Stereoregularity is also influenced by the size of the initiator. This has been demonstrated using triethylaluminum and equimolar amounts of different alcohols as the initiating system for polymerizing acetaldehyde.[4] These reagents react according to reaction (11.10). As the size of the alcohol is increased, the degree of isotactic placement increases, varying from as low as 14% with ethyl alcohol to as high as 92% with *t*-butyl alcohol.

$$
Al(C_2H_5)_3 + ROH \rightarrow (C_2H_5)_2AlOR + C_2H_6 \qquad (11.10)
$$

11.2.3 Polymerization of Cyclic Ethers

The history of cyclic ether polymerization dates back to some early experiments by Wurtz on ethylene oxide in the last century.[13] Earlier in this century Staudinger did considerable work on polyether synthesis using both homogeneous and heterogeneous initiator systems.[14,15] Much of the impetus for research in this area in recent years has resulted from the development of polyurethane foams, which are made from hydroxyl-terminated polyethers and diisocyanates.[16] (Polyurethane chemistry is discussed in Chapter 13.)

Trioxane (**1**), used in commercial production of polyoxymethylene under cationic conditions, is one of the most important cyclic ethers.[17] The probable polymerization mechanism, illustrated for the BF_3–H_2O initiating system, involves initial protonation and ring-opening to form a resonance-stabilized cation (11.11), and propagation by successive addition (11.12) and ring-opening (11.13) steps. The counterion (not shown) in these reactions is $HOBF_3^-$. The commercial product (Celanese trade name Celcon) is actually a random copolymer of trioxane and ethylene oxide. The —CH_2CH_2O— units arising from the latter prevent the polymer from unzipping at elevated temperatures.

$$\text{(dioxolane)} \xrightarrow{\text{F}_3\bar{\text{B}}-\overset{+}{\text{O}}\text{H}_2} \text{HO}\overset{+}{\text{(dioxolane)}} \longrightarrow \text{HOCH}_2\text{OCH}_2\overset{+}{\text{OCH}}_2 \qquad (11.11)$$

$$\xrightarrow{\text{Trioxane}} \text{HOCH}_2\text{OCH}_2\text{OCH}_2\overset{+}{\text{O}}\text{(dioxolane)} \qquad (11.12)$$

$$\longrightarrow \text{HOCH}_2\text{OCH}_2\text{OCH}_2\text{OCH}_2\text{OCH}_2\overset{+}{\text{OCH}}_2 \qquad (11.13)$$

Apart from trioxane, the most important cyclic ethers are ethylene oxide[18] (**10**), propylene oxide (**11**), hexafluoropropylene oxide (**12**), 3,3-di(chloromethyl)oxacyclobutane (**13**), and tetrahydrofuran (oxacyclopentane) (**14**).

$$\underset{\textbf{10}}{\text{CH}_2\overset{\text{O}}{-}\text{CH}_2} \qquad \underset{\textbf{11}}{\text{CH}_3\text{CH}\overset{\text{O}}{-}\text{CH}_2} \qquad \underset{\textbf{12}}{\text{CF}_3\text{CF}\overset{\text{O}}{-}\text{CF}_2}$$

$$\underset{\textbf{13}}{\text{O}\underset{\text{CH}_2\text{Cl}}{\overset{\text{CH}_2\text{Cl}}{\diamond}}} \qquad \underset{\textbf{14}}{\text{(oxolane)}}$$

Structures of the corresponding polymers are given in Table 11.1. Although the four- and five-membered rings are occasionally called 1,3-epoxypropane and 1,4-epoxybutane, respectively, the term *epoxide* is more commonly reserved for the three-membered ring, and this is how it is used in this book. The terms *oxirane* and *oxetane* are also used to denote ethylene oxide and oxacyclobutane, respectively.

Depending on molecular weight, poly(ethylene oxide) and poly(propylene oxide) range in properties from liquids to waxy solids. A significant difference between the two polymers is that the former is water soluble and the latter is insoluble. Poly(hexafluoropropylene oxide) has the advantage of retaining flexibility at temperatures as low as $-50°C$ and exhibiting excellent high temperature stability.[19] Polyethers derived from **13** and **14** are semicrystalline with T_ms of 177 and 45°C, respectively.

Because of the strain inherent in the three-membered ring, epoxides are particularly amenable to ring-opening polymerization and can be converted to polyether with cationic, anionic, or coordination initiators. Free radical initiation occurs with difficulty and is complicated by competing hydrogen abstraction reactions. The less reactive larger ring cyclic ethers polymerize well only under cationic conditions. The chemistry of these various processes follows. Stereospecific polymerization of epoxides is covered in the next section.

11.2.3.1 Cationic Polymerization of Epoxides

Various types of protonic and Lewis acid systems initiate the polymerization of epoxides.[20,21] Mechanisms involving the latter type are frequently complex and not well understood; some, in fact, are believed to react with monomer to form metal alkoxides that cause polymerization to occur via an *anionic* coordination mechanism. In the case of proton donors the mechanism is straightforward, but not very high molecular weights are obtainable. The initial step is formation of an oxonium ion (11.14). Propagation (11.15) involves ring-opening attack of monomer to form a new oxonium ion that competes with chain termination by reaction with water (11.16).

$$\underset{\substack{\diagup \ \ \diagdown \\ CH_2-CH_2}}{O} + HA \ \rightleftharpoons \ \underset{\substack{\diagup \ \diagdown \\ CH_2-CH_2}}{\overset{H}{\underset{+}{O}} \, A^-} \tag{11.14}$$

$$\underset{\substack{\diagup \diagdown \\ CH_2-CH_2}}{\overset{H}{\underset{+}{O}} \, A^-} + \underset{\substack{\diagup \diagdown \\ CH_2-CH_2}}{O} \ \rightleftharpoons \ HOCH_2CH_2\overset{+}{O}\underset{\diagdown CH_2}{\overset{\diagup CH_2}{\big|}} A^- \tag{11.15}$$

$$H{\leftarrow}OCH_2CH_2{\rightarrow}_n \overset{+}{O}\underset{\diagdown CH_2}{\overset{\diagup CH_2}{\big|}} A^- + H_2O \ \longrightarrow \ H{\leftarrow}OCH_2CH_2{\rightarrow}_n OCH_2CH_2OH + HA \tag{11.16}$$

Propylene oxide and other substituted epoxides undergo similar reactions except that ring opening is regiospecific at the carbon bearing the substituent group (11.17).

$$\underset{\substack{\diagup \ \ \diagdown \\ CH_2-CHCH_3}}{\overset{H}{\underset{+}{O}} \, A^-} + \underset{\substack{\diagup \ \ \diagdown \\ CH_2-CHCH_3}}{O} \ \longrightarrow \ HOCH_2\underset{\underset{CH_3}{|}}{CH}\overset{+}{O}\underset{\diagdown CH_2}{\overset{\diagup CHCH_3}{\big|}} A^- \tag{11.17}$$

Higher-molecular-weight polymer is obtained using stannic chloride initiator, but the mechanism appears to be quite different. Kinetics studies indicate that each stannic chloride molecule initiates two polymer chains, which has led to a suggestion[4,17] that initiation proceeds according to reaction (11.18). Such a scheme has the disadvantage of requiring considerable charge separation in the growing chains unless chain coiling allows some ion–counterion interaction. More work is necessary to elucidate these mechanisms.

$$SnCl_4 + 4\underset{\substack{\diagup \ \ \diagdown \\ CH_2-CH_2}}{O} \ \longrightarrow$$

$$\underset{\underset{CH_2}{|}}{\overset{CH_2}{|}}\!\!\!\diagup\!\!\overset{+}{O}CH_2CH_2O{-}(SnCl_4)^{2-}{-}OCH_2CH_2\overset{+}{O}\underset{\diagdown CH_2}{\overset{\diagup CH_2}{\big|}} \tag{11.18}$$

11.2.3.2 Cationic Polymerization of Other Cyclic Ethers

Oxacyclobutanes and tetrahydrofuran can be polymerized by a variety of cationic initiating systems, and in each case the initiating species is the corresponding oxonium ion.[21–23] Propagation undoubtedly involves nucleophilic attack of monomer on the α-carbon with simultaneous ring opening (11.19, 11.20), a reaction analogous to the normal S_N2 displacement of oxonium ions.

$$\underset{\square}{\overset{+}{O}}{\diagup}^H + \underset{\square}{\overset{+}{O}} \ \longrightarrow \ \underset{\square}{\overset{+}{O}}\diagdown\diagup\diagdown OH \quad X^- \tag{11.19}$$

$$\text{(11.20)}$$

As expected from considerations of ring strain, these monomers are not as reactive as epoxides, although the initiation rate can be increased by addition of an epoxide "promoter." Substituted oxetanes such as **13** polymerize readily. Substituted tetrahydrofurans, on the other hand, resist polymerization, with the exception of some bicyclic ethers such as 7-oxabicyclo[2.2.1]heptane (**15**). This monomer, which may be considered a disubstituted tetrahydrofuran, owes its reactivity (11.21) to strain associated with the bicyclic ring system.[24]

$$\text{(11.21)}$$

15

11.2.3.3 Anionic Polymerization of Epoxides

Low-molecular-weight polyethers can be prepared with basic catalysts such as sodium or potassium hydroxides or alkoxides. Mechanistic studies indicate that initiation (11.22) involves S_N2 displacement to form alkoxide ion and that propagation occurs by nucleophilic displacement involving the new alkoxide ion (11.23). The low molecular weight is due to the fact that initiation occurs rapidly in competition with propagation. As with other anionic polymerizations, the chain end remains active, and higher molecular weights can be obtained by addition of more monomer.

$$RO^- + CH_2\!-\!CH_2 \longrightarrow ROCH_2CH_2O^- \qquad \text{(11.22)}$$

$$ROCH_2CH_2O^- + CH_2\!-\!CH_2 \longrightarrow ROCH_2CH_2OCH_2CH_2O^- \qquad \text{(11.23)}$$

Kinetics studies suggest that protonation of the epoxide or possibly hydrogen bonding involving the alcohol catalyzes ring opening.

In the case of propylene oxide, nucleophilic substitution occurs preferentially at the less hindered carbon (11.24), and the resultant polymer, having pendant methyl groups, is capa-

$$RO^- + CH_2\!-\!CHCH_3 \longrightarrow ROCH_2CHO^- \longrightarrow RO\!\!\left[\!CH_2CHO\!\right]\!\!\underset{CH_3}{} \qquad \text{(11.24)}$$

ble of exhibiting stereoregularity. Again, only low-molecular-weight polymers are formed, and this, together with the fact that rather broad molecular weight distributions are observed, suggests that chain transfer processes occur easily. Observations of both allyl ether and propenyl ether end groups from infrared spectra of poly(propylene oxide) suggest two types of chain transfer, one (11.25) involving an E2-type elimination from monomer and the other (11.26) an intramolecular proton transfer.

$$RO^- + CH_3CH\overset{O}{\overset{\diagup\diagdown}{-}}CH_2 \longrightarrow ROH + CH_2=CH\overset{O^-}{\overset{|}{C}}H_2 \longrightarrow$$

$$CH_2=CHCH_2O\!\!-\!\!\left[CH_2CHO\right]_n\!\!-\!\!CH_2CHO^-$$
$$\qquad\qquad\qquad\quad \overset{|}{CH_3}\quad\;\; \overset{|}{CH_3}$$

(11.25)

$$CH_2=CHCH_2 \;\; \overset{\diagdown}{O}CHCH_3 \longrightarrow$$
$$\qquad\quad \overset{\diagdown}{O}\!\!\sim\!\!\sim\!\!\sim\!\!\overset{|}{C}H_2$$

$$\left[CH_2=CH\overset{\overset{\displaystyle -}{\cdot\cdot}}{C}H \;\; HO\overset{|}{C}HCH_3 \longleftrightarrow \;:CH_2\!\!-\!\!CH=CH \;\; HO\overset{|}{C}HCH_3\right] \longrightarrow$$
$$\qquad\qquad \overset{\diagdown}{O}\!\!\sim\!\!\sim\!\!\overset{|}{C}H_2 \qquad\qquad\qquad\qquad \overset{\diagdown}{O}\!\!\sim\!\!\sim\!\!\overset{|}{C}H_2$$

(11.26)

$$CH_3\!\!-\!\!CH=CH \;\; \overset{\diagdown}{O}CHCH_3$$
$$\qquad\qquad\quad \overset{\diagdown}{O}\!\!\sim\!\!\overset{|}{C}H_2$$

11.2.3.4 Polymerization of Epoxides with Coordination Catalysts

The highest molecular weights have been achieved with coordination catalysts, and the different catalyst systems that have been investigated are far too numerous to list here. They can, however, be broken down into three general types: (1) alkaline earth (group II) compounds; (2) mixtures of transition metal compounds with compounds of neighboring metals; and (3) metal–porphyrin complexes. Examples of the first type[4,25] include alkaline earth oxides, carbonates, sulfates, alkoxides, amides, carboxylates, and chelates, with the most reactive being amides or amide–alkoxides such as **16** and **17**. Chelates of β-keto esters, β-diketones, o-hydroxyphenyl ketones, and β-keto amides have the advantage of being soluble in monomer. The mechanism of

$$\text{Ca(NH}_2)_2 \qquad\qquad \text{Ca(NH}_2)\text{OC}_2\text{H}_5$$
$$\textbf{16} \qquad\qquad\qquad \textbf{17}$$

polymerization by these initiators is not known, but it may involve propagation by coordination of monomer with a complex alkoxide ion-pair chain end followed by anionic attack of the alkoxide ion on the ring:

$$\sim\!\!\sim\!\!O^-M^{++} + CH_2\overset{O}{\overset{\diagup\diagdown}{-}}CH_2 \longrightarrow \sim\!\!\sim\!\!\bar{O} \quad M^{++} \longrightarrow \sim\!\!\sim\!\!OCH_2CH_2O^-M^{++}$$

(11.27)

Examples of the second type[4,25] include typical Ziegler–Natta systems: aluminum, magnesium, zinc, and ferric alkoxides; aluminum and zinc alkyls; and ferric chloride–propylene oxide complex. Polymerization with metal alkyls presumably proceeds by a mechanism similar to that in reaction (11.27). The ferric chloride–propylene oxide complex, an active catalyst that gives high-molecular-weight crystalline polymer with propylene oxide, is believed to have the structure **18**, where $m + n = 4$ or 5.

$$ClFe \begin{cases} [OCH_2CH(CH_3)]_m Cl \\ [OCH_2CH(CH_3)]_n Cl \end{cases}$$

18

The metal alkoxides and metal alkyls are generally much more reactive in the presence of water or alcohols. The function of these coreactants is not simply that of proton donor as it is in normal cationic polymerization. Instead they react to modify the organometallic compound. For the diethylzinc–water and trimethylaluminum–water systems, the most effective ratios of catalyst to water are 1:1 and 1:1.5, respectively. The active species, which are believed to be **19** and **20**, yield very-high-molecular-weight polymer. A proposed pathway for propylene oxide propagation[25] with **19** involves intramolecular rearrangement (11.28) of a catalyst–monomer complex.

$$C_2H_5ZnOZnC_2H_5 \qquad (CH_3)_2AlOAl(CH_3)_2$$

19 **20**

$$C_2H_5ZnO\overset{\delta-}{Zn}C_2H_5 \longrightarrow C_2H_5ZnOCH_2\underset{CH_3}{CHOZnC_2H_5} \tag{11.28}$$

$$\underset{CH_2-CHCH_3}{\overset{O^{\delta+}}{\diagup\diagdown}}$$

Metalloporphyrin initiators, such as tetraphenylporphinatoaluminum chloride (**21**), were developed more recently.[26–30] They polymerize epoxides to very-low-polydispersity ($\overline{M}_w/\overline{M}_n = 1.1$)

21

living polymer, with the propagating chain being the complex aluminum alkoxide, as shown for propylene oxide in reaction (11.29). The metalloporphyrin end group is not fixed, but un-

$$\text{Al}-Cl + CH_2-\overset{O}{\diagdown}CHCH_3 \longrightarrow \text{Al}-O\underset{CH_3}{CHCH_2}\left[O\underset{CH_3}{CHCH_2}\right]Cl \tag{11.29}$$

dergoes rapid exchange with other polymer molecules. Particularly interesting is the fact that even in the presence of chain transfer agents such as alcohols, the low polydispersity is retained even as the molecular weight is reduced. This results from the reversible exchange of alkoxide (11.30), which is much faster than propagation. Thus the forward reaction (transfer)

$$\left(Al-O\text{wwww} + ROH \rightleftharpoons \left(Al-OR + HO\text{wwww}\right.\right. \tag{11.30}$$

kills the chain, but the reverse reaction revives it, a phenomenon that led to the resultant polymer being characterized as *immortal!* The exchange reaction works equally well with such alcohols as ethylene glycol or hydroxyethyl methacrylate to yield polymers having well-defined molecular weights containing reactive end groups.

11.2.4 Stereochemistry of Epoxide Polymerization

Simple cationic or anionic polymerization of propylene oxide does not lead to stereoregular polyether, although a crystalline, optically active polymer has been obtained from levorotatory propylene oxide using potassium hydroxide as initiator.[3,4] In this case the stereoregularity arises because the S_N2 displacement at the less hindered carbon (11.31) does not disturb the chiral center (designated with an asterisk). Potassium hydroxide itself is not capable of inducing stereoregularity, as is shown by the fact that *d,l*-propylene oxide forms amorphous, inactive polymer under the same conditions. Infrared studies of polyethers prepared from *cis*- and *trans*-1,2-dideuteroethylene oxide have shown quite conclusively that inversion always occurs at the carbon undergoing nucleophilic attack.

$$HO^- + \overset{O}{CH_2-\underset{*}{C}HCH_3} \longrightarrow HOCH_2\overset{CH_3}{\underset{*}{C}HO^-} \tag{11.31}$$

Only with coordination catalysts have stereoregular high-molecular-weight polymers been obtained,[25,31] and the mechanism of their formation has been the subject of much speculation. Presumably, the stereoregular placement is influenced by the nature of the active sites on the catalyst surface, as it is with Ziegler–Natta polymerization of vinyl monomers (Chapter 8).

Another epoxide that has been considered for market development is 1,4-dichloro-2,3-epoxybutane (**22**). Both *cis* and *trans* isomers of **22** undergo stereospecific cationic polymerization to yield semicrystalline polymer, the former *racemic*-diisotactic (all R or all S), and the latter *meso*-diisotactic (RS–RS sequences). It has been suggested that coordination between the propagating oxonium ion and the chlorine of the β-chloromethyl group of the last chain unit plays a role in steric control.[32]

$$ClCH_2\overset{}{C}H-\overset{}{C}HCH_2Cl$$
$$\diagdown O \diagup$$

22

11.3   Preparation of Polyethers by Step-Reaction Polymerization

11.3.1   Synthesis of Polyethers from Glycols and Bisphenols

In the preceding chapter we learned that self-condensation reactions of glycols to form polyethers such as that from 1,3-propanediol (11.32) are complicated by the tendency to form cyclic ethers. Dehydration to form alkene is another limiting factor.

$$n\text{HOCH}_2\text{CH}_2\text{CH}_2\text{OH} \xrightarrow{(-\text{H}_2\text{O})} \text{H}\text{—}[\text{OCH}_2\text{CH}_2\text{CH}_2]_n\text{—OH} \qquad (11.32)$$

Condensation of benzylic glycols such as **23** occurs more readily because the reaction (11.33) involves resonance-stabilized carbocation intermediates. Even so, the polymer (**24**) obtained is of relatively low molecular weight, having a $\overline{\text{DP}}$ of about 50.

$$\qquad (11.33)$$

| 23 | 24 |

The best way to prepare polyethers from dihydroxy compounds is by nucleophilic displacement reactions. An example is the reaction of a bisphenol with dichloromethane under basic conditions, which leads to high-molecular-weight aromatic polyformal[33,34] (11.34). The function of the base is to convert the bisphenol to the nucleophilic alkoxide ($^-\text{O–Ar–O}^-$).

$$\text{HO—Ar—OH} + \text{CH}_2\text{Cl}_2 \xrightarrow{\text{base}} \text{—}[\text{OArOCH}_2]\text{—} \qquad (11.34)$$

From the commercial standpoint, however, the greatest success has been realized with nucleophilic aromatic substitution (11.35), in which the leaving group (X) is activated by strongly electron-attracting groups in Ar'. The resultant polymers thus combine ether with other backbone functionalities.

$$\text{HO—Ar—OH} + \text{X—Ar—X} \xrightarrow{\text{Base}} \text{—}[\text{O—Ar—O—Ar'}]\text{—} + 2\text{ X}^- \qquad (11.35)$$

Three engineering plastics are manufactured this way: *polysulfone*[35] (**25**), *polyetherimide* (PEI),[36,37] (**26**), and *polyetheretherketone* (PEEK)[38,39] (**27**). (A related polymer, polyetherketone, is featured in exercise 8 at the end of the chapter.) PEEK and other ether–ketone polymers are notable in that they generate little smoke upon combustion. The leaving groups in the synthesis of the three polymers are Cl, NO_2, and F, respectively. Trade names are Udel (Union Carbide) for polysulfone, Ultem (General Electric) for PEI, and Victrex (ICI) for PEEK. Polymers **25** and **26** are amorphous with T_gs of about 220 and

215°C, respectively, whereas **27** is semicrystalline with a T_g of about 144°C and a T_m of about 334°C.

25

26

27

Phase-transfer methods have also been applied to polyether synthesis.[40] An example is the reaction (11.36) of 4,4'-dihydroxy-α-methylstilbene (**28**) with α,ω-dibromoalkanes catalyzed by tetrabutylammonium hydrogen sulfate (TBAH) to yield thermotropic liquid crystalline polymer (**29**), in which the methylene groups serve as flexible spacers between the mesogenic stilbene groups.[41]

28

(11.36)

29

11.3.2 Polyacetals and Polyketals

Apart from the chain-reaction polymerization of aldehydes and dichloromethane displacement reactions described earlier, there are three general methods for synthesizing polymers containing acetal or ketal groups: (1) reaction of diols with carbonyl compounds, an extension of the classical acetal (or ketal) synthesis, (2) acetal (or ketal) exchange, and (3) addition of diols to dialkenes.

Formation of acetals and ketals by the acid-catalyzed reaction of excess alcohol with aldehydes and ketones, respectively, is commonly employed in organic chemistry to protect the carbonyl function. By reacting equimolar amounts of aldehyde and diol and removing byproduct water by azeotropic distillation, one forms polyacetals (11.37). A serious limitation is the tendency for cyclic acetals also to be formed; nevertheless, high-melting polyformals (**30** and **31**) have been prepared by heating the corresponding cyclic diols with paraformaldehyde in the presence of acid catalysts.[40]

$$RCHO + HO-R'-OH \xrightarrow{H^+} \left[\begin{array}{c} CH-O-R'-O \\ | \\ R \end{array} \right] + H_2O \qquad (11.37)$$

$$\left[O-\underset{CH_3CH_3}{\overset{CH_3CH_3}{C}}-OCH_2 \right] \qquad \left[O-\langle \rangle-OCH_2 \right]$$

30   **31**

The tendency to form cyclic acetals can be used to advantage by reacting tetrols such as pentaerythritol (**32**) with dialdehydes to form cyclic polyacetals (**33**) (11.38). The polymer prepared from glutaraldehyde (R = —CH$_2$CH$_2$CH$_2$—), for example, has been investigated for fiber and wire coating applications. Particularly interesting is the reaction of **32** with 1,4-cyclohexanedione (**34**) to form *spiro* polyketal (**35**) (11.39).[43] Spiro polymers (i.e., polymers having recurring ring units joined through a common carbon atom) exhibit the same type of backbone rigidity and thermal stability as ladder polymers.

$$\begin{array}{c} HOCH_2 \\ \diagdown \\ C \\ \diagup \\ HOCH_2 \end{array} \begin{array}{c} CH_2OH \\ \diagup \\ \\ \diagdown \\ CH_2OH \end{array} + OCH-R-CHO \xrightarrow{H^+}$$

32

$$\qquad (11.38)$$

$$\left[CH \underset{OCH_2}{\overset{OCH_2}{<}} C \underset{CH_2O}{\overset{CH_2O}{>}} CH-R \right] + 2H_2O$$

33

$$32 + O=\langle \rangle=O \xrightarrow{H^+} \left[\text{spiro polyketal structure} \right] + 2H_2O \qquad (11.39)$$

34   **35**

Acid-catalyzed acetal or ketal exchange reactions give rise to the same types of polyacetals and polyketals shown above, except that alcohol is the byproduct. Polymer **33**, for example, may be prepared by reacting tetrol **32** with diacetal (R'O)$_2$CH—R—CH(OR')$_2$ (prepared from the dialdehyde) in the presence of acid. Four equivalents of byproduct R'OH would be formed per repeating unit.

Linear polyformals can also be prepared by an entirely different route, the acid-catalyzed stepwise addition polymerization of glycols to divinyl ethers[3] (11.40). Polyacetals of this type have sparked interest as bioerodable polymers for the controlled release of drugs.[44]

$$CH_2=CHOROCH=CH_2 + HO—R'—OH \xrightarrow{H^+} \begin{bmatrix} CHOROCHOR'O \\ | \quad\quad | \\ CH_3 \quad CH_3 \end{bmatrix} \quad (11.40)$$

11.3.3 Poly(phenylene oxide)s

Aromatic polyethers have received considerable attention in recent years because of their thermal stability. The most successful method of preparation involves oxidative coupling of 2,6-disubstituted phenols.[35,45] If the substituents are methyl groups (reaction (11.41), the

$$(11.41)$$

36

polymerization can be done at room temperature by bubbling oxygen through a solution of the phenol in the presence of a cooper(I) salt–amine complex. Pyridine is commonly used as amine and solvent, and copper(I) chloride as the copper salt. Somewhat higher temperatures are required for halogen-substituted phenols. The polymer (36) obtained from 2,6-dimethylphenol, known commercially as poly(phenylene oxide) (PPO), has good thermal and hydrolytic stability. It may be recalled from Chapter 3 that 36 is used primarily as an engineering plastic (Noryl) in a homogeneous blend with polystyrene. Copolymers with 2,3,6-trimethylphenol, which exhibit a higher T_g and greater strength than the homopolymer, are also used commercially as polyblends. If the substituent alkyl groups are sufficiently bulky, carbon–carbon rather than carbon–oxygen coupling occurs; 2,6-di-t-butylphenol, for example, gives the diphenoquinone (37) rather than polymer.

37

The mechanism of polymerization has been investigated in some detail. It is believed that the active catalyst is a basic copper(II) salt (resulting from oxidation of copper(I) chloride) complexed with two amine groups. Initiation might involve reaction between catalyst and phenol according to reaction (11.42). Electron transfer to copper would generate phe-

$$—\overset{|}{\underset{|}{Cu}}—OH + ArOH \longrightarrow —\overset{|}{\underset{|}{Cu}}—OAr + H_2O \quad (11.42)$$

noxy radicals that might be expected to lead to chain-growth polymerization by successive radical additions at the *para* position; however, the reaction has the characteristics of a step-growth rather than a chain-growth process, as evidenced by the fact that high molecular weight is achieved only at high degrees of polymerization. It is now believed that radical coupling to give quinol ethers occurs (11.43) followed by dissociation to give the original aryloxy radicals or two new radicals (11.44), which, in turn, could couple. Chain buildup thus occurs in a stepwise fashion, and a broad molecular weight distribution results from hydrogen transfer reactions.

$$\text{(11.43)}$$

$$\text{(11.44)}$$

Block copolymers have been prepared from 2,6-diphenylphenol and 2,6-dimethylphenol by oxidative coupling.[46] The diphenyl derivative is polymerized first; then the dimethyl derivative is added to the reaction mixture. If the order is reversed, or if a mixture is used, a random copolymer results.

Reactions (11.45) through (11.48) illustrate additional methods of synthesizing poly-(phenylene oxide)s,[35] none of which has been developed commercially.

$$\text{(11.45)}$$

$$\text{(11.46)}$$

$$\text{(11.47)}$$

$$\text{(11.48)}$$

11.3.4 Epoxy Resins

Epoxy resins are a special type of polyether. They are synthesized first as oligomers with epoxide end groups that are capable of undergoing further polymerization to form structurally complex network polymers. The fact that a number of books have been devoted to epoxy resins[47–52] attests to their importance. They are widely used as adhesives and structural polymers, especially in composites.

The oligomers are prepared by the step-reaction polymerization of a diphenol, most commonly bisphenol A, and an epoxide-containing compound in the presence of base. Epichlorohydrin (38) is the most widely used comonomer. In its reactions with phenols, 38 reacts in the manner of a diepoxide. The reaction sequence involves the attack by phenolate anion at the less hindered carbon of the epoxide ring, followed by intramolecular displacement of chloride to regenerate an epoxide (39) (11.49). Further reaction (11.50) of 39 with diphenolate results in a product (40) having a pendant hydroxy group.

$$^-O-Ar-O^- + \overset{O}{\overset{\diagup\diagdown}{CH_2-CH}}CH_2Cl \longrightarrow \ ^-O-Ar-OCH_2\overset{O^-}{\underset{|}{CH}}-CH_2Cl \qquad (11.49)$$

$$\mathbf{38}$$

$$\longrightarrow \ ^-O-Ar-OCH_2\overset{O}{\overset{\diagup\diagdown}{CH}}-CH_2 + Cl^-$$

$$\mathbf{39}$$

$$\mathbf{39} + \ ^-O-Ar-O^- \longrightarrow \ ^-O-Ar-OCH_2\overset{O^-}{\underset{|}{CH}}CH_2O-Ar-O^- \qquad (11.50)$$

$$\xrightarrow{H_2O} \ ^-O-Ar-OCH_2\overset{OH}{\underset{|}{CH}}CH_2O-Ar-O^-$$

$$\mathbf{40}$$

Reaction with excess 38 leads eventually to an epoxide-terminated oligomer (41). The end groups are commonly referred to as *glycidyl ether* groups. The molecular weight of the oligomer is varied by suitable adjustment of the epichlorohydrin excess, and may vary from a viscous liquid to a solid ($n = 2$–25) depending on the degree of polymerization. Optimum properties depend on the application.

$$\overset{O}{\overset{\diagup\diagdown}{CH_2-CH}}CH_2 \!\!\left[O-Ar-OCH_2\overset{OH}{\underset{|}{CH}}CH_2 \right]_{\!n} \!\!O-Ar-O\overset{O}{\overset{\diagup\diagdown}{CH}}-CH_2$$

$$\mathbf{41}$$

Besides bisphenol A, diphenols such as hydroquinone and resorcinol (1,4- and 1,3-dihydroxybenzene, respectively) have also been used in epoxy formulations, as have certain phenol–formaldehyde condensation products, including novolacs (discussed in Chapter 14).

While the glycidyl ethers represent the most common type of epoxy prepolymer, other epoxide structures are used commercially, among them multifunctional glycidyl amines such as **42** and **43**, cycloaliphatic compounds such as vinylcylohexene dioxide (**44**) and dicyclopentadiene dioxide (**45**), epoxidized diene polymers, and epoxidized oils.

$$CH_2\text{—}CHCH_2O\text{—}\langle \bigcirc \rangle\text{—}N(CH_2CH\text{—}CH_2)_2$$

42

$$(CH_2\text{—}CHCH_2)_2N\text{—}\langle \bigcirc \rangle\text{—}CH_2\text{—}\langle \bigcirc \rangle\text{—}N(CH_2CH\text{—}CH_2)_2$$

43

44   **45**

Further polymerization of prepolymer to form crosslinked polymer (curing) can be accomplished several ways, although amines and carboxylic acid anhydrides are most commonly used. Amines react by nucleophilic addition to the epoxide ring (11.51). Diamines and polyamines are also used as curing agents and, in some applications, the epoxy resin is blended with other polymers such as melamine– or urea–formaldehyde resins, which cause crosslinking by essentially the same reaction to give network copolymer. Reaction with amines is exothermic and is known to be catalyzed by proton donors.

$$RNH_2 + CH_2\text{—}CH\text{\tiny wwww} \longrightarrow RNHCH_2\overset{OH}{\underset{|}{CH}}\text{\tiny wwww} \longrightarrow RNCH_2\overset{OH}{\underset{|}{CH}}\text{\tiny wwww etc.} \quad (11.51)$$

$$\underset{\underset{OH}{\overset{|}{CH_2CH}\text{\tiny wwww}}}{}$$

Partially cured epoxy resins that are stable in storage but cure further on heating can be prepared[53] by adding hydrazides of structure **46**. Resins of this type are called *B-stage* resins (A-stage refers to uncured resins, B-stage to partially cured, and C-stage to fully cured.)

$$RNHCH_2CH_2\overset{O}{\overset{\|}{C}}NHNH_2$$

46

Anhydrides—for example, phthalic anhydride (**47**)—react with pendant hydroxyl groups to give ester acids (11.52). The acid group then reacts with epoxide groups (11.53) or other hydroxyls (11.54) to form more ester groups. Ether groups also form by the acid-catalyzed reaction between hydroxyl and epoxide groups. Obviously crosslinking of epoxy resins is a very complex process, and it is impossible to give a definitive structure for the final product.

(11.52)

47

(11.53)

(11.54)

Curing with anhydrides is usually catalyzed by amine. Some of the more common anhydrides in commercial use include phthalic (**47**), maleic (**48**), and chlorendic (**49**); the liquid anhydrides, dodecenylsuccinic anhydride (**50**) and nadic methyl anhydride* (**51**); and pyromellitic dianhydride (**52**). Liquid Diels–Alder adducts derived from maleic anhydride and terpenes have also been reported.[54] It appears that Diels–Alder adducts improve the high-temperature stability of epoxy resins because the retrograde Diels–Alder reaction generates alkenes in the polymeric network, which crosslink further and cause more char formation.[55] The advantages of a dianhydride such as **52** are that the cure rate is higher and the final product is more densely crosslinked. Whichever curing agent is used, there is a preponderance of free hydroxy groups in the cured polymer, which gives rise to the superior adhesive properties of epoxies.

| **48** | **49** | **50** | **51** | **52** |

*Nadic methyl anhydride is the common name for methylbicyclo[2.2.1]hept-5-ene-2,3-dicarboxylic anhydride, the Diels–Alder adduct of the mixed methyl-substituted cyclopentadienes with maleic anhydride.

Because of the reactive end groups, epoxy resins are easily grafted onto other polymers. Thermosetting acrylic polymers have been prepared, for example, by reaction of an epoxy resin with an acrylic to form a graft copolymer (11.55) capable of being crosslinked. For certain coatings applications, epoxies are also reacted in analogous fashion with drying oil fatty acids to take advantage of the free radical crosslinking propensity of the fatty acid. (Drying oil polymerization is discussed in the next chapter.)

$$\text{mCH}_2\underset{\underset{\text{CO}_2\text{H}}{|}}{\overset{\overset{\text{CH}_3}{|}}{\text{C}}}\text{m} \quad + \quad \text{CH}_2\text{—CH}\text{m} \quad \longrightarrow \quad \text{mCH}_2\underset{\underset{\underset{\parallel}{\text{COCH}_2\text{CH}}\text{m}}{\overset{\text{O} \quad \text{OH}}{}}}{\overset{\overset{\text{CH}_3}{|}}{\text{C}}}\text{m} \tag{11.55}$$

11.4 Polysulfides, Poly(alkylene polysulfide)s, and Polysulfones

11.4.1 Polysulfides

Polysulfides, or poly(alkylene sulfide)s as they are sometimes called, are the sulfur analogs of polyethers. The polysulfide that has achieved the greatest commercial success is poly(*p*-phenylene sulfide) (PPS) (**53**).[56] An engineering plastic, PPS is manufactured by Phillips Petroleum (trade name Ryton) by the reaction of *p*-dichlorobenzene with sodium sulfide (11.56) in a polar solvent at about 250°C. The polymer is highly crystalline with a T_m of about 290°C, and it can be made electrically conducting by the addition of dopant (*cf.* Chapter 4). When first made, PPS has a relatively low molecular weight, but heating in air causes chain extension and some crosslinking. The heat-cured polymer is used as an engineering plastic for applications requiring good chemical and heat resistance, while the low-molecular-weight PPS is used as a coating.

$$\text{Cl—}\underset{}{\underset{}{\bigcirc}}\text{—Cl} + \text{Na}_2\text{S} \longrightarrow \left[\underset{}{\bigcirc}\text{—S}\right] + 2\text{ NaCl} \tag{11.56}$$

53

Several other routes to PPS have been reported. Two having possible commercial utility are the oxidative coupling of diphenyl disulfide (or thiophenol) with a vanadium catalyst (11.57, acac = acetylacetonate)[57,58] and the melt condensation of *p*-iodobenzene with sulfur (11.58).[59,60] The latter, which forms product with some disulfide units in the backbone, can be processed to a prepolymer stage, then further polymerization can be effected in the solid state.

$$\underset{}{\bigcirc}\text{-SS-}\underset{}{\bigcirc} \quad \xrightarrow[\text{O}_2]{\text{VO(acac)}_2} \quad \left[\underset{}{\bigcirc}\text{-S}\right] \tag{11.57}$$

$$\text{I-}\underset{}{\bigcirc}\text{-I} \quad \xrightarrow[\text{(Melt)}]{\text{S}} \quad \left[\underset{}{\bigcirc}\text{-S}\right]\left[\underset{}{\bigcirc}\text{-SS}\right] \tag{11.58}$$

A variety of methods have been developed for synthesizing aliphatic polysulfides, many analogous to those used for preparing polyethers. The following are among the more interesting:

1. Polymerization of thiocarbonyl compounds.[61] Thiocarbonyl fluoride (**54**), for example, polymerizes at $-78°C$ with free radical or anionic initiators (11.59) to yield a tough high-molecular-weight elastomeric polymer.

$$F_2C{=}S \rightarrow {+}CF_2{+}$$ (11.59)
$$\underset{\text{54}}{}$$

2. Ring-opening polymerization of episulfides.[62] Like epoxides, the somewhat more reactive episulfides, such as ethylene sulfide (**55**), polymerize under the influence of cationic, anionic, and coordination initiators:

$$\overset{\displaystyle S}{\underset{\text{55}}{CH_2{-}CH_2}} \xrightarrow{\text{Initiator}} {+}CH_2CH_2S{+}$$ (11.60)

3. The base-catalyzed reaction of dihalides with dithiols.[63] This reaction (11.61) is analogous to the Williamson ether synthesis. Thiols being more acidic than alcohols, the base readily converts the dithiol to the dithiolate salt ($^-S{-}R{-}S^-$).

$$HS{-}R{-}SH + Br{-}R'{-}Br \xrightarrow{\text{Base}} {+}S{-}R{-}S{-}R'{+}$$ (11.61)

4. Addition of dithiols to dienes.[64] Addition occurs under both free radical (11.62) and ionic (11.63) conditions, the former by anti-Markovnikov and the latter by Markovnikov addition. Reaction (11.62) is unusual in that it combines a free radical chain reaction

$$\xrightarrow{\text{Peroxide}} {+}S{-}R{-}SCH_2CH_2{-}R'{-}CH_2CH_2{+}$$ (11.62)

$$\begin{array}{c} HS{-}R{-}SH \\ + \\ CH_2{=}CH{-}R'{-}CH{=}CH_2 \end{array} \Bigg|_{H^+ \text{ or } OH^-}$$

$$\xrightarrow{H^+ \text{ or } OH^-} {+}S{-}R{-}\underset{\underset{CH_3}{|}}{S}CH{-}R'{-}\underset{\underset{CH_3}{|}}{CH}{+}$$ (11.63)

with step-reaction polymerization (see review exercise 18). Molecular weights of 20,000–60,000 are obtained. Uncatalyzed Michael addition occurs if the carbon–carbon double bonds are conjugated with carbonyl groups (11.64).[65–67]

$$HS{-}R{-}SH + \underset{O}{\overset{O}{N{-}R'{-}N}} \longrightarrow {+}S{-}R{-}S{-}\Bigg[\underset{O}{\overset{O}{N{-}R'{-}N}} \Bigg]$$ (11.64)

5. Formation of polythioacetals and polythioketals.[64] The reaction of dithiols with aldehydes (11.65) and the reaction of cyclic diketones with tetrathiols (11.66) lead to

polythioacetals and spiro polythioketals, respectively. As was the case with poly-acetals, acetal exchange reactions may also be used to make the thio polymers.

$$R-CHO + HS-R'-SH \longrightarrow \left[\begin{array}{c} CH-S-R'-S \\ | \\ R \end{array}\right] + H_2O \qquad (11.65)$$

$$\qquad (11.66)$$

11.4.2 Poly(alkylene polysulfide)s

Related to the polysulfides are the poly(alkylene polysulfide)s prepared from a dihalide and a sodium polysulfide.[68] The most common dihalides are 1,2-dichloroethane and bis(2-chloroethyl)formal (**56**).

Typically the sodium polysulfide is prepared from sodium hydroxide and sulfur at elevated temperatures. The polysulfide is then reacted with dihalide in water to give a dispersion of polymer (11.67). In these polymers, x usually averages between 2 and 4.

$$ClCH_2CH_2OCH_2OCH_2CH_2Cl + Na_2S_x \rightarrow$$
$$\underset{\textbf{56}}{\qquad\qquad} +CH_2CH_2OCH_2OCH_2CH_2S_x + 2NaCl \quad (11.67)$$

Primary halides give the best yield of polymer, which is to be expected because the mechanism involves nucleophilic displacement of halide by polysulfide anion. Secondary and tertiary halides, particularly the latter, tend to undergo elimination rather than substitution.

Poly(alkylene polysulfide)s (sometimes referred to as *Thiokol rubbers*) are useful elastomers. Their properties can be modified by varying the number of carbon atoms in the repeating unit or the number of sulfur atoms. Increasing either will increase the elastomeric quality of the polymer. Crosslinkable polymers can be prepared by adding small amounts of polyhalides such as trichloropropane.

The most important method of manufacturing curable elastomers involves first forming the crosslinked polymer using a polyhalide monomer, then degrading the product to a thiol-terminated fusible polymer. This is accomplished by reduction with sodium hydrosulfide in the presence of sodium sulfite (11.68). The average molecular weight of the degraded polymer depends on the amount of sodium hydrosulfide used.

$$+RSSR + NaSH + Na_2SO_3 \rightarrow +RSH + NaSR + Na_2S_2O_3 \qquad (11.68)$$

Numerous oxidizing agents are used to cure the thiol-terminated polymer, including lead or zinc peroxides, cumene hydroperoxide, or even air in thin-film applications. Nonoxidative curing may be accomplished with zinc oxide or other suitable oxides that form stable thiol salts, or with aldehydes such as furfural that cure by thioacetal formation. The polymers can also be cured by blending with other types of polymers, including polyesters, epoxy resins, and phenolic resins.[69]

Poly(alkylene polysulfide)s are among the oldest synthetic polymers, having been in commercial use since 1930. They lack the strength of most other synthetic elastomers and tend to have unpleasant odors because of the presence of low-molecular-weight thiols and disulfides. Nevertheless, they are exceptionally resistant to oils and solvents and to weathering, and hence are widely used in gaskets, gasoline hoses, and sealants. Large volumes are now being used as binders for solid rocket propellants.

11.4.3 Polysulfones

Polysulfones are derivatives of polysulfides and can be prepared from them by oxidation—for example, with hydrogen peroxide (11.69). Introduction of the sulfone group normally results in a much higher polymer melt temperature relative to the polysulfide.

$$\{RS\} \xrightarrow{\text{H}_2\text{O}_2} \left\{ \begin{array}{c} \text{O} \\ \| \\ \text{RS} \\ \| \\ \text{O} \end{array} \right\} \qquad (11.69)$$

In recent years attention has centered on aromatic polysulfones as thermally stable engineering plastics.[70–73] One method of preparing them by a nucleophilic displacement reaction was described earlier (structure **25**). Another commercial route to polysulfones involves a Friedel–Crafts-type reaction using sulfonyl chlorides (11.70). The product in this case (**57**) is referred to in the plastics industry as a *polyethersulfone*, whereas that described earlier (**25**) is known simply as *polysulfone* (regardless of the fact that ether formation is the focal point of its synthesis). If diphenyl ether is replaced with biphenyl in reaction (11.70), the product (**58**) is called a *polyarylsulfone*. Polysulfones **57** and **58** are manufactured by ICI (trade name Victrex) and 3M (Astrel), respectively.

(11.70)

57

58

An additional route to polysulfones is the alternating copolymerization of sulfur dioxide[74] with vinyl monomers (11.71). The resultant polymers exhibit relatively low ceiling temperatures (~100°C when R = H, ~0°C when R = CH$_3$). Depolymerization occurs on heating or on exposure to ionizing radiation. As a consequence, such polysulfones are useful in electron beam resist applications.

$$\text{CH}_2{=}\text{CR}_2 + \text{SO}_2 \xrightarrow{\text{free radical initiator}} \{\text{CH}_2\text{CR}_2\text{SO}_2\} \qquad (11.71)$$

References

1. A. A. Miller and F. R. Mayo, *J. Am. Chem. Soc.*, **78**, 1017 (1956).
2. K. Kishore and T. Mukundan, *Nature*, **324**, 130 (1986).
3. N. G. Gaylord (Ed.), *Polyethers*, Part 1, Wiley-Interscience, New York, 1963.
4. J. Furukawa and T. Saegusa, *Polymerization of Aldehydes and Oxides*, Wiley-Interscience, New York, 1963.
5. R. B. Akin, *Acetal Resins*, Reinhold, New York, 1962.
6. S. J. Barker and M. B. Bruce, *Polyacetals*, Elsevier, New York, 1970.
7. K. J. Persak and R. A. Fleming, in *High Performance Polymers: Their Origin and Development* (R. B. Seymour and G. S. Kirshenbaum, Eds.), Elsevier, New York, 1986, pp. 105–14.
8. O. Vogl, H. C. Miller, and W. H. Sharkey, *Macromolecules*, **5**, 658 (1972).
9. K. Neeld and O. Vogl, *Macromol. Rev.*, **16**, 1 (1981).
10. M. Chang, A. Rembaum, and C. J. McDonald, *J. Polym. Chem., Polym. Lett. Ed.*, **22**, 279 (1984).
11. E. Brown, A. Racois, M. Grimaud, J. C. Lecog, R. Tixien, and M. Corgier, *Makromol Chem.*, **179**, 2887 (1978).
12. G. F. Pregalia and M. Binaghi, in *The Stereochemistry of Macromolecules*, Vol. 2 (A. D. Ketley, Ed.), Marcel Dekker, New York, 1967, Chap. 3.
13. A. Wurtz, *Ann. Chim. Phys.*, **69**, 330, 334 (1863); *Chem. Ber.*, **10**, 90 (1879).
14. H. Staudinger and O. Schweitzer, *Chem. Ber.*, **62**, 2395 (1929).
15. H. Staudinger and H. Lehmann, *Liebigs Ann. Chem.*, **505**, 41 (1933).
16. F. E. Bailey, Jr., and J. V. Koliske, *Alkylene Oxides and Their Polymers*, Marcel Dekker, New York, 1990.
17. Y. Ishii and S. Sakai, in *Ring-Opening Polymerization* (K. C. Frisch and S. L. Reegen, Eds.), Marcel Dekker, New York, 1969, Chap. 1.
18. F. E. Bailey and J. V. Koleske, *Poly(ethylene oxide)*, Academic Press, New York, 1976.
19. J. T. Hill, *J. Macromol. Sci.-Chem.*, **8**, 499 (1974).
20. P. Dreyfuss and M. P. Dreyfuss, in *Ring-Opening Polymerization* (K. C. Frisch and S. L. Reegen, Eds.), Marcel Dekker, New York, 1969, Chap. 2.
21. S. Penczek, P. Kubisa, and K. Matyjaszewski, *Adv. Polym. Sci.*, **37**, 1 (1980).
22. P. Dreyfuss, *Poly(tetrahydrofuran)*, Gordon and Breach, New York, 1982.
23. P. Dreyfuss and M. P. Dreyfuss, *Adv. Polym. Sci.*, **4**, 528 (1967).
24. Y. Yokoyama and H. K. Hall, Jr., *Adv. Polym. Sci.*, **42**, 107 (1982).
25. T. Tsuruta, in *The Stereochemistry of Macromolecules*, Vol. 2 (A. D. Ketley, Ed.), Marcel Dekker, New York, 1967, Chap. 4.
26. T. Aida, Y. Maekawa, S. Asano, and S. Inoue, *Macromolecules*, **21**, 1195 (1988).
27. S. Inoue, *J. Macromol. Sci-Chem.*, **A25**, 571 (1988).
28. S. Inoue and T. Aida, in *New Methods for Polymer Synthesis* (W. J. Mijs, Ed.), Plenum, New York, 1992.
29. S. Inoue and T. Aida, *Chemtech*, **24**(5), 28 (1994).
30. T. Aida and S. Inoue, *Acc. Chem. Res.*, **29**, 39 (1996).
31. T. Tsuruta and Y. Kawakami, in *Comprehensive Polymer Science*, Vol. 3 (G. C. Eastman, A. Ledwith, S. Russo, and P. Sigwalt, Eds.), Pergamon, Oxford, 1989, Chap. 33.
32. E. J. Vandenberg, *J. Polym. Sci., Polym. Chem. Ed.*, **13**, 2221 (1975).
33. A. S. Hay et al. *J. Polym. Sci., Polym. Lett. Ed.*, **21**, 449 (1983).
34. T. J. Shea and G. M. Loucks, *Polymer Preprints*, **28**(2), 354 (1987).
35. A. S. Hay, *Adv. Polym. Sci.*, **4**, 496 (1967).
36. R. O. Johnson and H. S. Burlhis, *J. Polym. Sci., Polym. Symp.*, **70**, 129 (1983).
37. D. M. White et al., *J. Polym. Sci., Polym. Chem. Ed.*, **19**, 1635 (1981).
38. D. J. Blundell and B. N. Osborn, *Polymer*, **24**, 953 (1983).
39. T. E. Attwood, P. C. Dawson, J. L. Freeman, L. R. J. Hoy, J. B. Rose, and P. A. Staniland, *Polymer*, **22**, 1096 (1981).
40. Y. Imai, *J. Macromol. Sci.-Chem.*, **A15**, 833 (1981).

41. V. Percec, T. D. Shaffer, and H. Nava, *J. Polym. Sci., Polym. Lett. Ed.*, **22**, 637 (1984).
42. W. J. Jackson and J. R. Caldwell, in *Polymerization and Polycondensation Processes* (R. F. Gould, Ed.), Adv. Chem. Ser. 34, American Chemical Society, Washington, D.C., 1962, pp. 200ff.
43. W. J. Bailey and A. A. Volpe, *J. Polym. Sci. A-1*, **8**, 2109 (1970).
44. J. Heller, D. W. H. Penhale, and R. F. Helwing, *J. Polym. Sci., Polym. Lett. Ed.*, **18**, 293 (1980).
45. A. S. Hay, *Polym. Eng. Sci.*, **16**, 1 (1976).
46. J. G. Bennett Jr., and G. D. Cooper, *Macromolecules*, **3**, 101 (1970).
47. W. G. Potter, *Epoxide Resins*, Springer-Verlag, New York, 1970.
48. I. Skeist, *Epoxy Resins*, Reinhold, New York, 1958.
49. C. A. May (Ed.), *Epoxy Resins: Chemistry and Technology*, 2nd ed., Marcel Dekker, New York, 1988.
50. P. F. Bruins, *Epoxy Resin Technology*, Wiley-Interscience, New York, 1968.
51. B. Sedláček and J. Kahovek (Eds.), *Crosslinked Epoxies*, de Gruyler, New York, 1986.
52. R. S. Bauer (Ed.), *Epoxy Resin Chemistry*, 2 vols., ACS Symp. Ser. 114 and 221, American Chemical Society, Washington, D.C., 1979 and 1983.
53. D. Aelony, *J. Appl. Polym. Sci.*, **13**, 227 (1969).
54. T. Matynia, *J. Appl. Polym. Sci.*, **25**, 1 (1980).
55. G. J. Fleming, *J. Appl. Polym. Sci.*, **13**, 989 (1969).
56. L. C. Lopez and G. L. Wilkes, *J. Macromol. Sci.-Rev. Macromol. Chem. Phys.*, **C29**, 83, (1989).
57. E. Tsuchida, K. Yamamoto, M. Jikei, and H. Nishidi, *Macromolecules*, **22**, 4138 (1989).
58. E. Tsuchida, K. Yamamoto, M. Jikei, E. Shouji, and H. Nishida, *J. Macromal. Sci.-Chem.*, **A28**, 1285 (1991).
59. M. Rule, D. R. Fagerburg, J. J. Watkins, and P. B. Lawrence, *Makromol. Chem., Rapid Commun.*, **12**, 221 (1991).
60. M. Rule, D. R. Fagerburg, J. J. Watkins, R. B. Lawrence, R. L. Zimmerman, and J. D. Cloyd, *Makromol. Chem., Macromol. Symp.*, **54/55**, 233 (1992).
61. W. H. Sharkey, in *Polyaldehydes* (O. Vogl, Ed.), Marcel Dekker, New York, 1967, pp. 91ff.
62. F. Lautenschlaeger and R. T. Woodhaus, in *Cyclic Monomers* (K. C. Frisch, Ed.), Wiley-Interscience, New York, 1973, Chap. 5.
63. V. Percec, H. Nava, and B. C. Auman, *Polym. J.* **16**, 681 (1984).
64. F. O. Davis and E. M. Fettes, in *Polyethers*, Part 3 (N. G. Gaylord, Ed.), Wiley-Interscience, New York, 1962, Chap. 12.
65. J. E. White and M. D. Scaia, *Polymer*, **25**, 850 (1984).
66. V. A. Sergeyev, V. I. Nedel'kin, and S. S. Arustamyan, *Polymer*, **26**, 365 (1985).
67. N. D. Ghatge and R. A. N. Murphy, *Polymer*, **22**, 1250 (1981).
68. M. B. Berenbaum, in *Polyethers*, Part 3 (N. G. Gaylord, Ed.), Wiley-Interscience, New York, 1962, Chap. 13.
69. J. R. Panek, in *Polyethers*, Part 3 (N. G. Gaylord, Ed.), Wiley-Interscience, New York, 1962, Chap. 14.
70. R. A. Clendenning, A. G. Farnham, and R. N. Johnson, in *High Performance Polymers: Their Origin and Development* (R. B. Seymour and G. S. Kirshenbaum, Eds.), Elsevier, New York, 1986, pp. 149–58.
71. K. J. Ivin and J. B. Rose, *Adv. Macromol. Chem.*, **1**, 336 (1968).
72. C. Arnold, *J. Polym. Sci., Macromol. Rev.*, **14**, 265 (1979).
73. H. Ulrich, *Introduction to Industrial Polymers*, 2nd ed., Hanser, New York, 1993, pp. 153–55.
74. A. H. Fawcett, in *Encyclopedia of Polymer Science and Engineering*, 2nd ed., Vol. 10 (H. F. Mark, N. M. Bikales, C. G. Overberger, G. Menges, and J. I. Kroschwitz, Eds.), Wiley-Interscience, New York, 1988, pp. 408–432.

Review Exercises

1. The following polyethersulfide is prepared by reacting diphenyl ether with SCl_2 in chloroform. Give the IUPAC name of the polymer. (T. Fujisawa and M. Kakatuni, *J. Polym. Sci., B*, **8**, 19, 1970)

2. Assuming that the polymerization reactions are feasible, write repeating units for the polymers prepared from the following monomers:

(a) $(CH_3)_2CHCHO$ (b) ϕCCH_3 (with O double bonded above C)   (c) (structure: oxolane ring with CH_3 and CH_3 substituents)   (d) (structure: bicyclic with O)

3. Predict the structure of the soluble polymer prepared by polymerization of *o*-phthalaldehyde. (C. Aso and S. Tagami, *Macromolecules*, **2**, 414, 1969)

4. Write the repeating units of the polyethers prepared under basic conditions
   (a) from bisphenol A and

   $$ClCH_2-\text{(benzene ring with CH}_3\text{, CH}_3\text{ top and CH}_3\text{, CH}_3\text{ bottom)}-CH_2Cl$$

   (b) from

   (structure: dichloro dibenzoyl benzene)   and   (structure: spiro bisphenol with HO and OH groups)

   In part (b), what advantage might be gained by using this particular bisphenol? (You may wish to refer to Chapter 4.) [(a) ref. 6, p. 168; (b) P. M. Hergenrother and B. J. Jensen, *Polym. Preprints*, **26**(2), 174, 1985]

5. Write a reaction showing the synthesis of spiro polymer **35** by a ketal exchange reaction.

6. Acetals and ketals are usually stable to base but undergo rapid hydrolysis in the presence of dilute acid. Polyformal **30**, however, is quite stable to acid. A polymer film began to disintegrate in 10% HCl only after 4 hours at 100°C. Rationalize the polymer's good acid resistance. (ref. 40)

7. Review the discussion of polysulfone, PEI, and PEEK in Section 11.3.1, then write reactions for their synthesis.

8. A polymer related in structure to PEEK is PEK (polyetherketone), also called PAEK (polyaryletherketone).

   (structure: repeating unit of PEK)

   (a) Suggest two possible reactions leading to PEK from the information given in Section 11.3.1.
   (b) Write IUPAC names for PEEK and PEK.

9. Give the structure of the epoxy prepolymer that would be formed from (a) bisphenol A and excess epichlorohydrin; (b) resorcinol and excess epichlorohydrin; and (c) bisphenol A and excess diepoxide **44**. (More than one structure is possible; write a representative one.)

10. An adhesives chemist prepared an epoxy oligomer from 119 g of bisphenol A and 60.2 g of epichloro-hydrin by reacting them together under basic conditions until thin-layer chromatography showed that the bisphenol A had reacted completely. What was the average molecular weight of the oligomer? (*Hint:* Consider the stoichiometric imbalance.)

11. What would be the most likely reason for using chlorendic anhydride (**48**) as a curing agent for epoxy resins?

12. A commercially available polyethersulfone (Union Carbide's Radel) not described in this chapter has the structure

    Which two compounds are most likely used in its synthesis?

13. In the Friedel–Crafts-type synthesis of polysulfones (reaction 11.70), crosslinking is not a problem, even though the ether group is *ortho, para* directing. Explain why. Suggest a single monomer that would yield the same polysulfone as that shown in the reaction.

14. A polysulfone exhibiting no olefinic absorption in its IR spectrum results from the alternating free radical copolymerization of SO_2 with such diallyl monomers as

    Propose possible structures for the polymers. (S. Harada et al., *Makromol. Chem.*, **90**, 177, 1966; **107**, 64, 78, 1967; **176**, 1289, 1975)

15. Phase-transfer catalysis has been used to synthesize the polymer

    A methylene chloride–water system was used.
    (a) Give structures for the most likely monomers used.
    (b) Given the reaction conditions, what side reaction might have been expected? (Show the product.)
       Why is the side reaction not favored? (Y. Imai et al., *J. Polym. Sci., Polym. Lett. Ed.*, **17**, 85, 1979)

16. Percec and coworkers (ref. 63) have prepared sequential (+AB+) block copolymers under phase-transfer conditions by reacting bromine-terminated polysulfides with phenol-terminated polyethers. Choosing any combination of monomers, write reactions illustrating such a synthesis.

17. Give the structure of the polyetherimide prepared from

This synthesis is less satisfactory than the one used to prepare **26**. What factors might be responsible? (S. Maiti and B. K. Mandel, *Makromol. Chem., Rapid Commun.*, **6**, 841, 1985).

18. Write plausible mechanisms for the reaction of 1,6-hexanedithiol with 1,5-hexadiene under acidic, basic, and free radical conditions. (See reactions 11.62 and 11.63.)

19. The following polymer (a polyestersulfideamine) was prepared without added initiator or catalyst.

Which monomers were most likely used? (*Hint:* See reaction 11.64.) (P. Ferruti and E. Ranucci, *J. Polym. Sci. C: Polym. Lett.*, **26**, 357, 1988)

20. Block copolymers of poly(alkylene disulfide)s and vinyl polymers have been prepared by reaction of the former with vinyl monomers under free radical conditions. Propose a mechanism. (M. L. Hallensleben, *Eur. Polym. J.*, **13**, 437, 1977)

12

POLYESTERS

12.1 Introduction

Polyesters[1-3] are among the more versatile synthetic polymers in that they find wide commercial use as fibers, plastics, and coatings. The common methods of synthesizing simple esters are used to make polyesters. These include direct esterification (12.1), transesterification (12.2), and the reaction of alcohols with acyl chlorides (12.3) or anhydrides (12.4).

$$RCO_2H + R'OH \rightleftharpoons RCO_2R' + H_2O \qquad (12.1)$$

$$RCO_2R'' + R'OH \rightleftharpoons RCO_2R' + R''OH \qquad (12.2)$$

$$RCOCl + R'OH \rightarrow RCO_2R' + HCl \qquad (12.3)$$

$$(RCO)_2O + R'OH \rightarrow RCO_2R' + RCO_2H \qquad (12.4)$$

These reactions are mechanistically well understood. Each involves nucleophilic addition to the carbonyl group, addition that is facilitated by the polar nature of the carbon–oxygen double bond, the ability of the carbonyl oxygen atom to assume a formal negative charge, and the planar configuration of the trigonal carbon that minimizes steric interference. A general mechanism may be written (reaction 12.5: Y=OH, OR", Cl, or OCOR) showing each step as reversible, although it should be recognized that in the case of acid chlorides and anhydrides the overall process is, for practical purposes, irreversible.

$$(12.5)$$

Less traditional methods are also used in polyester manufacturing, among them *acidolysis* (12.6) (a variation of transesterification used primarily to make aromatic polyesters), the

$$RCO_2H + R'CO_2R'' \rightleftharpoons RCO_2R'' + R'CO_2H \qquad (12.6)$$

reaction of carboxylic acids with epoxides (12.7), nucleophilic displacement (12.8), and ring-opening reactions of cyclic esters. Application of these reactions to the synthesis of linear polyesters is covered in Section 12.2 of this chapter. Hyperbranched polyesters are discussed briefly in Section 12.3, and Section 12.4 is concerned with polyesters that are crosslinked by use of polyfunctional monomers, or by incorporation of reactive double bonds into the polymer backbone.

$$RCO_2H + CH_2\overset{\displaystyle O}{\overbrace{}}CH_2 \rightarrow RCO_2CH_2CH_2OH \qquad (12.7)$$

$$RCO_2^- + R'Br \rightarrow RCO_2R' + Br^- \qquad (12.8)$$

Commercially important linear polyesters are shown in Table 12.1. No attempt is made to list applications of these polymers, because they are so diverse. Suffice it to say that the polyesters containing aromatic rings are used primarily as engineering plastics with the exception of poly(trimethylene terephthalate) and poly(ethylene terephthalate), the former being used as a textile fiber and the latter as both fiber and plastic. Indeed, poly(ethylene terephthalate) (PET) may be considered the workhorse of the polyester industry, having enjoyed

TABLE 12.1. Commercially Available Linear Polyesters[a]

Common Name	Principal Structure
Poly(ethylene terephthalate) (PET)	
Poly(trimethylene terephthalate)	
Poly(butylene terephthalate)[b] (PBT)	
Poly(ethylene naphthalene-2,6-dicarboxylate) (PEN)	
Poly(4-hydroxybenzoate)[c]	

continued

TABLE 12.1. (continued)

Common Name	Principal Structure
Poly(bisphenol A terephthalate/isophthalate)[d]	![structure]
Polycarbonate[b]	![structure]
Polycaprolactone	$\left[\!\!\begin{array}{c}O\\ \|\\ C(CH_2)_5O\end{array}\!\!\right]$
Poly(3-hydroxybutyrate/valerate) (PHBV)[e]	$\left[\begin{array}{c}O\quad CH_3\\ \|\quad \|\\ CCH_2CHO\end{array}\right]\left[\begin{array}{c}O\quad CH_2CH_3\\ \|\quad \|\\ CCH_2CHO\end{array}\right]$
Poly(glycolic acid)	$\left[\!\!\begin{array}{c}O\\ \|\\ CCH_2O\end{array}\!\!\right]$
Poly(lactic acid)	$\left[\begin{array}{c}O\\ \|\\ CCHO\\ \|\\ CH_3\end{array}\right]$

[a] Copolymers and/or polyblends are available for most of the polymers listed.
[b] A heterogeneous polyblend of PBT and polycarbonate (50:50 wt %) is commercially available.
[c] Primarily copolymers in liquid crystal polymer (LCP) formulations.
[d] Homopolymers and copolymers are available.
[e] Bacterial metabolism product of carbohydrates.

widespread commercial use since the 1940s.[4] Well-known trade names are Dacron and Mylar (both du Pont) for the fiber and plastic, respectively. Poly(trimethylene terephthalate), on the other hand, is a relative newcomer, being first introduced by Shell Chemical (Corterra) in 1996. Another relatively new polyester, poly(ethylene naphthalene-2,6-dicarboxylate) (PEN), is being used in blends with PET for improved bottle formulations.

Another polyester fiber (Eastman's Kodel) that was in commercial production until a few years ago is poly(1,4-dihydroxymethylcyclohexyl terephthalate) (**1**). Kodel was discontinued because it could not compete economically with PET. The aliphatic polyesters are of principal interest because of their degradability. As indicated in the footnotes of Table 12.1, a variety of copolyesters are also available. Not included in Table 12.1 are hydroxy-terminated aliphatic polyesters used in the manufacture of polyurethanes (Chapter 13).

1

12.2 Linear Polyesters

12.2.1 Preparation of Polyesters by Polycondensation Reactions

In principle any dibasic acid will condense with any glycol or any hydroxy acid will self-condense to form linear polyester.[5] In practice very few have achieved commercial importance, the major limitations being availability of monomer and chemical and physical properties of the finished product.

The most widely used linear polyester is poly(ethylene terephthalate) (**2**). It is prepared by direct esterification of terephthalic acid with ethylene glycol or—more commonly because of the acid's high melting point and low solubility—by transesterification (alcoholysis) of dimethyl terephthalate with ethylene glycol (12.9). The reaction is run in bulk or in solution using an excess of ethylene glycol to increase the reaction rate. This leads initially to a mixture of oligomers and bis(hydroxyethyl) terephthalate (**3**), which is subsequently transesterified with removal of the excess ethylene glycol to produce polymer. Salts of manganese or zinc are commonly added to catalyze the reaction, and a vacuum is used to facilitate removal of the glycol. Other aromatic–aliphatic polyesters are prepared similarly.

$$CH_3O_2C-\langle\bigcirc\rangle-CO_2CH_3 + HOCH_2CH_2OH \longrightarrow$$

$$\left[\begin{array}{c} O \\ \| \\ C-\langle\bigcirc\rangle-\overset{O}{\overset{\|}{C}}OCH_2CH_2O \end{array}\right] + 2CH_3OH \qquad (12.9)$$

2

$$HOCH_2CH_2O\overset{O}{\overset{\|}{C}}-\langle\bigcirc\rangle-\overset{O}{\overset{\|}{C}}OCH_2CH_2OH$$

3

More recently a process based on ethylene oxide instead of ethylene glycol has been commercialized in Japan. Ethylene oxide reacts rapidly with terephthalic acid (12.10) to form intermediate **3**, which then undergoes alcoholysis as described above to form polyester.

$$HO_2C-\langle\bigcirc\rangle-CO_2H + 2CH_2-CH_2 \longrightarrow \mathbf{3} \qquad (12.10)$$

Polyesters prepared from phenolic monomers are commonly called *polyarylates*, a term coined by Russian workers who have been particularly active in this field.[6] Polyarylates are difficult to prepare by direct polycondensation with diacids because the high temperatures necessary for achieving practical molecular weights frequently lead to decarboxylation or carboxyl-catalyzed ether formation. Consequently, they are usually made by transesterification methods[7] and, less commonly (because of high cost), by interfacial or solution polymerization of the acid chlorides.[8] The commercially important poly(4-hydroxybenzoate) (**4**), for example, is made by alcoholysis of the phenyl ester (12.11) or acidolysis of the acetyl derivative (12.12). One example of a commercial polyarylate copolymer (**5**), marketed by Hoechst Celanese (trade name Vectra), contains **4** and 6-hydroxynaphthalene-2-carboxylate repeating units. Polyarylates characteristically exhibit good thermal and chemical resistance. They are used in oven cookware, chemical processing equipment, automotive parts, and electrical and electronics equipment. The synthesis of poly(bisphenol A isophthalate) (**6**) (12.13) illustrates an interfacial synthesis of a polyarylate.

$$HO-\!\!\bigcirc\!\!-CO_2Ph \xrightarrow{\Delta} \left[O-\!\!\bigcirc\!\!-\overset{\overset{O}{\|}}{C} \right] + PhOH \qquad (12.11)$$

4

$$CH_3CO_2-\!\!\bigcirc\!\!-CO_2H \xrightarrow{\Delta} 4 + CH_3CO_2H \qquad (12.12)$$

5

$$(12.13)$$

6

A number of attempts have been made to prepare polyesters directly from dicarboxylic acids and dihydroxy compounds at low to moderate temperatures using activating agents.[9] Poly(ethylene terephthalate), for example, may be synthesized at room temperature from terephthalic acid and ethylene glycol in the presence of picryl chloride[10] (**7**), whereas polyarylates are formed directly at about 120°C using diphenyl chlorophosphate[11] (**8**) or *p*-toluenesulfonyl chloride[12] (**9**). Pyridine is used as a solvent or cosolvent in each case to remove byproduct HCl or to assist in the activation process. These reactions contrast with direct polycondensation or transesterification reactions, which are normally run at temperatures in excess of 200°C. The role of the activating agent is to convert the acid into a highly reactive ester intermediate. While such methods are useful in laboratory syntheses, they are too costly for industrial-scale processes.

Structures with labels **7**, **8**, **9** (Cl-P(OPh)₂ shown).

Polyester synthesis by nucleophilic displacement reactions are much less common because of the low nucleophilicity of carboxylate ion. One such reaction, the synthesis of the biodegradable poly(glycolic acid) (**10**), has been accomplished by reacting bromoacetic acid with an equimolar amount of triethylamine in nitromethane or ether at ambient temperatures (12.14).[13] Phase-transfer catalysis facilitates this type of reaction. Phase-transfer methods have also been used to make sulfonate (**11**) and phosphonate (**12**) polyesters[14] from the corresponding sulfonyl chlorides (12.15) and phenylphosphonic dichloride (12.16).

$$BrCH_2CO_2H + (C_2H_5)_3N \longrightarrow \left[OCH_2\overset{\displaystyle O}{\overset{\displaystyle \|}{C}} \right] + (C_2H_5)_3\overset{+}{N}H\ Br^- \qquad (12.14)$$

$$\mathbf{10}$$

$$HO-Ar-OH + ClSO_2-Ar'-SO_2Cl \longrightarrow \left[O-Ar-OSO_2-Ar'-SO_2 \right] \qquad (12.15)$$

$$\mathbf{11}$$

$$HO-Ar-OH + Cl-\underset{\underset{\displaystyle Ph}{|}}{\overset{\overset{\displaystyle O}{\|}}{P}}-Cl \longrightarrow \left[O-Ar-O-\underset{\underset{\displaystyle Ph}{|}}{\overset{\overset{\displaystyle O}{\|}}{P}} \right] \qquad (12.16)$$

$$\mathbf{12}$$

Anhydrides may also be used as monomers for polyesters. Succinic anhydride and ethylene glycol react initially to form hydroxy acid, which undergoes subsequent esterification to form poly(ethylene succinate) (12.17). Such reactions are used most commonly in the preparation of polyester resins (Section 12.4), where high degrees of polymerization are not so important. Under very mild conditions, dianhydrides form base-soluble polyesters (12.18), which lend themselves to crosslinking or grafting through the free carboxyl groups. A wide variety of block copolymers of polyesters have been reported.[15] The reaction of hydroxyl-terminated polyester with an epoxide-terminated polyether (12.19) to form an [AB] multi-block copolymer is an example.

$$\begin{matrix} CH_2-C \\ | \quad\quad\ \ \\ CH_2-C \end{matrix} \begin{matrix} O \\ O \\ O \end{matrix} + HOCH_2CH_2OH \xrightarrow{\ fast\ } HOCH_2CH_2O\overset{O}{\overset{\|}{C}}CH_2CH_2\overset{O}{\overset{\|}{C}}OH \xrightarrow{\ slow\ }$$

$$\left[OCH_2CH_2O\overset{O}{\overset{\|}{C}}CH_2CH_2\overset{O}{\overset{\|}{C}} \right] + H_2O \qquad (12.17)$$

$$(12.18)$$

Polyester   Polyether   Polyester

$$(12.19)$$

Block copolymer

The effect of structure on polyester properties is evident in the melting points presented in Table 12.2.[16,17] The first four entries illustrate why aromatic units in the backbone are usually necessary to give the polymers other than elastomeric properties. The fifth shows the chain-stiffening effect of naphthalene units in the backbone.[18] The four indicate how different functionalities influence thermal properties. The particularly high melting point resulting

TABLE 12.2. Representative Melting Points of Linear Polyesters[a]

Structure	Approximate Melting Temperature (T_m) (°C)
	50
	265
	228
	232–267

continued

TABLE 12.2. (continued)

Structure	Approximate Melting Temperature (T_m) (°C)
[naphthalene diacyl] —OCH$_2$CH$_2$O—	337
—C(=O)—⬡—(CH$_2$)$_4$—⬡—C(=O)OCH$_2$CH$_2$O—	170
—C(=O)—⬡—O(CH$_2$)$_2$O—⬡—C(=O)OCH$_2$CH$_2$O—	240
—C(=O)—⬡—NH(CH$_2$)$_2$NH—⬡—C(=O)OCH$_2$CH$_2$O—	273
—C(=O)—⬡—CH=CH—⬡—C(=O)OCH$_2$CH$_2$O—	420
—C(=O)—⬡—C(=O)—O—⬡—SO$_2$—⬡—O—	330
—C(=O)—⬡(CH$_2$(CH$_2$)$_3$CH$_3$)—C(=O)—O—⬡—SO$_2$—⬡—O—	246
—C(=O)—⬡—O—	610
—C(=O)—⬡—C(=O)—O—⬡—O—	600

[a]Data from Wilfong[16]; Evers, and Fisch[17]; Jackson[19]; Cheng and Wunderlich[18]; *Modern Plastics Encyclopedia* 1985, McGraw-Hill, New York, 1984; and *Chem. Engin. News*, **73**, 14 (July 17, 1995).

from incorporation of a double bond between the *p*-phenylene units reflects the significance of restricting rotation. The next two polyesters show the effect of including a "plasticizing" five-carbon chain in the backbone.

The remaining two polyesters in Table 12.2 represent extreme examples of chain stiffening. Because the melting points are too high for most molding applications, high-temperature powder sintering techniques are used to fabricate the polymers. Attempts to improve the solubility and lower the melting point include copolymerization with other polyarylates.[19]

12.2.2 Polycarbonates

Polycarbonates[20–22] are polyesters of carbonic acid. Although carbonic acid itself is not a stable compound, its derivatives (phosgene, urea, carbonates) are commercially available. The reaction of gaseous phosgene with bisphenols, notably bisphenol A (12.20), was developed into commercial production independently by Farbenfabriken Bayer in Germany and General Electric in the United States. The product (**13**) is marketed under such trade names as Lexan (General Electric) and Merlon (Mobay). Interfacial polymerization is the preferred method, with the aqueous phase containing stoichiometric amounts of NaOH to neutralize the byproduct HCl. A tertiary amine catalyst is usually added, along with a monofunctional phenol to control the chain length.

$$(12.20)$$

13

While any dihydroxy compound can, in principle, form a polycarbonate with phosgene, only bisphenol A has gained commercial importance, although other bisphenols are sometimes incorporated as property modifiers. The polymer is partially crystalline, is transparent, and has unusually high impact strength (see Chapter 4, Table 4.1), which makes it ideal for laboratory safety shields, bullet-proof windows, safety helmets, and the like. Polycarbonate is also thermally stable (it can be molded at temperatures as high as 550 to 600°C) and self-extinguishing. Other uses include gears, bushings, automotive parts, tableware, food containers, machine housings, medical appliances, and telephone and electronics parts.[23] Several polyblends of polycarbonate have been developed as engineering plastics, the most important being those with poly(butylene terephthalate) and acrylonitrile–butadiene–styrene (ABS) copolymer.[23,24]

Table 12.3 compares the glass transition temperatures of polycarbonates of varying structure. Interestingly, replacing the methyl groups of bisphenol A polycarbonate has only a marginal effect on the T_g except where a five- or six-membered ring is incorporated. The great-

TABLE 12.3. Glass Transition Temperatures (T_gs) of Representative Polycarbonates[a]

R_1	R_2	R_3	T_g (°C)
CH_2	H	H	147
$C(CH_3)_2$	H	H	149
$C(CH_3)_2$	CH_3	H	125
$C(CH_3)_2$	CH_3	CH_3	210
SO_2	CH_3	CH_3	260
$C(CH_2CH_2CH_3)_2$	H	H	148
$C(C_6H_5)_2$	H	H	121
	H	H	167
	H	H	171

[a]Data from Schnell[21] and Vernaleken.[22]

est effect occurs when all ring positions *ortho* to the carbonate group are substituted, and when the isopropylidene group is replaced by sulfonyl, both of which severely restrict rotational freedom.

Although interfacial polymerization is the preferred method of synthesis, transesterification (12.21) is sometimes used, with diphenyl carbonate being the ester of choice. Transesterification generally results in lower-molecular-weight polymer because of difficulties in removing phenol from the viscous polymer mass. Furthermore, the polymerization temperature must be carefully controlled to avoid rearrangement (12.22) that could result in crosslinking or chain branching.

(12.21)

$$\text{(structure)} \xrightarrow{\Delta} \text{(structure)} \qquad (12.22)$$

Polycarbonates have also been synthesized from dihydroxy compounds and bischloro-formates[21] (12.23) and by alternating copolymerization of epoxides with carbon dioxide (12.24).[25,26] A considerable amount of developmental work has been devoted to the latter,[27]

$$\text{ClCO—R—OCCl} + \text{HO—R′—OH} \longrightarrow$$

$$\left[\text{CO—R—OCO—R′—O} \right] + 2\,\text{HCl} \qquad (12.23)$$

$$\underset{\text{O}}{\overset{\text{CH}_2—\text{CHR}}{\diagdown}} + \text{CO}_2 \xrightarrow{\text{Zn(C}_2\text{H}_5)_2\cdot\text{H}_2\text{O}} \left[\text{CH}_2\text{CHOCO} \right] \qquad (12.24)$$

not surprising considering the obvious economic advantage of low monomer cost. Alternating copolycarbonates have also been reported[28] from the reaction of carbon dioxide, diols and dihalides under basic conditions (12.25).

$$\text{HO—R—OH} + \text{CO}_2 + \text{Cl—R′—Cl} \xrightarrow{\text{K}_2\text{CO}_3}$$

$$\left[\text{CO—R—OCO—R′—O} \right] \qquad (12.25)$$

Carbon dioxide may also be used in the form of a carbonate salt, as shown in reaction (12.26).[29] In this instance, a crown ether catalyst is used to bind the potassium ions, thus increasing the nucleophilicity of carbonate ion. Additionally, polythiocarbonates may be prepared using carbon disulfide as a comonomer.[30] Polycarbonate synthesis from cyclic oligomers is covered in the next section.

$$\text{BrCH}_2\text{—(ring)—CH}_2\text{Br} + \text{K}_2\text{CO}_3 \xrightarrow{\text{18-crown-6}} \left[\text{CH}_2\text{—(ring)—CH}_2\text{OCO} \right] \qquad (12.26)$$

12.2.3 Preparation of Polyesters by Ring-Opening Polymerization

Ring-opening polymerization of cyclic esters (lactones) is brought about by cationic, anionic, and complex coordination initiators.[31–33] The mechanism of cationic polymerization appears in most instances to involve an intermediate acylium ion with each propagation step proceeding with acyl–oxygen bond cleavage (12.27, Z^+ = initiator). An exception is β,β-

dimethyl-β-propiolactone (**14**), which undergoes alkyl–oxygen cleavage to give the preferred tertiary carbocation intermediate (12.28).

$$\text{(CH}_2)_x\overset{\displaystyle C\!\!=\!\!O}{\underset{O}{\big|}} \quad \xrightarrow{Z^+} \quad ZO(CH_2)_x\overset{\displaystyle O}{\overset{\|}{C}} + \quad \longrightarrow$$

(12.27)

$$ZO(CH_2)_x\overset{O}{\overset{\|}{C}}\!\!\left[O(CH_2)_x\overset{O}{\overset{\|}{C}}\right]\!\!O(CH_2)_x\overset{O}{\overset{\|}{C}} +$$

$$\underset{\textbf{14}}{\overset{\displaystyle CH_3\ CH_3}{\underset{\displaystyle O}{\underset{C}{\big|}}}\quad \xrightarrow{Z^+}\quad ZO \quad \longrightarrow\quad ZOCCH_2\overset{CH_3}{\underset{CH_3}{\big|}} + \quad\longrightarrow$$

(12.28)

$$ZOCCH_2\overset{CH_3}{\underset{CH_3}{C}}\!\!\left[OCCH_2\overset{CH_3}{\underset{CH_3}{C}}\right]\!\!OCCH_2\overset{CH_3}{\underset{CH_3}{C}} +$$

Two mechanisms have been suggested for anionic polymerization of lactones. Nucleophilic addition to the carbonyl group followed by acyl–oxygen cleavage (analogous to ordinary ester saponification) leads to propagation through alkoxide ion chain ends (12.29, B^- = initiator). As in the case of epoxide polymerization, living polymers may be formed. There is, however, a tendency for the growing chain to depropagate via a back-biting mechanism to form cyclic oligomers. The resultant chain-ring equilibrium, which may severely limit the molecular weight, is less prevalent when Li^+ is the counterion, and high-molecular-weight polyesters have been prepared with alkyllithium and lithium alkoxide initiators.[34]

$$(CH_2)_x\overset{\displaystyle C\!\!=\!\!O}{\underset{O}{\big|}}\ \xrightarrow{B^-}\ (CH_2)_x\overset{\displaystyle C\overset{B}{\underset{O^-}{\big\diagup}}}{\underset{O}{\big|}}\ \longrightarrow\ BC(CH_2)_xO^-\ \longrightarrow$$

(12.29)

$$BC(CH_2)_xO\!\!\left[\overset{O}{\overset{\|}{C}}(CH_2)_xO\right]\!\!\overset{O}{\overset{\|}{C}}(CH_2)_xO^-$$

The other anionic mechanism involves nucleophilic attack on the carbon adjacent to the ring oxygen, giving alkyl–oxygen cleavage and propagating carboxylate ion chain ends. Such a mechanism (12.30) applies to the more highly strained β-propiolactone (**15**).* This reac-

*Supporting evidence is derived from studies of the base-catalyzed hydrolysis of β-lactones in ^{18}O-enriched water.

$$\overset{\displaystyle CH_2\!\!-\!\!C\!\!=\!\!O}{\underset{R}{\overset{\diagdown}{\underset{CH\!\!-\!\!O}}}}\ \xrightarrow[^{18}OH^-]{H_2{}^{18}O}\ RCHCH_2CO_2H\ \underset{^{18}OH}{\big|}$$

Nucleophilic attack at the carbonyl carbon would result in the ^{18}O in the carboxylic acid group.

tion is subject to chain transfer reactions resulting in unsaturated chain ends (12.31) and low molecular weights.

$$
\text{B}^- \quad \underset{\overset{|}{\text{CH}_2-\text{O}}}{\text{CH}_2-\text{C}}\!\!\diagup\!\!\overset{\text{O}}{} \longrightarrow \text{BCH}_2\text{CH}_2\overset{\text{O}}{\overset{\|}{\text{C}}}\text{O}^- \longrightarrow
$$

15

$$
\text{BCH}_2\text{CH}_2\overset{\text{O}}{\overset{\|}{\text{C}}}\text{O}\!\!\left[\!\text{CH}_2\text{CH}_2\overset{\text{O}}{\overset{\|}{\text{C}}}\text{O}\!\right]\!\text{CH}_2\text{CH}_2\overset{\text{O}}{\overset{\|}{\text{C}}}\text{O}^-
$$

(12.30)

$$
\text{\tiny ww}\overset{\text{O}}{\overset{\|}{\text{C}}}\text{O}^-\!,\ \underset{\overset{|}{\text{CH}_2-\text{O}}}{\text{H}-\text{CH}-\text{C}}\!\!\diagup\!\!\overset{\text{O}}{} \longrightarrow \text{\tiny ww}\overset{\text{O}}{\overset{\|}{\text{C}}}\text{OH} + \underset{\overset{|}{\text{CH}_2}}{\text{CH}-\text{C}}\!\!\diagup\!\!\overset{\text{O}}{\diagdown\text{O}^-} \longrightarrow
$$

(12.31)

$$
\text{CH}_2\!\!=\!\!\text{CHCO}\!\!\left[\!\text{CH}_2\text{CH}_2\overset{\text{O}}{\overset{\|}{\text{C}}}\text{O}\!\right]\!\text{CH}_2\text{CH}_2\overset{\text{O}}{\overset{\|}{\text{C}}}\text{O}^-
$$

Polymerization of β-lactones may also be initiated with metalloporphyrin initiators of the type encountered in the previous chapter.[35,36] Ring opening takes place at the alkyl–oxygen bond (12.32) and—as was the case with epoxides—the result is an "immortal" polymer suitable for forming block copolymers. Another example of complex coordination catalysis is the ring-opening polymerization of ϵ-caprolactone (**16**) using the dibutylzinc–triisobutyl-aluminum system (12.33).[37] Polycaprolactone (**17**) is used in the medical field for bioerodable surgical sutures and postoperative support pins, and as a delivery system for controlled drug release.[38] With the exception of the five-membered γ-butyrolactone, most lactones appear to form polyester.

$$
\left(\!\!\diagdown\!\!\text{Al}-\text{Cl}\right. + \mathbf{15} \longrightarrow \left(\!\!\diagdown\!\!\text{Al}\!\!-\!\!\text{O}\overset{\text{O}}{\overset{\|}{\text{C}}}\text{CH}_2\text{CH}_2\right)\!\!-\!\!\text{Cl}
$$

(12.32)

$$
\left(\!\!\diagdown\!\!\overset{\text{O}}{\diagup}\!\!=\!\!\text{O}\right) \xrightarrow{(n\text{-}C_4H_9)_2Zn\text{---}(i\text{-}C_4H_9)_3Al} \left[\!\!-\!\!(\text{CH}_2)_5\overset{\text{O}}{\overset{\|}{\text{C}}}\text{---O}\!\!-\!\!\right]
$$

(12.33)

16   **17**

Two other important biodegradable polyesters made from cyclic esters are the previously mentioned poly(glycolic acid) (**10**) and poly(lactic acid) (**20**). They are prepared from glycolide (**18**) and lactide (**19**), respectively, using Lewis acid catalysts (reactions 12.34 and 12.35). Polymers **10** and **20**, which are also called polyglycolid and polylactide, are used in similar medical applications.[39,40] Polyglycolide (trade name Dexon) and block copolymers of glycolide and lactide are manufactured by American Cyanamide. Poly(lactic acid) has also been suggested as a biodegradable packaging film.[41]

$$
\text{O}\!\!=\!\!\left<\!\!\overset{\overset{\text{O}}{\diagup}}{\diagdown\text{O}}\!\!\right>\!\!=\!\!\text{O} \xrightarrow{\text{SbF}_3} \left[\!\!-\!\!\text{CH}_2\overset{\text{O}}{\overset{\|}{\text{C}}}\text{O}\!\!-\!\!\right]
$$

(12.34)

18   **10**

$$\text{(12.35)}$$

19 **20**

More recent studies have been directed towards ring-opening polymerization of ketene acetals, which yield polyesters under both free radical[43,44] and cationic conditions.[45] The peroxide-initiated polymerization (12.36) of 2-methylene-1,3-dioxolane (**21**) to form polymer (**22**) is an example. The mechanism involves addition of peroxy radical to **21** to form radical **23**, followed by ring opening (12.37) to form radical **24**, which, in turn, initiates further ring opening (12.38). Vinyl polymerization (nonring opening) is a competing reaction, especially at lower temperatures (about 50:50 at 60°C); but with the more hindered 2-methylene-1,3-dioxepane (**25**), almost pure polycaprolactone (**17**) is formed. Steric effects also favor ring opening under cationic conditions.[45] (It should be noted that free radical ring-opening polymerization[46] is by no means common.)

$$\xrightarrow[\Delta]{\text{peroxide}} \quad \text{(12.36)}$$

21 **22**

$$\longrightarrow \quad \text{(12.37)}$$

23 **24**

$$\textbf{24} + \textbf{21} \longrightarrow \text{RO} \quad \longrightarrow \text{etc.} \quad \text{(12.38)}$$

$$\xrightarrow[\Delta]{\text{peroxide}} \textbf{17} \quad \text{(12.39)}$$

25

Under free radical conditions ketene acetal monomers copolymerize readily with vinyl monomers. Introduction of ester units from **21** into polyethylene yields a copolymer (**26**) that is biodegradable by initial breakdown to form fatty acid-type residues.

$$+CH_2CH_2\overline{}_xCH_2\overset{O}{\overset{\|}{C}}OCH_2CH_2\overline{}CH_2CH_2\overline{}_y$$

26

Another significant development in ring-opening polymerization involves the synthesis and polymerization of cyclic carbonate oligomers[47,48] to yield the commercially important bisphenol A polycarbonate (**13**). To prepare the cyclic oligomers, bisphenol A is first converted to the bis(chloroformate) (**27**) with phosgene. A dichloromethane solution is then added slowly to a mixture of bisphenol A, triethylamine, and sodium hydroxide to yield the oligomers (**28**, n = 2–20) in about 85% yield (12.40). The slow addition maintains the con-

(12.40)

centration of **27** at a low level to minimize formation of linear polymer. Heating with a basic catalyst converts **28** to the linear polymer (**13**) having molecular weights in the range of 50,000 to 700,000. Chain length is controlled by the addition of monofunctional phenols or diphenyl carbonate. Advantages of the cyclic oligomer process are that no byproducts are formed in the polymerization step, the oligomer mixture has a low viscosity, and the highly corrosive phosgene is eliminated from the polymer processing equipment. Cycloaliphatic carbonates also undergo ring-opening polymerization (see exercise 8), but the products lack the properties to make them commercially viable. Polyarylates can also be made from cyclic oligomers.[49,50]

12.2.4 Microbial Polyesters

Polyesters are commonly found as the reserve energy material in bacteria, just as starch is the reserve material of plants. The bacterial polyesters, first characterized in the 1920s,[51] are known collectively as poly(hydroxy alkanoate)s (PHAs), the simplest and most common of which is poly(3-hydroxybutyrate) (PHB) (**29**). Longer side chains, up to nine carbon atoms, have also been identified. All have the isotactic configuration. There is considerable interest in these polyesters because of their biocompatibility and biodegradability.[52–58]

Bacterial PHB is highly crystalline and has a T_g of 5°C and a T_m of 180°C. Its properties resemble those of polypropylene, except that it is more brittle and has better resistance to ultraviolet radiation. Early attempts to exploit PHB commercially were unsuccessful, until ICI in the 1980s began production of a random copolymer, poly(3-hydroxy butyrate-*co*-3-hydroxy valerate) (PHBV), using the bacterium *Alcaligenes eutrophus*. The copolymer structure is given in Table 12.1. Marketed under the trade name Biopal, PHBV is used to make compostable bottles, but degradable diapers, packaging material, and agricultural mulch film, as well as biomedical applications, are other possibilities. Little has been done with the longer-chain bacterial polyesters, which characteristically exhibit properties of thermoplastic elastomers.

One of the interesting aspects of bacterial polyesters is that their structure can be influenced by what is fed to the bacteria. Thus double bonds may be introduced using unsaturated fatty acid nutrients,[59] as can aromatic groups using phenyl-substituted carboxylic acid nutrients.[60] Another interesting development, which might well determine the commercial viability of PHAs, involves recombinant gene technology.[61] The genetic information needed to make the polyesters has already been spliced into *Escherichia coli*, which produces much higher yields of polymer than *A. eutrophus* by fermentation. Even more promising for longer-term development, the required genes have been transferred to plants for agricultural production of polyesters.[61] Tobacco and canola are two likely candidates.

12.3 Hyperbranched Polyesters

All the polyesters described up to this point are linear. There is growing interest, however, in hyperbranched polyesters because of their solubility, low viscosity, and high degree of functionality.[62–66] The high solubility arises from the dendritic branching which prevents close chain packing, as well as from the high functionality. While they may lack the beautiful symmetry of the dendrimers, hyperbranched polymers can be synthesized in a single, uncontrolled, divergent polymerization reaction, rather than the controlled multiple steps, divergent or convergent, needed to make dendrimers. It may be recalled that A_xB-type monomers are required, where x is greater than one. Representative examples for polyesters are **30**, **31**, and **32**. Monomer **30** polymerizes by an acidolysis reaction with acetic acid being the byproduct.

At the high temperature needed to maintain a practical rate of propagation, crosslinking occurs by the formation of anhydride units between the carboxyl groups, but these are readily hydrolyzed once the polymerization is complete. Crosslinking is not a problem with **31**, which undergoes direct esterification at a lower temperature in the presence of an organotin catalyst. Nor is crosslinking a problem with **32**, which yields trimethylsilyl chloride byproduct.

Individual polymer structures are much too complex to show here, but representative structural units arising from **31** include terminal (**33**), linear (**34**), and dendritic (**35**). The degree of branching (DB) is defined[67] as

$$DB = \frac{\text{dendritic} + \text{terminal units}}{\text{dendritic} + \text{terminal} + \text{linear units}}$$

The percent branching is DB $\times$ 100. Monomer **32**, for example, yields polyester having $\overline{M}_w$ and $\overline{M}_n$ of 164,000 and 55,000, respectively, with 50–60% branching.

33	**34**	**35**

12.4 Crosslinked Polyesters

There are two types of crosslinked or network polyesters:

1. Those prepared from polyfunctional monomers so that crosslinking occurs during the polyesterification reaction. (These are referred to below as *saturated polyester resins*.)
2. Those crosslinked by a separate addition polymerization reaction through reactive double bonds incorporated into the polyester backbone. (These are commonly called *unsaturated polyesters*.)

Both types are referred to as *alkyds* or *polyester resins*, and both are widely used in a variety of commercial applications. Their chemistry will be discussed in some detail.

12.4.1 Saturated Polyester Resins

Some of the common polyfunctional alcohols and acids or acid derivatives used to make saturated alkyd resins[68] include glycerol, pentaerythritol, sorbitol, citric acid, trimellitic acid, and pyromellitic dianhydride. Phthalic anhydride is by far the most common dibasic acid derivative used.

The polyester prepared from phthalic anhydride and glycerol is called *glyptal*, and its network structure typifies this type of resin (reaction 12.41). Structures are complex and highly crosslinked, the degree of crosslinking depending on the reactants employed and the extent of reaction. Normally, polymerization is carried to the point of forming a viscous liquid that can be hardened later to a more fully crosslinked product, for example, in a mold. In production the process is monitored by periodically checking the viscosity and acid number. (See Chapter 2, Section 2.3.)

$$(12.41)$$

Obviously one can vary the structure at will by changing the type and concentration of reactant, or by adding some diol or monofunctional acid or alcohol. These types of alkyds have little use other than as adhesives or as additives in certain types of coatings, such as nitrocellulose. They are, however, extremely important when modified with natural or synthetic oils (triglycerides). Such products are called *oil-modified alkyds*.

Prior to latex emulsion paints, oil-modified alkyds were the most important type of coating. Oils are incorporated into polyester resins by two general methods:

1. Polyol and oil are transesterified (12.42) so that mixing occurs; the mixed product is then reacted with dibasic acid or anhydride in a second step (12.43) to give the final complex crosslinked polymer. (R represents the long carbon chain of the fatty acid in these reactions.)

2. Dibasic acid or anhydride, polyol, and oil fatty acids (called "split oils") are reacted together in a single reaction.

$$(12.42)$$

$$(12.43)$$

The net effect of either process is to incorporate the long-chain oil structure into the polyester. If the oil itself is reacted with polyol and dibasic acid in one step, a two-phase product results. After initial esterification, the viscosity is increased by heating to a high temperature. This process, known as *bodying*, results from a combination of further esterification and polymerization through the double bonds of the oil.

The type of oil used may contain varying amounts of unsaturation depending on the application. Examples of the more unsaturated oils (*drying oils*) are linseed oil, which contains principally linoleic (**36**) and linolenic (**37**) acids, and tung oil, a major constituent of which is elaeostearic acid (**38**). Dehydrated castor oil, a synthetic oil, also contains linoleic acid. Olive oil and castor oil are examples of nondrying oils.

$$CH_3(CH_2)_4CH{=}CHCH_2CH{=}CH(CH_2)_7CO_2H$$
36

$$CH_3(CH_2CH{=}CH)_3(CH_2)_7CO_2H$$
37

$$CH_3(CH_2)_3CH{=}CH{-}CH{=}CH{-}CH{=}CH(CH_2)_7CO_2H$$
38

The oil-modified alkyds, dissolved in suitable solvents (the choice depending usually on the amount of oil used), together with appropriate additives such as pigments and metal naphthenates (driers), constitute *alkyd paints*. The function of the oil is to cause *drying* by a free radical crosslinking reaction involving allylic radicals. (Oils "dry" by themselves when spread as a film.) The reaction depends on the presence of atmospheric oxygen and is speeded up by the presence of the driers, often cobalt or manganese naphthenates.*

The rapidity of drying depends on the amount of unsaturation and the amount of oil incorporated into the alkyd (referred to as *oil length*). Oil lengths are commonly designated as follows:

> Short-oil alkyds, 30 to 50% oil
> Medium-oil alkyds, 50 to 65% oil
> Long-oil alkyds, 65 to 80% oil

The short-oil alkyds usually have to be baked (baking enamels) to effect drying in a reasonable time, although the coating is generally more durable. Alkyds modified with nondrying oils will always remain tacky; they are usually used for blending with other types of resins.

Oil-modified alkyds are sometimes modified further by addition of styrene. The styrene is copolymerized with the oil portion of the alkyd to give a tougher, more durable film by reactions similar to those discussed in the next section. The products are referred to as *styrenated alkyds*.

12.4.2   Unsaturated Polyesters

Unsaturated polyesters[69,70] are among the most common polymers used in conjunction with glass fiber reinforcing. They are prepared from difunctional monomers, one of which contains a double bond that is capable of undergoing addition polymerization in a subsequent crosslinking reaction. Typically, the linear unsaturated polyester is processed to a relatively

---

*Naphthenic acids are petroleum-derived complex mixtures consisting primarily of fused cyclopentane rings with aliphatic side chains and carboxylic acid groups. Metal naphthenates are viscous liquids.

low molecular weight; then it is dissolved in a monomer such as styrene to form a viscous solution. Crosslinking, usually initiated with free radical initiators, is thus a vinyl copolymerization between the polyester and the solvent monomer. Styrene is by far the most commonly used solvent monomer, although one can use such others as vinyl acetate or methyl methacrylate or, for improved flame-resistance, halogenated monomers such as *ortho-* and *para*-bromostyrene.[71]

The only ingredients of commercial significance for introducing unsaturation into the polymer backbone are maleic anhydride and fumaric acid, particularly the former because of its lower cost.[72] If only unsaturated acid and glycol are used, the final product is too highly crosslinked and brittle to be useful. For this reason, copolyesters are normally prepared containing both unsaturated acid and a noncrosslinkable acid. Reaction (12.44) shows a typical

(12.44)

unsaturated polyester synthesis from maleic and phthalic anhydrides (in a 1:1 molar ratio) and diethylene glycol. The phthalic and maleic units would be randomly distributed. Crosslinking with styrene is depicted in reaction (12.45).

(12.45)

Properties of unsaturated polyesters can be readily varied by using different types of diacids and glycols or different reactant ratios. Solvent monomer may also be varied in type as well as concentration.

Some of the dibasic acids or anhydrides commonly used in addition to those mentioned above are isophthalic acid, adipic acid, and succinic anhydride. Work has also been done on polyesters prepared from 2,2-diphenylpropane dioxyacetic acid (**39**), a diacid prepared from bisphenol A and chloroacetic acid.[73] Common glycols include ethylene, propylene, diethylene, and dipropylene glycols. To obtain maximum hardness and flexural strength, "rigid" monomers such as phthalic anhydride, isophthalic acid, and propylene glycol are normally used. Greater flexibility and higher impact strengths are obtained from the longer chain monomers. Flame resistance is imparted by halogenated monomers such as chlorendic anhydride and tetrabromophthalic anhydride. (Refer to the discussion in Section 4.5, Chapter 4.) Polyesters with good chemical resistance are prepared with glycols that affords steric hindrance to the ester linkage. (Section 4.6, Chapter 4.) It may be recalled from Chapter 8 that unsaturated polyesters have also been synthesized by ADMET polymerization (reaction 8.40).

$$HO_2CCH_2O-\!\!\langle\bigcirc\rangle\!\!-\overset{\overset{\displaystyle CH_3}{|}}{\underset{\underset{\displaystyle CH_3}{|}}{C}}-\!\!\langle\bigcirc\rangle\!\!-OCH_2CO_2H$$

39

Most unsaturated polyesters are not suitable as coatings because atmospheric oxygen inhibits the crosslinking reaction. Two approaches have been used to circumvent this difficulty. One is to mix wax into the polyester-styrene solution. Not being completely soluble, the wax rises to the surface and excludes oxygen and, incidentally, reduces loss of styrene by evaporation. Wax has the disadvantage of imparting a dull finish to the film.

The other approach is to incorporate into the polyester backbone monomers such as tetrahydrophthalic anhydride (**40**) and dodecenylsuccinic anhydride (**41**), which contain isolated double bonds. These allow crosslinking by the same oxygen-induced mechanism that causes crosslinking of oil-modified alkyds.

$$CH_3(CH_2)_8CH\!=\!CHCH_2-$$

40   **41**

A number of photochemical methods have been developed to cure unsaturated polyester coatings or transparent articles, among them the use of disulfides[74] or dyes[75] to initiate photocrosslinking, or the incorporation of photosensitive groups such as chalconyl or cinnamyl into the polymer[76] (see Table 9.1, Chapter 9). Photochemical curing not only allows the formulation of *one-pot* systems, but B-stage (partially cured) resins can be formed simply by removing the light source once the appropriate viscosity is reached.

Because unsaturated polyesters are so widely used, a number of strategies have been suggested for recycling. One recent method[77] involves the alcoholysis of waste resin with propylene glycol, followed by re-esterification with additional maleic anhydride. More styrene is then added to the reconstituted polymer for crosslinking.

References

1. Bjorksten Research Laboratories, *Polyesters and Their Applications*, Reinhold, New York, 1956.
2. V. V. Korshak and S. V. Vinogradova, *Polyesters*, Pergamon, New York, 1965.
3. I. Goodman and J. A. Rhys, *Polyesters*, Vol. 1, *Saturated Polymers*, Iliffe, London, 1965.
4. B. V. Petukhov, *The Technology of Polyester Fibres*, Macmillan, New York, 1963.
5. P. W. Morgan, *Condensation Polymers: By Interfacial and Solution Methods*, Wiley-Interscience, New York, 1965, Chap. 8.
6. V. V. Korshak, *Heat-Resistant Polymers*, Halsted, New York, 1972.
7. J. P. Critchley, G. J. Knight, and W. W. Wright, *Heat-Resistant Polymers*, Plenum, New York, 1983, Chap. 4.
8. S. C. Temin, in *Interfacial Synthesis*, Vol. 2 (F. Millich and C. E. Carraher, Jr., Eds.), Marcel Dekker, New York, 1977, Chap. 12.
9. F. Higashi, in *New Methods for Polymer Synthesis* (W. J. Mijs, Ed.), Plenum, New York, 1992, Chap. 7.
10. H. Tanaka, Y. Iwanaga, G. Wu, K. Sanui, and N. Ogata, *Polym. J.*, **14**, 648 (1982).
11. F. Higashi, A. Hoshio, Y. Yamada, and M. Ozawa, *J. Polym. Sci., Polym. Chem. Ed.*, **23**, 69 (1985).
12. F. Higashi, N. Akiyama, I. Takahashi, and T. Koyama, *J. Polym. Sci., Poly. Chem. Ed.*, **22**, 1653 (1984).
13. A. G. Pinkus and R. Subramanyam, *J. Polym. Sci., Polym. Chem. Ed.*, **22**, 1131 (1984).
14. Y. Imai and M. Ueda, in *Crown Ethers and Phase Transfer Catalysis in Polymer Science* (L. J. Mathias and C. E. Carraher, Jr., Eds.), Plenum, New York, 1984, pp. 121–137.
15. A. Noshay and J. E. McGrath, *Block Copolymers: Overview and Critical Survey*, Academic, New York, 1977, Chap. 7.
16. R. E. Wilfong, *J. Polym. Sci.*, **54**, 385 (1961).
17. G. F. L. Ehlers, R. C. Evers, and K. R. Fisch, *J. Polym. Sci., A-1*, **7**, 3413 (1969).
18. S. Z. D. Cheng and B. Wunderlich, *Macromolecules*, **21**, 789 (1988).
19. W. J. Jackson Jr., *Br. Polym. J.*, **12**, 154 (1980).
20. W. F. Christopher and D. W. Fox, *Polycarbonates*, Reinhold, New York, 1962.
21. H. Schnell, *Chemistry and Physics of Polycarbonates*, Wiley-Interscience, New York, 1964.
22. H. Vernaleken, in *Interfacial Synthesis*, Vol. 2 (F. Millich and C. E. Carraher, Jr., Eds.), Marcel Dekker, New York, 1977, Chap. 13.
23. S. K. Sikdar, *Chemtech.*, **17**, 112 (1987).
24. B. D. Hanrahan, S. R. Angeli, and J. Runt, *Polym. Bull.*, **14**, 399 (1985).
25. M. Kobayashi, S. Inoue, and T. Tsuruta, *J. Polym. Sci., Polym. Chem. Ed.*, **11**, 2383 (1973).
26. R. Rokicki and W. Kuran, *J. Macromol. Sci., Rev. Macromol. Chem.*, **C21**, 135 (1981).
27. S. Inoue, *Chemtech*, **6**, 588 (1976).
28. S. Oi, K. Nemoto, S. Matsumo, and Y. Inoue, *Macromol. Rapid Commun.*, **15**, 133 (1994).
29. K. Soga, S. Hosoda, and S. Ikeda, *J. Polym. Sci., Polym. Chem. Ed.*, **17**, 517 (1979).
30. G. Rokicki, W. Kuran, and J. Kieliewicz, *Makromol. Chem.*, **180**, 2779 (1979).
31. G. L. Brode and J. V. Koleske, in *Polymerizations of Heterocyclics*, (O. Vogl and J. Furukawa, Eds.), Marcel Dekker, New York, 1973, pp. 117–152.
32. J. O. Brash and D. J. Lyman, in *Cyclic Monomers*, (K. C. Frisch, Ed.), Wiley-Interscience, New York, 1972, Chap. 2.
33. R. D. Lundberg and E. F. Cox, in *Ring-Opening Polymerization* (K. C. Frisch and S. L. Reegen, Eds.), Marcel Dekker, New York, 1969, Chap. 6.
34. M. Morton and M. Wu, in *Ring-Opening Polymerization* (J. E. McGrath, Ed.), ACS Symp. Ser. 286, American Chemical Society, Washington, D.C., 1985, pp. 175–82.
35. S. Inoue and T. Aida, in *Ring-Opening Polymerization: Mechanism, Catalysis Structure, Utility* (D. J. Brunelle, Ed.), Hanser, New York, 1993, Chap. 6.
36. S. Inoue, *Makromol. Chem., Macromol. Symp.*, **3**, 295 (1986).
37. R. D. Lundberg, J. V. Koleske, and K. B. Wischmann, *J. Polym. Sci. A-1*, **7**, 2915 (1969).
38. H. J. Sanders, *Chem. Eng. News*, **63**(13), 30 (1985).
39. K. S. Devi and P. Vasudevan, *J. Macromol. Sci., Rev. Macromol. Chem. Phys.*, **C25**, 315 (1985).
40. S. W. Shalaby and R. A. Johnson, in *Biomedical Polymers: Designed-to-Degrade-Systems* (S. W. Shalaby, Ed.), Hanser, New York, 1994, Chap. 1.

41. M. Chasin and R. Langer (Eds.), *Biodegradable Polymers as Drug Delivery Systems*, Marcel Dekker, New York, 1990.
42. R. G. Sinclair, *J. Macromol. Sci.—Pure Appl. Chem.*, **A33**, 585 (1996).
43. W. J. Bailey et al., *J. Macromol. Sci.-Chem.*, **A25**, 781 (1988).
44. W. J. Bailey, in *Ring-Opening Polymerization* (J. E. McGrath, Ed.), ACS Symp. Ser. 286, American Chemical Society, Washington, D.C., 1985, pp. 47–65.
45. J. V. Crivello, R. Malik, and Y.-L. Lai, *J. Polym. Sci., Part A: Polym. Chem.*, **34**, 3103 (1996).
46. T. Endo and T. Yokozawa, in *New Methods for Polymer Synthesis* (W. J. Mijs, Ed.), Plenum, New York, 1992, Chap. 5.
47. D. J. Brunelle and T. G. Shannon, *Macromolecules*, **24**, 3035 (1991).
48. D. J. Brunelle, in *Ring-Opening Polymerization: Mechanism Catalysis Structure, Utility* (D. J. Brunelle, Ed.), Hanser, New York, 1993, Chap. 11.
49. D. J. Brunelle, T. L. Evans, T. G. Shannon, E. P. Boden, K. R. Steward, L. P. Fontana, and K. D. Bonauto, *Polym. Prepr.*, **30**(2), 569 (1989).
50. H. Jiang, T. Chen, and J. Xu, *Macromolecules*, **30**, 2839 (1997).
51. M. Lemoigne, *Ann. Inst. Pasteur (Paris)*, **39**, 144 (1925); *Bull. Soc. Chim. Biol.*, **8**, 770 (1926).
52. Y. Doi, *Microbial Polyesters*, VCH, New York, 1990.
53. P. J. Hocking and R. H. Marchessault, in *Chemistry and Technology of Biodegradable Polymers* (G. J. L. Griffin, Ed.), Blackie, London, 1994, Chap. 4.
54. R. H. Marchessault, *Trends Polym. Sci.*, **4**, 169 (1996).
55. P. Kurcok, M. Kowalczuk, G. Adamus, Z. Jedlinski, and R. W. Lenz, in *Degradable Polymers, Recycling, and Plastics Waste Management* (A.-C. Albertsson and S. J. Huang, Eds.), Marcel Dekker, New York, 1995.
56. R. W. Lenz, *Adv. Polym. Sci.*, **107**, 1 (1993).
57. M. K. Cox, *J. Macromol. Sci.—Pure Appl. Chem.*, **A32**, 607 (1995).
58. R. Sharma and A. R. Ray, *J. Macromol. Sci.-Rev. Macromol. Chem. Phys.*, **C35**, 327 (1995).
59. R. A. Gross, in *Biomedical Polymers: Designed-to-Degrade-Systems* (S. W. Shalaby, Ed.), Hanser, New York, 1994, Chap. 7.
60. B. Hazer, R. W. Lenz, and R. C. Fuller, *Polymer*, **37**, 5951 (1996).
61. S. F. Williams and O. P. Peoples, *Chemtech*, **26**(9), 38 (1996).
62. B. I. Voit, *Acta Polymerica*, **46**, 87 (1995).
63. M. Johansson, E. Malmoström, and A. Hult, *J. Polym. Sci., Part A: Polym. Chem.*, **31**, 619 (1993).
64. C. J. Hawker, R. Lee, and J. M. J. Fréchet, *J. Am. Chem. Soc.*, **113**, 4583 (1991).
65. H. R. Kricheldorf and O. Stöber, *Macromol. Rapid Commun.*, **15**, 87 (1994).
66. S. R. Turner, F. Walter, B. I. Voit, and T. H. Mourey, *Macromolecules*, **27**, 167 (1994).
67. C. J. Hawker and J. M. J. Fréchet, *J. Am. Chem. Soc.*, **112**, 7638 (1990).
68. C. R. Martens, *Alkyd Resins*, Reinhold, New York, 1961.
69. H. V. Boenig, *Unsaturated Polyesters: Structure and Properties*, Elsevier, Amsterdam, 1964.
70. B. Parkyn, F. Lamb, and B. V. Clitton, *Polyesters*, Vol. 2, *Unsaturated Polyesters and Polyester Plasticizers*, Elsevier, New York, 1967.
71. D. Alsheh and G. Marom, *J. Appl. Polym. Sci.*, **22**, 3177 (1978).
72. B. C. Trivedi and B. M. Culbertson, *Maleic Anhydride*, Plenum Press, New York, 1982, Chap. 12.
73. R. S. Lenk, *J. Appl. Polym. Sci.*, **15**, 2211 (1971).
74. Neth. Pat. Appl. 6,517,086 (to Farbenfabriken Bayer, A. G.), 1966 (*Chem. Abstr.*, **66**, 11427h, 1967).
75. I. J. Alexander and R. J. Scott, *Br. Polym. J.*, **15**, 30 (1983).
76. D. S. Sadafule and S. P. Panda, *J. Appl. Polym. Sci.*, **24**, 511 (1979).
77. K. H. Yoon, A. T. DiBenedetto, and S. J. Huang, *Polymer*, **38**, 2281 (1997).

Review Exercises

1. Write IUPAC names for the polyesters in Table 12.1.

2. Low-molecular-weight polyesters used in the manufacture of polyurethanes (Chapter 13) are poly(ethylene adipate), poly(butylene adipate), and poly(diethylene glycol adipate). Write the repeating units of the polymers.

3. Write equations for three ways to synthesize poly(hydroquinone terephthalate),

$$\left[\overset{\displaystyle O}{\underset{\displaystyle \|}{C}} - \bigcirc - \overset{\displaystyle O}{\underset{\displaystyle \|}{C}} - O - \bigcirc - O \right]$$

4. The nucleic acids DNA and RNA (see Chapter 18) are polyesters of phosphoric acid (i.e., polyphosphates). Write a reaction for a laboratory preparation of a polyphosphate.

5. Although γ-butyrolactone resists polymerization, the following five-membered ring lactones polymerize readily. Explain. (H. K. Hall, Jr., et al., *Macromolecules*, **4**, 142, 1971; *J. Am. Chem. Soc.*, **80**, 6412, 1958)

6. The following lactones undergo base-catalyzed polymerization to yield high-molecular-weight polyester. Write the repeating units, including end groups, when $^{18}OH^-$ is used as initiator. Why is chain transfer not a problem with these monomers? [R. P. J. G. Thiebaut et al., *Ind. Plastiques Mod.* (Paris), **14**(2), 13, 1962; *Chem. Abstr.*, **57**, 6118e, 1962; H. K. Hall, Jr., et al., *Macromolecules* **2**, 475, 1969]

7. Monomer **21** (reaction 12.36) undergoes some vinyl polymerization in competition with free radical ring-opening polymerization (W. J. Bailey et al., *J. Macromol. Sci.-Chem.*, **A21**, 1611, 1984). (a) Give the structure of the expected vinyl polymer. (b) If one of the ring hydrogens of **21** is replaced with phenyl, the monomer undergoes only ring-opening polymerization. Give a plausible explanation. (c) Under what other conditions might you expect polymerization to occur? Explain. (d) In principle, the polyester formed in reaction (12.36) could be prepared by step-reaction polymerization of 4-hydroxybutanoic acid. Can you foresee any difficulties with this approach? Explain.

8. Give the structure of the polymer expected from the anionic polymerization of

Write a mechanism for the polymerization. (ref. 21)

9. Metalloporphyrin initiators bring about the copolymerization of epoxides with carbon dioxide to form polycarbonates (see reaction 12.24). Additionally they initiate alternating copolymerization of epoxides with cyclic anhydrides (refs. 34, 36). What type of polymer would be formed in the latter case? Explain.

10. The bacterial polyester PHB (**29**) is isotactic. How might one synthesize atactic PHB? (ref. 58)

11. Write the repeating units expected from the following polymerization reactions:

(a) $HOCH_2 - \bigcirc - CO_2CH_3 \overset{\Delta}{\longrightarrow}$

(b)

(c)

(d)

(e)

(M. Okada et al., *Macromolecules*, **19**, 503, 1986)

(f)

(R. H. Hasek et al., *J. Org. Chem.*, **27**, 60, 1962)

(g)

(F. M. Houlihan et al., *Macromolecules*, **19**, 13, 1986)

(h)

(G. Schwarz and H. R. Kricheldorf, *Makromol. Chem., Rapid Commun.*, **9**, 717, 1988)

(i)

(A. L. Cimecioglu, et al., *J. Polym. Sci. A: Polym. Chem.*, **26**, 2129, 1988)

(j)

(J. M. Mikroyannidis, *Eur. Polym. J.*, **22**, 185, 1986)

(k) $CH_2{=}CHCH_2CH_2O\overset{O}{\overset{\|}{C}}{-}\langle\bigcirc\rangle{-}\overset{O}{\overset{\|}{C}}OCH_2CH_2CH{=}CH_2$ $\xrightarrow{\text{Mo catalyst}}$

(K. B. Wagener et al., *Macromolecules*, **25**, 3862, 1992)

(l) $\overset{S}{\overset{\diagup\diagdown}{CH_2{-}CH_2}} + CS_2 \xrightarrow{ZnEt_2{\cdot}H_2O}$ (Ref. 30)

(m) $\langle\text{cyclohexene oxide}\rangle O + CO_2 \xrightarrow{\text{Zn catalyst}}$

(M. Super et al., *Macromolecules*, **30**, 368, 1997)

(n) $\xleftarrow{\text{structure}} \xrightarrow[\Delta]{\text{Peroxide}}$

[W. J. Bailey and L.-L. Zhou, *Polym. Prepr.*, **30**(1), 195, 1989]

12. Referring to exercise 11(e), what stereochemistry would be expected in the polymer if ring opening involved (a) acyl–oxygen cleavage, and (b) alkyl–oxygen cleavage of the ester group?

13. The *t*-butyloxycarbonyl (BOC) group is commonly used to protect amines (e.g., in proteins) and alcohols. The BOC group may be removed by heating:

$$RX\overset{O}{\overset{\|}{C}}OC(CH_3)_3 \xrightarrow{\Delta} RXH + CO_2 + CH_2{=}C(CH_3)_2$$
$$(X = NH \text{ or } O)$$

   (a) Predict the outcome of heating the polymer prepared in exercise 11 (g).
   (b) Does this exercise suggest a possible alternate synthesis of poly(vinyl alcohol)? (See Chapter 9.) Explain. (Y. Yiang et al., *Polym. Bull.*, **17**, 1, 1987; H. W. Gibson and P. R. Kurek, *Polym. Commun.*, **28**, 97, 1987)

14. Polyester $\text{+CH}_2CH_2CO_2\text{+}$ has been synthesized by treating acrylic acid ($CH_2{=}CHCO_2H$) with potassium acetate (KOAc) in the presence of a crown ether catalyst (B. Yamada et al., *J. Polym. Sci., Polym. Lett. Ed.*, **14**, 277, 1976). (a) Write a plausible mechanism for the reaction. (b) Suggest two alternative syntheses.

15. A practical route to unsaturated polyesters has been developed that makes use of scrap poly(ethylene terephthalate) bottles. The method involves digestion of the bottles with propylene glycol followed by reaction with maleic anhydride. Write reactions for the process.

CHAPTER 13

POLYAMIDES AND RELATED POLYMERS

13.1 Introduction

Historically and commercially, polyamides[1-4] occupy an important place in the world of polymers. The early work of Carothers at du Pont dealt with both polyesters and polyamides, and emphasis on the latter led to the first commercial development of synthetic fibers, known commonly as nylons. Nylon fibers characteristically have good tensile strength and elasticity and are widely used in the manufacture of clothing, tire cord, carpeting, and rope. In plastic form polyamides are used in making such items as gears, piping, wire insulation, zippers, and brush bristles.

Table 13.1 lists the numerous polyamides for fibers or plastics applications that have been made available commercially. As impressive as the list is, however, it should be stressed that a large percentage of the total production (about 80%) comprises just two polymers, nylons 6 and 66, the remainder being reserved for more specialized applications. (Refer to Chapter 1 for the nylon numbering system.) Not all those listed enjoy worldwide production; nylons 7 and 9, for example, have been produced only in Russia, and nylon 8 has never gained a sound commercial foothold.

Nylons represent roughly 25% of total fibers production and about 40% of all engineering plastics. Of all the nylon polymers produced, between 80 and 90% are used as fibers.

Ordinary polyamides, having the functional group

$$\underset{\text{—C—NH—}}{\overset{\overset{\displaystyle O}{\parallel}}{}}$$

in the polymer backbone are covered in Sections 13.2 and 13.3. Section 13.4 describes polyureas, which, like polycarbonates, are derived from the simplest dibasic acid, carbonic acid. Natural polyamides (proteins) are discussed in Chapter 18.

$$\underset{\text{—NH—C—NH—}}{\overset{\overset{\displaystyle O}{\parallel}}{}} \qquad \text{Urea}$$

TABLE 13.1. Polyamides Developed for Commercial Use[a]

Structure	Generic and/or Common Name[b]	Type[c]
$[\text{NH(CH}_2)_5\text{C}]$	Nylon 6 (polycaprolactam)	F, P
$[\text{NH(CH}_2)_6\text{C}]$	Nylon 7 [poly(7-heptanamide)]	F, P
$[\text{NH(CH}_2)_7\text{C}]$	Nylon 8 (polycapryllactam)	F, P
$[\text{NH(CH}_2)_8\text{C}]$	Nylon 9 [poly(9-nonanamide)]	F
$[\text{NH(CH}_2)_{10}\text{C}]$	Nylon 11 [poly(11-undecanamide)]	P
$[\text{NH(CH}_2)_{11}\text{C}]$	Nylon 12 (polylauryllactam)	P
$[\text{NH(CH}_2)_4\text{NHC(CH}_2)_4\text{C}]$	Nylon 46 [poly(tetramethylene adipamide)]	F, P
$[\text{NH(CH}_2)_6\text{NHC(CH}_2)_4\text{C}]$	Nylon 66 [poly(hexamethylene adipamide)]	F, P
$[\text{NH(CH}_2)_6\text{NHC(CH}_2)_7\text{C}]$	Nylon 69 [poly(hexamethylene azelamide)]	P
$[\text{NH(CH}_2)_6\text{NHC(CH}_2)_8\text{C}]$	Nylon 610 [poly(hexamethylene sebacamide)]	P
$[\text{NH(CH}_2)_6\text{NHC(CH}_2)_{10}\text{C}]$	Nylon 612 [poly(hexamethylene dodecanediamide)]	P
$[\text{NH}\text{—cyclohexylene—CH}_2\text{—cyclohexylene—NHC(CH}_2)_{10}\text{C}]$	Poly(methylene-4,4'-dicyclohexylene dodecanediamide)	F
$[\text{NHCH}_2\text{—cyclohexylene—CH}_2\text{NHC(CH}_2)_6\text{C}]$	Poly(1,4-cyclohexylenedimethylene suberamide)	F
$[\text{NH—phenylene—NHC—phenylene—C}]$	Poly(*m*-phenylene isophthalamide)	F
$[\text{NH—phenylene—NHC—phenylene—C}]$	Poly(*p*-phenylene terephthalamide)	F

continued

TABLE 13.1. (continued)

Structure	Generic and/or Common Name[b]	Type[c]
$\left[\begin{array}{c} \overset{O}{\underset{\|}{C}}-\bigcirc-\overset{O}{\underset{\|}{C}}-NHCH_2\overset{CH_3}{\underset{\|}{CH}}CH_2\overset{CH_3}{\underset{\|}{\overset{\|}{C}}}CH_2CH_2NH \\ \qquad\qquad\qquad\qquad\overset{\|}{CH_3} \end{array}\right]$	Poly(2,4,4-trimethylhexamethylene terephthalamide)[d]	P

[a] Excluding copolymers.
[b] Nylons prepared from lactams are named accordingly; others are prepared by polycondensation.
[c] F = fiber. P = plastic.
[d] The 2,2,4-trimethyl isomer is also used.

A number of polymers related in structure to polyamides are discussed in Sections 13.5–13.7, including polyhydrazides, polyurethanes, and polyimides. The characteristic functional groups of these polymers are as follows:

$$-\overset{O}{\underset{\|}{C}}-NH-NH- \qquad \text{Hydrazide}$$

$$-O-\overset{O}{\underset{\|}{C}}-NH- \qquad \text{Urethane}$$

$$-\overset{O}{\underset{\|}{C}}-N-\overset{O}{\underset{\|}{C}}- \qquad \text{Imide}$$

13.2   Polyamides

13.2.1   Preparation of Polyamides by Polycondensation Reactions

There are three condensation reactions employed in synthesis of amides: dehydration of amine salts of carboxylic acids (13.1), the reaction of acid chlorides with amines (13.2), and, less commonly, aminolysis of esters (13.3). All three methods have been extended to polyamide synthesis.

$$R-\overset{O}{\underset{\|}{C}}-O^-\overset{+}{N}H_2R_2' \xrightarrow{\text{heat}} R-\overset{O}{\underset{\|}{C}}-NR_2' + H_2O \qquad (13.1)$$

$$R-\overset{O}{\underset{\|}{C}}-Cl + R_2'NH \longrightarrow R-\overset{O}{\underset{\|}{C}}-NR_2' + HCl \qquad (13.2)$$

$$R-\overset{\overset{\displaystyle O}{\|}}{C}-OR'' + R_2'NH \longrightarrow R-\overset{\overset{\displaystyle O}{\|}}{C}-NR_2' + R''OH \qquad (13.3)$$

Direct amidation between dibasic acid and diamine requires the usual stoichiometric balance, but since the reaction proceeds via the intermediate ammonium salt, purification of the salt guarantees this balance. Nylon 66 (**1**, $x = 4$, $y = 6$), originally developed by Carothers, is manufactured this way (13.4), as is the more recently developed nylon 46[5] (**1**, $x = 4$, $y = 4$). Polymerization temperatures in the range of 260 to 280°C are used to obtain practical molecular weights. This method is also applied to amino acids as long as the two functional groups are separated by at least six carbon atoms to prevent cyclization to lactam. An example is the formation of nylon 11 (**2**) from 11-aminoundecanoic acid (13.5).* Nylons 7 and 9 are made similarly.

$$HO_2C(CH_2)_xCO_2H + H_2N(CH_2)_yNH_2 \longrightarrow$$

$$[^-O_2C(CH_2)_xCO_2^-][H_3\overset{+}{N}(CH_2)_y\overset{+}{N}H_3] \xrightarrow{\text{heat}}$$

$$\left[\overset{\overset{\displaystyle O}{\|}}{C}(CH_2)_x\overset{\overset{\displaystyle O}{\|}}{C}NH(CH_2)_yNH\right] + 2H_2O$$

1

$$(13.4)$$

$$H_2N(CH_2)_{10}CO_2H \xrightarrow{\text{heat}} \left[NH(CH_2)_{10}\overset{\overset{\displaystyle O}{\|}}{C}\right] + H_2O \qquad (13.5)$$

2

Low-temperature solution methods have been developed to prepare polyamides from diacids and diamines or from amino acids, among them the important *phosphorylation* reaction[6] (13.6). It has been proposed that an intermediate phosphorus complex (**3**) is formed, which undergoes rapid displacement by amine. The more basic aliphatic diamines are not suitable for this reaction because they displace pyridine in the complex. As was the case with the activated synthesis of polyesters (Chapter 12), such reactions are not practical for large-

---

*Nylon 11 is produced by the French company Aquitaine-Organico. The monomer, 11-aminoundecanoic acid, is prepared by an interesting process that begins with castor oil. Ricinoleic acid, obtained by hydrolysis of the oil, is thermally degraded to undecylenic acid; the latter is converted to 11-bromoundecanoic acid by free radical hydrobromination, and this, in turn, is converted to the corresponding amino acid with ammonia.

$$\underset{\underset{\displaystyle |}{\overset{\displaystyle OH}{}}}{CH_3(CH_2)_5CHCH}{=}CH(CH_2)_7CO_2H \xrightarrow{\text{heat}} CH_2{=}CH(CH_2)_8CO_2H + C_6H_{13}CHO$$

$$\downarrow \text{HBr, peroxide}$$

$$H_2N(CH_2)_{10}CO_2H \xleftarrow{\text{NH}_3} BrCH_2CH_2(CH_2)_8CO_2H$$

The polymer is sold under the trade name Rilsan 11.

scale industrial production. Microwave radiation has also been used to increase polyamidation rates.[7]

$$HO_2C—Ar—CO_2H + H_2N—Ar'—NH_2 \xrightarrow[\text{LiCl}]{\text{(PhO)}_3\text{P, pyridine}}$$ (13.6)

$$\left[\overset{\overset{\text{O}}{\|}}{C}—Ar—\overset{\overset{\text{O}}{\|}}{C}—NH—Ar'—NH \right] + PhOH + HOP(OPh)_2$$

$$\text{NP(OPh)}_2\text{OCArCOP(OPh)}_2\text{N}$$

3

Although the acid chloride route (13.2) is most versatile for preparing polyamides,[8,9] cost considerations have limited its use primarily to aromatic polyamides,[10–13] or *aramids*, as their fibers are called. Both solution and interfacial techniques are used. A case in point is the synthesis (13.7) of poly(*m*-phenylene isophthalamide) (**4**) (marketed under the du Pont trade name Nomex). It may be recalled from Chapter 3 (Section 3.7) that the *para,para* isomer of **4** (du Pont's Kevlar) is spun into extremely high-modulus fibers from a lyotropic liquid crystalline solution in sulfuric acid.

$$H_2N \underset{+}{\bigcirc} NH_2 \quad Cl—\overset{\overset{\text{O}}{\|}}{C} \bigcirc \overset{\overset{\text{O}}{\|}}{C}—Cl \longrightarrow$$ (13.7)

$$\left[-NH \bigcirc NH\overset{\overset{\text{O}}{\|}}{C} \bigcirc \overset{\overset{\text{O}}{\|}}{C} - \right] + 2HCl$$

4

Polysulfonamides and polyphosphonamides are also synthesized from the corresponding acid chlorides,[8] as shown for poly(hexamethylene-*m*-benzenedisulfonamide) (**5**) (13.8) and poly(hexamethylene phosphonamide) (**6**) (13.9). In general, the polyphosphonamide synthesis is less successful because of differences in reactivity between the two chlorine atoms and because of a tendency to form cyclic structures.

$$H_2N(CH_2)_6NH_2 + ClO_2S \bigcirc SO_2Cl \longrightarrow$$ (13.8)

$$\left[NH(CH_2)_6NHSO_2 \bigcirc SO_2 \right] + 2HCl$$

5

$$H_2N(CH_2)_6NH_2 + Cl\overset{\overset{O}{\|}}{-P-}Cl \longrightarrow \left[\overset{\overset{O}{\|}}{-P-}NH(CH_2)_6NH-\right] + 2HCl \quad (13.9)$$

6

The aminolysis reaction (13.3) is not used as frequently to make polyamides, although the commercially available poly(2,4,4-trimethylhexamethylene terephthalamide) and its 2,2,4 isomer (see Table 13.1) is synthesized by the reaction of dimethyl terephthalate with a mixture of 2,4,4- and 2,2,4-trimethyl-1,6-hexanediamine.

13.2.2 Polymerization of Lactams

Ring-opening polymerization of lactams[14,15] is a commercially important process, more so than the analogous polymerization of lactones to give polyesters. Most important is the polymerization of caprolactam (**7**) (13.10) to give nylon 6. The reaction can be brought about with cationic or anionic initiators or with water.

$$\overset{O}{\underset{NH}{\bigcirc}} \longrightarrow \left[\overset{\overset{O}{\|}}{-C(CH_2)_5NH-}\right] \quad (13.10)$$

7

Cationic polymerization[16,17] with acid, HA, is believed to involve initial protonation to give both O-protonated and N-protonated amide (13.11), mainly the former because of resonance stabilization. Nucleophilic attack by monomer on the more reactive N-protonated acid (13.12), followed by successive ring-opening acylation of the primary amine function, results in formation of polyamide (13.13).

$$\mathbf{7} + HA \rightleftharpoons \left(\overset{\overset{+}{O}H}{\underset{NH}{\bigcirc}} \longleftrightarrow \overset{OH}{\underset{\overset{NH}{+}}{\bigcirc}}\right) \rightleftharpoons \overset{O}{\underset{\overset{NH_2}{+}}{\bigcirc}} \quad (13.11)$$

$$\mathbf{7} + \overset{O}{\underset{\overset{NH_2}{+}}{\bigcirc}} \longrightarrow \overset{O}{\underset{N}{\bigcirc}}-\overset{\overset{O}{\|}}{C(CH_2)_5\overset{+}{N}H_3} \quad (13.12)$$

$$\overset{O}{\underset{N}{\bigcirc}}-\overset{\overset{O}{\|}}{C(CH_2)_5\overset{+}{N}H_3} + n \ \mathbf{7} \longrightarrow$$

$$\overset{O}{\underset{N}{\bigcirc}}-\left[\overset{\overset{O}{\|}}{C(CH_2)_5NH}\right]_{n-1}-\overset{\overset{O}{\|}}{C(CH_2)_5\overset{+}{N}H_3} \quad (13.13)$$

Generally, high molecular weights are not obtainable under cationic conditions, possibly because of intramolecular cyclization reactions involving terminal amine groups (for example, 13.14).

$$
\text{wwwNH}\overset{\overset{\displaystyle O}{\|}}{C}(CH_2)_5\overset{+}{N}H_3 \longrightarrow \text{wwwNHC} \Big\langle \begin{array}{c} \overset{+}{N}H-CH_2 \\ \\ CH_2-CH_2 \end{array} \Big\rangle \begin{array}{c} CH_2 \\ | \\ CH_2 \end{array} + H_2O \qquad (13.14)
$$

Strong bases such as sodium hydride, sodium amide, or alkali metals are effective anionic initiators. The mechanism, using base B:⁻, involves initial formation of resonance-stabilized anion (13.15) which reacts with monomer in a slow step (13.16), slow because the new anion is not stabilized by resonance. Once formed, however, it undergoes rapid proton transfer (13.17) and propagation (13.18) by successive formation of resonance-stabilized anions.

$$
\mathbf{7 + B:^-} \; \rightleftharpoons \; \left(\; \bigcirc\!\!\!\overset{O}{\underset{N:\_}{}} \; \longleftrightarrow \; \bigcirc\!\!\!\overset{O^-}{\underset{N}{}} \; \right) + BH \qquad (13.15)
$$

$$
\bigcirc\!\!\!\overset{O}{\underset{N:^-}{}} + \mathbf{7} \; \xrightarrow{\text{slow}} \; \bigcirc\!\!\!\overset{O}{\underset{N}{}}\!\!-\!\!\overset{\overset{\displaystyle O}{\|}}{C}(CH_2)_5\overset{..}{N}H \qquad (13.16)
$$

$$
\overset{\mathbf{7}}{\underset{\text{fast}}{\rightleftharpoons}} \; \bigcirc\!\!\!\overset{O}{\underset{N}{}}\!\!-\!\!\overset{\overset{\displaystyle O}{\|}}{C}(CH_2)_5NH_2 \; + \; \bigcirc\!\!\!\overset{O}{\underset{N:^-}{}} \qquad (13.17)
$$

$$
\xrightarrow{\text{fast}} \; \bigcirc\!\!\!\overset{O}{\underset{N}{}}\!\!-\!\!\overset{\overset{\displaystyle O}{\|}}{C}(CH_2)_5\overset{..}{N}\!\!-\!\!\overset{\overset{\displaystyle O}{\|}}{C}(CH_2)_5NH_2 \qquad (13.18)
$$

Because the second step is slow, this type of ring-opening polymerization is characterized by an induction period. The induction period can, however, be eliminated by addition of coinitiators (called *activators*) such as acyl halides or anhydrides that ensure formation of a stabilized anion in the second step. This is illustrated in reactions (13.19) and (13.20) using an acid chloride.

$$
\bigcirc\!\!\!\overset{O}{\underset{N:^-}{}} + R\!\!-\!\!\overset{\overset{\displaystyle O}{\|}}{C}\!\!-\!\!Cl \longrightarrow \bigcirc\!\!\!\overset{O}{\underset{N}{}}\!\!-\!\!\overset{\overset{\displaystyle O}{\|}}{C}\!\!-\!\!R + Cl^- \qquad (13.19)
$$

$$
\bigcirc\!\!\!\overset{O}{\underset{N}{}}\!\!-\!\!\overset{\overset{\displaystyle O}{\|}}{C}\!\!-\!\!R + \bigcirc\!\!\!\overset{O}{\underset{N:^-}{}} \longrightarrow \bigcirc\!\!\!\overset{O}{\underset{N}{}}\!\!-\!\!\overset{\overset{\displaystyle O}{\|}}{C}(CH_2)_5\overset{..}{N}\overset{\overset{\displaystyle O}{\|}}{C}R \qquad (13.20)
$$

One of the difficulties associated with anionic lactam polymerization is that proton transfer can occur with the polymer chain, particularly at low monomer concentrations. This results in chain branching and a polydisperse polymer.

Water is also used to initiate polymerization, a process referred to as *hydrolytic polymerization*. It is believed that initially hydrolysis occurs to form amino acid (13.21) which reacts with lactam by ring cleavage (13.22). Although some polymerization undoubtedly occurs through the free amino and carboxyl groups of the intermediate amino acid, the primary propagation process is ring opening.

$$7 + H_2O \longrightarrow HO_2C(CH_2)_5NH_2 \tag{13.21}$$

$$HO_2C(CH_2)_5NH_2 + \text{(lactam)} \longrightarrow HO_2C(CH_2)_5NHC(CH_2)_5NH_2 \tag{13.22}$$

Besides nylon 6 from anionic or hydrolytic polymerization of caprolactam,* nylon 8 and nylon 12 are made by ring-opening polymerization of capryllactam (**8**) and lauryllactam (**9**).† The five-membered butyrolactam (**10**) (commonly called 2-pyrrolidone) polymerizes slowly under the usual anionic conditions to yield nylon 4; however, nylon 4 having $\overline{M}_w$ greater than 1 million is formed rapidly under conditions of phase-transfer catalysis.[18] Valerolactam (**11**) (also called α-piperidone) forms relatively low-molecular-weight nylon 5 slowly under basic conditions. Both nylons 4 and 5 revert to monomer at elevated temperatures.

Some other ring-opening reactions related to lactam polymerization include the base-catalyzed reaction of N-carboxy-α-amino acid anhydrides[19–21] (**12**) (prepared from α-amino acid and phosgene) to give the corresponding nylon 2 (**13**) (13.23), and the preparation of a polysulfonamide **15** from propiosultam (**14**) (13.24).[22] The former reaction (13.23) is of par-

---

*Caprolactam is made cheaply from phenol by the following reaction sequence:

The last step is an example of a Beckmann rearrangement. Other methods involve converting cyclohexanone to caprolactone using peracetic acid, followed by reaction of the cyclic ester with ammonia to give lactam; forming cyclohexanone oxime directly from cyclohexane by a light-catalyzed process; and forming caprolactam by an acetyl exchange reaction between cyclohexanone oxime and acetyl caprolactam.

†Capryllactam and lauryllactam are made from cyclooctanone oxime and cyclododecanone oxime, respectively. Lauryllactam is produced by the French company Aquitaine-Organico, which markets nylon 12 under the trade name Rilsan 12. The Japanese firm, Mitsubishi Petrochemical Company, produces nylon 12 by a similar route.

ticular interest to protein chemists because the anhydrides, when prepared from naturally occurring α-amino acids, yield useful models for protein studies.

$$\underset{\textbf{12}}{\begin{matrix} HN-C \diagup^O \\ | \qquad\diagdown O \\ R_2C-C\diagdown \\ \qquad\quad {}^O \end{matrix}} \xrightarrow{\text{base}} \underset{\textbf{13}}{\left[\begin{matrix} O \quad R \\ \| \quad | \\ C-C-NH \\ \qquad | \\ \qquad R \end{matrix} \right]} + CO_2 \qquad (13.23)$$

$$\underset{\textbf{14}}{\left[\begin{matrix} \diagdown_N{}^{\diagup SO_2} \\ \ | \\ \ H \end{matrix} \right]} \xrightarrow{\text{base}} \underset{\textbf{15}}{-\!\!\left[SO_2CH_2CH_2CH_2NH\right]\!\!-} \qquad (13.24)$$

13.2.3 Miscellaneous Methods of Preparing Polyamides

Besides the polycondensation and ring-opening methods described above, a variety of other
routes to polyamides have been described, several of which are given in Table 13.2. One of
the most useful is the reaction of diisocyanates with dicarboxylic acids[23] (13.25); it is the only
one of those listed that enjoys widespread commercial use. The carboxylic acid adds across
the carbon–nitrogen double bond to give initially an unstable polymeric anhydride (**16**) which

$$\underset{\textbf{16}}{\left[\begin{matrix} O & & & O & & O & & O \\ \| & & & \| & & \| & & \| \\ C-NH-R-NH-\!\!\!& C-O-& C-R'-& C-O \end{matrix} \right]}$$

undergoes decarboxylation to form polyamide. If diisothiocyanate, SCN—R—NCS, is used
instead of diisocyanate, carbon oxysulfide is evolved; hence loss of carbon dioxide (or car-

TABLE 13.2. Miscellaneous Methods of Synthesizing Polyamides

Reaction	Number
OCN—R—NCO + HO$_2$C—R'—CO$_2$H $\longrightarrow$	
$\left[\text{NH-R-NH-} \overset{\displaystyle O}{\overset{\displaystyle \|}{C}} \text{-R'-NH} \right] + 2CO_2$	(13.25)
OCN—R—NCO + CH$_2$=C(NR'$_2$)$_2$ $\longrightarrow$ $\left[\overset{O}{\overset{\|}{C}}\text{NH-R-NH}\overset{O}{\overset{\|}{C}}\text{-C}\!-\!\!\underset{\displaystyle C(NR'_2)_2}{} \right]$	(13.26)
NC—R—CN + HO—R'—OH^a $\longrightarrow$ $\left[\text{NH-}\overset{O}{\overset{\|}{C}}\text{-R-}\overset{O}{\overset{\|}{C}}\text{-NH-R'} \right]$	(13.27)
NC—R—CN + CH$_2$O^b $\longrightarrow$ $\left[\text{NH-}\overset{O}{\overset{\|}{C}}\text{-R-}\overset{O}{\overset{\|}{C}}\text{-NH-CH}_2 \right]$	(13.28)
	continued

TABLE 13.2. (continued)

Reaction	Number

$$RN{=}C{=}O \xrightarrow{\text{base}^c} \left[\begin{array}{cc} R & O \\ | & \| \\ N & C \end{array}\right]$$ (13.29)

$$CH_2{=}CHCNH_2 \xrightarrow{\text{base}} \left[CH_2CH_2CNH\right]$$ (13.30)

$$H_2N{-}R{-}NH_2 + ArCOC{-}R'{-}COAr^d \longrightarrow \left[C{-}R'{-}C{-}NH{-}R{-}NH\right] + 2ArCO_2H$$ (13.31)

$$I{-}Ar{-}I + H_2N{-}Ar'{-}NH_2 + 2CO \xrightarrow[R_3N]{\text{Pd catalyst}} \left[C{-}Ar{-}C{-}NH{-}Ar'{-}NH\right]$$ (13.32)

---

[a] Both OH groups are tertiary.
[b] Trioxane is used as the source of CH_2O.
[c] Typical bases: RLi, ROLi, RNa, NaCN. Crystalline polymer is obtained at $-80°C$.
[d] Ar = 2,4,6-trimethylphenyl.

bon oxysulfide) must involve the carbon attached to nitrogen. Use is made of this reaction in generating foams by treating carboxyl-terminated low-molecular-weight polyesters with diisocyanates such as 2,4-toluenediisocyanate (**17**) (13.33). The evolved carbon dioxide is trapped as bubbles in the viscous medium, and thereby produces the foam. Once set, such foams are flexible and find use in cushions and other upholstery. Sometimes they are referred to erroneously as "urethane" foams. More is said on this subject later in the chapter.

$$HO_2C\text{\wwwww}CO_2H + \text{ } \underset{\textbf{17}}{\text{[OCN—NCO, CH}_3\text{]}} \longrightarrow$$

polyester

$$\left[\underset{\text{CH}_3}{C{-}NH{-}\text{—}{-}NH{-}C}\right] + 2CO_2$$ (13.33)

The second reaction (13.26) involves a nucleophilic addition to the isocyanate to yield a product, **18**, which is capable of further addition by the same mechanism.[24] The nucleophilic monomer (a *ketene aminal*) behaves like an enamine, but is highly reactive by virtue of having two nitrogen atoms to increase the nucleophilicity of the double bond.

$$\begin{array}{c} R_2'N \\ \diagdown \\ C{=}CH{-}C{-}NH{-}R{-}NCO \\ \diagup \\ R_2'N \end{array}$$

18

The acid-catalyzed polyaddition of tertiary diols to dinitriles (13.27) (an extension of what is known as the Ritter reaction)[8] provides a route to polyamides having substituents on the carbon adjacent to the amide nitrogen. The mechanism involves addition of an alcohol-derived carbocation to the carbon–nitrogen triple bond, hence any species capable of generating a carbocation (alcohol, alkene, aldehyde, alkyl halide) can in principle undergo this reaction. The following reaction (13.28) is a case in point. Reaction (13.28) leads to relatively high-molecular-weight polyamide of the inherently unstable diaminomethane.

Nylon 1 derivatives are obtained by the base-catalyzed homopolymerization of isocyanates (13.29).[25–27] Very-high-molecular-weight crystalline polymers are obtainable by this reaction; however, the polymers tend to depolymerize at elevated temperatures.

Nylon 3 is obtained by the base-catalyzed step-growth addition polymerization of polyacrylamide (13.30).[28,29] This Michael addition reaction is analogous to the formation of polyester from acrylic acid (see Chapter 12, exercise 14). The polycondensation of diamine with dianhydride (13.31) is related to the acid chloride synthesis, and is likewise adaptable to both solution and interfacial techniques.[30] The last reaction in Table 13.2 (13.32) represents a novel and potentially attractive carbonylation route to aromatic polyamides.[21,22]

13.3 Properties of Polyamides

Like polyesters, polamides lend themselves to a wide variety of structural variations. They melt at much higher temperatures than polyesters, however, because of intermolecular hydrogen bonding. Nylon 66, for example, melts at about 265°C, whereas the corresponding poly(hexamethylene adipate) melts below 60°C. Table 13.3 lists the approximate melting

TABLE 13.3. Approximate Melting Points of Aliphatic Polyamides[a]

Nylon	Melting Point (°C)	Nylon	Melting Point (°C)
3	>320	46	295
4	265	56	230
5	250–270	66	265
6	233	76	220
7	233	68	240
8	200	69	205
9	188	410	236
10	188	510	186
11	190	610	225
12	179	612	212

[a]Data from J. Zimmerman, in *Encyclopedia of Polymer Science and Engineering*, Vol. 2 (H. F. Mark, N. M. Bikales, C. G. Overberger, G. Menges, and J. I. Kroschwitz, Eds.), Wiley-Interscience, New York, 1988, pp. 515–581.

TABLE 13.4. Approximate Melting Temperatures of Terephthalamides[a]

Diamine	Melting Temperature (°C)
Ethylene	455
Trimethylene	399
Tetramethylene	436
Pentamethylene	353
Hexamethylene	371
Heptamethylene	341

[a] Data from Frazer.[33]

points of several linear aliphatic polyamides, the first set representing those having one amide group per repeating unit, and the second set, two. Both series exhibit an odd–even relationship typical of homologous series. Polyamides having an even number of carbon atoms between amide groups tend to be higher melting, a phenomenon reflecting the better packing ability within the crystalline lattice. Similar behavior is seen with the partially aromatic poly(alkylene terephthalamides) in Table 13.4. Fully aromatic polyamides[10,11,33] containing *para-* or *meta-*phenylene linkages do not melt below 400°C, whereas those having *ortho* arrangements are considerably lower melting. The *ortho-*substituted polymers are also significantly less thermally stable, exhibiting up to 90% weight loss between 200 and 300°C, compared with the *m-* and *p-*substituted polymers, which are stable up to 400°C.[34]

A major problem with wholly aromatic polyamides having high softening temperatures (particularly those prepared with *p-*phenylene units) is their low solubility and resultant difficulty in fabrication. One approach to solving this has been the development of highly regular copolyamides, referred to as *ordered copolyamides*,[35] which are synthesized by treating diamine monomers containing one type of repeating unit with diacid chlorides of another. Reaction (13.34), for example, illustrates the formation of an ordered copolyamide containing *m-*benzamide and isophthalamide units. The corresponding random copolymer has a melting point (about 300°C) about 100°C lower than that of the ordered copolymer. Compared with polyamide **4** or its *para* analog, the ordered copolyamides are generally more soluble and less crystalline. Films and fibers of ordered copolyamides can be oriented, however, to induce greater crystallinity.

(13.34)

Substitutent groups on the polymer backbone reduce the melting point of polyamides significantly. Nylon 66 containing α-methyl groups on the carboxyl portion of the polymer melts at about 166°C, compared with 265°C for the unsubstituted polymer. Also, substituting about 50% of the hydrogen atoms attached to nitrogen with isobutyl groups reduces the melting point of nylon 610 more than 70°C. If the substituent groups are sufficiently bulky, the polymers are amorphous.

Substituent groups can be introduced at the nitrogen atoms by such reactions as methoxymethylation (13.35), alkylation (13.36), and polyetherification (13.37). Methoxymethylation of 36% of the nitrogens reduces the melting point of nylon 66 by about 115°C and increases its solubility. Considerable work has been done in these areas, as well as with radiation grafting, to improve flexibility or solubility and, in the case of fibers, elasticity and wash-and-wear properties.[36,37] Random copolymerization is another strategy employed commercially for reducing crystallinity.

$$
\text{www NHC www} \quad
\begin{cases}
\xrightarrow[\text{CH}_3\text{OH}]{\text{CH}_2\text{O, OH}^-} \quad \text{www N}\!-\!\text{C www} \quad \text{CH}_2\text{OCH}_3 & (13.35) \\[2em]
\xrightarrow[\text{2. RBr}]{\text{1. NaH}} \quad \text{www N}\!-\!\text{C www} \quad \text{R} & (13.36) \\[2em]
\xrightarrow{\text{CH}_2\!-\!\text{CH}_2} \quad \text{www N}\!-\!\text{C www} \quad [\text{CH}_2\text{CH}_2\text{O}]\!-\!\text{H} & (13.37)
\end{cases}
$$

As was the case with aliphatic polyesters, biodegradability and biocompatibility are properties that have been incorporated into polyamides, with particular emphasis in the biomedical areas.[38,39] If natural α-amino acids are introduced into a nylon 66 formulation, for example, the polymer becomes biodegradable.[40] Another interesting development is the combination of degradable polyester and polyamide sequences in a graftable copolymer of L-lactic acid and L-glycine (**19**) that has potential as an artificial mammalian tissue matrix.[41]

$$
\underset{\text{CH}_3}{\text{www OCHC www}} \quad \underset{(\text{CH}_2)_4\text{NH}_2}{\text{www NHCHC www}}
$$

19

Another important characteristic of polyamides is a propensity for absorbing moisture from their immediate environment, which generally has a plasticizing effect.[42] Because properties may thus vary with climate, moisture absorption must be taken into account in commercial applications. This is of less concern with the more hydrophobic polyamides such as nylons 11 and 12.

13.4 Polyureas

Polyureas[8] are polyamides of carbonic acid, just as polycarbonates are polyesters of carbonic acid. Like the more conventional polyamides, they are tough, high melting, and suitable for elastomer and fiber applications. They are usually higher melting than polyamides of like structure.

The best method of synthesizing polyureas is to react diamines with diisocyanates. This is an exothermic reaction that can be done in solution or by the interfacial technique. The reaction of 2,4-toluenediisocyanate (**17**) with 4,4′-diaminodiphenyl ether (**19**) (13.38), for example, gives a polyurea **20** melting at about 320°C. Because it is a step-growth addition reaction (addition of amine across the carbon–nitrogen double bond), there is no byproduct, which simplifies the polymerization process.

(13.38)

The reaction of diamines with phosgene, analogous to the polycarbonate synthesis with diols and phosgene, is another way to make polyureas. Polyhexamethyleneurea (**21**), prepared interfacially from 1,6-hexanediamine and phosgene (13.39), melts at about 295°C, which is significantly higher than nylon 6 (215°C) or nylon 66 (265°C), which have comparable spacing between the amide groups.

$$H_2N(CH_2)_6NH_2 + Cl-\overset{O}{\overset{\|}{C}}-Cl \longrightarrow \overline{}NH(CH_2)_6NH\overset{O}{\overset{\|}{C}}\overline{} \qquad (13.39)$$

21

Miscellaneous methods of polyurea synthesis, shown in Table 13.5, include the reaction of diamines with carbonate esters[8] (13.40); with biscarbamyl chlorides[6] (13.41); and with carbon dioxide[42] (13.42).

Despite their generally good properties, economic factors have limited the commercial development of polyurea plastics to specialty applications, such as reaction injection molding (RIM) systems for automobile parts using amine-terminated prepolymers in conjunction with aromatic diisocyanates.[43] Other applications include encapsulation of pharmaceuticals, inks and dyes, and the modification of wool fibers by interfacial grafting techniques.[44]

Another type of urea polymer, prepared by condensing urea with formaldehyde, is covered in Chapter 14.

TABLE 13.5. Miscellaneous Methods of Synthesizing Polyureas

Reaction	*Number*
H_2N—R—NH_2 + PhOCOPh ⟶ $\left[NH-R-NHC \right]$ + 2 PhOH	(13.40)
H_2N—R—NH_2 + ClCN⟨ ⟩NCCl ⟶ $\left[NH-R-NHCN⟨ ⟩NC \right]$ + 2 HCl	(13.41)
H_2N—R—NH_2 + CO_2 $\xrightarrow[\text{pyridine}]{O=PH(OPh)_2}$ $\left[NH-R-NHC \right]$ + H_2O	(13.42)

13.5 Polyurethanes

Like polyureas, polyurethanes are carbonic acid derivatives. An older term for them is *poly-carbamate*, from carbamic acid ($RNHCO_2H$). Polyurethanes are used in a wide variety of products, including fibers (particularly the elastic variety), elastomers, foams, coatings, and adhesives. A number of books have been written on the subject.[45–48]

There are two principal methods of forming polyurethanes: the reaction of bischlorofor-mates (**22**) (made from dihydroxy compounds with excess phosgene) with diamines (13.43), and, more important from the industrial perspective, the reaction of diisocyanates with dihydroxy compounds (13.44), which has the advantage of no byproducts. Bischloroformates

$$\text{ClCO—R—OCCl} + H_2N\text{—R'—}NH_2 \longrightarrow$$
22

$$\left[CO-R-OCNH-R'-NH \right] + 2\ \text{HCl} \tag{13.43}$$

$$\text{OCN—R—NCO} + \text{HO—R'—OH} \longrightarrow \left[CNH-R-NHCO-R'-O \right] \tag{13.44}$$

are less reactive than ordinary acid chlorides; nevertheless they react with diamines at low temperatures under interfacial conditions.[8,49] Polyurethane **23** prepared this way, for example, melts at about 180°C, compared with 295°C for nylon 46. Reaction (13.44) is actually more complex because the urethane groups, once formed, can react further with free isocyanate (13.45) to form *allophonate* groups, which leads to chain branching and crosslinking.

$$\left[CO(CH_2)_2OCNH(CH_2)_6NH \right]$$
23

$$\text{wNHCOw} + \text{wNCO} \xrightarrow{\Delta} \text{wNCOw} \qquad (13.45)$$

with the product NCO group bearing C=O and NHw.

Industrial polyurethane production is based primarily on low-molecular-weight aliphatic hydroxy-terminated polyethers such as poly(propylene glycol) (**24**), or polyesters such as poly(diethylene glycol adipate) (**25**). Low-molecular-weight diols such as 1,4-butanediol and 1,6-hexanediol are also used as chain extenders. The most important isocyanates are a mixture of 2,4-toluenediisocyanate (**17**) and its 2,6 isomer (principally the former), methylene-4,4'-diphenyldiisocyanate (**26**) (MDI), and "polymeric isocyanate" (**27**), which is derived from the reaction product of aniline with formaldehyde. Aliphatic diisocyanates such as 1,6-hexanediisocyanate and the hydrogenation product of MDI are used primarily in light-stable coatings.

$$\left[\text{CH}_2\underset{\underset{\text{CH}_3}{|}}{\text{CHO}} \right] \qquad \left[\overset{O}{\overset{||}{\text{C}}}(\text{CH}_2)_4\overset{O}{\overset{||}{\text{C}}}\text{OCH}_2\text{CH}_2\text{OCH}_2\text{CH}_2\text{O} \right]$$

$$\mathbf{24} \qquad\qquad\qquad \mathbf{25}$$

$$\text{OCN}—⟨\bigcirc⟩—\text{CH}_2—⟨\bigcirc⟩—\text{NCO}$$

$$\mathbf{26}$$

$$\mathbf{27}$$

Polyurethane **28** was one of the early fibers produced in Germany as an alternative to the patent-protected nylon fibers. Made from 1,6-hexanediisocyanate and 1,4-butanediol (13.45), it was marketed by Bayer under the trade name Perlon U. Today the most important polyurethane fibers consist of elastomeric $+\text{AB}+$ block copolymers containing alternating "soft" and "hard" segments.[50,51] (Du Pont's Spandex is an example.) Typically, an

$$\left[\overset{O}{\overset{||}{\text{C}}}\text{NH}(\text{CH}_2)_4\text{NH}\overset{O}{\overset{||}{\text{C}}}\text{O}(\text{CH}_2)_4\text{O} \right]$$

$$\mathbf{28}$$

aromatic diisocyanate is reacted in excess with a hydroxy-terminated polyether or polyester having an average molecular weight of 2000 to 3000 to yield an isocyanate-terminated polymer that is, in turn, reacted with a diamine chain extender. The sequence of steps is shown in Scheme 13.1. Note that chain extension occurs via urea rather than urethane linkages.

Elastomeric fibers exhibit morphological characteristics similar to those of thermoplastic elastomers. The hard segments associate into crystalline microdomains while the pre-

$$HO\text{\small wwww}OH \quad + \quad OCN\text{—}Ar\text{—}NCO$$

(Polyether or polyester)   (Excess)

$$OCN\text{—}Ar\text{—}NH\overset{O}{\overset{\|}{C}}\text{—}\!\!\left[\!O\text{\small wwww}O\overset{O}{\overset{\|}{C}}NH\text{—}Ar\text{—}NH\overset{O}{\overset{\|}{C}}\!\right]\!\!\text{—}O\text{\small wwww}O\overset{O}{\overset{\|}{C}}NH\text{—}Ar\text{—}NCO$$

                 OCN—Ar—NCO H_2N—R—NH_2
                   (unreacted)

$$\text{—}\!\!\left[\!\!\left(\!\overset{O}{\overset{\|}{C}}NH\text{—}Ar\text{—}NH\overset{O}{\overset{\|}{C}}NH\text{—}R\text{—}NH\!\right)\!\!\left(\!\overset{O}{\overset{\|}{C}}NH\text{—}Ar\text{—}NH\overset{O}{\overset{\|}{C}}O\text{\small wwww}O\!\right)\!\!\right]\!\!\text{—}$$

        "Hard" segment   "Soft" segment

SCHEME 13.1. Synthesis of an elastomeric polyurethane for fiber applications.

dominant soft segments, which constitute the continuous phase, remain amorphous and randomly coiled. Upon stretching, the soft segments become elongated and crystalline, but when the tension is removed they rapidly revert to the amorphous state.

A variety of conventional elastomers and thermoplastic elastomers are also marketed. The high reactivity of the isocyanate group makes polyurethanes ideally suited to RIM technology.[52] Large-scale objects such as automobile bumpers are made this way.

Polyurethane foams[53] are made from isocyanate-terminated prepolymers having a molecular weight of about 3000 prepared according to reaction (13.44) using excess diisocyanate. Water is added, which causes an increase in molecular weight by formation of urea groups with simultaneous loss of carbon dioxide (13.46). As the evolved gas causes the polymer to foam, the polymerization reaction increases the viscosity and sets the foam before it collapses. A low-boiling liquid such as dichloromethane, which is boiled by the reaction exotherm, is usually added to augment the foaming action of the carbon dioxide.

$$OCN\text{\small wwww}NCO + H_2O \longrightarrow OCN\text{\small wwww}NH\overset{O}{\overset{\|}{C}}OH \xrightarrow{(-CO_2)}$$

$$(13.46)$$

$$OCN\text{\small wwww}NH_2 \longrightarrow \text{\small wwww}\!\left[\!NH\overset{O}{\overset{\|}{C}}NH\!\right]\!\text{\small wwww}$$

Flexible foams are usually prepared from dihydroxy polyesters or polyethers, rigid foams from polyhydroxy prepolymers. Rigid foams are sometimes prepared without water by reacting a hydroxyl-terminated prepolymer with diisocyanate in the presence of a blowing agent. In this case, molecular weight buildup is via urethane linkages. Such foams are particularly effective insulators because the blowing agents, trapped in the foam cells, have very low thermal conductivity. Metal salts such as stannous 2-ethylhexanoate and tertiary amines, particularly diazabicyclo[2.2.2]octane (**29**) (commonly called DABCO), are normally used to catalyze urethane formation. DABCO is a more effective catalyst than most other tertiary amines, possibly because the rigid cyclic system minimizes steric interference.

$$\text{wNHCOw} + \text{wNCO} \xrightarrow{\Delta} \text{wNCOw} \qquad (13.45)$$

with the carbonyl O above NHCO, and the product showing the structure:

Industrial polyurethane production is based primarily on low-molecular-weight aliphatic hydroxy-terminated polyethers such as poly(propylene glycol) (**24**), or polyesters such as poly(diethylene glycol) adipate) (**25**). Low-molecular-weight diols such as 1,4-butanediol and 1,6-hexanediol are also used as chain extenders. The most important isocyanates are a mixture of 2,4-toluenediisocyanate (**17**) and its 2,6 isomer (principally the former), methylene-4,4'-diphenyldiisocyanate (**26**) (MDI), and "polymeric isocyanate" (**27**), which is derived from the reaction product of aniline with formaldehyde. Aliphatic diisocyanates such as 1,6-hexanediisocyanate and the hydrogenation product of MDI are used primarily in light-stable coatings.

$$\left[CH_2\underset{\underset{CH_3}{|}}{CH}O\right] \qquad \left[\underset{O}{\overset{O}{\|}}C(CH_2)_4\overset{O}{\overset{\|}{C}}OCH_2CH_2OCH_2CH_2O\right]$$

      24   **25**

$$OCN—\langle\text{ring}\rangle—CH_2—\langle\text{ring}\rangle—NCO$$

26

27

Polyurethane **28** was one of the early fibers produced in Germany as an alternative to the patent-protected nylon fibers. Made from 1,6-hexanediisocyanate and 1,4-butanediol (13.45), it was marketed by Bayer under the trade name Perlon U. Today the most important polyurethane fibers consist of elastomeric +AB+ block copolymers containing alternating "soft" and "hard" segments.[50,51] (Du Pont's Spandex is an example.) Typically, an

$$\left[\overset{O}{\overset{\|}{C}}NH(CH_2)_4NH\overset{O}{\overset{\|}{C}}O(CH_2)_4O\right]$$

28

aromatic diisocyanate is reacted in excess with a hydroxy-terminated polyether or polyester having an average molecular weight of 2000 to 3000 to yield an isocyanate-terminated polymer that is, in turn, reacted with a diamine chain extender. The sequence of steps is shown in Scheme 13.1. Note that chain extension occurs via urea rather than urethane linkages.

Elastomeric fibers exhibit morphological characteristics similar to those of thermoplastic elastomers. The hard segments associate into crystalline microdomains while the pre-

$$HO\text{\small\textcolor{black}{$\sim\!\!\sim\!\!\sim$}}OH \quad + \; OCN\text{—}Ar\text{—}NCO$$

(Polyether or polyester)   (Excess)

$$OCN\text{—}Ar\text{—}NHC\underset{O}{\overset{O}{\parallel}}\Big[O\text{\small$\sim\!\!\sim$}OCNH\text{—}Ar\text{—}NHC\Big]O\text{\small$\sim\!\!\sim$}OCNH\text{—}Ar\text{—}NCO$$

OCN—Ar—NCO | H₂N—R—NH₂
(unreacted)

$$-\Big[\Big(\overset{O}{\underset{\parallel}{C}}NH\text{—}Ar\text{—}NH\overset{O}{\underset{\parallel}{C}}NH\text{—}R\text{—}NH\Big)\Big(\overset{O}{\underset{\parallel}{C}}NH\text{—}Ar\text{—}NH\overset{O}{\underset{\parallel}{C}}O\text{\small$\sim\!\!\sim$}O\Big)\Big]-$$

"Hard" segment   "Soft" segment

SCHEME 13.1. Synthesis of an elastomeric polyurethane for fiber applications.

dominant soft segments, which constitute the continuous phase, remain amorphous and randomly coiled. Upon stretching, the soft segments become elongated and crystalline, but when the tension is removed they rapidly revert to the amorphous state.

A variety of conventional elastomers and thermoplastic elastomers are also marketed. The high reactivity of the isocyanate group makes polyurethanes ideally suited to RIM technology.[52] Large-scale objects such as automobile bumpers are made this way.

Polyurethane foams[53] are made from isocyanate-terminated prepolymers having a molecular weight of about 3000 prepared according to reaction (13.44) using excess diisocyanate. Water is added, which causes an increase in molecular weight by formation of urea groups with simultaneous loss of carbon dioxide (13.46). As the evolved gas causes the polymer to foam, the polymerization reaction increases the viscosity and sets the foam before it collapses. A low-boiling liquid such as dichloromethane, which is boiled by the reaction exotherm, is usually added to augment the foaming action of the carbon dioxide.

$$OCN\text{\small$\sim\!\!\sim$}NCO + H_2O \longrightarrow OCN\text{\small$\sim\!\!\sim$}NH\overset{O}{\underset{\parallel}{C}}OH \xrightarrow{(-CO_2)}$$

$$OCN\text{\small$\sim\!\!\sim$}NH_2 \longrightarrow \Big[\text{\small$\sim\!\!\sim$}NH\overset{O}{\underset{\parallel}{C}}NH\text{\small$\sim\!\!\sim$}\Big] \tag{13.46}$$

Flexible foams are usually prepared from dihydroxy polyesters or polyethers, rigid foams from polyhydroxy prepolymers. Rigid foams are sometimes prepared without water by reacting a hydroxyl-terminated prepolymer with diisocyanate in the presence of a blowing agent. In this case, molecular weight buildup is via urethane linkages. Such foams are particularly effective insulators because the blowing agents, trapped in the foam cells, have very low thermal conductivity. Metal salts such as stannous 2-ethylhexanoate and tertiary amines, particularly diazabicyclo[2.2.2]octane (**29**) (commonly called DABCO), are normally used to catalyze urethane formation. DABCO is a more effective catalyst than most other tertiary amines, possibly because the rigid cyclic system minimizes steric interference.

29

The reaction of diisocyanates with carboxyl-terminated polyesters (reaction 13.33), as discussed earlier, is also used to make flexible foams, in this case by amide formation. It is not uncommon to see such reactions included under urethane foams even though no urethane chemistry is involved. In other words, all foam manufacture that uses diisocyanates tends to be grouped under polyurethanes.

Flexible foams are used as insulators, including textile laminates for cold-weather clothing; automobile crash panels; upholstery; bedding; carpet underlays; synthetic sponges; and other miscellaneous uses. Rigid foams are most commonly used in insulated construction panels, for packaging delicate items, for lightweight furniture, and for marine flotation equipment. The use of these materials in construction has prompted attempts at preparing nonflammable polyurethanes. Phosphorus-containing polyurethanes of type **30**, for example, are self-extinguishing.[54]

30

A different strategy for reducing flammability of foams is to convert terminal and pendant isocyanate groups to *polyisocyanurates*.[55] The trimerization reaction (13.47) is facilitated by alkali metal alkoxide, phenoxide, and carboxylate catalysts. The isocyanurate moiety (**31**) serves the dual purpose of reducing flammability and increasing thermal stability.

$$3 \text{\textasciitilde\textasciitilde\textasciitilde NCO} \xrightarrow{\text{catalyst}} \qquad (13.47)$$

31

For "one-pot" coatings formulations, polyurethanes typically prepared with polyesters containing polyfunctional alcohols are treated with excess diisocyanate. Upon exposure to atmospheric moisture, the isocyanate-terminated polymer cures by forming urea linkages according to reaction (13.46). Reaction of the urea groups with unreacted isocyanate leads to further crosslinking via *biuret* groups (13.48). In other one-pot coatings, the isocyanate groups are reacted with a "blocking group" such as phenol, which dissociates on heating to regenerate the isocyanate groups.

$$\text{\textasciitilde\textasciitilde\textasciitilde NHCNH\textasciitilde\textasciitilde\textasciitilde} + \text{\textasciitilde\textasciitilde\textasciitilde NCO} \longrightarrow \text{\textasciitilde\textasciitilde\textasciitilde NHCN\textasciitilde\textasciitilde\textasciitilde} \qquad (13.48)$$

13.6 Polyhydrazides

There are three general methods of preparing polyhydrazides[8]; the reaction of a diacid chloride with hydrazine or hydrazine hydrate (13.49), with a dihydrazine (13.50), and with a dihydrazide (13.51). Of the three, the dihydrazide route (13.51) is preferred because the monomers are easily prepared from diesters and hydrazine and are easily purified. Hydrazine is readily available, but it is so hygroscopic that it is difficult to achieve an exact stoichiometric balance. Furthermore, one amino group of hydrazine is sufficiently less reactive than the other that condensations of this type (13.49) must be run in solution and cannot be done by interfacial polymerization.

$$
\begin{array}{cc}
\xrightarrow[(-2\,HCl)]{H_2NNH_2} & \left[\!\!\begin{array}{c}\overset{O}{\overset{\|}{C}}-R-NHNH\end{array}\!\!\right] \quad (13.49)
\end{array}
$$

$$
\underset{ClC-R-CCl}{\overset{O\quad\;\;O}{\overset{\|\qquad\|}{}}}
$$

$$
\xrightarrow[(-2\,HCl)]{H_2NNH-R'-NHNH_2} \left[\!\!\begin{array}{c}\overset{O}{\overset{\|}{C}}-R-\overset{O}{\overset{\|}{C}}NHNH-R'-NHNH\end{array}\!\!\right] \quad (13.50)
$$

$$
\xrightarrow[(-2\,HCl)]{H_2NNHC-R'-CNHNH_2} \left[\!\!\begin{array}{c}\overset{O}{\overset{\|}{C}}-R-\overset{O}{\overset{\|}{C}}NHNH\overset{O}{\overset{\|}{C}}-R'-NHNH\end{array}\!\!\right] \quad (13.51)
$$

These polymers have not achieved commercial importance, although there has been interest in aromatic polyhydrazides as high-modulus, thermally stable fibers[32] and because they may exhibit liquid crystalline properties.[56] Tire cord represents a potential market for materials of this type. Polyamidehydrazides such as **32** are also of interest.[35] In Chapter 16 we shall see how polyhydrazides may be converted to heterocyclic polymers.

32

13.7 Polyimides

Although imides may have an open-chain structure, it is the cyclic imides that are of interest in polymer chemistry. Such imides are most commonly prepared from cyclic anhydrides and ammonia or amines via the intermediate amide-acid (13.52).

Polyimides have experienced extremely rapid development in recent years,[57–62] the major emphasis being on engineering applications. High-strength composites, thermally stable films, molding compounds, and adhesives are numbered among the products.

29

The reaction of diisocyanates with carboxyl-terminated polyesters (reaction 13.33), as discussed earlier, is also used to make flexible foams, in this case by amide formation. It is not uncommon to see such reactions included under urethane foams even though no urethane chemistry is involved. In other words, all foam manufacture that uses diisocyanates tends to be grouped under polyurethanes.

Flexible foams are used as insulators, including textile laminates for cold-weather clothing; automobile crash panels; upholstery; bedding; carpet underlays; synthetic sponges; and other miscellaneous uses. Rigid foams are most commonly used in insulated construction panels, for packaging delicate items, for lightweight furniture, and for marine flotation equipment. The use of these materials in construction has prompted attempts at preparing nonflammable polyurethanes. Phosphorus-containing polyurethanes of type **30**, for example, are self-extinguishing.[54]

30

A different strategy for reducing flammability of foams is to convert terminal and pendant isocyanate groups to *polyisocyanurates*.[55] The trimerization reaction (13.47) is facilitated by alkali metal alkoxide, phenoxide, and carboxylate catalysts. The isocyanurate moiety (**31**) serves the dual purpose of reducing flammability and increasing thermal stability.

$$3\text{wwwNCO} \xrightarrow{\text{catalyst}} \qquad (13.47)$$

31

For "one-pot" coatings formulations, polyurethanes typically prepared with polyesters containing polyfunctional alcohols are treated with excess diisocyanate. Upon exposure to atmospheric moisture, the isocyanate-terminated polymer cures by forming urea linkages according to reaction (13.46). Reaction of the urea groups with unreacted isocyanate leads to further crosslinking via *biuret* groups (13.48). In other one-pot coatings, the isocyanate groups are reacted with a "blocking group" such as phenol, which dissociates on heating to regenerate the isocyanate groups.

$$\text{wwwNHCNHwww} + \text{wwwNCO} \longrightarrow \text{wwwNHCNwww} \qquad (13.48)$$

13.6 Polyhydrazides

There are three general methods of preparing polyhydrazides[8]; the reaction of a diacid chloride with hydrazine or hydrazine hydrate (13.49), with a dihydrazine (13.50), and with a dihydrazide (13.51). Of the three, the dihydrazide route (13.51) is preferred because the monomers are easily prepared from diesters and hydrazine and are easily purified. Hydrazine is readily available, but it is so hygroscopic that it is difficult to achieve an exact stoichiometric balance. Furthermore, one amino group of hydrazine is sufficiently less reactive than the other that condensations of this type (13.49) must be run in solution and cannot be done by interfacial polymerization.

$$\begin{array}{c} \xrightarrow[(-2\,HCl)]{H_2NNH_2} \end{array} \quad \begin{bmatrix} O \\ \| \\ C-R-NHNH \end{bmatrix} \qquad (13.49)$$

$$\begin{matrix} O & O \\ \| & \| \\ ClC-R-CCl \end{matrix} \xrightarrow[(-2\,HCl)]{H_2NNH-R'-NHNH_2} \begin{bmatrix} O & O \\ \| & \| \\ C-R-CNHNH-R'-NHNH \end{bmatrix} \qquad (13.50)$$

$$\xrightarrow[(-2\,HCl)]{\overset{O\ \ \ \ O}{\overset{\|\ \ \ \ \|}{H_2NNHC-R'-CNHNH_2}}} \begin{bmatrix} O & O & O \\ \| & \| & \| \\ C-R-CNHNHC-R'-NHNH \end{bmatrix} \qquad (13.51)$$

These polymers have not achieved commercial importance, although there has been interest in aromatic polyhydrazides as high-modulus, thermally stable fibers[32] and because they may exhibit liquid crystalline properties.[56] Tire cord represents a potential market for materials of this type. Polyamidehydrazides such as **32** are also of interest.[35] In Chapter 16 we shall see how polyhydrazides may be converted to heterocyclic polymers.

32

13.7 Polyimides

Although imides may have an open-chain structure, it is the cyclic imides that are of interest in polymer chemistry. Such imides are most commonly prepared from cyclic anhydrides and ammonia or amines via the intermediate amide-acid (13.52).

Polyimides have experienced extremely rapid development in recent years,[57–62] the major emphasis being on engineering applications. High-strength composites, thermally stable films, molding compounds, and adhesives are numbered among the products.

Oxidative and hydrolytic stability, in addition to thermal stability, are characteristic of polyimides.

$$(13.52)$$

There are two types of polyimides:

1. *Condensation polyimides*, in which the imide functional group or its precursor is formed during the polymerization reaction.

2. *Addition polyimides*, which consist of low-molecular-weight, usually oligomeric imide-containing, compounds or mixtures of compounds having reactive terminal groups that upon heating undergo addition polymerization to yield network polymers.

Condensation polyimides were the first to be developed.[62–64] The reaction, illustrated with pyromellitic dianhydride (**33**), involves initial formation of polyamide (13.53) followed by ring closure to form polyimide (13.54). The formation of a stable five-membered ring is the driving force for forming linear rather than crosslinked polymer.

33

$$(13.53)$$

$$(13.54)$$

If R is aliphatic in the above equations, the two steps may be run in one operation. If R is aromatic, however, the final product is infusible and insoluble, and the second step must be carried out in the solid state—for example, with a cast film of intermediate polyamide or an impregnated fiber for manufacturing composites. Formation of byproduct water in the cyclization step is of concern because it may cause voids that structurally weaken the finished product.

The half ester, formed by reacting **33** with two equivalents of alcohol, may be used in place of the dianhydride, in which case alcohol is a byproduct in the formation of intermediate polyamide—or *poly(amic acid)*, as it is commonly called.

Among the important polyimides prepared by the dianhydride-diamine route is **34** (du Pont's Kapton). The polymer melts in excess of 600°C and exhibits little weight loss up to

34

500°C in an inert atmosphere. Extensive structural studies have been conducted on polyimides[63,65,66] and, not unexpectedly, it has been found that increasing the level of *meta* substitution in the aromatic diamine causes a reduction in both viscosity and glass transition temperature. Interestingly, polyimide **35**, which contains no hydrogen, exhibits greater thermal stability than the analogous polymer prepared from **33**, which suggests that carbon–hydrogen bonds are the sites for oxidative decomposition.[67]

35

An alternative one-step route to polyimides is the reaction of dianhydrides with diisocyanates[68–70] (13.55). With carbon dioxide as a byproduct, the reaction has obvious implications for the manufacture of foams. Polymerization takes place at elevated temperatures or at moderate temperatures (50–60°C) in polar aprotic solvents if water or tetracarboxylic

$$33 + OCN-R-NCO \longrightarrow \quad + 2CO_2 \qquad (13.55)$$

acid is present. Amines catalyze the reaction. The mechanism is complex and probably involves initial hydrolysis of isocyanate or anhydride (or both), and perhaps the intercession of an unstable seven-membered ring intermediate (**36**).

36

Although a large number of dianhydrides have been applied to polyimide synthesis, the only one in widespread use besides **33** is benzophenone-3,3′,4,4′-tetracarboxylic dianhydride (**37**). The fluorinated anhydride **38** finds occasional use in polyimides where high moisture resistance is a necessity (see Chapter 4, Section 4.6). One formulation that makes use of dianhydride **37** is a *polyimidesulfone* (**39**) developed by the National Aeronautics and Space Administration and marketed as an engineering plastic.

37   **38**

39

Two varieties of addition polyimides that have achieved commercial importance are the bismaleimides (**40**) and bisnadimides* (**41**).[71,72] The maleimide and nadimide groups are among the few monomers that undergo facile thermal addition polymerization. The R groups in the structures may be aliphatic or aromatic, but are more commonly the latter. Bismaleimides (BMIs) are synthesized from maleic anhydride and a monomeric or oligomeric diamine, and they are cured simply by heating at 200–400°C (13.56) to yield a highly crosslinked polymer. To reduce brittleness arising from the high crosslink density, some commercial BMIs are copolymerized with other monomers, for example, *o,o′*-diallylbisphenol A (**42**).

40   **41**

Nadimide is a generic term for the 5-norbornene-2,3-dicarboximide system. It is synthesized from the corresponding nadic anhydride or acid-ester. Nadic anhydride is the Diels–Alder adduct of maleic anhydride and cyclopentadiene.

$$40 \xrightarrow{200\text{--}400°C} \qquad (13.56)$$

42

Commercial products also make use of the Michael addition of primary or secondary diamines to effect chain extension (13.57) prior to thermal curing.[71–73] Chain extension reduces the crosslink density and brittleness of the finished product. An example is Kerimid 601 (Rhone-Poulenc trade name), which employs 4,4′-diaminodiphenylmethane (**43**) as the feedstock for both the bismaleimide and diamine according to reaction (13.57). The Michael addition reaction is also used to prepare linear *polyaspartimides** (**44**) from equimolar quantities of bismaleimide and secondary diamines[74,75] (13.58). In Chapter 17 we shall see how bismaleimide monomers may be used for preparing polyimides by cycloaddition reactions.

$$H_2N\text{---}\langle\ \rangle\text{---}CH_2\text{---}\langle\ \rangle\text{---}NH_2$$

43

$$2\ \mathbf{40} + H_2N\text{---}R'\text{---}NH_2 \longrightarrow \qquad (13.57)$$

$$\mathbf{40} + H\text{---}N\underset{\smile}{\frown}N\text{---}H \longrightarrow \qquad (13.58)$$

44

For the bisnadimide type (**41**), an amine-terminated imide prepolymer or a simple diamine is end-capped with nadimide groups by reaction with nadic anhydride or ester-acid.[76–78] Thermal curing at 270 to 310°C involves an initial retrograde Diels–Alder reaction followed

---

*The term *polyaspartimide* is derived from the naturally occurring amino acid aspartic acid, $HO_2CCH_2CH(NH_2)CO_2H$.

Imide oligomer or monomer

SCHEME 13.2. Polymerization of a PMR-type nadimide end-capped prepolymer.

by copolymerization of the resultant cyclopentadiene and maleimide end groups, as illustrated in Scheme 13.2. Curing is more complex than that shown, because homopolymerization of maleimide or copolymerization of unreacted nadimide groups may also occur.[79,80] The curing cycle must be controlled carefully to preclude voids arising from the highly volatile cyclopentadiene. Polymers of this type have been designated PMR (for *polymerization of monomeric reactants*).[76]

45

Other thermally labile end groups were first mentioned in Chapter 4 (see Table 4.4) as one strategy for improving the processability of thermally stable polymers. An example is the ethynyl-capped polyimide oligomer **45**, marketed by Gulf Oil Chemicals as a high-temperature adhesive.

References

1. R. M. Moncrieff, *Man-Made Fibers*, Butterworth, London, 1975, Chaps. 19–22.
2. M. I. Kohan (Ed.), *Nylon Plastics*, Wiley-Interscience, New York, 1973.
3. D. B. Jacobs and J. Zimmerman, in *Polymerization Processes* (C. E. Schildknecht, Ed.), Wiley-Interscience, New York, 1977, Chap. 12.
4. H. K. Livingston, M. S. Sioshansi, and M. D. Glick, *J. Macromol. Sci.-Revs., Macromol. Chem.*, **C6**, 29 (1971).
5. E. Roerdink and J. M. M. Warnier, *Polymer*, **26**, 1582 (1985).
6. J. Preston and W. L. Hofferbert, *J. Polym. Sci., Poly. Symp. Ed.*, **65**, 13 (1978).
7. K. H. Park, S. Watanabe, M. Kakimoto, and Y. Imai, *Polym. J.*, **25**, 209 (1993).
8. P. W. Morgan, *Condensation Polymers: By Interfacial and Solution Methods*, Wiley-Interscience, New York, 1965.
9. V. Z. Nikonov and V. M. Savinov, in *Interfacial Synthesis*, Vol. 2 (F. Millich and C. E. Carraher, Eds.), Marcel Dekker, New York, 1977, Chap. 15.
10. V. V. Korshak, *Heat-Resistant Polymers*, Halsted, New York, 1972.
11. J. P. Critchley, G. J. Knight, and W. W. Wright, *Heat-Resistant Polymers*, Plenum, New York, 1983.
12. R. S. Lenk, *Macromol. Rev.*, **13**, 355 (1978).
13. M. Jaffe and R. S. Jones, in *High Technology Fibers*, Part A (M. Lewin and J. Preston, Eds.), Marcel Dekker, New York, 1985, Chap. 9.
14. H. K. Reimschuessel, in *Ring-Opening Polymerization* (K. C. Frisch and S. L. Reegen, Eds.), Marcel Dekker, New York, 1969, Chap. 7.
15. F. Millich and K. V. Seshadri, in *Cyclic Monomers* (K. C. Frisch, Ed.), Wiley-Interscience, New York, 1972, Chap. 3.
16. J. Šebenda, *J. Macromol. Sci.-Chem.*, **A6**, 1145 (1972).
17. M. Rothe and G. Bertalan, in *Ring-Opening Polymerization* (T. Saegusa and E. Goethals, Eds.), ACS Symp. Ser. 59, American Chemical Society, Washington, D.C., 1977.
18. R. Bacskai, in *Crown Ethers and Phase Transfer Catalysts in Polymer Science* (L. J. Mathias and C. E. Carraher, Jr., Eds.), Plenum, New York, 1984, pp. 183–199.
19. M. Swarc, *Adv. Polym. Sci.*, **4**, 1 (1965).
20. H. R. Kricheldorf, *Alpha-Amino Acid-N-Carboxy Anhydride and Related Heterocycles*, Springer-Verlag, New York, 1987.
21. H. J. Harwood, in *Ring-Opening Polymerization* (J. E. McGrath, Ed.), ACS Symp. Ser. 286, American Chemical Society, Washington, D.C., 1985, pp. 67–85.
22. W. R. Sorenson and T. W. Campbell, *Preparative Methods of Polymer Chemistry*, 2nd ed., Wiley-Interscience, New York, 1968, 349 ff.
23. J. H. Saunders and K. C. Frisch, *Polyurethanes: Chemistry and Technology*, Part 1, Wiley-Interscience, New York, 1962, p. 79.
24. D. F. Regelman, L. M. Alberino, and R. J. Lockwood, in *Reaction Injection Molding* (J. E. Kresta, Ed.), ACS Symp. Ser. 270, American Chemical Society, Washington, D.C., 1985, pp. 125–134.
25. V. E. Shashoua, W. Sweeny, and R. F. Tietz, *J. Am. Chem. Soc.*, **82**, 866 (1960).
26. G. Natta, D. Pietro, and M. Cambini, *Makromol. Chem.*, **56**, 200 (1962).
27. T. E. Patten and B. M. Novak, *J. Am. Chem. Soc.*, **113**, 5065 (1991).
28. T. Otsu, B. Yamada, M. Itahashi, and T. Mori, *J. Polym. Sci., Polym. Chem. Ed.*, **14**, 1347 (1976).
29. L. W. Bush and D. S. Breslow, *Macromolecules*, **1**, 189 (1968).
30. M. Ueda, O. Hara, A. Sato, and Y. Imai, *J. Polym. Sci., Polym. Chem. Ed.*, **17**, 769 (1979).
31. M. Yoneyama, M. Kakimoto, and Y. Imai, *Macromolecules*, **21**, 1908 (1988).
32. R. H. Perry, *Chemtech*, **24**(2), 18 (1994).
33. A. H. Frazer, *High Temperature Resistant Polymers*, Wiley-Interscience, New York, 1968, Chap. 3.
34. R. A. Dine-Hart, B. J. C. Moore, and W. W. Wright, *J. Polym. Sci. B*, **2**, 369 (1964).
35. W. B. Black and J. Preston (Eds.), *High-Modulus Wholly Aromatic Fibers*, Marcel Dekker, New York, 1973.

36. A. K. Mukherjee and H. R. Goel, *J. Macromol. Sci., Rev. Macromol. Chem.*, **C25**, 99 (1985).
37. N. A. Platé and V. P. Shibaev, *J. Polym. Sci. Macromol. Rev.*, **8**, 117 (1974).
38. A. Nathan and J. Kohn, in *Biomedical Polymers: Designed-to-Degrade Systems* (S. W. Shalaby, Ed.), Hanser, New York, 1994, Chap. 5.
39. H. Sekiguchi and B. Coutin, in *The Polymeric Materials Encyclopedia: Synthesis, Properties and Applications* (J. C. Salamone, Ed.), CRC, Boca Raton, 1996.
40. I. Arvanitoyannis, E. Psomiadou, N. Yamamoto, E. Nikolaou, and J. M. V. Blanchard, *Polymer*, **36**, 2957 (1995).
41. D. A. Barrera, E. Zylstra, P. T. Lansbury, Jr., and R. Langer, *J. Am. Chem. Soc.*, **115**, 11010 (1993).
42. T. S. Ellis, *J. Appl. Polym. Sci.*, **36**, 451 (1988).
43. H. Ulrich, *Introduction to Industrial Polymers*, 2nd ed., Hanser, New York, 1993, Chap. 12.
44. K. C. Stueben and A. E. Barnabeo, in *Interfacial Synthesis*, Vol. 2, (F. Millich and C. E. Carraher, Eds.), Marcel Dekker, New York, 1977, Chap. 18.
45. J. H. Saunders and K. C. Frisch, *Polyurethanes: Chemistry and Technology*, Wiley-Interscience, New York, Part 1, 1962; Part 2, 1964.
46. L. N. Phillips and D. B. Parker, *Polyurethanes—Chemistry, Technology, and Properties*, Gordon and Breach, New York, 1964.
47. K. N. Edwards (Ed.), *Urethane Chemistry and Applications*, ACS Symp. Ser. 172, American Chemical Society, Washington, D.C., 1981.
48. E. N. Doyle, *The Development and Use of Polyurethane Products*, McGraw-Hill, New York, 1969.
49. T. Tanaka and Y. Yokoyama, in *Interfacial Synthesis*, Vol. 2 (F. Millich and C. E. Carraher, Eds.), Marcel Dekker, New York, 1977, Chap. 17.
50. M. Couper, in *High Technology Fibers*, Part A (M. Lewin and J. Preston, Eds.), Marcel Dekker, New York, 1985, Chap. 2.
51. D. C. Allport and A. A. Mohajer, in *Block Copolymers* (D. C. Allport and W. H. Janes, Eds.), Halsted, New York, 1973, Chap. 5.
52. H. F. Hespe, in *Engineering Plastics Handbook, Vol. 2: Engineering Plastics* (J. N. Epel et al., Eds.), ASM International, Metals Park, Ohio, 1988, pp. 257–64.
53. T. T. Healey (Ed.), *Polyurethane Foams*, Gordon and Breach, New York, 1968.
54. E. Dyer and R. A. Dunbar, *J. Polym. Sci. A-1*, **8**, 629 (1970).
55. J. A. Brydson, *Plastics Materials*, 3rd ed., Butterworth, London, 1975, Chap. 27.
56. P. W. Morgan, *J. Polym. Sci., Polym. Symp.*, **65**, 1 (1978).
57. K. L. Mittal (Ed.), *Polyimides*, Vols. 1 and 2, Plenum, New York, 1984.
58. M. I. Beesonov, M. M. Koton, V. V. Kudryavtsev, and L. A. Laius, *Polyimides—Thermally Stable Polymers*, Plenum, New York, 1987.
59. D. Wilson, H. D. Stenzenberger, and P. Hergenrother (Eds.) *Polyimides*, Blackie, London, 1990.
60. D. A. Scola and J. H. Vontell, *Chemtech*, **13**(2), 112 (1989).
61. C. E. Sroog, *Progr. Polym. Sci.*, **16**, 561 (1994).
62. C. E. Sroog, A. L. Endrey, S. V. Abramo, C. E. Berr, W. M. Edwards, and K. L. Olivier, *J. Polym. Sci., Part A*, **3**, 1373 (1965).
63. C. E. Sroog, *Macromol. Rev.*, **11**, 161 (1976).
64. W. Volksen, *Adv. Polym. Sci.*, **117**, 111 (1994).
65. I. K. Varma, R. N. Goel, and D. S. Varma, *J. Polym. Sci., Polym. Chem. Ed.*, **17**, 703 (1979).
66. V. L. Bell, K. E. M. Hett, and G. M. Stokes, *J. Appl. Polym. Sci.*, **26**, 3805 (1981).
67. S. S. Hirsch, *J. Polym. Sci. A-1*, **7**, 15 (1969).
68. P. S. Carleton, W. J. Farrissey Jr., and J. S. Rose, *J. Appl. Polym. Sci.*, **16**, 2983 (1972).
69. W. M. Alvino and L. E. Edelman, *J. Appl. Polym. Sci.*, **19**, 2961 (1975).
70. N. D. Ghatge and U. P. Mulik, *J. Polym. Sci., Polym. Chem. Ed.*, **18**, 1905 (1980).
71. H. D. Stenzenberger, M. Herzog, W. Romer, R. Scheiblich, and N. J. Reeves, *Br. Polym. J.*, **15**, 2 (1983).
72. H. D. Stenzenberger, *Adv. Polym. Sci.*, **117**, 165 (1994).
73. I. K. Varma and S. Sharma, *Eur. Polym. J.*, **20**, 1101 (1984).
74. J. E. White, M. D. Scaia, and D. A. Snider, *J. Appl. Polym. Sci.*, **29**, 891 (1984).
75. V. L. Bell and P. R. Young, *J. Polym. Sci., Polym. Chem. Ed.*, **24**, 2647 (1986).

76. T. T. Serafini, P. Delvigs, and G. R. Lightsey, *J. Appl. Polym. Sci.*, **16**, 905 (1972).
77. D. A. Scola and M. P. Stevens, *J. Appl. Polym. Sci.*, **26**, 231 (1981).
78. S. E. Delos, R. K. Schellenberg, J. E. Smedley, and D. E. Kranbuel, *J. Appl. Polym. Sci.*, **27**, 4295 (1982).
79. N. G. Gaylord and M. Martan, in *Cyclopolymerization and Polymers with Ring-Chain Structures* (G. Butler and J. E. Kresta, Eds.), ACS Symp. Ser 195, American Chemical Society, Washington, D.C., 1982, pp. 97–105.
80. M. A. B. Meador, J. C. Johnson, and P. J. Cavano, *Macromolecules*, **30**, 515 (1997).
81. H. Stenzenberger, *Br. Polym. J.*, **20**, 389 (1988).

Review Exercises

1. Write IUPAC names for as many of the polyamides in Table 13.1 as you can.

2. Ignoring steric effects, which would you expect to have the greater effect on a polyamide's melting point, substitution on nitrogen or substitution on carbon? Explain.

3. Give plausible explanations for the following facts:
   (a) Most commercially important polyamides are aliphatic, whereas commercial polyesters are mostly aromatic.
   (b) Polycondensation of α-amino acids is not a practical way to obtain nylons 4 and 5.
   (c) Polyurethanes used for organ implantation are made with polyether glycols, not the less expensive polyester glycols.
   (d) The percent crystallinity and T_m of nylon 46 are significantly higher than those of nylon 66, more so than one might expect on the basis of chain lengths alone.
   (e) Polyureas are generally higher melting than polyamides of like structure.
   (f) When primary diamines are used to make polyaspartimides (reaction 13.58), gelation occurs. [J. E. White et al., *Polym. Preprints*, **26**(1), 132, 1985]
   (g) Polyurethanes can only be made from AA + BB monomers, not from the AB type.

4. Write a mechanism for the anionic polymerization of 2-pyrrolidone (**10**). Explain the function of the phase-transfer catalyst in increasing the rate. (ref. 18)

5. Polysulfonamide **5** (reaction 13.8) is soluble in base. Explain why. In basic solution the polymer can be methylated with dimethyl sulfate. Write the structure for the methylated polymer. How might the same polymer be formed directly?

6. Nylon 1313 has been investigated by the U.S. Department of Agriculture as a potential engineering plastic. Starting point for the synthesis is erucic acid, $CH_3(CH_2)_7CH{=}CH(CH_2)_{11}CO_2H$, obtainable from the seeds of crambe, a potential cattle feed. Beginning with erucic acid as the only source of carbon compounds, suggest a route to nylon 1313. (H. J. Nieslag, *Ind. Eng. Chem. Prod. Res. Dev.*, **16**, 101, 1977)

7. Write a plausible mechanism for the synthesis of nylon 3 from acrylamide (reaction 13.30, Table 13.2). Write reactions for two other ways that could, in principle, result in the same polymer.

8. What other reaction could, in principle, occur in reaction (13.31) (Table 13.2)? Why doesn't it? (Note the structure of the Ar groups.)

9. From the information given in Section 13.2, write the reactions for the synthesis of poly(2,4,4-trimethyl-hexamethylene terephthalamide) (Table 13.1).

10. Write a reaction illustrating the synthesis of the polyamidehydrazide **32**. (ref. 35)

11. Two commercial engineering plastics are the *polyimidesulfone* **39** and the *polyamideimide* of structure

Write reactions illustrating their synthesis.

12. The commercially important polyetherimide described in Chapter 11 (structure **26**) is made by a displacement reaction. Suggest an alternative synthesis. [T. Takekoshi et al., *Polym. Preprints*, **24**(2), 312, 1983]

13. In Chapter 11 (reaction 11.64), the synthesis of polyimidesulfides by Michael addition was described. Suggest an alternative route that does not entail prior synthesis of bismaleimides. (V. A. Sergeyev et al., *Polym. Commun.*, **26**, 365, 1985)

14. Substitute diisocyanate (OCN—R—NCO) for diacid chloride in reactions (13.49)–(13.51) and write the expected repeating units of the polymers. (The polymers are called polybiureas, polysemicarbazides, and polyacylsemicarbazides, respectively.)

15. When D,L-aspartic acid [$HO_2CCH(NH_2)CH_2CO_2H$] is heated at 200–300°C, a polymer having a repeating unit with molecular formula $C_4H_3NO_2$ is formed. What is the structure of the repeating unit? (*Chem. Engr. News*, July 20, 1992)

16. A polyurea exhibiting nonlinear optical (NLO) properties has the structure

What monomers are most likely used in its synthesis?
(D. M. Burland et al., *Chem. Rev.*, **94**, 31, 1994)

17. Predict the repeating unit of the polymers formed in each of the following reactions:

(a)

(R. C. Evers, E. J. Moore, and T. Abraham, *J. Polym. Sci. A: Polym. Chem.*, **26**, 3213, 1988)

(b)

(V. V. Korshak et al., *Izv. Akad. Nauk SSSR. Sci. Khim.*, **6**, 1402, 1969; *Chem. Abstr.*, **71**, 81784, 1969)

(c)

$$\xrightarrow[(-CO_2)]{\Delta}$$

(H. Ulrich and R. Richter, *J. Org. Chem.*, **38**, 2557, 1973)

(d)

$+ CH_2{=}CHSO_2CH{=}CH_2 \xrightarrow{\text{base}}$

(M. Russo and L. Mortillaro, *J. Polym. Sci. A-1*, **7**, 3337, 1969)

(e)

$+ H_2N{-}\bigcirc{-}O{-}\bigcirc{-}NH_2 \xrightarrow[CO_2 \text{ base}]{\text{Pd catalyst}}$

(R. J. Perry et al., *Macromolecules*, **28**, 2607, 1995)

(f)

$+ H_2N{-}\bigcirc{-}O{-}\bigcirc{-}NH_2 \xrightarrow[CO_2 \text{ base}]{\text{Pd catalyst}}$

(R. J. Perry et al., *Macromolecules*, **28**, 3509, 1995)

(g) $\quad H_2N(CH_2)_6NH\overset{\overset{\displaystyle O}{\|}}{C}(CH_2)_4CO_2CH_3 \xrightarrow{\Delta}$

(F. Joufret and P.-J. Madec, *J. Polym. Sci., Part A: Polym. Chem.*, **34**, 2363, 1996)

(h)

$+ H_2N{-}N{=}CH{-}\bigcirc{-}CH{=}N{-}NH_2 \longrightarrow \xrightarrow[\text{pyridine}]{Ac_2O}$

(D. B. Sobanski et al., *J. Polym. Sci., Polym. Chem. Ed.*, **A23**, 189, 1986)

(i)

$+ CH_3NH(CH_2)_6NHCH_3 \longrightarrow$

[J. E. White et al., *Polym. Preprints*, **26**(1), 132, 1985]

(j)

(H. K. Reimschuessel et al., *Macromolecules*, **2**, 567, 1967)

(k)

$$\underset{\text{NCO}}{\overset{\mid}{\text{RCHCH}_2\text{NCO}}} \xrightarrow{\text{base}} \text{(linear polymer)}$$

(ref. 25) (*Hint*: See reaction 13.29)

(l) $\text{H}_2\text{N}\text{—R—NH}_2 + \text{CS}_2 \xrightarrow{(\text{H}_2\text{S})}$

(ref. 31 and G. J. M. van der Kerk et al., *Rec. Trav. Chim.*, **74**, 1301, 1955)

(m)

(L. J. Mathias and K. Al-Jumah, *J. Polym. Sci., Polym. Chem. Ed.*, **18**, 2911, 1980)

(n)

(*Hint*: See Section 12.2.3 and Ref. 44 in Chap. 12.)

(o)

(S. A. Pooley et al., *Eur. Polym. J.*, **31**, 547, 1995)

(p)

(*Note*: A Curtius rearrangement with loss of N_2 is involved in this reaction.) (J. F. Kinstle and L. E. Sepulveda, *J. Polym. Sci., Polym. Lett. Ed.*, **15**, 467, 1977)

(q)

(ref. 24) (*Hint*: See reaction 13.26)

18. Give the structure of the intermediate poly(amic acid) formed in the reaction shown in exercise 17(h).

19. When the polymer formed by the reaction in exercise 17(p) was heated to 200°C, considerable weight loss occurred. A solid that condensed on the cooler surfaces of the thermolysis apparatus was found to be identical to the polymer being heated, indicating that the polymer had apparently "sublimed." What is a more plausible explanation?

20. One approach to improving the tractability of aromatic polyamides has been to synthesize polymers of structure

which, upon prolonged heating at 200 to 300°C under nitrogen, expel ethylene by the retrograde Diels–Alder reaction to form the anthracene system. (A. H. Frazer et al., *J. Polym. Sci., Polym. Chem. Ed.*, **23**, 2779, 2791, 1985)
   (a) Explain the rationale for this approach. (You may wish to review the discussion of thermal stability in Chapter 4.)
   (b) Suggest a potential problem that might arise in processing (other than the usual problems of availability of monomers). (*Hint*: Polyimide processing has a similar problem.)

21. When aromatic amides substituted in the ortho position with alkynyl groups are heated, they undergo a cyclization reaction:

(isomer mixture)

Show what would happen if the polymer formed in exercise 17(a) were to be similarly heated. What effect would you expect this modification to have on (a) the polymer's glass transition temperature? (b) the polymer's processability?

14

PHENOL–, UREA–, AND MELAMINE–FORMALDEHYDE POLYMERS

14.1 Introduction

Phenol (**1**), urea (**2**), and melamine (**3**) are three compounds of seemingly disparate structure; yet they are related in terms of how they react with formaldehyde, as well as in the processing and applications of the resultant polymers.

Phenol–formaldehyde condensation polymers, often referred to as *phenolic resins*, were the first true synthetic polymers to gain commercial acceptance. This occurred early in this century, and they have maintained a prominent position in the polymer market to the present time. Urea–formaldehyde and melamine–formaldehyde resins appeared a few years later (in the early 1920s and 1930s, respectively). About 70% of all thermosetting polymers produced is made up of phenol–, urea–, and melamine–formaldehyde polymers, with phenolics enjoying the lion's share. Because the three polymer types exhibit some similarities in their chemistry, but more important, because their processing and end-use applications are related, they are treated together in this chapter.

Phenol–formaldehyde resins[1–6] are normally prepared by two different methods. One involves a base catalyst with an excess of formaldehyde over phenol. The initially formed product (called a *resole*) can be cured to a thermosetting polymer simply by heating; as such, it constitutes a one-component system. The other method utilizes an excess of phenol over formaldehyde in the presence of an acidic catalyst. In this case, the initial product, called a *novolac* (also spelled *novolak*), requires the addition of more formaldehyde to effect curing. The chemistry of these two processes is discussed below.

Phenolic resins are widely used as lacquers and varnishes, molding compounds, laminates (particularly for decorative wall panels and table tops), and adhesives (notably for plywood and particle board).[7] Flameproof fibers based on crosslinked novolacs (Carborundum trade name Kynol) are used for thermal insulation and protective clothing.[8] Urea and melamine resins[7,9,10] (often referred to as *amino resins*) are used in similar applications and also in the treatment of textiles for improved crease and shrink resistance and for improving the wet strength of paper. Phenolic polymers are sometimes used as ion-exchange resins when other functional groups are present.

14.2   Phenol–Formaldehyde Polymers: Resoles

Resoles are the product of the reaction between phenol and excess formaldehyde in the presence of base. Under these conditions phenol is present as the resonance-stabilized anion (**4**).

4

The first step in the polymerization involves addition of the anion to formaldehyde to give *ortho*- and *para*-substituted methylolphenols. This is shown (reaction 14.1) for the *ortho* product.* (Although reactions involving the *ortho* positions only are given below, it is understood that analogous reactions also occur at *para* positions.)

(14.1)

Because phenol is very reactive, simple monoaddition reactions seldom occur; instead, a mixture of monomethylolphenols, dimethylolphenols, and trimethylolphenols is formed with substitution occurring almost exclusively at the *ortho* and *para* positions. The initially formed methylolphenols condense on heating to give *resoles*, which are, in effect, low-molecular-weight prepolymers. They are soluble in base and contain a large number of free methylol groups. Resoles are, of course, complex mixtures of compounds, but a representative structure (**5**) is given. The methylene bridges linking the benzene rings result from condensation between methylolphenols and available *ortho* or *para* positions, either by direct S_N2 displacement of the hydroxyl group (14.2) or by Michael addition (14.4) to an *ortho* or

---

*Addition does not occur at the oxygen atom because the product, a hemiacetal, is unstable. Furthermore, the Cannizzaro reaction of formaldehyde to give methanol and formate ion is not a significant side reaction because the rate is much lower than that of formaldehyde with phenolate anion.

para quinone methide structure that might be present in equilibrium (14.3) with the methylol anion.

5

(14.2)

(14.3)

Studies with model compounds such as *o*-hydroxybenzyl alcohol (also known as saligenin) show that formaldehyde is also evolved in methylene bridge formation, suggesting that initially the bridge can occur at a carbon bearing a methylol group (14.5).

(14.4)

(14.5)

Ether linkages may also form (14.6), if the solution is very weakly basic or neutral.

$$2 \quad \text{(phenolate with } CH_2OH\text{)} \longrightarrow \text{(ether-linked phenolates } CH_2OCH_2\text{)} + H_2O \quad (14.6)$$

In commercial production, resoles are normally processed to a workable viscosity; then subsequent polymerization to high-molecular-weight network polymer (called *resite*) can be effected simply by heating. Depending on the application, this can be done directly on the base solution of resole or on neutral or slightly acidic solution. As a typical example, plywood adhesives are prepared by mixing appropriate additives such as wood flour with the basic resole solution. The mixture is then spread on the surface of the wood veneers prior to placing them in a hot press. The heat of the press not only causes polymerization to occur but also steams off the water. Excellent bonding is achieved by reaction between the resin and phenolic constituents of the wood. Thin sheets of paper impregnated with resin are often employed in place of the basic adhesive solution.

Curing of phenolic resins involves methylene bridge formation as described above. In acidic or neutral media, ether formation (14.6) is a significant crosslinking reaction, particularly at lower temperatures. This is because protonation of the alcohol groups by phenolic protons or by protons from added acid renders the groups much more susceptible to either S_N1 or S_N2 displacement by neighboring alcohol groups (14.7).

$$\xrightarrow{S_N1 \text{ or } S_N2} \quad (14.7)$$

Methylene bridges also form in acidic media by electrophilic substitution involving benzylic carbocations and available ring positions (14.8). At about 150°C formaldehyde is evolved from dissociation of benzylic alcohol groups (14.9), thus making available more sites for methylene bridge formation.

$$\longrightarrow \quad + H^+ \quad (14.8)$$

$$\xrightarrow{150°C} \quad + CH_2O \quad (14.9)$$

A reaction that occurs at about 180°C and that contributes to crosslinking is dehydration to form the highly reactive *ortho* or *para* quinone methides (14.10), which undergo a variety of reactions including cycloaddition with other quinone methides to give chroman groups

(14.11) or dimerization to give unsaturated linkages (14.12). Phenolic resins have been modified by reaction with drying oils to incorporate some flexibility into the otherwise very

$$ \text{(14.10)} $$

rigid structure. This modification presumably involves a Diels–Alder reaction between quinone methide groups and the double bonds of the drying oil.

$$ \text{(14.11)} $$

$$ \text{(14.12)} $$

It is apparent, then, that high-molecular-weight phenolic resins have extremely complex structures. It should be borne in mind that network polymers are formed because phenol is polyfunctional. If phenol is substituted in the *para* or in one *ortho* position, the functionality is reduced to two and it is possible to form linear polymer; however, high-molecular-weight polymers have not been prepared this way.

14.3 Phenol–Formaldehyde Polymers: Novolacs

Acid catalysis with excess phenol leads to a phenol–formaldehyde condensation product quite different from that obtained by base catalysis. The mechanism involves protonation of the carbonyl group (14.13) followed by electrophilic aromatic substitution (14.14) at *ortho* or *para* positions.

$$ \text{H}-\overset{\text{O}}{\overset{\|}{\text{C}}}-\text{H} \;\underset{}{\overset{H^+}{\rightleftharpoons}}\; (\text{CH}_2{=}\overset{+}{\text{O}}\text{H} \;\longleftrightarrow\; \overset{+}{\text{CH}}_2{-}\text{OH}) \qquad \text{(14.13)} $$

$$ \text{(14.14)} $$

Under acidic conditions further reaction occurs to give methylene bridges as discussed in the preceding section (reactions 14.7 and 14.8). The net result is the formation, in the early stages of polymerizaton, of complex mixtures of low-molecular-weight polymers (for example, **6**) characterized by having random *para–para*, *ortho–ortho*, or *ortho–para* methylene linkages. It has been shown, again with model compounds, that the *para* position is more reactive with strong acid catalysts at pH less than 3 (the usual condensation conditions); hence

6

ortho–ortho linkages undoubtedly occur less frequently. The opposite is the case at pH 4.5 to 6 where condensation at *ortho* positions predominates. The use of divalent metal catalysts facilitates *ortho* condensation—not only the initial reaction of phenol with formaldehyde, but also the subsequent formation of methylene bridges. (The products are referred to as *o,o'-*novolacs.) This directive effect of the catalyst probably results from chelation involving the phenolic hydroxyl group.[11] A possible mechanism involving hydrated formaldehyde and a zinc salt is given in reactions (14.15) to (14.17). The significance of having substitution predominantly in the *ortho* positions is that curing (described below) occurs more rapidly because the more reactive *para* positions remain available for crosslinking.

$$Zn^{2+} + HOCH_2OH \rightleftharpoons {}^+ZnOCH_2OH + H^+ \qquad (14.15)$$

$$(14.16)$$

$$(14.17)$$

TABLE 14.1. Effect of Reactant Ratio on Molecular Weight[a]

Moles Formaldehyde/10 Moles Phenol	Average Molecular Weight	Resin Type
1	229	Dihydroxydiphenylmethane
2	256	Novolac
3	291	Novolac
4	334	Novolac
5	371	Novolac
6	437	Novolac
7	638	Novolac
8	850	Novolac
9	1000	Novolac
12	—	Resite

[a] From Carswell,[6] courtesy of Wiley-Interscience.

Organotin compounds[12] and metal phenolates[13] are also effective in promoting *ortho* substitution. The reaction of phenoxymagnesium bromide with formaldehyde, for example, results in low-molecular-weight linear novolacs (**7**). The reaction of phenol with para-formaldehyde in anhydrous xylene at 190°C yields similar products.[14]

7

Unless excess phenol is used, the condensation reaction proceeds to high-molecular-weight infusible resin (resite), so in practice less than an equivalent of formaldehyde is reacted with phenol. The resultant product, a novolac, is fusible and has an average molecular weight dependent in large measure on the phenol/formaldehyde ratio, as shown in Table 14.1. Novolacs can also be prepared under basic conditions, but the reaction is complicated by a tendency toward chain branching and gelation. In the presence of divalent metal compounds, *ortho* condensation is preferred, as described above.

The basic difference between resoles and novolacs is that the latter contain no hydroxymethyl groups for all practical purposes and hence cannot be converted to network high polymer simply by heating. Crosslinking is brought about by adding additional formaldehyde or, more commonly, by adding paraformaldehyde or hexamethylenetetramine (**8**), a

8

high-melting (>230°C) solid having an interesting cage structure that is obtained by the reaction of formaldehyde with ammonia. Formaldehyde is released from these compounds under the influence of heat and pressure, as in a molding operation, and crosslinking occurs
again by formation of methylene bridges. Some benzylamine linkages (**9**) are also formed
when hexamethylenetetramine is used. Other types of additives commonly used as reinforcers
or extenders in molding formulas include wood flour, chopped rags, and glass fibers.
Asbestos, once commonly used as a reinforcer, is now out of favor because of health concerns. Pigments are also added, as are stearates or oils to facilitate releasing from the mold.
Phenolic resins are sometimes blended with synthetic rubber to enhance flexibility.

9

   Commercial production of both resoles and novolacs involves two basic steps. The
monomers and catalyst are reacted in aqueous solution to a relatively low viscosity; then water is removed under vacuum, and the solid product is ground to a powder. At this point the
polymer is of relatively low molecular weight, soluble, and fusible. It is referred to as the
A-stage. (As described earlier, aqueous solutions of resoles, with appropriate additives, are
used directly as plywood glues or for impregnating paper.) The A-stage resin is mixed with
additives (including hexamethylenetetramine or paraformaldehyde in the case of novolacs),
then heated further to a higher-molecular-weight B-stage. The polymer is converted to resite
(C-stage) in a final molding operation.

14.4   Chemical Modifications of Phenolic Resins

The only aldehyde of any significance, apart from formaldehyde, of use in phenolic resins
is furfural (**10**), obtainable from corncobs and oat hulls. It is frequently used in the prepara-

10

tion of molding resins because it improves the polymer's flow properties. Some other aldehydes that have been studied are acrolein, acetaldehyde, and butyraldehyde, although reactions of the latter two are complicated by their tendency to undergo aldol condensations,
particularly under basic conditions. Starch hydrolysates, which undergo dehydration to
form 5-hydroxymethylfurfural (**11**) under acidic conditions, have also been investigated as
a partial replacement for formaldehyde.[15] Related to **10** and **11** is furfuryl alcohol (**12**),
which undergoes acid-catalyzed condensation to form chemically resistant polymers called
furan resins,[16] which contain furan rings linked through methylene groups and thus are
structurally related to phenolic resins. The resins are used as tank and vat linings in chemical plants.

$$HOCH_2\underset{O}{\text{—}}CHO \qquad \underset{O}{\text{—}}CH_2OH$$

11   **12**

Considerably more work has been done on alternatives for phenol. Resorcinol (**13**) is much more reactive than phenol and is used to increase the rate of cure, particularly with cold set adhesives. Resorcinol's higher cost relative to that of phenol limits its use, however. The even higher-priced bisphenol A is used in certain coatings formulations for automotive body priming.[1] Bisphenol A–furfural resins have also been reported.[17] Monosubstituted phenols such as *o*- and *p*-cresol and the corresponding chlorophenols have been used to modify resin properties. Oil-soluble resins suitable for varnishes are prepared from *p-t*-butylphenol, *p*-phenylphenol, and *p-n*-octylphenol.

OH

OH

13

Unsaturated phenols are of interest because they provide further crosslinking by addition polymerization. Most important are those present in *cashew nut shell liquid* (CNSL).[18] CNSL, which is obtained from the shells by a process involving heating and decarboxylation, consists of mixtures of phenols with 15-carbon unsaturated side chains, the most important of which is cardanol (**14**). On reaction with formaldehyde, CNSL yields flexible resins suitable for brake lining and floor tile binders. Flexible phenolic resins are also obtained by using phenol modified with tung oil.[19] This modification involves ring substitution via the conjugated system of *α*-eleostearic acid (*cis,trans,trans*-9,11,13-octadecatrienoic acid), the principal fatty acid component of tung oil. Novolacs curable by addition polymerization have also been synthesized from 2-allylphenol (**15**) or by introducing allyl groups into a preformed novolac.[20]

OH

$(CH_2)_7CH=CH(CH_2)_5CH_3$

14

OH

$CH_2CH=CH_2$

15

Other naturally occurring phenols that are used as partial substitutes for phenol include *tannin*[21] and *lignin*.[22] The former, which is widely distributed in a variety of trees, consists of two types: hydrolyzable and condensed. Hydrolyzable tannins are mixtures of such phenolic compounds as pyrogallol (**16**), ellagic acid (**17**), and glucose esters of gallic acid (**18**) or condensed forms of gallic acid. The more abundant condensed tannins consist of a complex mix of flavonoid units linked together with carbohydrates. The increased use of tannins in phenolic resins has paralleled a decline in the leather tanning industries, once the sole domain of industrial tannins. Lignin, a major constituent of wood, is a complex phenolic polymer bearing a superficial resemblance to synthetic phenolic resins. Its structure is described in Chapter 18.

16   17   18

The flame resistance of phenolic resins may be enhanced with appropriate fillers or by the introduction of phosphate ester groups.[23] The latter is illustrated (14.18) with an *o,o'*-novolac.

$$(14.18)$$

Recent work has been directed toward production of aqueous dispersions or free-flowing solids of phenol–formaldehyde oligomers stabilized with a protective colloidal polysaccharide coating.[24,25] This technology yields prepolymers that are easier to manipulate and process to resites than are conventional phenolic resins.

14.5 Urea–Formaldehyde Polymers

Urea (**2**), prepared commercially from ammonia and carbon dioxide, undergoes reactions with formaldehyde that are analogous, at least in the initial stages, to those of phenol. Unlike other amides, urea is basic because the single carbonyl group is not sufficient to compensate for two amino groups. Urea undergoes nucleophilic addition to formaldehyde to give methylol derivatives (14.19). The solution is kept mildly basic by careful pH control because the methylol derivatives condense rapidly under acidic conditions.

$$\mathbf{2} + \text{H}-\overset{\text{O}}{\overset{\|}{\text{C}}}-\text{H} \longrightarrow \text{H}_2\text{NCNHCH}_2\text{OH} + \text{HOCH}_2\text{NHCNHCH}_2\text{OH} \qquad (14.19)$$

The similarity between urea and phenolic resins, at least as far as their chemistry is concerned, ends at this point.

There has been considerable speculation recently concerning the structure of urea–formaldehyde polymers. It had long been assumed that further condensation involved cyclic intermediates formed by initial dehydration of the protonated methylol derivatives to give imines (14.20), which then trimerize (14.21). The corresponding cyclic compound having methylol groups attached to the amide nitrogen atoms would be formed from dimethylolurea.

$$\underset{H}{H_2N\overset{O}{\overset{\|}{C}}NHCH_2\overset{+}{O}H} \rightleftharpoons H_2N\overset{O}{\overset{\|}{C}}N{=}CH_2 + H_3O^+ \tag{14.20}$$

$$3H_2N\overset{O}{\overset{\|}{C}}N{=}CH_2 \longrightarrow \tag{14.21}$$

Subsequent condensation would then occur between methylol and amide groups (14.22) to give network polymer (**19**). Support for this mechanism stems from known cyclization reactions of other amides and also from the nitrogen content, which approximates that expected for a polymer formed by this mechanism.

$$\overset{O}{\overset{\|}{N\text{C}NHCH_2OH}} + H_2N\overset{O}{\overset{\|}{C}} \longrightarrow \overset{O}{\overset{\|}{N\text{C}NHCH_2NHC\text{N}}} + H_2O \tag{14.22}$$

19

More recent work[26,27] suggests that network polymers are *not* formed, particularly with the relatively low formaldehyde–urea ratios employed in many industrial applications. Instead, it has been proposed that linear oligomeric urea–formaldehyde condensates (**20**) are

$$\sim\!\!\sim\!\!CH_2NH\overset{O}{\overset{\|}{C}}NHCH_2NH\overset{O}{\overset{\|}{C}}NH\!\sim\!\!\sim$$

20

formed that then separate as a colloidal dispersion stabilized by association with excess formaldehyde. "Curing" results from agglomeration of the colloidal particles with concomitant release of formaldehyde. Support for this viewpoint is based on several observations: (1) The polymers are soluble in formaldehyde and concentrated sulfuric acid, and can be recovered from the latter unchanged. (2) A plot of the logarithm of solution viscosity versus time during polymerization exhibits a sharp break, unlike similar plots for phenol–formaldehyde polymers, which show a continuous increase in viscosity with development of the polymer network. (3) X-ray diffraction patterns and laser Raman spectra of the cured resins indicate the presence of crystalline regions and an absence of water, respectively, both consistent with hydrogen-bonded proteinlike or nylonlike close chain packing. And (4) scanning electron micrographs of the cured resins show surface characteristics more in common with those of coalesced colloids than high-molecular-weight network polymers. The structure of urea–formaldehyde resins may thus resemble that of proteins more than that of phenolic resins. FTIR studies,[28] on the other hand, suggest that some methylene and ether crosslinks are probably present, as are cyclic ether units, and that the true structure of urea–formaldehyde polymers is a function of both feed ratio and processing pH.

As was mentioned earlier, urea–formaldehyde resins are used in applications similar to those of phenol–formaldehyde resins—for molding, laminating, and adhesives. One advantage of the urea resins is that they are light in color, and hence are more suitable for decorative use. (The dark color of phenolic resins is due to a combination of oxidation and quinone methide formation.) Also, decorative interior plywood is normally glued with urea resin because the dark-colored phenolic resins can stain the veneer. Exterior plywood, however, is glued with phenolic resin because it has better weather resistance. In the coatings field, urea–formaldehyde resins are sometimes blended with alkyd baking enamels (Chapter 12) to improve hardness. Urea resins are also used to impart crease and shrink resistance to fabrics through crosslinking reactions. Urea–formaldehyde foams were once marketed as home insulation; however, these were discontinued because of health concerns arising from release of formaldehyde vapors.

14.6   Melamine–Formaldehyde Polymers

Melamine (**3**) is an aromatic, heterocyclic compound prepared by the thermal trimerization of cyanamide (14.23). Its systematic name is 2,4,6-triamino-1,3,5-triazine.

$$3H_2N-C\equiv N \xrightarrow{\text{heat}} \quad \mathbf{3} \qquad (14.23)$$

Melamine undergoes condensation reactions with formaldehyde in a manner analogous to those of urea. A major difference is that, in the initial nucleophilic addition to formaldehyde, each amino group may form a dimethylol derivative (14.24). In practice, prepolymers are prepared having some monomethylol-substituted amino groups, and subsequent curing is mainly through methylene bridge formation by condensation between methylol and amino groups (14.25). The prepolymers are water soluble but can be made soluble in organic solvents by etherification with such alcohols as 1-butanol.

$$3 + CH_2O \longrightarrow (HOCH_2)_2N \underset{\text{(excess)}}{\quad} \text{...} \quad N(CH_2OH)_2 \qquad (14.24)$$

$$(14.25)$$

Melamine resins are generally harder and more moisture resistant than urea resins. They are used in molding and laminating formulations and, like urea resins, as alkyd enamel modifiers and textile fiber finishes. Under the trade name Formica they are used for laminated counter and table tops. Their largest use is in the manufacture of decorative dinnerware. A modified melamine–formaldehyde resin is marketed as a reinforcing fiber by BASF (trade name Basofil).

References

1. A. Knop and L. A. Pilato, *Phenolic Resins*, Springer-Verlag, New York, 1979.
2. A. A. K. Whitehouse, E. G. K. Pritchett, and G. Barnett, *Phenolic Resins*, Iliffe, London, 1967.
3. N. J. L. Megson, *Phenolic Resin Chemistry*, Academic Press, New York, 1958.
4. R. W. Martin, *The Chemistry of Phenolic Resins*, Wiley-Interscience, New York, 1956.
5. D. F. Gould, *Phenolic Resins*, Reinhold, New York, 1959.
6. T. S. Carswell, *Phenoplasts*, Wiley-Interscience, New York, 1947.
7. A. Pizzi (Ed.), *Wood Adhesives: Chemistry and Technology*, Marcel Dekker, New York, 1983, Chaps. 2 and 3.
8. J. Economy, L. C. Wohrer, F. J. Frechette, and G. Y. Lei, *Appl. Polym. Symp.*, **21**, 81 (1973).
9. L. L. Williams, I. H. Updegraff, and J. C. Petropolous, in *Applied Polymer Science* (R. W. Tess and G. W. Poehlein, Eds.), ACS Symp. Ser. 285, American Chemical Society, Washington, D.C., 1985, Chap. 45.
10. B. Meyer, *Urea–Formaldehyde Resins*, Addison-Wesley, Menlo Park, Calif., 1979.
11. M. F. Drumm and J. R. LeBlanc, in *Step-Growth Polymerizations* (D. H. Solomon, Ed.), Marcel Dekker, New York, 1973, Chap. 5.
12. A. Ninagawa, Y. Ohnishi, H. Takeuchi, and H. Matsuda, *Makromol. Chem., Rapid Commun.*, **6**, 793 (1985).
13. G. Casiraghi, G. Sartori, F. Bigi, M. Cornia, E. Dradi, and G. Casnati, *Makromol. Chem.*, **182**, 2151 (1981).
14. G. Casiraghi, G. Casnati, M. Cornia, G. Sartori, and F. Bigi, *Makromol. Chem.*, **182**, 2973 (1981).

15. H. Koch and J. Pein, *Polym. Bull.*, **13**, 525 (1985).
16. J. A. Brydson, *Plastics Materials*, Butterworth, London, 1975, Chap. 28.
17. R. B. Jambusaria and S. P. Potnis, *Polym. J.*, **6**, 333 (1974).
18. P. A. Mahanwar and D. D. Kale, *J. Appl. Polym. Sci.*, **61**, 2107 (1996).
19. Y. Yoshimura, *J. Appl. Polym. Sci.*, **29**, 1063 (1984).
20. K. M. Hui and L. C. Yip, *J. Polym. Sci., Polym. Chem. Ed.*, **14**, 2323 (1976).
21. A. Pizzi, in *Wood Adhesives: Chemistry and Technology* (A. Pizzi, Ed.), Marcel Dekker, New York, 1983, Chap. 4.
22. H. N. Nimz, in *Wood Adhesives: Chemistry and Technology* (A. Pizzi, Ed.), Marcel Dekker, New York, 1983, Chap. 5; G. H. van der Klashorst, *ibid.*, Vol. 2, Chap. 6.
23. B. F. Dannels and A. F. Shepard, *J. Polym. Sci. A-1*, **6**, 2051 (1968).
24. G. L. Brode, T. R. Jones, and S. W. Chow, *Chemtech*, **13**, 630 (1983).
25. G. L. Brode, *J. Macromol. Sci.-Chem.*, **A22**, 897 (1985).
26. T. J. Pratt, W. E. Johns, R. M. Rammon, and W. L. Plagemann, *J. Adhesion*, **17**, 275 (1985).
27. A. K. Dunker, W. E. Johns, R. Rammon, B. Farmer, and S. J. Johns, *J. Adhesion*, **19**, 153 (1986).
28. S. S. Jada, *J. Appl. Polym. Sci.*, **35**, 1573 (1988).

Review Exercises

1. The ratio of *ortho* to *para* substitution in the reaction of phenol with formaldehyde is greatest at about pH 6. *Ortho* substitution is also favored in aprotic nonpolar solvents. Give a plausible explanation for this behavior.

2. Besides urea and melamine, aniline also reacts with formaldehyde to form thermosetting resins (rarely used today). Their chemistry resembles that of both phenolic and melamine resins. Write plausible reactions for the formation of aniline–formaldehyde resins under acidic, basic, and neutral conditions. (ref. 7)

3. Urea– and melamine–formaldehyde resins are widely used as curing agents for coatings such as alkyds (see Chapter 12). In such cases the resins are often etherified with, for example, 1-butanol. Also, a melamine substitute such as "benzoguanamine" (2-phenyl-4,6-diamino-1,3,5-triazine) is sometimes used (although less frequently because of increasing cost). (a) What advantage might such formulations have over conventional urea or melamine resins? (b) Show mechanistically how a urea or melamine resin may be butylated under acidic conditions. (c) What reactions might be involved in curing a glyptal-type coating resin (see Chapter 12) with urea– or melamine–formaldehyde resin, and what advantage would this type of modification have from the standpoint of processing? (ref. 9)

4. Show mechanistically how furan resins are formed from furfuryl alcohol (**12**) under acidic conditions. Crosslinking is accompanied by loss of unsaturation. Suggest a possible crosslinking reaction. (ref. 16)

5. Another ingredient of CNSL is cardol, which is structurally similar to cardanol (**14**), but with an extra OH at the free *meta* position. What polymers other than phenolic resins could be made from cardol?

C H A P T E R

15

HETEROCYCLIC
POLYMERS

15.1 Introduction

In Chapter 4, where the concept of thermal stability was introduced, we learned that aromatic polymers exhibit better heat resistance than nonaromatic polymers. To meet the needs of the aerospace industries for light-weight materials that retain their properties for prolonged periods at elevated temperatures, a tremendous amount of research was devoted to the synthesis and evaluation of aromatic heterocyclic polymers from the 1960s through the 1980s, with the result that virtually every known heterocyclic system has appeared in a polymer formulation.[1-5] Relatively few have seen commercial development, partly because of the high cost of starting materials, but probably more because of the difficulty in processing polymers that have very high glass transition temperatures. Some of the strategies for dealing with processing problems were addressed in Chapter 4. In this chapter we will explore just a few of the numerous synthetic methods for making heterocyclic polymers, with particular emphasis on polymers that are used commercially or have been considered for commercial development.

A remarkable number of heterocyclic polymers are prepared by reactions that lead to intermediate polyamides or related polymers that subsequently undergo cyclization leading to the heterocyclic ring systems. Examples are given in Table 15.1. In general, four types of reactions are used to make heterocyclic polymers:

1. Oxidative polymerization of an existing heterocyclic compound. The conducting polymer polypyrrole (cf. Chapter 4) is an example. These are discussed in the next section.

2. Homopolymerization of monomers containing a single type of functional group capable of undergoing a cyclization reaction. The commercially important polycyanurate resins (Section 15.3) are representative.

3. Cyclization of precursor polymers. Table 15.1 includes, for example, heterocyclic polymers prepared from polyhydrazides. These are discussed in Section 15.4.

TABLE 15.1. Heterocyclic Polymers Derived from Polyamides and Related Polymers

Ring System	Polymer Type	Functional Groups Required	Intermediate Stage
	Polybenzimidazole	Phenyl ester + diamine	Polyimine
	Polybenzoxazole	Phenyl ester + o-aminophenol	Polyamide
	Polybenzothiazole	Phenyl ester + o-aminothiophenol	Polyamide
	Polyimidazopyrrolone	Anhydride + diamine	Polyamide
	Poly(1,3,4-oxadiazole)	Hydrazide	Polyhydrazide
	Poly(1,2,4-triazole)	Hydrazide + amine	Polyhydrazide
	Poly(1,3,4-thiadiazole)	Thiohydrazide	Polythiohydrazide
	Polyhydantoin	Isocyanate + α-amino acid	Polyurea
	Poly(parabanic acid)	Isocyanate + HCN	Polyurea
	Polythiazoline	Thiourea + alcohol	Polythiourea
	Polyimidine	Lactone + amine	Polyamide
	Polybenzoxazinone	Acid chloride + o-aminobenzoic acid	Polyamide

continued

TABLE 15.1. (continued)

Ring System	Polymer Type	Functional Groups Required	Intermediate Stage
	Polybenzoxazinedione	Amine + *o*-hydroxy-benzoate ester	Polyamide
	Polyquinazolinedione	Isocyanate + *o*-amino-benzoic acid	Polyurea
	Polyisoindoloquina-zolinedione	Anhydride + *o*-amino-benzamide	Polyamide
	Polytetraazopyrene	Tetraamine + diphenyl ester	Polyamide

4. Reaction of di- or polyfunctional monomers that undergo polycondensation reactions. These constitute the majority, and most of the polymers listed in Table 15.1 are of this type. Condensation polyimides, discussed in Chapter 13, are another example. (Note that polyimides are, in fact, heterocyclic polymers; they were included in Chapter 13 because of their structural relationship with polyamides.)

Polyfunctional monomers most commonly used are dicarboxylic acids or their derivatives (esters, acid chlorides, and nitriles) in conjunction with such monomers as bis(*o*-phenylenediamine)s (**1**), diamines substituted in the *ortho* position with other nucleophilic groups (**2**), and dihydrazidines (also called bisamidrazones) (**3**). Examples of heterocyclic polymers prepared with these monomers are covered in Section 15.5.

1

2

(X = OH, SH, or carbonyl-containing group)

$$ \underset{3}{H_2NNH-\overset{\overset{\displaystyle NH}{\|}}{C}-Ar-\overset{\overset{\displaystyle NH}{\|}}{C}-NHNH_2} $$

15.2 Polypyrrole, Polyfuran, and Polythiophene

Polypyrrole (**4**) is of principal interest because, when doped, it exhibits electrical conductivities comparable to those of polyacetylene and poly(sulfur nitride) (see Chapter 4) but is more readily processable. The polymer is prepared by electrolytic oxidation of pyrrole,[6,7] a reaction that apparently proceeds via three steps: (1) formation of a radical cation, (2) radical combination, and (3) deprotonation. These steps are illustrated in Scheme 15.1 for formation of dimer and trimer.

4

Electrolysis of an acetonitrile solution of pyrrole in the presence of tetraethylammonium fluoroborate results in precipitation of highly conducting BF_4^--doped films of polypyrrole on the anode.[8] Chemical oxidizing agents such as iron(III) chloride can also be used to make polypyrrole,[9] but the conductivity is usually inferior to that obtained by electropolymerization.

Furan[10,11] and thiophene[12,13] undergo polymerization in the presence of Lewis acids or mineral acids. The polymer structures are relatively complex, however, and cannot be represented in an idealized form comparable to that of polypyrrole. On the basis of spectroscopic evidence and elemental analyses, the principal structures present in polyfuran and polythiophene appear to be **5** and **6**, respectively. (Polyfuran should not be confused with the furan resins described in Chapter 14.)

5   **6**

SCHEME 15.1. Proposed mechanism for the electrolytic polymerization of pyrrole.

Electrolytic oxidation has also been used to prepare a form of polythiophene that is conducting.[14] In this case the structure is best represented in the idealized from as **7**.

7

15.3 Polycyanurate and Polyphthalocyanine Resins

Polycyanurate[15,16] resins are formed by the thermal homopolymerization of compounds containing two or more cyanate ester (—O—C≡N) groups. The principal curing reaction is trimerization to form a network structure of 1,3,5-triazine rings (15.1). As such the polymers are also called *poly(1,3,5-triazine)*s. Several cyanate esters are used commercially, including one derived from bisphenol A (**8**) and a cycloaliphatic dicyanate (**9**) from dicyclopentadiene. The network polymers derived from **8** and **9** have glass transition temperatures of 289 and 244°C, respectively. Some applications of polycyanurates include circuit boards, brake linings, grinding wheels, aircraft parts, and radome cones.

$$\text{NCO—Ar—OCN} \xrightarrow{\Delta}$$ (15.1)

8

9

Polyphthalocyanines,[17] synthesized by the thermal polymerization of bisphthalonitriles (**10**), are of interest as matrix resins for high-temperature composites. The phthalocyanine ring system is like that found in metal porphyrin compounds (see reaction 16.32 in the next chapter), but without the coordinated metal. The polymers are very complex in nature with triazine and possibly other ring systems present besides phthalocyanine.

10

15.4 Heterocyclic Polymers Derived from Precursor Polymers

Polyhydrazides (**11**), first introduced in Chapter 13, can be converted to *poly(1,3,4-oxadiazole)*s (**12**) by dehydration (15.2) at about 280°C with about 95% conversion.[18] Using a precursor polymer in this fashion is an often convenient way to overcome processing difficulties associated with a high T_g or intractability. (Recall the precursor route to polyacetylene discussed in Chapter 9.) Thus articles fabricated from the soluble thermoplastic polyhydrazide can be transformed to poly(1,3,4-oxadiazole) simply by heating. Interestingly, polymer **12** can also be synthesized directly by heating a dicarboxylic acid or its derivative, or a dihydrazide, in solution with concentrated sulfuric acid or PPA.[19] Presumably the polyhydrazide is an intermediate in this reaction.

$$
\left[Ar\!-\!\overset{O}{\overset{\|}{C}}\!-\!NH\!-\!NH\!-\!\overset{O}{\overset{\|}{C}} \right] \xrightarrow[\;(-H_2O)\;]{\Delta} \left[Ar\underset{N-N}{\overset{O}{<\!\!\!>}} \right]
$$

$$
\mathbf{11} \qquad\qquad\qquad\qquad\qquad\qquad \mathbf{12} \tag{15.2}
$$

Polyhydrazide **11** can also be converted to phenyl-substituted *poly(1,2,4-triazole)* (**13**) by prolonged heating with aniline in the presence of polyphosphoric acid (PPA) (15.3). Conversion is not as high, however, and some chain degradation occurs.[20] As might be expected, the sulfur analog of **12**, *poly(1,3,4-thiadiazole)* (**15**), is prepared by dehydrosulfurization (15.4) of the less accessible polydithiahydrazide (**14**). All three polymers (**12**, **13**, and **15**) exhibit good thermal and hydrolytic stability, with decomposition temperatures in the range of 450–500°C. There has been considerable interest in poly(1,3,4-oxadiazole)s (acronym POD) for high-modulus fiber applications,[21,22] and one type (**12**, Ar = *p*-phenylene) has been marketed in Russia under the Oksalon trade name.

$$
\mathbf{11} \xrightarrow[\text{PPA}, \Delta]{\text{PhNH}_2} \left[Ar\underset{N-N}{\overset{\overset{\displaystyle Ph}{\overset{|}{N}}}{<\!\!\!>}} \right] \tag{15.3}
$$

$$
\mathbf{13}
$$

$$
\left[Ar\!-\!\overset{S}{\overset{\|}{C}}\!-\!NH\!-\!NH\!-\!\overset{S}{\overset{\|}{C}} \right] \xrightarrow[\;(-H_2S)\;]{\Delta} \left[Ar\underset{N-N}{\overset{S}{<\!\!\!>}} \right] \tag{15.4}
$$

$$
\mathbf{14} \qquad\qquad\qquad\qquad\qquad\qquad \mathbf{15}
$$

Another precursor polymer that cyclizes to give backbone oxadiazole and triazole units is the poly(*N*-acylamidrazone) **16**, prepared from monomer **3** and a diacid chloride.[23] Depending on the cyclization conditions, **16** loses ammonia to form poly(1,3,4-oxadiazole) (15.5) or water to yield poly(1,2,4-triazole) (15.6), or both.

$$\left[\begin{array}{c} \overset{O}{\underset{\|}{C}}NHNH\overset{NH}{\underset{\|}{C}}-Ar-\overset{NH}{\underset{\|}{C}}NHNH\overset{O}{\underset{\|}{C}}-Ar'- \end{array}\right] \quad \underset{16}{} \quad \xrightarrow{-NH_3} \left[\begin{array}{c} \overset{O}{\underset{N-N}{\diagup}}-Ar-\overset{O}{\underset{N-N}{\diagup}}-Ar'- \end{array}\right] \quad (15.5)$$

$$\xrightarrow{-H_2O} \left[\begin{array}{c} \overset{NH}{\underset{N-N}{\diagup}}-Ar-\overset{NH}{\underset{N-N}{\diagup}}-Ar'- \end{array}\right] \quad (15.6)$$

Polymers containing conjugated triple bonds have also been converted to polymers containing heterocyclic rings (pyrrole, thiophene, and pyrazole); these reactions are covered in Chapter 17.

15.5 Heterocyclic Polymers from Polyfunctional Monomers

15.5.1 Polybenzimidazoles

These were among the first and most successful of the wholly aromatic polymers,[24,25] an outgrowth of the pioneering work of Carl Marvel.[26] Indeed, much of the flood of research into heterocyclic polymers was inspired by Marvel's success with polybenzimidazoles. They are normally prepared from esters of dibasic acids (particularly phenyl esters) and tetramines (reaction 15.7) or from diamino esters (15.8). The former reaction is used to prepare the commercially important polybenzimidazole **17** (PBI) in which R is *m*-phenylene (structure **1**, Chapter 4). Esters rather than acids are used because the latter tend to decarboxylate at the high temperatures necessary for benzimidazole formation. Polybenzimidazoles can, however, be prepared from diacids at moderate temperatures (about 140°C) when phosphorus pentoxide in methanesulfonic acid is used as a condensing agent.[27]

$$\begin{array}{c} H_2N \quad\quad NH_2 \\ \\ H_2N \quad\quad NH_2 \end{array} + \overset{O}{\underset{\|}{PhOC}}-R-\overset{O}{\underset{\|}{COPh}} \longrightarrow \quad\quad (15.7)$$

$$\left[\begin{array}{c} HN \quad\quad NH \\ \\ N \quad\quad N \end{array}-R-\right] + 2H_2O + 2PhOH$$

$$\underset{17}{}$$

$$H_2N-\overset{O}{\underset{\|}{\diagdown}}-COPh \quad \underset{H_2N}{} \longrightarrow \left[\begin{array}{c} HN \\ \\ N \end{array}\right] + H_2O + PhOH \quad (15.8)$$

The more conventional synthetic procedure is to heat diester and tetraamine at temperatures below 300°C to form a solid prepolymer, with the final polymerization being done in the solid state at temperatures up to 400°C. It is tempting to write a mechanism involving nucleophilic attack of amine on the ester carbonyl followed by loss of phenol to give polyamide intermediate. Subsequent attack by the second amino group at the amide carbonyl followed by loss of water would then give imidazole. Interestingly, however, it has been shown[28] that water is evolved *before* phenol, suggesting a more plausible mechanism via an intermediate imine (15.9). Addition of the second amino group to the carbon–nitrogen double bond followed by rate-determining loss of phenol gives imidazole (15.10).

$$(15.9)$$

$$(15.10)$$

Polybenzimidazoles can also be made by the reaction of tetramines with dialdehydes[29] or dialdehyde bisulfite addition compounds.[30] This leads to an intermediate polyimine (**18**) that undergoes cyclization on reaction with oxygen (15.11).

$$(15.11)$$

Polybenzimidazoles are highly colored (yellow to black) and exhibit good flame resistance. The limiting oxygen index (LOI) of **17** (R = *m*-phenylene) is 28, compared with values of 17 or 18 for aramids. The polymers are porous because of loss of the more bulky phenol in the later stages of polymerization, and they retain moisture strongly, even above 100°C. Nevertheless, they exhibit very high hydrolytic stability. At elevated temperatures in the absence of air, polybenzimidazoles decompose to yield phenazine- and carbazole-type chars (**19** and **20**, respectively), but retain up to 80% of their mass up to 800°C. In air they undergo catastrophic weight loss between 450 and 500°C.

          19   **20**

If dianhydrides are reacted with tetraamines, ladder *polyimidazopyrrolones* (also called *ladder pyrrones*) are formed that combine the structural features of imidazole and pyrrolone rings.[31–33] Thermoplastic polyamide, which can be cast or molded prior to cyclization, is formed at room temperature (15.12). Subsequent heating results in a two-reaction sequence (15.13, 15.14) to yield blood-red thermosetting polymer (**21**). The center imidazopyrrolone structure of **21** would be inverted if the isomeric polyamide structure were to form in reaction (15 12). Polyimidazopyrrolones exhibit excellent thermal stability and radiation resistance.

(15.12)

(15.13)

(15.14)

21

15.5.2 Polybenzoxazoles and Polybenzothiazoles

These polymers are related in structure to polybenzimidazoles and are prepared by analogous reactions using bis(*o*-aminophenol)s for the former and bis(*o*-aminothiophenol)s for the latter.[21,34–37] Reaction (15.15) illustrates the synthesis of the polybenzoxazole (PBO) and polybenzothiazole (PBT) polymers (**22** and **23**, respectively) that were developed for commercial use, primarily as high-modulus fibers to compete with aramids. Both polymers have since been discontinued.

X = O or S

(15.15)

22 (X = O)
23 (X = S)

15.5.3 Polyhydantoins

Several routes are available for the synthesis of polyhydantoins,[38] among them the reaction of di-isocyanates with bis(α-amino acid)s (15.16) to form polymer **24**. Initially a polyurea is formed by addition of the amino group to isocyanate, which is then followed by cyclization with loss of water. Characteristically, polyhydantoins exhibit high glass transition temperatures and good chemical resistance. They have been used as wire coatings and insulating films for electric motors.[39]

$$
\text{HO}_2\text{CCNH}-\text{Ar}-\text{NHCCO}_2\text{H} + \text{OCN}-\text{Ar}'-\text{NCO} \longrightarrow
$$

(15.16)

24

15.5.4 Poly(parabanic acid)s

Also called *poly(1,3-imidazolidine-2,4,5-trione)s*, these polymers are manufactured by the reaction of diisocyanates with HCN (15.17) to yield an intermediate poly(5-imino-1,3-imidazolidinedione) (**25**), which is hydrolyzed (15.18) in a separate step to the poly(parabanic acid) (**26**).[40] The polymers may be thought of as dicarbonyl derivatives of polyureas; indeed, the parabanic acid unit can be prepared by the reaction of polyureas with oxalyl chloride[41] (15.19).

$$
\text{OCN}-\text{R}-\text{NCO} + \text{HCN} \longrightarrow
$$

(15.17)

25

$$
\mathbf{25} \xrightarrow{\text{H}^+ \ \text{H}_2\text{O}}
$$

(15.18)

26

(15.19)

Poly(parabanic acid)s were originally considered lower cost substitutes for polyimides, but development work was discontinued.

15.5.5 Polyquinoxalines and Polypyrazines

Polyquinoxalines[23,42] are synthesized from tetracarbonyl compounds and tetraamines. An example is the ladder polymer **27** shown in reaction (15.20). Polyquinoxalines have generated considerable interest as heat-resistant resins for aerospace applications. An example of an intramolecularly curable polyquinoxaline is given in Chapter 4 (Scheme 4.1).

(15.20)

27

The polymerization reactions can be carried out in solution or in the melt, but generally postpolymerization of the isolated polymer at about 350 to 400°C is necessary to achieve high molecular weights. Polyquinoxalines are stable in nitrogen at temperatures up to 400 to 500°C; then weight loss occurs gradually at temperatures up to 800 to 900°C. In air they decompose rapidly at 400 to 500°C. Ladder polyquinoxalines (**27**) reportedly show little weight loss in nitrogen at temperatures below 600 to 700°C.

The quinoxaline unit may also be referred to as a *benzopyrazine* group. Polymers containing free pyrazine groups (*polypyrazines*, **28**)[43] are prepared from bis(α-haloketones) by reaction with ammonia under oxidative conditions (15.21). Presumably, the corresponding bis(α-aminoketone) is intermediate in the reaction. If the polymer is heated with hydrogen peroxide, some of the nitrogen atoms are converted to the polar N-oxide groups, which enhances polymer solubility.

(15.21)

28

15.5.6 Polypyrazoles and Polyimidazoles

The pyrazole ring is most conveniently prepared from 1,3-dicarbonyl compounds and hydrazine derivatives. The extension of this reaction to bishydrazines and tetracarbonyl compounds (15.22) gives polypyrazoles[44] (**29**). Aromatic polypyrazoles of the type shown are black, infusible materials exhibiting good thermal stability.

(15.22)

29

The imidazole ring is isomeric with the pyrazole ring. Almost the entire emphasis in the area of imidazole polymers has been with polybenzimidazoles. Polymers of relatively low molecular weight having nonbenzo imidazole linkages (**30**) have been synthesized from tetraketones, dialdehydes, and ammonia[45] (15.23).

(15.23)

30

15.5.7 Poly(as-triazine)s and Polytriazolines

Poly(*as*-triazine)s (**31**) and polytriazolines (**32**) are synthesized from dihydrazidines by reaction with bisdicarbonyl compounds (15.24) and dialdehydes (15.25), respectively.[23,46] As might be expected, the nonaromatic polytriazolines are not thermally stable; however, poly(*as*-triazine)s are stable up to about 380°C, at which point the nitrogen–nitrogen bond breaks; the backbone remains intact up to about 530°C.

(15.24)

32

31

(15.25)

32

15.5.8 Polyquinolines and Polyanthrazolines

The quinoline and anthrazoline rings are structurally related and, as might be expected, polyquinolines (**33**) and polyanthrazolines (**34**) are synthesized in essentially the same manner.[47–49] Bis(*o*-aminocarbonyl) compounds and diacetyl compounds, reacted together in the presence of an acid catalyst, are the monomers in each case (15.26, 15.27). If R is phenyl in these reactions, the polymers are significantly more soluble than if R is hydrogen. Both polymer types are stable to above 500°C.

(15.26)

33

(15.27)

34

References

1. P. E. Cassidy, *Thermally Stable Polymers*, Marcel Dekker, New York, 1980, Chaps. 6–9.
2. V. Rossbach and G. Oberlein, in *Handbook of Polymer Synthesis*, Part B (H. R. Kricheldorf, Ed.) Marcel Dekker, 1992, Chap. 19.
3. V. V. Korshak, *Heat-Resistant Polymers*, Halstead, New York, 1972.
4. J. P. Critchley, G. J. Knight, and W. W. Wright, *Heat-Resistant Polymers*, Plenum, New York, 1983.
5. A. H. Frazer, *High Temperature Resistant Polymers*, Interscience, New York, 1968.
6. G. Wegner, *Angew. Chem. Int. Ed. Engl.*, **20**, 361 (1981).
7. M. Satch and E. Hasegawa, *Macromol. Symp.*, **105**, 211 (1996).
8. A. F. Diaz, K. K. Kanazawa, and G. P. Gandini, *J. Chem. Soc., Chem. Commun.*, 635 (1979).
9. N. Toshima and S. Hara, *Prog. Polym. Sci.*, **20**, 153 (1995).
10. A. Gandini, *Adv. Polym. Sci.*, **25**, 47 (1977).
11. B. S. Lamb and P. Kovacic, *J. Polym. Sci., Polym. Chem. Ed.*, **18**, 2423 (1980).
12. J. Roncali, *Chem. Rev.*, **92**, 711 (1992).
13. G. Schopf and G. Kossmehl, *Polythiophenes: Electrically Conducting Polymers*, Springer-Verlag, Berlin, 1997.
14. R. J. Walman and J. Bargon, *Can. J. Chem.*, **64**, 76 (1986).
15. I. Hamerton (Ed.), *Chemistry and Technology of Cyanate Ester Resins*, Chapman and Hall, 1994.
16. D. A. Shimp, in *Engineered Materials Handbook, Vol. 2: Engineering Plastics* (J. N. Epel et al., Eds.), ASM International, Metals Park, Ohio, 1987, pp. 232–39.
17. A. W. Snow and J. R. Griffith, in *Encyclopedia of Polymer Science and Engineering*, 2nd ed., Vol. 11

(H. F. Mark, N. M. Bikales, C. G. Overberger, G. Menges, and J. I. Kroschwitz, Eds.), Wiley-Interscience, New York, 1988, pp. 212–25.

18. A. H. Frazer and F. T. Wallenberger, *J. Polym. Sci. A*, **2**, 1157 (1964).
19. Y. Iwakura, K. Uno, and S. Hara, *J. Polym. Sci. A*, **3**, 45 (1965); *Makromol. Chem.*, **94**, 103 (1966); *Makromol. Chem.*, **95**, 248 (1966).
20. J. R. Holsten and M. R. Lilyquist, *J. Polym. Sci. A*, **3**, 3905 (1965).
21. H. H. Yang, *Aromatic High-Strength Fibers*, Wiley, New York, 1989, Chap. 7.
22. W. B. Black and J. Preston (Eds), *High-Modulus Wholly Aromatic Fibers*, Marcel Dekker, New York, 1973.
23. P. M. Hergenrother, in *Step-Growth Polymerizations* (D. H. Solomon, Ed.), Marcel Dekker, New York, 1972, Chap. 4.
24. E. W. Neuse, *Adv. Polym. Sci.*, **47**, 1 (1982).
25. A. B. Conciotori, A. Buckley, and D. E. Stuetz, in *High Technology Fibers, Part A* (M. Lewin and J. Preston, Eds.), Marcel Dekker, New York, 1985, Chap. 6.
26. H. Vogel and C. S. Marvel, *J. Polym. Sci.*, **50**, 511 (1961).
27. M. Ueda, M. Sato, and A. Mochizuki, *Macromolecules*, **18**, 2723 (1985).
28. W. Wrasidlo and H. H. Levine, *J. Polym. Sci. A*, **2**, 4795 (1964).
29. E. W. Neuse and M. S. Loonat, *Macromolecules*, **16**, 128 (1985).
30. J. Higgins and C. S. Marvel, *J. Polym. Sci. A-1*, **8**, 171 (1970).
31. C. S. Marvel, in *High-Temperature Polymers* (C. L. Segal, Ed.), Marcel Dekker, New York, 1967, pp. 7 ff.
32. B. Nartissov, *J. Macromol. Sci., Rev. Macromol. Chem.*, **C11**, 143 (1974).
33. V. V. Korshak and A. L. Rusanov, *J. Macromol. Sci., Rev. Macromol. Chem.*, **C21**, 275 (1982).
34. J. F. Wolfe, B. H. Loo, and F. E. Arnold, *Macromolecules*, **14**, 95 (1981).
35. J. F. Wolfe and F. E. Arnold, *Macromolecules*, **14**, 909 (1981).
36. E. W. Choe, and S. N. Kim, *Macromolecules*, **14**, 920 (1981).
37. R. C. Evers, F. E. Arnold, and T. E. Helminiak, *Macromolecules*, **14**, 925 (1981).
38. R. Merten, *Angew. Chem. Int. Ed. Engl.*, **10**, 294 (1971).
39. J. A. Brydson, *Plastics Materials*, 3rd ed., Butterworth, London, 1975, pp. 483–84.
40. T. L. Patton, *Polym. Prepr.*, **12**(1), 162 (1971).
41. G. Caraculacu, E. Scortanu, and A. A. Caraculacu, *Eur. Polym. J.*, **19**, 142 (1983).
42. P. M. Hergenrother, *J. Macromol. Sci., Rev. Macromol. Chem.*, **C6**, 1 (1971).
43. J. Higgins, J. F. Jones, and A. Thornburgh, *Macromolecules*, **2**, 558 (1969); *J. Polym. Sci. A-1*, **9**, 763 (1971).
44. E. W. Neuse, *Macromolecules*, **1**, 171 (1968).
45. B. Krieg and G. Manecke, *Makromol. Chem.*, **108**, 210 (1967).
46. P. M. Hergenrother, *J. Polym. Sci., Polym. Chem. Ed.*, **12**, 2857 (1974).
47. W. H. Beever and J. K. Stille, *J. Polym. Sci., Polym. Symp.*, **65**, 41 (1978).
48. J. K. Stille, *Macromolecules*, **14**, 870 (1981).
49. Y. Imai, E. F. Johnson, T. Katto, M. Kurihara, and J. K. Stille, *J. Polym. Sci., Polym. Chem. Ed.*, **13**, 2233 (1975).

Review Exercises

1. Write structures of the monomers that would react to form

   (a) the polyquinoxaline

   (ref. 23)

(b) the polyisoanthrazoline

(ref. 49)

(c) the poly(benzoxazinone)

(N. Yoda, *Polym. Sci. Technol.*, **10**, 682, 1969)

2. Give the structure of A and B in the following reaction sequence:

$$H_2N-NH-\overset{O}{\underset{||}{C}}-\overset{O}{\underset{||}{C}}-NH-NH_2 \ +$$

$\longrightarrow$ A $\xrightarrow{\text{PPA}, \Delta}$ B

(G. Caraculacu et al., *Eur. Polym. J.*, **29**, 1143, 1993)

3. Complete the following reactions:

(a)

$\xrightarrow[(-PhOH, -CO_2)]{\Delta}$

(b) H_2NNH $NHNH_2 + OCH$——$CHO \xrightarrow{\Delta}$

(P. M. Hergenrother, *J. Polym. Sci. A-1*, **7**, 945, 1969)

(c)

$+$ $\longrightarrow$ $\xrightarrow{O_2}$

(ref. 1)

(d)

$+$ $\xrightarrow[(-6H_2O)]{\Delta}$

(J. Szitza et al., *J. Polym. Sci. A-1*, **9**, 691, 1971)

(e)

(E. W. Neuse, *Macromolecules*, **1**, 171, 1968)

4. When the monomer

is reacted with $H_2NNH-Ar-NHNH_2$, a polymer is formed in high yield with loss of 4 equivalents of H_2O per repeating unit. Suggest a likely structure and explain its formation from a mechanistic viewpoint. (Y. Imai et al., *J. Polym. Sci., Polym. Chem. Ed.*, **14**, 2797, 1976)

5. When 4-amino-3-nitrobenzil (shown) was heated with iron filings in acetic acid and *m*-cresol to reduce the nitro group to amine, a thermally stable polymer was isolated when the reaction mixture was poured into methanol. What is the structure of the polymer? (T. Fang, *Macromolecules*, **24**, 444, 1991)

16

INORGANIC AND PARTIALLY
INORGANIC POLYMERS

16.1   Introduction

Major emphases in polymer development over the years have been on organic polymers. The only polymers in commercial use for an extended period that have inorganic backbones are the *polysiloxanes*, which date back to the 1940s. Nevertheless a considerable volume of work dealing with inorganic and partially inorganic polymers has been published,[1–7] although most of the products developed remain laboratory curiosities. In more recent years some other mixed inorganic–organic polymers have been marketed, including *polyphosphazenes*, *polysilanes*, and polymers containing carborane units in the backbone. These polymers and their applications are listed in Table 16.1. Other major areas of activity include *organometallic polymers*, particularly the ferrocene type, and *coordination polymers*, which have coordinate covalent bonds to metals as an integral part of the polymer backbone.

Each of the polymers described above are partially inorganic. There are, however, some interesting wholly inorganic polymers, the simplest of which is *polymeric sulfur*, consisting of unsubstituted chains of sulfur atoms. Polymeric sulfur is formed by heating the familiar rhombic sulfur, which is made up of eight-membered rings, to about 160°C.[8] The reaction is free radical in nature. Further heating to 175°C or higher results in depolymerization. When the polymer, plasticized with some residual S_8 rings, is cooled to room temperature, it exhibits elastomeric properties with a T_g of about -30°C. Polymeric sulfur is not stable; it reverts gradually to rhombic sulfur at room temperature. Selenium and tellurium, which lie below sulfur in the periodic table, behave similarly.

Poly(sulfur nitride), another wholly inorganic polymer first mentioned in Chapter 4 in the context of conducting polymers, is discussed in the next section. Poly(sulfur nitride) is by no means the only wholly inorganic polymer to have attracted attention. Boron nitride (BN), an inorganic polymer having structures analogous to those of graphite and diamond, has been manufactured into refractory fibers by the high-temperature (>1500°C) reaction of boric oxide filaments with urea or ammonia,[9] or by pyrolysis of borazine ($B_3N_3H_6$)-based polymers.[10] The BN fibers, in turn, have been similarly reacted with metal halides to form

TABLE 16.1. Partially Inorganic Polymers Developed for Commercial Use

Type	Principal Structure	Typical Applications
Polysiloxane	$\begin{array}{c} CH_3 \\ \mid \\ -Si-O- \\ \mid \\ CH_3 \end{array}$	Elastomers, sealants, lubricating oils, greases, hydraulic fluids, biomedical devices, cosmetics
Polysilane[b]	$\begin{array}{c} R \\ \mid \\ -Si- \\ \mid \\ R \end{array}$	Silicon–carbide ceramic precursors
Polyphosphazene[a]	$\begin{array}{c} OR \\ \mid \\ -P=N- \\ \mid \\ OR \end{array}$	Oil-resistant hose, gaskets, O-rings
Polycarboranesiloxane	$\begin{array}{c} CH_3 \\ \mid \\ -C(B_{10}H_{10})C-Si-O- \\ \mid \\ CH_3 \end{array}$	High-temperature elastomers, greases, lubricants

[a] R = mixed fluoroalkyl groups.
[b] R = methyl or phenyl primarily.

such nitride polymers as NbN and TiN.[11] Similar methods have been employed to make metal carbide fibers from carbon fiber. There is considerable interest in these materials as strengthening agents for ceramics.

It is not possible to cover all aspects of inorganic–organic polymers in a single chapter because so much has been published in the field, but some of the more significant and timely topics are included. The reader should keep in mind, however, that this is an area of very active research; current literature should be consulted for recent advances. It should not be forgotten, too, that some polymeric silicates, notably glass and asbestos, have been in service far longer than any of the polymers described in this chapter.

16.2 Poly(sulfur nitride)

Also called *polythiazyl*, poly(sulfur nitride)[12,13] (**2**) is prepared by the solid-state polymerization (16.1) of disulfur dinitride (**1**) at room temperature. The dimer **1**, which is, itself, formed by heating a cyclic tetramer, polymerizes slowly to yield lustrous, gold-colored monoclinic crystals of **2**. The polymer is a most unusual material, being the first reported example of a nonmetallic covalent polymer exhibiting electrical conductivity comparable to that of metals (see Chapter 4, Section 4.8); indeed, it even exhibits superconductivity at 0.26 K. Apart from its conductivity, **2** exhibits high reflectivity and malleability, both properties characteristic of metals. It is assumed that the high conductivity arises not only from delocalization along the chains, but also by S–S, N–N, and S–N orbital overlap between adjacent parallel chains. The polymer is relatively stable at room temperature, but decomposes over prolonged periods or on heating.

$$S_2N_2 \longrightarrow \{S{=}N\} \tag{16.1}$$
$$\quad\mathbf{1} \qquad\qquad \mathbf{2}$$

Poly(sulfur nitride) reacts with bromine to form polymers having the composition $\{SNBr_x\}$, where x is in the range 0.3 to 0.4.[14] The brominated polymers have higher conductivities than **2**.

At the present time, poly(sulfur nitride) and its derivatives are of more theoretical than practical interest.

16.3 Polysiloxanes

Despite the position of silicon directly below carbon in group IV of the periodic table, the properties of the two elements are quite different. While its normal oxidation state (like that of carbon) is 4, silicon can expand its valence shell by using available $3d$ orbitals for bonding.

A significant difference between the two elements is that the Si—Si bond energy is considerably less (about 30 kcal/mol) than that of the C—C bond; hence silanes, analogous to alkanes, having the general formula Si_nH_{2n+2} are less stable than their alkane counterparts. On the other hand, Si—O bonds are more stable (about 22 kcal/mol) than C—O bonds, and polymers having recurring Si—O linkages (polysiloxanes) are very important. There is evidence from bond length and dipole moment data of some double-bond character in Si—O bonds arising from overlap of the p orbitals of oxygen with vacant d orbitals of silicon.

Polysiloxanes,[15–17] frequently referred to as *silicones* (a rather misleading name), are prepared by hydrolysis of alkylsilicon or arylsilicon halides. The starting materials can be made from silicon and alkyl or aryl halides by heating at 250 to 280°C in the presence of copper (reaction 16.2). This is the preferred procedure for preparing methylchlorosilanes and phenylchlorosilanes.

$$Si + RCl \xrightarrow[250-280°C]{Cu} SiCl_4 + RSiCl_3 + R_2SiCl_2 + R_3SiCl \tag{16.2}$$

Hydrolysis of the halides gives the corresponding unstable silanols, which undergo condensation to form the siloxane linkage (16.3). Hydrolysis of dichlorides leads to linear polymers—for example, the preparation of polydimethylsiloxane **3** from dimethyldichlorosilane (16.4)—whereas trichlorides give crosslinked polysiloxanes.

$$-\overset{|}{\underset{|}{Si}}{-}Cl \xrightarrow{H_2O} \left[-\overset{|}{\underset{|}{Si}}{-}OH\right] \longrightarrow -\overset{|}{\underset{|}{Si}}{-}O{-}\overset{|}{\underset{|}{Si}}{-} + H_2O \tag{16.3}$$

$$Cl{-}\overset{\overset{\displaystyle CH_3}{|}}{\underset{\underset{\displaystyle CH_3}{|}}{Si}}{-}Cl \xrightarrow{H_2O} \left[\overset{\overset{\displaystyle CH_3}{|}}{\underset{\underset{\displaystyle CH_3}{|}}{Si}}{-}O\right] \tag{16.4}$$
$$\mathbf{3}$$

As a general method of preparing polysiloxanes, hydrolysis of the halides is not satisfactory because of the tendency to form cyclic siloxanes (mainly trimers or tetramers) under the conditions of hydrolysis, although some siloxane resins are prepared this way. To obtain high-molecular-weight polymer, the cyclic products are purified by distillation and polymerized by acid- or base-catalyzed ring-opening reactions.[18]

In the presence of sulfuric acid, ring opening and further condensation take place to give linear siloxane polymers. *Silicone oils* are prepared by adding some hexamethyldisiloxane during the polymerization reaction. This leads to polysiloxanes with unreactive trimethyl-siloxyl end groups (16.5). The molecular weight, and hence the viscosity, of the oils can be controlled by the amount of hexamethyldisiloxane added. Characteristically, silicone oils have good thermal stability and demonstrate little change in viscosity with temperature; hence they are ideal as lubricants and hydraulic fluids.

$$n\left[\begin{array}{c} CH_3 \\ | \\ -Si-O- \\ | \\ CH_3 \end{array}\right]_x \xrightarrow[H_2SO_4]{(CH_3)_3SiOSi(CH_3)_3} (CH_3)_3SiO\left[\begin{array}{c} CH_3 \\ | \\ Si-O \\ | \\ CH_3 \end{array}\right]_{xn}Si(CH_3)_3 \qquad (16.5)$$

Base-catalyzed ring-opening polymerization of cyclic siloxanes results in high-molecular-weight linear polymer with rubberlike properties. Crosslinking can be effected by cohydrolysis with alkyltrichlorosilanes or by treatment with peroxides or oxygen. Replacing some of the methyl groups with vinyl groups results in more effective curing. One-component, room-temperature-curable silicone elastomers consist of silanol-terminated polymer packaged with a crosslinking agent precursor such as methyltriacetoxysilane and a catalyst such as dibutyltin dilaurate. When exposed to atmospheric moisture, the precursor is hydrolyzed to silanol (16.6), which then reacts with the polysiloxane chain ends to yield crosslinked polymer (16.7). The term *RTV rubber* (room-temperature vulcanization) is used for elastomers of this type.

$$CH_3Si(O_2CCH_3)_3 \xrightarrow{H_2O} CH_3Si(OH)_3 + 3\ CH_3CO_2H \qquad (16.6)$$

$$\begin{array}{c} CH_3 \\ | \\ \sim\!\!\sim\!\!Si-OH + CH_3Si(OH)_3 \\ | \\ CH_3 \end{array} \longrightarrow \begin{array}{c} CH_3\quad CH_3 \\ |\qquad | \\ \sim\!\!\sim\!\!Si-O-Si-O\!\!\sim\!\!\sim \\ |\qquad | \\ CH_3\quad O \end{array} \qquad (16.7)$$

Among the more remarkable properties of polysiloxanes are a very low glass transition temperature (about $-127°C$), considering the high degree of substitution on the polymer backbone. (Compare data in Tables 3.1 and 3.2, Chapter 3.) This much greater flexibility arises from the larger bond angles and bond lengths associated with the siloxane backbone relative to those of vinyl polymers or polyethers (Table 16.2), which allow much greater conformational variability.[19] This, and other properties, have led to a broad range of applications. In addition to the silicone oil lubricants and hydraulic fluids mentioned earlier, *silicone rubber* is useful as a potting compound for encapsulation of electrical parts because of

TABLE 16.2. Bond Lengths and Bond Angles in Polymer Backbones[a]

Backbone Unit	Bond Length ($\overset{\circ}{A}$)	Bond Angle
C—C—C	1.54	112°
C—O—C	1.42	111°
Si—O—Si	1.63	130°

[a]Data from Owen.[19]

a combination of good thermal and oxidative stability, chemical resistance, and dielectric properties. Polysiloxanes also find use as wire coatings, gaskets, adhesives, caulking compounds, and paper-release coatings for such applications as pressure-sensitive labels or floor tiles. Because of their physiological inertness, polysiloxanes are an attractive material for construction of artificial organs for surgical implantation. Silicone resins are also used for water-repellent coatings, mold-release agents, antifoaming agents, foam stabilizers for polyurethanes, and nonstick surfaces for cooking utensils.

Modification of polysiloxanes is a fruitful area of research. One of the more interesting developments is the preparation of soluble ladder polymer by the base-catalyzed hydrolysis of phenyltrichlorosilane (16.8). The reaction is unusual in that the polymer, known as *polyphenylsilsequioxane*[20,21] (**4**), has a regular *cis*-syndiotactic structure with a *cis-anti-cis* arrangement of fused rings.

$$C_6H_5SiCl_3 \xrightarrow[OH^-]{H_2O} \left[\begin{array}{c} \text{structure 4} \end{array} \right] \tag{16.8}$$

4

Other modifications include incorporation of aryl groups (16.9) or metal atoms such as titanium (16.10) into the backbone[22,23]; synthesis of polysiloxane comb polymers by the macromer technique[24] (16.11); and formation of copolymers of siloxanes with other polymer types. An example is the formation of silicone–ether block copolymers (16.12) for use as surfactants.[25]

$$R_2SiCl_2 + HO-Ar-OH \longrightarrow \left[\begin{array}{c} R \\ | \\ Si-O-Ar-O \\ | \\ R \end{array} \right] \tag{16.9}$$

$$R_2SiCl_2 + Ti(OR')_4 \longrightarrow \begin{bmatrix} R \\ | \\ O-Si-O \\ | \\ R \end{bmatrix}\begin{bmatrix} OR' \\ | \\ Ti-O \\ | \\ OR' \end{bmatrix} \tag{16.10}$$

$$\text{CH}_2=\!\!\!\left\langle\!\!\bigcirc\!\!\right\rangle\!\!-\begin{bmatrix} CH_3 \\ | \\ SiO \\ | \\ CH_3 \end{bmatrix}\!\!-Si(CH_3)_3 \xrightarrow{\text{initiator}} \begin{matrix} CH_2 \\ | \\ CH \end{matrix}\!\!-\left\langle\!\!\bigcirc\!\!\right\rangle\!\!-\begin{bmatrix} CH_3 \\ | \\ SiO \\ | \\ CH_3 \end{bmatrix}\!\!-Si(CH_3)_3 \tag{16.11}$$

$$\begin{bmatrix} CH_3 \\ | \\ Si-O \\ | \\ CH_3 \end{bmatrix}\begin{matrix} CH_3 \\ | \\ Si-OC_2H_5 \\ | \\ CH_3 \end{matrix} + HO\begin{bmatrix} R \\ | \\ CH_2CHO \end{bmatrix} \xrightarrow{(-C_2H_5OH)}$$

$$\tag{16.12}$$

$$\begin{bmatrix} CH_3 \\ | \\ Si-O \\ | \\ CH_3 \end{bmatrix}\begin{matrix} CH_3 \\ | \\ Si-O \\ | \\ CH_3 \end{matrix}\begin{bmatrix} R \\ | \\ CH_2CHO \end{bmatrix}$$

Particularly interesting are modifications of inorganic glasses with polysiloxanes.[26] The procedure involves hydrolyzing silicon alkoxides (16.13), then copolymerizing the hydrolyzate with a hydroxyl-terminated polysiloxane (16.14). Pyrolysis of the hydrolyzate alone (the *sol-gel process*) yields the unmodified ceramic. Incorporation of the polymer is a potential route to reducing brittleness. The copolymers, which represent the marriage of a ceramic with a conventional polymer, have been appropriately designated *ceramers*.

$$SiOR_4 + 4H_2O \xrightarrow{H^+} Si(OH)_4 + 4ROH \tag{16.13}$$

$$Si(OH)_4 + HO\begin{bmatrix} CH_3 \\ | \\ SiO \\ | \\ CH_3 \end{bmatrix}\begin{matrix} CH_3 \\ | \\ SiOH \\ | \\ CH_3 \end{matrix} \xrightarrow[(-H_2O)]{\Delta} -SiO\begin{bmatrix} CH_3 \\ | \\ SiO \\ | \\ CH_3 \end{bmatrix}\begin{matrix} CH_3 & O \\ | & | \\ Si-O-Si-O- \\ | & | \\ CH_3 & O \end{matrix} \tag{16.14}$$

16.4 Polysilanes

Although polysilanes (**5**) (polymers containing silicon–silicon bonds in the backbone) have been known for a long time, linear high-molecular-weight tractable polysilanes are of relatively recent vintage.[27–30] They are formed in moderate yield by the condensation of a dichloro organosilane monomer in the presence of finely dispersed sodium in refluxing toluene (16.15). The high degree of backbone substitution is permitted by the much longer Si—Si bond length (2.34 Å) relative to that of carbon.

$$RR'SiCl_2 \xrightarrow[\text{toluene}]{Na, \Delta} \left[\begin{array}{c} R \\ | \\ -Si- \\ | \\ R' \end{array}\right] + 2NaCl \tag{16.15}$$

5

One polysilane that has attracted considerable interest is **6**, which contains equal numbers of dimethylsilane and methylphenylsilane units. Polymer **6** has been dubbed *polysilastyrene* by analogy with polystyrene; however, the methylphenylsilane units are randomly dispersed, not alternating as is the case with the phenyl groups of polystyrene. Polysilane **6** undergoes photodegradation under the influence of UV radiation and is therefore of interest in UV-active photoresists. If doped with arsenic pentafluoride, **6** exhibits semiconducting properties, which indicates that there is considerable delocalization of the sigma electrons along the backbone. This differs, of course, from carbon–chain polymers, which require conjugated double bonds for delocalization.

$$\begin{array}{cc} CH_3 & CH_3 \\ | & | \\ \text{wwwSiwwwwSiwww} \\ | & | \\ CH_3 & Ph \end{array}$$

6

When polysilanes are heated at 1200 to 1400°C, side-chain carbon atoms insert themselves into the backbone, and the polymers eventually degrade to silicon carbide. Controlled pyrolysis of polysilane fibers thus leads to SiC fibers that, like boron nitride, are of interest for ceramics applications.

An alternative synthesis of polysilane **5** is the dehydrogenative coupling of dialkylsilanes ($RR'SiH_2$) in the presence of a metallocene catalyst.[31] A similar procedure has been used to prepare high-molecular-weight *polystannanes* (polymers containing recurring tin atoms in the backbone).[32]

16.5 Polyphosphazenes

Polyphosphazenes, or *phosphonitrilic* polymers, as they are also called, are among the more interesting and commercially promising of the inorganic-type polymers developed in recent years.[33–35] Starting material for their preparation is usually hexachlorocyclotriphosphazene (**7**) (phosphonitrilic chloride trimer), a commercially available monomer synthesized from phosphorus pentachloride and ammonium chloride. The monomer undergoes ring-opening polymerization to give linear polydichlorophosphazene (**8**) on heating (16.16), with some crosslinking occurring on prolonged heating.

$$\begin{array}{c} \text{(structure of 7)} \xrightarrow{230-300°C} \left[\begin{array}{c} Cl \\ | \\ -N=P- \\ | \\ Cl \end{array}\right] \end{array} \tag{16.16}$$

7   **8**

Fluoro, bromo, and isothiocyano derivatives of cyclotriphosphazene give analogous linear polymers. The mechanism of polymerization is believed to involve initial ionization of a phosphorus–chlorine bond, followed by attack on the resultant cation by an electron-rich nitrogen atom (16.17) to initiate ring-opening polymerization. Chain branching might then occur by a similar mechanism involving ionization of phosphorus–chlorine bonds in the polymer backbone.

$$\text{(16.17)}$$

Polydichlorophosphazene, a true inorganic polymer with a backbone isoelectronic with that of siloxane polymers, is of no practical value as such because it is extremely sensitive to atmospheric moisture, undergoing ready hydrolysis to phosphate, ammonia, and hydrochloric acid. It also depolymerizes to trimer, tetramer, and other oligomers at temperatures above 350°C. It can, however, be converted to much more stable derivatives by replacing the chlorine atoms with alkoxy, aryloxy, or amino groups. Quantitative replacement of the chlorines can be accomplished by refluxing in an ether solvent for several hours with sodium alkoxide or aryloxide (16.18) or with primary or secondary amine (16.19). (The correspondingly substituted cyclotriphosphazenes are quite stable and resist polymerization.) Reaction of the difluoro analog with alkyllithium or dialkylmagnesium compounds (16.20) yields polyphosphazenes containing phosphorus–carbon bonds.[36]

$$\text{(16.18)}$$

$$\text{(16.18)}$$

$$\text{(16.20)}$$

An examination of the phosphonitrilic backbone might lead one to predict properties similar to those of the highly crystalline polyacetylene or poly(sulfur nitride). In fact, the opposite is the case; polyphosphazenes form amorphous, clear, colorless films with elastomeric properties despite the alternation of single and double bonds. Polyphosphazenes are also nonconducting. The high conformational mobility can be explained in terms of the $(p–d)\pi$ overlap, which is less restrictive relative to $(p-p)\pi$ bonding because of the availability of five d orbitals on each P atom. The lack of color or conductivity, on the other hand, suggests that electron delocalization is minimal. An orbital picture (**9**) to account for the four valence electrons not involved in sigma bonding (three on N and one on P) places two electrons in a lone-pair orbital on N, while the remaining two forms a $(2p-3d)\pi$ bond. This leads to an orbital node on each P atom, thus limiting delocalization to three backbone atoms.

9

Phosphonitrilic polymers are used in a variety of applications. In the area of elastomers, fluoroalkoxy-substituted polymers exhibit an ideal combination of chemical resistance and thermal stability. As an example, the copolymer **10**, prepared by refluxing **8** with a mixture of sodium trifluoroethoxide and heptafluorobutoxide, retains its flexibility at temperatures as low as $-77°C$ and is stable up to $300°C$. The low-temperature flexibility makes the elastomer attractive for fuel lines and sealants in arctic environments.

10

Phosphonitrilic polymers are also useful as fire-retardant materials, not only because they are self-extinguishing but because they burn with only moderate evolution of smoke. As was discussed in Chapter 4 (Section 4.7), phosphonitrilic polymers are also of interest in the medical field both as structural materials for artificial organs and as delivery vehicles for controlled release of medicinal agents.[37-39]

Phosphonitrilic polymers, both linear and crosslinked, have also been prepared by linking cyclic phosphazene rings together. This type of work is normally directed toward synthesis of thermally stable polymers, taking advantage of the inherent stability of the cyclotriphosphazene and cyclotetraphosphazene rings. Typical of the methods reported are the reaction of a dichlorocyclophosphazene (**11**) with a dihydroxy compound (16.21), or the analogous reaction of **7** to yield network polymer. Flame- and heat-resistant polyimide composites have also been prepared from maleimide-substituted phosphazenes such as **12**.[40,41]

(16.21)

11

12

Polythiophosphazenes[42] (**14**), which are structurally related to polyphosphazenes, have been prepared by heating cyclic thiophosphazene **13** (16.22). Like polymer **8**, **14** is moisture sensitive. Replacement of the chlorine atoms with phenoxy groups using sodium phenoxide only marginally reduces the hydrolytic instability, presumably because the sulfur–nitrogen double bond is more exposed than the phosphorus–nitrogen double bonds of substituted polyphosphazenes. A more stable polythiophosphazene can be prepared with more sterically hindered aryloxy substituents, however. If sulfur is replaced with carbon in **13**, the resultant polymer is a *polycarbophosphazene*.[43]

$$\text{13} \xrightarrow{90°C} \text{14} \tag{16.22}$$

13   **14**

16.6   Carborane Polymers

Apart from the boron nitride work described earlier, the most successful approach to synthesizing well-characterized boron-containing polymers has been to incorporate carborane units into the polymer.[44–48]

It is outside the scope of this book to delve into the fascinating and complex area of carborane chemistry. Most of the polymer work has been concerned with only two of these structures—the icosahedral $C_2B_{10}H_{12}$ and the pentagonal bipyramidal $C_2B_5H_7$, particularly the former. Both have a cage structure and are prepared from the corresponding boranes, decaborane ($B_{10}H_{14}$) and pentaborane (B_5H_9), by reaction with acetylene. The $C_2B_{10}H_{12}$ carborane exists in three isomeric forms, named *ortho* (**15**), *meta* (**16**), and *para* (**17**). (Only the skeletal arrangement is shown. One hydrogen is attached to each boron and carbon atom.) Reaction of decarborane with acetylene gives **15**, which can be converted to **16** by heating to 475 to 600°C; **16**, in turn, is converted to **17**, by heating to 630 to 700°C. The great sta-

15   **16**   **17**

○ = boron   ● = carbon

bility of these compounds may be attributed to the complete delocalization of the electrons, giving them pseudoaromatic character.

Because the hydrogens attacked to the carbon atoms are more acidic than those attached to boron, they can be replaced, for example, with lithium by reaction with butyllithium; hence a variety of carborane derivatives have been synthesized. One of these is vinylcarborane, which can be polymerized through the vinyl group or copolymerized to give high-molecular-weight polymers. Some examples of nonvinyl polymer-forming reactions involving the icosahedral carborane nucleus ($C_2B_{10}H_{10}$) are given in reactions (16.23) through (16.26).

$$\text{Cl}-\overset{\overset{\displaystyle O}{\|}}{C}-C(B_{10}H_{10})C-\overset{\overset{\displaystyle O}{\|}}{C}-\text{Cl}$$

$$\xrightarrow{\text{HO}-\text{R}-\text{OH}} \left[\overset{\overset{\displaystyle O}{\|}}{C}-C(B_{10}H_{10})C-\overset{\overset{\displaystyle O}{\|}}{C}-\text{R}-\text{O}\right] \quad (16.23)$$

$$\xrightarrow{\text{H}_2\text{N}-\text{R}-\text{NH}_2} \left[\overset{\overset{\displaystyle O}{\|}}{C}-C(B_{10}H_{10})C-\overset{\overset{\displaystyle O}{\|}}{C}-\text{R}-\text{NH}\right] \quad (16.24)$$

$$\text{HOCH}_2-C(B_{10}H_{10})C-\text{CH}_2\text{OH} + \text{CH}_3\text{CH}-\text{CH}_2 \longrightarrow$$
$$\overset{\diagdown O \diagup}{}$$

$$\text{H}\left[\text{OCH}_2\underset{\underset{\displaystyle CH_3}{|}}{\text{CH}}\right]_x \text{OCH}_2-C(B_{10}H_{10})C-\text{CH}_2\text{O}\left[\underset{\underset{\displaystyle CH_3}{|}}{\text{CHCH}_2\text{O}}\right]_y \text{H} \quad (16.25)$$

$$\text{Li}-C(B_{10}H_{10})C-\text{Li} + 2\text{Cl}-\underset{\underset{\displaystyle CH_3}{|}}{\overset{\overset{\displaystyle CH_3}{|}}{\text{Si}}}-\text{O}-\underset{\underset{\displaystyle CH_3}{|}}{\overset{\overset{\displaystyle CH_3}{|}}{\text{Si}}}-\text{Cl} \longrightarrow$$

$$\text{Cl}-\underset{\underset{\displaystyle CH_3}{|}}{\overset{\overset{\displaystyle CH_3}{|}}{\text{Si}}}-\text{O}-\underset{\underset{\displaystyle CH_3}{|}}{\overset{\overset{\displaystyle CH_3}{|}}{\text{Si}}}-C(B_{10}H_{10})C-\underset{\underset{\displaystyle CH_3}{|}}{\overset{\overset{\displaystyle CH_3}{|}}{\text{Si}}}-\text{O}-\underset{\underset{\displaystyle CH_3}{|}}{\overset{\overset{\displaystyle CH_3}{|}}{\text{Si}}}-\text{Cl} \xrightarrow{\text{H}_2\text{O}} \quad (16.26)$$

$$\left[\underset{\underset{\displaystyle CH_3}{|}}{\overset{\overset{\displaystyle CH_3}{|}}{\text{Si}}}-\text{O}-\underset{\underset{\displaystyle CH_3}{|}}{\overset{\overset{\displaystyle CH_3}{|}}{\text{Si}}}-C(B_{10}H_{10})C-\underset{\underset{\displaystyle CH_3}{|}}{\overset{\overset{\displaystyle CH_3}{|}}{\text{Si}}}-\text{O}-\underset{\underset{\displaystyle CH_3}{|}}{\overset{\overset{\displaystyle CH_3}{|}}{\text{Si}}}-\text{O}\right]$$

Reaction (16.26) and other related reactions have been used to synthesize a series of *m*-carborane–siloxane polymers that have been marketed under the trade names Dexsil (Olin) and Ucarsil (Union Carbide). They exhibit elastomeric properties with excellent thermal and oxidative stability, but their high cost is a drawback. One such polymer has been used as a stationary liquid phase in gas chromatography for performing separations at temperatures as high as 500°C.

Siloxane–carborane polymers having similar properties have been synthesized using the pentagonal bipyramidal $C_2B_5H_7$ carborane (**18**), although these usually contain other carboranes as minor ingredients.[44]

18

16.7 Organometallic Polymers

Organometallic polymers are, by definition, polymers containing carbon–metal bonds. They represent a fruitful area of research with a rich variety of structures.[49–53] Many have unusual properties: intense ultraviolet absorption, for example, which makes them attractive for exterior coatings. Some are of interest for their biocidal properties.[54] Others have potential in the biomedical field.[55] Some organometallic polymers have found commercial use as catalysts. The high cost of the starting materials limits commercial development, however.

The ferrocene moiety has been the focus of much of the research because it is readily available and easily functionalized. Vinylferrocene (**19**) is converted to polyvinylferrocene with free radical, cationic, or complex coordination initiators, and ferrocenylmethyl acrylate (**20**) behaves similarly to the more conventional acrylic monomers. In the case of free radical polymerization, azo initiators are normally employed because peroxides oxidize the iron to Fe(III). While the topic of vinyl polymerization was covered in Part II, it is appropriate here to consider the polymerization of these vinyl metallocene monomers.

$$\text{(19)} \qquad \text{(20)} \quad -CH_2O_2CCH=CH_2$$

19   **20**

Despite its relationship to styrene, **19** exhibits atypical behavior toward free radical polymerization in benzene solution.[56] Termination, for example, involves intramolecular electron transfer (16.27) leading to Fe(III) chain ends rather than the more conventional bimolecular pathway. Furthermore, **19** has a much higher chain transfer constant than styrene, which leads to a more highly branched polymer. Monomer **20** follows normal free radical termination behavior, as does vinylruthenocene[57] (the ruthenium analog of **19**), which has an oxidation potential about 0.1 V more negative than that of **19**.

$$\text{\small{wCH}_2\dot{\text{C}}\text{H}} \longrightarrow \text{\small{wCH}_2\ddot{\text{C}}\text{H}^-} \xrightarrow{\text{electrophile}} \text{termination} \qquad (16.27)$$

To place the ferrocene unit directly into the backbone, both step-growth and ring-opening polymerization may be used. Polyesters and polyamides, for example, are readily synthesized from such monomers as **21** and **22** using solution or interfacial methods.[52,58,59]

21   **22**

Thermal ring-opening polymerization of bridged ferrocenes[7] is an effective method for producing soluble ferrocene polymers having number-average molecular weights in excess of 10^5. Examples are the formation of poly(ferrocenylsilane)s[60] (**23**) (16.28) and poly(ferrocenylethylene)[61] (**24**) (16.29). For such reactions to proceed at a practical rate, the bridging group in the monomer must be short enough to introduce ring strain, the alleviation of which provides the driving force for the reaction. The cyclopentadiene rings of the strained monomers are tilted with respect to one another from the normal parallel arrangement in unstrained ferrocenes.* Polymer **23** exhibits semiconducting properties when doped with iodine. Pyrolysis of **23** yields iron–silicon–carbon ceramics.

$$(16.28)$$

23

$$(16.29)$$

24

Examples of nonferrocene organometallic vinyl monomers[56] are **25–27**. Copolymers of **27** (R = *n*-butyl) with methyl methacrylate are used as biocidal paints for maintaining ships' hulls and offshore oil-drilling platforms free of barnacles and other marine organisms.[54]

25   **26**   **27**

---

*Bridged monomers of this type are termed [*n*]ferrocenophanes, where *n* denotes the number of atoms bridging the rings (1 and 2, respectively, in reactions 16.28 and 16.29). The general term for bridged ring systems is [*n*]cyclophane.

Aryl- or alkyl-substituted metal halides are also useful monomers for preparing organometallic polymers because the metal halides often exhibit the same type of reactivity as carboxylic acid chlorides. Solution and interfacial methods may thus be used to make metal-containing polyesters, polyamides, and related polymers.[59] Semimetal halides such as Ph_3SbCl_2 behave similarly. A general reaction (where M = tetravalent metal) is given (16.30).

$$R_2MCl_2 + {}^-O_2C-R'-CO_2^- \longrightarrow \left[M(R_2)-OC-\underset{\parallel}{\overset{O}{C}}-R'-\underset{\parallel}{\overset{O}{C}}O \right] \quad (16.30)$$

There are some other organometallic polymers that were not made deliberately but were found to be polymeric in nature on being characterized.[62] Examples are the nickel(O)–cyclooctatetraene complex (**28**), the norbornadiene–silver nitrate complex (**29**), and the arylethynyl copper complexes (**30**). Some of these, such as **30**, are of interest because they display semiconducting and photoconducting properties.

28

29   **30**

16.8 Coordination Polymers

Coordination polymers are polymers having coordinate covalent bonds in the repeating unit. As with organometallic polymers, considerable work has been done in this area.[51,52,62] and it still represents a very active field of research. Potential outlets for coordination polymers include thermally stable and conducting or semiconducting[63] materials, and water-based coatings, which are seen as a possible solution to the environmental problems associated with solvent-based coatings. While a large number of interesting coordination polymers have been prepared, their thermal properties are often no better, and frequently are worse, than those of the purely organic aromatic polymers.

There are generally three methods for preparing coordination polymers. Two of them yield true coordination polymers in that the coordination bonds are an integral part of the polymer backbone. The other method leads to products best described by the term *polymeric chelates*, because the chelated metal is only incidental to the overall polymer structure and breaking the coordination bonds would not rupture the polymer backbone.

The first method involves linking the polydentate ligands with metals, using metal salts, free metal, or coordination complexes (via ligand exchange) as coreactant. Examples leading to chelates through nitrogen ligands are the reaction of a bis(1,2-dioxime) with nickel acetate (16.31) and the reaction of tetracyanoethylene with copper to give a polymer having a phthalocyanine-type sheetlike structure (16.32). The phthalocyanine-type polymers have received considerable attention in recent years, and although their thermal stabilities have not always lived up to expectation, some network transition metal polyphthalocyanines linked through imide or benzimidazole groups have proved to be very stable indeed, exhibiting char yields of 80 to 90 percent at 800°C.[64] There is also considerable interest in cofacially linked polyphthalocyanines **31**, which, on doping, conduct electricity through the stacked phthalocyanine rings.[65,66]

$$(16.31)$$

$$(16.32)$$

M = metal or metalloid

31

Polymer formation through oxygen ligands has been accomplished with numerous types of compounds, with bis(β-diketones) being among the most successful as far as achieving high molecular weight is concerned. Ligand exchange between bis(β-diketone) and metal acetylacetonate (acac) (16.33) or tetrabutyl titanate to incorporate a hexacoordinate metal atom (16.34) typify this process. Another example, involving iron–oxygen backbone coordination in a degradable polymer, is described in Chapter 4 (Section 4.7) as a vehicle for releasing agricultural chemicals into farmland under controlled conditions.

$$(16.33)$$

$$(16.34)$$

One example of a polymer formed with both nitrogen and oxygen ligands is the product of the reaction of a bis(8-hydroxy-5-quinolyl) derivative with a metal acetylacetonate (16.35). Polymers with sulfur and mixed sulfur–nitrogen or sulfur–oxygen ligands have also been prepared.[62]

$$(16.35)$$

The second general method of forming coordination polymers is to polymerize a metal chelate through other functional groups. Chelates of thiopicolinamides having free amino groups (**32**) have been reacted with diacid chloride to form polyamides, for example; and dihydroxy derivatives of Schiff bases (imines) (**33**) have been converted to polycarbonate with phosgene.

32

33

The polymeric chelate type of polymer is normally prepared by complexing a polymer that has coordinating ligands as part of the backbone. One of the best studied examples involves the reaction of hydroxy-substituted polyimines (or polymeric Schiff bases, as they are also called) with metal salts, although not very high molecular weights have been achieved. The polymer prepared from 5,5′-methylenebis(salicylaldehyde) and *o*-phenylenediamine is shown (**34**) complexed with divalent metal. Polymeric chelates of a variety of transition met-

34

als have been precipitated by mixing polymer solution with soluble divalent metal salts. Interestingly, such polymers frequently have poorer thermal stability than the uncomplexed polymers. (Polyimines will be covered as a separate class of polymers in the next chapter.)

Crown ethers, which complex group I cations, have been extensively investigated as pendant chelating groups on vinyl polymers and as backbone chelating groups for condensation polymers.[67] Typical of the former is poly(4′-vinylbenzo-18-crown-6) (**35**) shown with complexed ion.[68] Crown ether monomers used to make condensation polymers include the diaminodibenzo derivative **36** and the diaza crown ether **37**.[69,70]

35

                                    36   37

Primarily because of steric effects, crown ether-containing polymers usually exhibit differing cation-binding characteristics than the monomers from which they are derived. In many instances, cations are bound more strongly by being sandwiched between two chelating rings.

References

1. J. Mark, H. R. Allcock, and R. West, *Inorganic Polymers*, Prentice Hall, Englewood Cliffs, New Jersey, 1992.
2. N. H. Ray, *Inorganic Polymers*, Academic, New York, 1978.
3. F. G. A. Stone and W. A. G. Graham (Eds.), *Inorganic Polymers*, Academic Press, New York, 1962.
4. D. N. Hunter, *Inorganic Polymers*, Wiley-Interscience, New York, 1963.
5. H. R. Allcock, *Heteroatom Ring Systems and Polymers*, Academic Press, New York, 1967.
6. M. Zeldin, K. J. Wynne, and H. R. Allcock (Eds.), *Inorganic and Organometallic Polymers*, ACS Symp. Ser. 360, American Chemical Society, Washington, D.C., 1987.
7. I. Manners, *Angew. Chem. Int. Ed. English*, **35**, 1602 (1996).
8. B. Meyer, *Chem. Rev.*, **76**, 367 (1976).
9. J. Economy, R. V. Anderson, and V. I. Matkovich, *J. Appl. Polym. Sci., Appl. Polym. Symp.*, **9**, 377 (1969).
10. R. T. Paine and L. G. Sneddon, *Chemtech*, **24**(7), 29 (1994).
11. J. Economy, W. D. Smith, R. Y. Lin, *J. Appl. Polym. Sci., Appl. Polym. Symp.*, **21**, 131 (1973).
12. C. M. Mikulski, P. J. Russo, M. S. Saran, A. G. MacDiarmid, A. F. Garito, and A. J. Heeger, *J. Am. Chem. Soc.*, **97**, 6358 (1975).
13. M. M. Labes, P. Love, and L. F. Nichols, *Chem. Rev.*, **79**, 1 (1979).
14. M. Akhtar, C. K. Chiang, M. J. Cohen, A. H. Heeger, J. Kleppinger, A. G. MacDiarmid, J. Milliken, M. J. Moran, and D. L. Peebles, in *Organometallic Polymers* (C. E. Carraher, Jr., J. F. Sheats, and C. U. Pittman, Eds.), Academic, New York, 1978, pp. 301–12.
15. E. G. Rochow, *Silicon and Silicones*, Springer-Verlag, New York, 1957.
16. G. Koerner, M. Schulze, and J. Weis, *Silicones: Chemistry and Technology*, CRC, Boca Raton, Florida, 1992.
17. S. J. Clarson and J. A. Semlyen (Eds.), *Siloxane Polymers*, Prentice Hall, Englewood Cliffs, New Jersey, 1991.
18. E. E. Bostick, in *Ring-Opening Polymerization* (K. C. Frisch and S. L. Reegen, Eds.), Marcel Dekker, New York, 1969, Chap. 8.
19. M. J. Owen, *Chemtech*, **11**, 288 (1981).
20. J. F. Brown Jr., L. H. Voght Jr., A. Katchman, J. W. Eustace, K. M. Kiser, and K. W. Krantz, *J. Am. Chem. Soc.*, **82**, 6194 (1960).
21. T. E. Helminiak and G. C. Berry, *J. Polym. Sci., Polym. Symp.*, **65**, 107 (1978).
22. A. H. Frazer, *High Temperature Resistant Polymers*, Wiley-Interscience, New York, 1968, Chap. 5.
23. K. A. Andrianov, *Metalorganic Polymers*, Wiley-Interscience, New York, 1965, Chap. 4.
24. Y. Kawakami, Y. Miki, T. Tsuda, R. A. N. Murthy, and Y. Yamashita, *Polym. J.*, **14**, 913 (1982).
25. B. Kanner, W. G. Reid, and I. H. Petersen, *Ind. Eng. Chem. Prod. Res. Develop.*, **6**, 88 (1967).
26. H.-H Huang, B. Orier, and G. L. Wilkes, *Polym. Bull.*, **14**, 557 (1985).
27. R. West and E. Carberry, *Science*, **189**, 179 (1975).

28. R. West, L. D. David, P. I. Djurovich, K. L. Stearly, K. S. V. Srinivasan, and H. Yu, *J. Am. Chem. Soc.*, **103**, 7352 (1981).
29. R. West and J. Maxka, in *Inorganic and Organometallic Polymers* (M. Zeldin, K. J. Wynne, and H. R. Allcock, Eds.), ACS Symp. Ser. 360, American Chemical Society, Washington, D.C., 1988, Chap. 2.
30. R. D. Miller and J. Michl, *Chem. Rev.*, **89**, 1359 (1989).
31. H.-G. Woo, J. F. Walzer, and T. D. Tilley, *J. Am. Chem. Soc.*, **114**, 7047 (1992).
32. T. Imori, V. Lu, H. Cai, and T. D. Tilley, *J. Am. Chem. Soc.*, **117**, 9931 (1995).
33. H. R. Allcock, *Chem. Rev.*, **72**, 315 (1972).
34. H. R. Allcock, *Phosphorus-Nitrogen Compounds*, Academic, New York, 1972.
35. H. R. Allcock, *Angew. Chem., Int. Ed. English*, **16**, 147 (1977).
36. T. L. Evans and H. R. Allcock, *J. Macromol. Sci.-Chem.*, **A16**, 409 (1981).
37. H. R. Allcock, *J. Polym. Sci., Polym. Symp.*, **70**, 71 (1983).
38. H. R. Allcock, in *Organometallic Polymers* (C. E. Carraher, Jr., J. F. Sheats, and C. U. Pittman, Jr., Eds.), Academic, New York, 1978, pp. 283–88.
39. A. G. Scopelianos, in *Biomedical Polymers: Designed-to-Degrade Systems*, S. W. Shalaby, Ed.), Hanser, New York, 1994, Chap. 6.
40. D. Kumar, *J. Polym. Sci., Polym. Chem. Ed.*, **23**, 1661, (1985).
41. D. Kumar, G. M. Fohlen, and J. A. Parker, *J. Polym. Sci., Polym. Chem. Ed.*, **22**, 927, 1141 (1984).
42. J. A. Dodge, I. Manners, and H. R. Allcock, *J. Am. Chem. Soc.*, **112**, 1268 (1990).
43. I. Manners, H. R. Allcock, G. Renner, and O. Nuyken, *J. Am. Chem. Soc.*, **111**, 5478 (1989).
44. R. N. Grimes, *Carboranes*, Academic, New York, 1970, Chap. 8.
45. R. E. Williams, *Pure Appl. Chem.*, **29**, 569 (1972).
46. D. J. Mangold, *J. Appl. Polym. Sci., Appl. Polym. Symp.*, **11**, 157 (1969).
47. H. A. Schroeder, *Inorg. Macromol. Rev.*, **1**, 45 (1970).
48. P. E. Cassidy, *Thermally Stable Polymers*, Marcel Dekker, New York, 1980, Chap. 10.
49. E. W. Neuse and H. Rosenberg, *Metallocene Polymers*, Marcel Dekker, New York, 1970.
50. E. W. Neuse, *Adv. Macromol. Chem.*, **1**, 1 (1970).
51. A. D. Pomogailo and V. S. Savostyanov, *J. Macromol. Sci., Rev. Macromol. Chem. Phys.*, **C25**, 375 (1985).
52. N. Hagihara, K. Sonogashira, and S. Takahashi, *Adv. Polym. Sci.*, **41**, 149 (1981).
53. E. W. Neuse, *J. Macromol. Sci.-Chem.*, **A16**, 3 (1981).
54. W. L. Yeager and V. J. Castelli, in *Organometallic Polymers* (C. E. Carraher, Jr., J. F. Sheats, and C. U. Pittman, Jr., Eds.), Academic, New York, 1978, pp. 175–80.
55. C. E. Carraher Jr. and C. G. Gebelein (Eds.), *Biological Activities of Polymers*, ACS Symp. Ser. 186, American Chemical Society, Washington, D.C., 1982.
56. C. U. Pittman, Jr., in *Organometallic Polymers* (C. E. Carraher, Jr., J. F. Sheats, and C. U. Pittman, Jr., Eds.), Academic, New York, 1978, pp. 1–11.
57. T. C. Willis and J. E. Sheats, *J. Polym. Sci., Polym. Chem. Ed.*, **22**, 1077 (1984).
58. K. Gonsalves, L. Zhan-Ru, and M. D. Rausch, *J. Am. Chem. Soc.*, **106**, 3862 (1984).
59. C. E. Carraher Jr., in *Interfacial Synthesis*, Vol. 2 (F. Millich and C. E. Carraher Jr., Eds.), Marcel Dekker, New York, 1977, Chap. 21.
60. D. A. Foucher, R. Ziembinski, B.-Z. Tang, P. M. Macdonald, J. Massey, C. R. Jaeger, G. J. Vancso, and I. Manners, *Macromolecules*, **26**, 2878 (1993).
61. D. Foucher, R. Ziembinski, R. Petersen, J. Pudelski, M. Edwards, Y. Ni, J. Massey, C. R. Jaeger, G. J. Vancso, and I. Manners, *Macromolecules*, **26**, 3992 (1994).
62. R. J. Cotter and M. Matzner, *Ring-Forming Polymerizations*, Part A, Academic, New York, 1969, Chaps. 5–9.
63. B. A. Bolto, in *Organic Semiconducting Polymers* (J. E. Katon, Ed.), Marcel Dekker, New York, 1968, Chap. 4.
64. B. N. Achar, G. M. Fohlen, and J. A. Parker, *J. Polym. Sci., Polym. Chem. Ed.*, **20**, 269, 2773, 2781 (1982); **23**, 801 (1985).
65. C. W. Dirk, T. Inabe, J. W. Lyding, K. F. Schoch, Jr., C. W. Kannewurf, and T. J. Marks, *J. Polym. Sci., Polym. Symp.*, **70**, 1 (1983).

66. P. M. Kuznesof, R. S. Nohr, K. J. Wynne, and M. E. Kenney, *J. Macromol. Sci.-Chem.*, **A16**, 299 (1981).
67. U. Tunca and Y. Yagci, *Progr. Polym. Sci.*, **19**, 233 (1994).
68. A. J. Varma, T. Majewicz, and J. Smid, *J. Polym. Sci., Polym. Chem. Ed.*, **17**, 1573 (1979).
69. A. Ricard and F. Lafuma, *Polymer*, **27**, 133 (1986).
70. L. J. Mathias and K. Al-Jumah, *J. Polym. Sci., Polym. Chem. Ed.*, **18**, 2911 (1980).

Review Exercises

1. Using either **21** or **22** and any other appropriate monomers, write reactions for the synthesis of a polyester, a polyamide, a polyurea, and a polyimide.

2. Suggest a way to make a polydimethylsiloxane–polycarbonate $+$AB$+$ block copolymer. (H. A. Vaughn, *J. Polym. Sci., B*, **7**, 569 1959)

3. Keeping in mind the mode of termination, derive a rate expression for the free radical polymerization of vinylferrocene in benzene solution. (See reaction 16.27.) (ref. 56)

4. Write repeating units of the polymers expected from the following reactions:

(a)

(H. Stuger and R. West, *Macromolecules*, **18**, 2349, 1985)

(b) $ClCC(B_{10}H_{10})CCCl + H_2N$——$NH_2 \longrightarrow$

(V. V. Korshak et al., *J. Polym. Sci. A-1*, **8**, 2351, 1970)

(c)

(L. Plummer and C. S. Marvel, *J. Polym. Sci. A-2*, 2559, 1964)

(d) $R_2Si(CH_2CH{=}CH_2)_2 \xrightarrow{AlBr_3}$

(N. C. Billingham et al., *J. Polym. Sci., Polym. Chem. Ed.*, **15**, 675, 1977)

(e) $(n\text{-}C_4H_9)_2GeCl_2 \xrightarrow[\Delta]{Na}$

(P. Trefonas and R. West, *J. Polym. Sci., Polym. Chem. Ed.*, **23**, 2099, 1985)

(f) $OCH{-}C(B_{10}H_{10})C{-}CHO +$ $\xrightarrow{\Delta}$

(E. W. Neuse et al., *Macromolecules*, **19**, 481, 1986)

(g) $(C_4H_9)_2SnCl_2 +$ $\longrightarrow$

(ref. 59)

(h) ZrCl$_2$ + Fe(cyclopentadienyl with $-CO_2^-$ groups) $\longrightarrow$

(ref. 59)

(i) Co$^+$PF$_6^-$ + Ph$_3$SbCl$_2$ $\longrightarrow$

(ref. 59)

(j) (salicylaldehyde-based bis-Schiff base) $-CH=N-\bigcirc-N=CH-$ with OH and HO groups + Co(OAc)$_2$ $\longrightarrow$

(P. Y. Maurya et al., *Makromol. Chem.*, **183**, 511, 1982)

(k) HO$-\bigcirc-\overset{CH_3}{\underset{CH_3}{C}}-\bigcirc$ with HO$_2$C, CO$_2$H, and $-$OH substituents + NiCl$_2$ $\longrightarrow$

(ref. 62)

(l) $-$CH$_2$CH$-$
    |
    C=O
    |
    NCH$_3$
(phenyl with N=CH and OH) + Cd(OAc)$_2$ $\longrightarrow$

(J. J. Laverty and Z. G. Gardlund, *J. Polym. Sci. A-1*, **9**, 243, 1971)

(m) Se(CH$_2$CH$_2$OH) + CH$_3-\bigcirc-$NCO with OCN substituent $\longrightarrow$

(H. Kroll and E. F. Bolton, *J. Appl. Polym. Sci.*, **14**, 2319, 1970)

(n)

$\xrightarrow{\Delta}$

(D. A. Foucher and I. Manners, *Makromol. Chem. Rapid. Commun.*, **14**, 63, 1993)

(o)

$+ \text{HO(CH}_2)_4\text{OH} \longrightarrow$

(C. U. Pittman et al., *Macromolecules*, **4**, 360, 1971)

(p)

$\xrightarrow{\Delta}$

(M. Liang and I. Manners, *J. Am. Chem. Soc.*, **113**, 4044, 1991)

(q) $(n\text{–}C_4H_9)_2SnH_2$ $\xrightarrow{\text{zirconocene catalyst}}$

(ref. 32)

(r)

$\xrightarrow{\text{W(O)(OR}_2)Cl_2}$

(P. S. Wolfe, F. J. Gómez, and K. B. Wagener, *Macromolecules*, **30**, 714, 1997)

5. The polymer formed in exercise 4(p) is a *polythionylphosphazene*. How would you expect the hydrolytic stability of a phenoxy-substituted polythionylphosphazene to compare with that of a similarly substituted polythiophosphazene (structure **14**)?

17

MISCELLANEOUS
ORGANIC POLYMERS

17.1 Introduction

This is the last chapter dealing with purely synthetic polymers. Included are nonvinyl polymers that do not fit into any of the significant commercial categories discussed previously. This is by no means to be considered an exhaustive compilation of "miscellaneous" organic polymers, but it does include (not necessarily in any logical sequence) most of the polymer types that have received fairly extensive treatment in the polymer literature. For the most part they are laboratory curiosities, although some, particularly the polyanhydrides, polycarbodiimides, and some of the aromatic polymers, have attracted considerable interest. Among those that have been developed commercially are polyphenylenes and poly(p-xylylene), as well as some related Friedel–Crafts polymers and polyamines.

17.2 Miscellaneous Unsaturated Polymers

Unsaturated polymers have been prepared by several different reactions. Already encountered in previous chapters are polydienes and metathesis polymers, which contain backbone or pendant ethylenic groups (or both), and conjugated polymers such as polyacetylene, poly(sulfur nitride), and polyphosphazenes. Although backbone-conjugated polymers having potential as conductors or semiconductors have received the most attention in recent years, the polycarbodiimides containing cumulated carbon–nitrogen bonds have also been investigated extensively.

17.2.1 Polycarbodiimides

Polycarbodiimides,[1,2] which contain the —N=C=N— unit in the backbone, are prepared from the readily available diisocyanates by a self-addition polymerization reaction catalyzed by organophosphorus compounds—in particular, phosphine oxides (R$_3$PO). The mechanism is considered to involve addition of catalyst across the carbon–nitrogen double bond (17.1) and loss of carbon dioxide to form a dipolar intermediate (17.2), which adds (17.3) to another isocyanate group, this time across a carbon–oxygen double bond. Decomposition of the cyclic intermediate (17.4) results in formation of the carbodiimide group. The overall polymerization reaction is illustrated for methylenebis(4-phenyl isocyanate) (17.5). Poly-

$$R\text{—}N\text{=}C\text{=}O + R_3'P \rightarrow O \longrightarrow \begin{array}{c} R\text{—}N\text{—}C\text{=}O \\ | \qquad | \\ R_3'P\text{—}O \end{array} \tag{17.1}$$

$$\begin{array}{c} R\text{—}N\text{—}C\text{=}O \\ | \qquad | \\ R_3'P\text{—}O \end{array} \longrightarrow R\text{—}\overset{-}{N}\text{—}\overset{+}{P}R_3' + CO_2 \tag{17.2}$$

$$R\text{—}\overset{-}{N}\text{—}\overset{+}{P}R_3' + R\text{—}N\text{=}C\text{=}O \longrightarrow \begin{array}{c} R\text{—}N\text{—}C\text{=}NR \\ | \qquad | \\ R_3'P\text{—}O \end{array} \tag{17.3}$$

$$\begin{array}{c} R\text{—}N\text{—}C\text{=}NR \\ | \qquad | \\ R_3'P\text{—}O \end{array} \longrightarrow RN\text{=}C\text{=}NR + R_3'P \rightarrow O \tag{17.4}$$

$$\tag{17.5}$$

carbodiimides prepared this way are generally of high molecular weight and exhibit properties comparable to those of some of the nylons. Depending on structure, they are semicrystalline, and are believed to be partially crosslinked because of reversible cycloaddition (17.6). Although highly resistant to chemicals in the solid state, polycarbodiimides are quite reactive in solution. For example, modification with amines (17.7) or water (17.8) leads to polymers having guanidine and urea groups, respectively. Reaction with hydrazoic acid (17.9) yields the heterocyclic polyaminotetrazole (**1**).[3]

$$\tag{17.6}$$

$$\tag{17.7}$$

$$\tag{17.8}$$

$$\tag{17.9}$$

1

17.2.2 Polyimines

Polyimines,[4,5] also called *azomethine* or *Schiff base polymers,* are another class of polymers containing carbon–nitrogen double bonds. The imine group is formed by addition of amines to carbonyl compounds followed by loss of water. The reaction is normally acid-catalyzed, although careful pH control is necessary because protonation of the amine reduces its nucleophilicity. Reaction of dialdehydes with diamines leads to polyimines by this step reaction polycondensation. Typically, conjugated polymers result from terephthaldehyde and *p*-phenylenediamine (17.10). High-strength fibers have been spun from liquid crystalline melts of aromatic polyimines, but they have not achieved the commercial success of the aramid fibers.

$$
\text{OCH}\!-\!\langle\bigcirc\rangle\!-\!\text{CHO} + \text{H}_2\text{N}\!-\!\langle\bigcirc\rangle\!-\!\text{NH}_2 \longrightarrow
$$

$$
\left[\langle\bigcirc\rangle\!-\!\text{CH}\!=\!\text{N}\!-\!\langle\bigcirc\rangle\!-\!\text{N}\!=\!\text{CH}\right] + 2\text{H}_2\text{O} \qquad (17.10)
$$

It may be recalled from the preceding chapter that appropriately substituted polyimines are useful chelating agents for coordination polymers. The extension of polyimine synthesis to tetracarbonyl compounds and tetraamines, which leads to the aromatic polyquinoxalines, is discussed in Chapter 15.

Related to the polyimines are the *polyamidines*[6–8] (**2**), which are prepared most directly from monocarboxylic acids and diamines by dehydration with poly(trimethylsilylphosphate); the *polyisocyanides* (**3**), prepared from isocyanides using nickel(II) catalysts[9] (17.12); and the *polycarbodiimides* (**4**), prepared by the living polymerization of carbodiimide monomers using a complex titanium catalyst[10] (17.13). These polycarbodiimides should not be confused with the previously described polymers that contain carbodiimide repeating units.

$$
\text{RCO}_2\text{H} + \text{H}_2\text{N}\!-\!\text{Ar}\!-\!\text{NH}_2 \xrightarrow{\text{poly(trimethylsilyl phosphate)}}
$$

$$
\left[\begin{array}{c} R \\ | \\ \text{N}\!=\!\text{C}\!-\!\text{NH}\!-\!\text{Ar} \end{array}\right] \qquad (17.11)
$$

2

$$
\text{RN}\!\equiv\!\text{C} \xrightarrow{\text{NiCl}_2 \cdot 6\text{H}_2\text{O}} \left[\begin{array}{c} \text{NR} \\ \| \\ \text{C} \end{array}\right] \qquad (17.12)
$$

3

$$
\text{RN}\!=\!\text{C}\!=\!\text{NR} \xrightarrow{\text{Ti complex}} \left[\begin{array}{c} \text{NR} \;\; R \\ \| \quad | \\ \text{C}\!-\!\text{N} \end{array}\right] \qquad (17.13)
$$

4

17.2.3 Polymers Containing Carbon–Carbon Double Bonds

The best route to polymers containing conjugated double bonds is the coordination polymerization of alkynes (Chapter 8). The well-known Wittig reaction between ylides and carbonyl compounds, as shown using AB (17.14) and AA + BB (17.15) monomers, represents an alternative synthesis, although the molecular weights of the resultant polymers are much lower.[11,12] Conjugated polymers (of low molecular weight) can also be prepared by condensing biscarbonyl compounds with monomers containing active methylene groups[13] (17.16).

$$\text{OCH}-\!\!\bigcirc\!\!-\text{CH}=\text{P}(C_6H_5)_3 \longrightarrow \left[\bigcirc\!-\text{CH}=\text{CH}\right] + (C_6H_5)_3\text{PO} \qquad (17.14)$$

$$\text{OCH}-\!\!\sideset{}{}{\mathop{\bigcirc}}_{S}\!\!-\text{CHO} + (C_6H_5)_3\text{P}=\text{CH}-\!\!\sideset{}{}{\mathop{\bigcirc}}_{S}\!\!-\text{CH}=\text{P}(C_6H_5)_3 \longrightarrow$$

$$\left[\sideset{}{}{\mathop{\bigcirc}}_{S}\!-\text{CH}=\text{CH}\right] + 2(C_6H_5)_3\text{PO} \qquad (17.15)$$

$$\text{NCCH}_2-\!\!\bigcirc\!\!-\text{CH}_2\text{CN} + \text{OCH}-\!\!\bigcirc\!\!-\text{CHO} \xrightarrow{\text{NaOC}_2\text{H}_5}$$

$$\left[\underset{\text{C}=\text{CH}}{\overset{\text{CN}}{|}}-\!\!\bigcirc\!\!-\text{CH}=\underset{}{\overset{\text{CN}}{\underset{|}{\text{C}}}}-\!\!\bigcirc\right] + 2H_2O \qquad (17.16)$$

17.2.4 Azo Polymers

Azo polymers[14] have nitrogen–nitrogen double bonds in the backbone. Several methods have been reported for their synthesis, with the highest molecular weights being obtained by oxidative coupling of aromatic diamines[15] (17.17) using catalysts similar to those used in preparing poly(phenylene oxide) by oxidative coupling of phenols (Chapter 11).

$$\text{H}_2\text{N}-\!\!\bigcirc\!\!-\text{NH}_2 \xrightarrow[O_2]{\text{Cu}_2\text{Cl}_2} \left[\text{N}=\text{N}-\!\!\bigcirc\right] \qquad (17.17)$$

Other methods include:

1. Photolysis of diazides[16] (17.18)

$$\text{N}_3-\text{Ar}-\text{N}_3 \xrightarrow{h\nu} \left[\text{Ar}-\text{N}=\text{N}\right] + 2\text{N}_2 \qquad (17.18)$$

2. Coupling of tetrazonium compounds with bisphenols[17] (17.19)

$$\bar{C}l\overset{+}{N_2}-Ar-\overset{+}{N_2}\bar{C}l + HO-\langle\bigcirc\rangle-R-\langle\bigcirc\rangle-OH \longrightarrow$$

$$\left[-N{=}N-Ar-N{=}N-\langle\bigcirc\rangle-R-\langle\bigcirc\rangle-\right] \quad (17.19)$$
$$HOOH$$

3. Decomposition of tetrazonium salts[18] (17.20)

$$\overset{+}{N_2}-Ar-\overset{+}{N_2} \xrightarrow{Cu_2Cl_2, \Delta} -[Ar-N{=}N]- \quad (17.20)$$

4. Reductive coupling of dinitro compounds[19] (17.21)

$$O_2N-Ar-NO_2 \xrightarrow[\text{triethylene glycol}]{KOH, Zn} -[Ar-N{=}N]- \quad (17.21)$$

17.2.5 Polymers Containing Carbon–Carbon Triple Bonds

Oxidative coupling is also used to prepare polyacetylenes,[20,21] which in this case are defined as polymers having backbone triple bonds. An example is the polymerization of *m*-diethynylbenzene (17.22), which leads to high-molecular-weight polymer if the monomer is

$$HC{\equiv}C\text{—}\langle\bigcirc\rangle\text{—}C{\equiv}CH \xrightarrow[O_2]{Cu_2Cl_2, R_3N} \left[\langle\bigcirc\rangle\text{—}C{\equiv}C-C{\equiv}C\right] \quad (17.22)$$

carefully purified. The regular zigzag structure of the polymer (**5**) results in high crystallinity and very low solubility. The incorporation of small amounts of *p*-diethynylbenzene, however, has the effect of increasing solubility by altering the regular structure, thus impeding crystallization, without changing the rigid character of the polymer backbone.

5

An alternative route to polyacetylenes involves the reaction of diacetylenes with aromatic dibromides (17.23) in the presence of a palladium catalyst.[22] The success of this process hinges on activation of the bromine atoms by electron-withdrawing groups. Conjugated diacetylenes can also be polymerized in the solid state using heat, light, or ionizing radiation[23,24] (17.24). This process, which involves free radical coupling reactions, results in high-molecular-weight polyeneynes that exhibit interesting color changes when they undergo phase transitions.

$$\text{Br—Ar—Br} + \text{HC}\equiv\text{C—Ar}'\text{—C}\equiv\text{CH} \xrightarrow[(-2\text{HBr})]{\text{Pd}}$$

$$+\!\!\text{Ar—C}\equiv\text{C—Ar}'\text{—C}\equiv\text{C}\!\!+ \tag{17.23}$$

$$\text{R—C}\equiv\text{C—C}\equiv\text{C—R} \xrightarrow[\text{ionizing radiation}]{\Delta, h\nu, \text{or}} \overset{}{=}\!\!\underset{\text{R}}{\overset{|}{\text{C}}}\text{—C}\equiv\text{C}\text{—}\underset{\text{R}}{\overset{|}{\text{C}}}\!\!= \tag{17.24}$$

Conjugated polyacetylenes are relatively unstable, but they can be converted to aromatic polymers having pyrrole (17.25), thiophene (17.26), or pyrazole (17.27) rings.[26]

$$(17.25)$$
$$(17.26)$$
$$(17.27)$$

17.3   Poly(*p*-phenylene) and Poly(*p*-xylylene)

17.3.1   Poly(p-phenylene)

The simplest of the fully aromatic polymers is poly(*p*-phenylene)[27,28] (**6**). It is prepared most conveniently by directly polymerizing benzene in the presence of combinations of Lewis acids and oxidizing agents. When aluminum chloride and cupric chloride are used (17.28), polymerization occurs rapidly at temperatures as low as 35 to 50°C.

$$(17.28)$$

6

The mechanism of benzene polymerization has been the subject of much debate, but on the basis of ESR studies there is general agreement that the initiating species is a benzene radical cation.[29] Theoretical calculations[30] and studies on polymerization of deuterated benzene[31] indicate a possible polymerization mechanism in which benzene rings are associated in a stacked, end-to-end arrangement (**7**), as shown in Scheme 17.1, with the radical cation delocalized over the entire chain. Chain buildup ceases when the radical cation concentration on the terminal phenyls becomes too small to promote further association. Subsequent

SCHEME 17.1. Proposed mechanism for benzene polymerization.

sigma bond formation results in **8**, which may undergo competing depropagation or oxidation and loss of H$^+$ to form **6**.

Poly(p-phenylene) has also been prepared from polymeric precursors, such as poly(1,3-cyclohexadiene) (**9**) by dehydrogenation and the diacetoxy derivative (**10**) by elimination of acetic acid[32]; and by palladium(0)-catalyzed coupling reactions[33–35] (17.29).

$$ \text{(17.29)} $$

As might be expected, poly(p-phenylene) has good thermal stability, but its very high melting point (it does not melt at temperatures below 530°C) and low solubility make it difficult to use in commercial applications. Nevertheless, *polyphenylenes* have been marketed (Monsanto) as laminating resins,[36a,37] although they bear little resemblance to pure **6**. Rather, they are synthesized from mixtures of biphenyl and *o*-, *m*-, and *p*-terphenyls, which are heated together with disulfonyl chlorides in the presence of a copper halide catalyst. Both HCl and SO$_2$ are split out in the process. Heat-curable acetylene-terminated polyphenylene prepolymers have also been marketed (Hercules). More recently, attention has focused on the conducting properties of **6** (see Chapter 4).

Poly(*p*-phenylene) and backbone-substituted poly(*p*-phenylene) have also been prepared by cycloaddition reactions. These are discussed in Section 17.5.

17.3.2 *Poly(p-xylylene)*

Poly(*p*-xylylene) (**11**) is not completely aromatic but contains benzene rings linked through two methylene groups. Considerable interest has been expressed in this polymer[27,28] because it can be synthesized by pyrolysis of the readily available *p*-xylene, a reaction (17.29a) believed to involve an intermediate quinodimethide. One of the byproducts of this synthesis is the bridged dimer, paracyclophane (**11a**), a compound that can also be pyrolyzed to poly(*p*-xylylene), either through the same quinodimethide intermediate or possibly via an intermediate diradical. The resultant polymer is generally formed in higher yield and with less chain branching than that prepared by *p*-xylene pyrolysis, and the paracyclophane route has the added advantage of being applicable to preparation of substituted poly(*p*-xylylene)s.

(17.29)

11

11a

Other routes to **11** or its chlorinated derivatives are given in reactions (17.30) through (17.32).[39]

(17.30)

(17.31)

(17.32)

Poly(*p*-xylylene) is a crystalline material having relatively poor thermal stability. It is more tractable than poly(*p*-phenylene) in that it can be fabricated into brittle fibers or tough films, the latter by vapor deposition from pyrolysis of **11a**. The polymer is of interest in medical applications because of its biocompatibility. Commercial poly(*p*-xylylene)s (Union Carbide's Parylene) include chlorinated and nonchlorinated varieties. (Vapor deposition of one-micron-thick coatings of Parylene onto paper is an excellent way to preserve old books and manuscripts.)[40] The analogous *ortho* and *meta* polyxylylenes have also been prepared, but they are very low-melting and not, therefore, as useful as poly(*p*-xylylene), which has a softening temperature in excess of 400°C. Some other polymers containing bridged benzene rings are described in the next section.

17.4   Friedel–Crafts Polymers

We first encountered Friedel–Crafts polymers in Chapter 11 in the preparation of polysulfones from disulfonyl chlorides. Similar reactions of dicarboxylic chlorides (17.33) leads to polyketones.[41] Dicarboxylic acids may in some instances be used in place of the acid chlorides in the presence of the strongly dehydrating P_2O_5-methanesulfonic acid.[42]

(17.33)

Benzyl chloride undergoes acid-catalyzed self-condensation (17.34) to yield highly branched low-molecular-weight polymers having repeating benzylic groups.[43] Similarly, α,α'-dichloro-*p*-xylene reacts with biphenyl (17.35) to give polymers of analogous structure.

(17.34)

(17.35)

Reactions of this type have led to commercial laminating resins and molding compounds.[36b] To circumvent the problem of HCl evolution during polymerization, benzylic ethers (**12**), which evolve methanol, have been developed as alternative Friedel–Crafts monomers. Reaction of **12** with phenol has also been used commercially to produce thermosetting poly-

mers that are related in structure to phenol–formaldehyde polymers, as well as to poly-
xylylene and polyphenylene resins.

$$CH_3OCH_2-\!\!\bigcirc\!\!-CH_2OCH_3$$

12

The Friedel–Crafts reaction has also been used to prepare crosslinked polystyrene.[44] A
mixture of styrene and p-di(chloromethyl)benzene (α,α'-dichloro-p-xylene) is polymerized
at low temperatures with a Lewis acid catalyst such as stannic chloride. Chain polymerization to form linear polystyrene occurs rapidly, while crosslinking (17.36), under the influence of the same catalyst, occurs relatively slowly, giving a product with a higher glass transition temperature than that of linear polystyrene.

$$\text{(17.36)}$$

17.5   Cycloaddition Polymerization

The two most widely studied cycloaddition reactions are the Diels–Alder reaction (2 + 4 cycloaddition) and the light-catalyzed 2 + 2 cycloaddition. Because of the simplicity of these
reactions and the absence of byproducts, they have been very popular with polymer chemists,
particularly the former, although there has been little commercial exploitation. The previously mentioned (Chapter 13) nadimide-terminated addition polyimides, which use reverse
(retrograde) Diels–Alder reactions in curing, and some photoresist resins that cure by 2 + 2
cycloaddition of cinnamic acid derivatives, represent two examples.

17.5.1   *Diels–Alder Polymerization*

In its simplest form the Diels–Alder reaction is the reaction of a diene with an alkene (called
the *dienophile*) to form a cyclohexene ring (17.37). In practice, dienophiles having electron-

$$\text{(17.37)}$$

attracting substituents on the double bond are most effective, while reactivity of the diene is
increased if the monomer is confined to the more reactive *s-cis* configuration, as is the case
with homoannular dienes. Illustrative of the latter is the facile dimerization or oligomeriza-

tion of cyclopentadiene (17.38). This type of reaction is not useful from the standpoint of polymer formation, because only low-molecular-weight materials are obtained. (Recall from Chapter 9 the reversible crosslinking of polymers by cyclopentadiene cycloaddition.)

$$\text{(structure)} \quad (17.38)$$

A large number of Diels–Alder polymerizations have been reported.[45–47] One of the earliest was the reaction of 2-vinyl-1,3-butadiene with *p*-benzoquinone to give a ladder polymer (17.39). In this case, each cycloaddition generates a new diene monomer.

$$\text{(structure)} \quad (17.39)$$

Some Diels–Alder polymerizations involve monomers having both diene and dienophile functions. Among these is the self-addition polymerization of 2-vinyl-1,3-butadiene (17.40). The structure of the polymer is more complex than that shown because different modes of addition or crosslinking by Diels–Alder addition are possible.

$$\text{(structure)} \quad (17.40)$$

Reaction of bisdienes with bisdienophiles has been most widely used in Diels–Alder polymerizations. An example is the reaction of a biscyclopentadiene with a bismaleimide (17.41) to form polyimide **13**. One factor that limits the molecular weight in reactions of this

$$\text{(structure)} \quad (17.41)$$

13

type is the tendency for the retrograde Diels–Alder reaction to occur at the high temperatures usually needed to obtain a high degree of polymerization. One way that this has been circumvented is by reacting biscyclopentadienones (**14**, also known as bistetracyclones) with diacetylenes. The Diels–Alder reaction (17.42) occurs to give the polymeric adduct (**15**), which loses carbon monoxide spontaneously to yield the wholly aromatic polymer **16**, thus rendering the reaction essentially irreversible. This represents an alternative synthesis of a poly(*p*-phenylene); however, unlike the unsubstituted polymer (**6**), this phenyl-substituted analog (**16**) is amorphous and more soluble, while exhibiting little weight loss in air or nitrogen below 550°C.

14

$$(17.42)$$

15

16

Certain compounds, known as *pseudobisdienes*, contain a single diene function but undergo decomposition to generate a new diene in the course of polymerization. An example is α-pyrone **17**, which reacts with bismaleimide (17.43) to form an intermediate (**18**) that loses carbon dioxide spontaneously (17.44) to form a new intermediate (**19**) containing both diene and dienophile units. This intermediate undergoes Diels–Alder polymerization (17.45). Other pseudobisdienes include thiophene-1,1-dioxide (**20**) and 2,5-dimethyl-3,4-diphenylcyclopentadieneone (**21**)* which react similarly, except that sulfur dioxide and carbon monoxide, respectively, are evolved.

$$(17.43)$$

17   **18**

*Compound **21** exists as a colorless dimer at room temperature, but dissociates to red monomer on heating.

$$18 \xrightarrow{(-CO_2)} \text{[structure 19]} \tag{17.44}$$

19

$$19 \longrightarrow \text{[structure]} \tag{17.45}$$

20 **21**

Many Diels–Alder polymers have interesting properties and can be fabricated into films or fibers. The preponderance of fused rings in many of the polymers results in high melting points and low solubility and, where the retrograde Diels–Alder reaction is not a problem, good thermal stability and chemical resistance.

17.5.2 2 + 2 Cycloaddition Polymerization

A major problem in preparing polymers by the light-catalyzed 2 + 2 cycloaddition results from side reactions caused by irradiation—for example, crosslinking, degradation, or rearrangement. Furthermore, the cycloaddition, shown (17.46) in simplified form, results in a cyclobutane ring, a structure that is inherently unstable because of ring strain.

$$\| + \| \xrightarrow{h\nu} \square \tag{17.46}$$

Most of the early work in this area dealt with solid-phase photopolymerization[48] of monomers such as 2,5-distyrylpyrazine (**22**), which forms highly crystalline stereoregular polymer (**23**) on irradiation (17.47). Stereoregularity is assured by the spatial requirements of the crystalline lattice. It should be noted that this type of reaction is a true photopolymerization process in that *each chain-propagating step involves a photochemical reaction,* as opposed to those reactions that are simply initiated by light.[49]

$$Ph-CH=CH-\text{[pyrazine]}-CH=CH-Ph \xrightarrow{h\nu} \text{[polymer structure]} \tag{17.47}$$

22 **23**

To achieve high molecular weights in solution processes, diene monomers must be selected that have little tendency to undergo vinyl polymerization or intramolecular cyclization.[50,51] Bismaleimides substituted at the double bonds with chlorine atoms (**24**) represent one type that forms linear cycloaddition polymers[49] (17.48). In this case, at least seven carbon atoms should separate the two maleimide groups to minimize intramolecular cycloaddition. Triplet sensitizers enhance the reaction.

$$(17.48)$$

$$(x > 7)$$

24

Polyimides have also been prepared by a combination of 2 + 2 and 2 + 4 cycloaddition reactions by copolymerizing bismaleimides photochemically with benzene[52] or alkylbenzenes.[53] In this case, the initial 2 + 2 cycloaddition to the benzene ring (17.49) results in a homoannular diene that subsequently undergoes Diels–Alder addition (17.50) to give polymer **25**. Only a head-to-tail structure is shown, but both head-to-head and tail-to-tail structures could presumably form, depending on whether the intermediate homoannular diene undergoes self-addition or reaction with bismaleimide. Polyimides prepared this way are insoluble, possibly because of crosslinking through the remaining double bond in the backbone. Also, they all tend to decompose by decyclization at about the same temperature (420–485°C) regardless of whether R′ is aliphatic or aromatic, thus demonstrating the sensitivity of the cyclobutane ring.

$$(17.49)$$

$$(17.50)$$

25

17.6 Polyanhydrides

Polymers with recurring anhydride groups[54–56] were among the first linear fiber-forming polymers to be investigated as part of the pioneering work of Carothers. They can be prepared from dicarboxylic acids by dehydration with reagents such as phosgene, acid chlorides, or acetic anhydride (17.51). Alternatively, polyanhydrides may be prepared by an anhydride exchange reaction (17.52) of a preformed mixed anhydride, with byproduct acetic anhydride being removed under reduced pressure.

$$HO_2C-R-CO_2H + (CH_3CO)_2O \xrightarrow{heat} \left[OC-R-C \right] + 2CH_3CO_2H \quad (17.51)$$

$$HO_2C-R-CO_2H + 2(CH_3CO)_2O \longrightarrow$$

$$CH_3COC-R-COCCH_3 \xrightarrow{heat} \left[OC-R-C \right] + (CH_3CO)_2O \quad (17.52)$$

As might be expected, butanedioic (succinic) and pentanedioic (glutaric) acids form mainly cyclic anhydrides rather than linear polymer under these conditions. Hexanedioic (adipic) acid forms a mixture of both. Propanedioic (malonic) acid and acids higher than adipic form polymer. Diacid chlorides may also be used as monomers for polyanhydrides, either by reaction (17.53) with dicarboxylate salts[55] (Z = ONa) or trimethylsilyl esters[57] (Z = OSiMe$_3$); or they may be converted to the diammonium salts followed by reaction with stoichiometric amounts of water as shown (17.54) for the polyanhydride (**26**) of terephthalic acid.[58]

$$Cl-C-R-C-Cl + Z-C-R'-C-Z \longrightarrow$$

$$\left[C-R-C-O-C-R'-C-O \right] \quad (17.53)$$

$$Cl-C-\bigcirc-C-Cl \xrightarrow{R_3N} Cl^- R_3NC-\bigcirc-CNR_3Cl^- \xrightarrow{H_2O}$$

$$\left[O-C-\bigcirc-C \right] \quad (17.54)$$

26

Aliphatic polyanhydrides are very sensitive to hydrolysis and degrade rapidly in the presence of moisture. Aromatic polyanhydrides, on the other hand, often exhibit surprisingly good hydrolytic stability, comparable to that of linear aromatic polyesters. Polymer **26**, for example, can be refluxed in 10M HCl for up to six hours with little degradation. This stability results primarily from a high degree of crystallinity. Degradable polyanhydrides are of current interest for controlled release applications, particularly in the biomedical field.[59,60]

17.7 Polyamines

The term *polyamine* refers here to polymers containing an amino group as an essential part of the backbone. It does not include heterocyclic amino polymers such as polyquinoxalines or polypyrazoles, nor does it include polymers having amino groups attached to the main chain or amino substituents on a pendant group.

TABLE 17.1. Polyaniline Structures

Name	Structure	Description
Leuco-emaraldine		Colorless, insulator, fully reduced
Emaraldine base		Green, insulator, partially oxidized
Emaraldine salt		Blue, conductor, partially oxidized
Pernigraniline		Purple, insulator, fully oxidized

One of the most important polyamines is *polyaniline*,[61] first mentioned in Chapter 4 because of its conducting properties. It is a commercial product with a variety of applications, including in batteries, computer devices, and electromagnetic shielding. As shown in Table 17.1, polyaniline exists in a variety of forms, each with a distinctive name and color representing varying oxidation states. Only the partially oxidized protonated emaraldine salt exhibits conducting properties; no additional dopants are needed. The polymer is made in aqueous solution by the reaction of aniline with an oxidizing agent such as ammonium persulfate[62] or an enzymatic peroxidase.[63] Electrochemical oxidation of aniline[64] is a convenient laboratory route to polyaniline films. Polyaniline owes its commercial success to its ease of synthesis and relatively low cost compared with other conducting polymers.

One of the best general methods of making polyamines is by Michael addition of primary or secondary diamines to activated double bonds, a reaction described in Chapter 13 for chain extension of bismaleimides. [Recall the synthesis of polyamidoamine (PAMAM) dendrimers by Michael addition in Chapter 10.] Typical is the reaction of piperazine with 1,4-diacrylpiperazine[65] (17.55). If primary amine (RNH$_2$) is substituted for piperazine in this reaction, polymer **27** is formed.

$$(17.55)$$

27

Nucleophilic substitution of primary dihalides with diamines is not a satisfactory way to make high-molecular-weight polyamines because salt formation between reactants and byproduct hydrogen halide upsets the stoichiometric balance. One way to circumvent this problem is to use disilylamines[66] (**28**) (prepared from diamine and a chlorosilane). Reaction with an active dihalide in the presence of catalytic amounts of ammonium chloride results in high-molecular-weight polymer with elimination of the halosilane. This is shown for the reaction of α,α'-dichloro-p-xylene with the disilyl derivative of piperazine (17.56).

$$R_3Si-N \quad N-SiR_3 + ClCH_2-\!\!\bigcirc\!\!-CH_2Cl \xrightarrow{NH_4Cl}$$

28

(17.56)

$$\left[N \quad N-CH_2-\!\!\bigcirc\!\!-CH_2\right] + 2R_3SiCl$$

Polyamines of low molecular weight may be prepared by the Mannich reaction involving secondary diamines or primary amines, formaldehyde, and compounds containing active hydrogen.[67–70] This is shown for pyrrole (**29**) (which reacts principally at ring positions 2 and 5) using piperazine (17.57) and benzylamine (17.58). Mannich-base polymers are used as complexing agents, flocculants, antioxidants, and coating binders.

(17.57)

(17.58)

Ring-opening polymerization of alkyleneimines[71] (aziridines) forms highly branched polyamines, and the product obtained from the simplest member of this monomer type, ethyleneimine, has been developed commercially (Dow trade name Montrek). The polymerization reaction is acid catalyzed and similar, mechanistically, to the polymerization of epoxides and episulfides. Polymerization is not initiated with basic catalysts unless an acyl group or other strongly electron-attracting substituent is attached to the nitrogen atom. Initiation is believed to involve S_N2 ring opening of a protonated alkyleneimine (17.59), with propagation (17.60) continuing by analogous reactions of the cyclic immonium end group.

(17.59)

(17.60)

Chain branching occurs by reaction of secondary amine groups in the backbone with immonium chain ends (17.61).

$$\text{mCH}_2\text{CH}_2\text{NHm} + \text{mCH}_2\text{CH}_2\overset{+}{\text{NH}} \underset{\text{CH}_2}{\overset{\text{CH}_2}{\diagdown}} \longrightarrow$$

$$\text{mCH}_2\text{CH}_2\overset{+}{\text{NH}}\text{m}$$
$$\underset{|}{\text{CH}_2\text{CH}_2\text{NHCH}_2\text{CH}_2\text{m}}$$

(17.61)

The four-membered ring analogs of aziridines (azetidines)—for example, conidine (**30**)—are also susceptible to this type of reaction (17.62).

(17.62)

30

Although aziridenes yield only branched polyamines, it is possible to prepare the linear polymer by ring-opening polymerization of 2-oxazolines[72,73] (**31**) to form the polymeric amide (**32**), followed by basic hydrolysis (17.63).

(17.63)

31   **32**

Polyaziridenes are used primarily in aqueous solution for treatment of textiles and paper. Aziridenes themselves are used for crosslinking carboxyl-containing polymers, such as acrylic emulsion coatings. In this instance, polyfunctional aziridenes (**33**) react with the carboxyl groups according to reaction (17.64).

(17.64)

33

17.8   Charge-Transfer Polymers

Charge-transfer complexing in polymeric systems has received considerable attention in recent years, particularly from the standpoint of the effect of complexation on electrical conductivity and photoconductivity.[74–76] In fact, it has been suggested that simple charge-transfer complexes in which donor and acceptor molecules alternate throughout a crystalline structure may be thought of as polymers.[77] We are concerned here, however, with covalently bonded polymers containing donor or acceptor groups.

There are basically three approaches to studying charge-transfer complexing in polymers: (1) complexing polymers containing donor groups or polymers containing acceptor groups with low-molecular-weight acceptors or donors, respectively; (2) mixing donor polymers with acceptor polymers; and (3) synthesizing polymers having both donor and acceptor groups. A lot of work has been done with donor polymers and low-molecular-weight acceptors, but the products are frequently brittle and intractable. Some polycarbonates and polyesters with good mechanical properties have been prepared[78] from aryliminodiethanol compounds such as **34** and **35**. These polymers form highly colored complexes with certain

$$\text{HOCH}_2\text{CH}_2\text{NCH}_2\text{CH}_2\text{OH}$$

$$\underset{\text{Ar}}{|}$$

34

$$\text{CH}_2\text{CH}_2\text{OH}$$

35

pi acceptor molecules with a resultant increase in conductivity. Table 17.2 shows the effect of adding equivalent amounts of 2,3-dichloro-5,6-dicyano-*p*-benzoquinone (**36**) to some polycarbonates (**37**) prepared from methoxy derivatives of **34** and bisphenol A dichloroformate. Addition of **36** to the *p*-methoxy derivative results in a thousandfold decrease in volume resistivity. Steric effects become evident, however, as the number of methoxy groups on the ring is increased.

36

TABLE 17.2. Effect of Equivalent Amounts of 2,3-dichloro-5,6-dicyano-*p*-benzoquinone on Volume Resistivities of Polycarbonate **52**[a]

Polymer Structure	51 Added	Volume Resistivity (ohm cm)
$R_1 = R_2 = R_4 = H; R_3 = OCH_3$	No	1.3×10^{16}
$R_1 = R_2 = R_4 = H; R_3 = OCH_3$	Yes	1×10^{13}
$R_1 = R_4 = OCH_3; R_2 = R_3 = H$	Yes	1×10^{11}
$R_1 = H; R_2 = R_3 = R_4 = OCH_3$	Yes	2×10^{11}

[a]Data from Sulzberg and Cotter.[78]

37

Acceptor polycarbonates (**38**) have been prepared from 5-nitroisophthalic acid ($R_1 = R_3 = H, R_2 = NO_2$) and 4,6-dinitroisophthalic acid ($R_1 = R_3 = NO_2, R_2 = H$) and from nitroterephthalic acid.[79] Blends of **37** and **38** are highly colored and have mechanical properties suitable for films, coatings, and fibers. Typically, a mixture of the *p*-methoxy derivative of **37** and the 5-nitro derivative of **38** is yellow-orange in color with a lowest energy absorption maximum of 398 nm compared with 298 and 330 nm, respectively, for the individual polymers.

38

Polyesters (**39**) of low molecular weight containing both donor and acceptor groups have been prepared[78] from monomer **34** with 5-nitroisophthalic acid or nitroterephthalic acid. Interestingly, **39** had a somewhat lower volume resistivity than blends of donor and acceptor polymers (**37** and **38**, respectively). It is not clear why only low-molecular-weight donor–acceptor polyesters could be prepared, since high-molecular-weight donor polymers and acceptor polymers were prepared from the same monomers. It has been suggested that loss of mobility or reactivity resulting from complexing during polymerization may upset the stoichiometric balance.[78]

39

Vinyl polymers containing donor and acceptor groups have also been the subject of considerable study.[80–82] An example is copolymer **40**, which exhibits a charge-transfer absorption enhanced an order of magnitude over that of mixtures of model compounds. Presumably, constraints in the polymer system causes a higher relative concentration of charge-transfer interactions.

40

17.9   Ionic Polymers

In Chapter 3 we learned that the presence of ionic groups in polymers can have interesting morphological effects, including the formation of ionic microdomains within the polymer matrix and a tendency for the polymers to exhibit thermoplastic elastomeric properties.[83–86] In Chapter 9 we encountered the commercially important ionomers—ionized copolymers of ethylene and methacrylic acid. In this section we are concerned with the synthesis of nonvinyl analogs of ionomers as well as polymers having ionic groups as part of the backbone structure.

Compared with the vinyl-based ionomers, nonvinyl ionomers have not been studied extensively. Among the few that have been reported are polyureas (**41**) prepared from diisocyanates and diaminobenzoates[87] and polyesters and polyurethanes containing tetra-*t*-butylammonium and -sulfonium groups from dihydroxy monomer **42**.[88] Related in structure

41

42

are polyurethane elastomers containing *backbone* ammonium ions[87,89] prepared by neutralization of amine-containing polyurethanes (17.65). Polymers of these types are of interest as coatings and adhesives and as finishing agents for textiles and leather.

Polymers containing backbone ammonium ions have also been prepared by the reaction of dihalides (preferably primary) with ditertiary amines such as 1,4-diazabicyclo[2.2.2]octane (DABCO) (17.66)[90] or with primary amines (17.67).[91] Polymers of this type, called *ionenes*,[92] are used commercially as flocculating agents for water treatment and as components of cosmetics and antimicrobial formulations. An ionene–polyether–ionene ABA block copolymer is of interest as a flocculating agent.[93]

$$\text{N} \quad \text{N} + \text{Br—R—Br} \longrightarrow \left[\begin{array}{c} \text{Br}^- \\ ^+\text{N} \quad \text{N}^+\text{—R} \\ \text{Br}^- \end{array} \right] \tag{17.66}$$

$$\text{RNH}_2 + \text{Cl—R}'\text{—Cl} \longrightarrow \left[\begin{array}{c} \text{Cl}^- \\ \text{R}'\text{—}^+\text{NH—} \\ \text{R} \end{array} \right] + \text{HCl} \tag{17.67}$$

Polymers containing ionic linkages as part of the backbone have been synthesized from dicarboxylic acids and divalent metal (M) acetates (17.68), or from dicarboxylate salts and metal halides (17.69). The products are called *halatopolymers*,[86,94] A halatomonomer,[95] **43**, prepared from divalent metal salts of mono(hydroxyethyl)phthalate, has been used to prepare polyurethanes and polyesters.[96–98] Halatopolymers are primarily of interest as adhesives.

$$\text{M(OCOCH}_3)_2 + \text{HO}_2\text{C—R—CO}_2\text{H} \rightarrow [\text{MO}_2\text{C—R—CO}_2]_x + 2\text{CH}_3\text{CO}_2\text{H} \tag{17.68}$$

$$\text{MCl}_2 + \text{Na}^{+-}\text{O}_2\text{C—R—CO}_2{}^-\text{Na}^+ \rightarrow [\text{MO}_2\text{C—R—CO}_2]_x + 2\text{NaCl} \tag{17.69}$$

$$\text{HOCH}_2\text{CH}_2\text{O}_2\text{C} \quad \text{CO}_2{}^-\text{M}^{2+-}\text{O}_2\text{C} \quad \text{CO}_2\text{CH}_2\text{CH}_2\text{OH}$$

43

Before we leave the subject of ionic polymers, it should be noted that *ion-exchange resins*[99] have a relatively long history of use in separations and water purification. The crosslinked polymers are classified as anionic or cationic exchange resins. The former contain basic groups capable of exchanging anions, such as styrene–divinylbenzene copolymers substituted with amine groups, or formaldehyde–aniline resins. Cationic exchange resins contain acidic groups capable of exchanging cations. Sulfonated phenol–formaldehyde or styrene–divinylbenzene copolymers, and crosslinked poly(acrylic acid)s, are typical examples. One type of cholesterol-reducing medicine is an anion exchange resin that functions by removing bile salt as it passes through the intestinal tract. To compensate, cholesterol is oxidized to bile acids, thereby reducing serum-cholesterol levels.

A variety of similar polymers of different molecular weight ranges are used as coagulating agents for removal of colloidal material in wastewater treatment. These polymers, known as *polyelectrolytes*,[100–102] range from cationic to nonionic to anionic. They are normally used in conjunction with inorganic coagulants such as activated silica, alum, or ferric chloride. Although used in very small amounts, they represent a large-volume market because of the extensive application of wastewater treatment. Typical polymers are polyacrylates, poly(vinylpyridinium salts), and poly(amino acids). Polyelectrolytes apparently aid coagulation by neutralizing the charge on colloidal particles, although there is evidence that their function may also be to bridge the charged particles, because coagulation can be brought about with polyelectrolytes having the same charge as the colloid.

Polyampholytes[103,104] are a type of polyelectrolyte or ionomer having both cationic and anionic groups, but no conventional counterions. An example is the copolymer of a methacrylate ester with ion-pair **44**.[105] Polyampholytes are useful chelating agents for removing trace metals from water. They are also used as ion-exchange membranes, flocculants, and soil conditioners and as ingredients in shampoos and hair conditioners.

$$\left[\begin{array}{c} CH_3 \\ | \\ CH_2{=}CCO_2CH_2CH_2\overset{+}{N}(CH_3)_3 \end{array} \right] \left[\begin{array}{c} CH_3 \\ | \\ CH_2{=}CCO_2CH_2CH_2SO_3^- \end{array} \right]$$

44

References

1. R. J. Cotter and M. Matzner, *Ring-Forming Polymerizations*, Part B1, Academic, New York, 1968, Chap. 1.
2. L. M. Alberino, W. J. Farrissey, and A. A. R. Sayigh, *J. Appl. Polym. Sci.*, **21**, 1999 (1977).
3. E. Dyer and P. E. Christie, *J. Polym. Sci A-1*, **6**, 729 (1968).
4. G. F. D'Alelio and R. K. Schoenig, *J. Macromol. Sci., Rev. Macromol. Chem.*, **3**, 105 (1969).
5. D. A. Tomalia, in *Handbook of Polymer Synthesis*, Part A (H. R. Kricheldorf, Ed.), Marcel Dekker, New York, 1992, Chap. 13.
6. S. Ogata, M. Kakimoto, and Y. Imai, *Makromol. Chem. Rapid Commun.*, **6**, 835 (1985).
7. L. J. Mathias and C. G. Overberger, *J. Polym. Sci., Polym. Chem. Ed.*, **17**, 1287 (1979).
8. K. Kurita, Y. Kusayama, and Y. Iwakura, *J. Polym. Sci., Polym. Chem. Ed.*, **15**, 2163 (1977).
9. R. J. M. Nolte and W. Drenth, in *New Methods for Polymer Synthesis* (W. J. Mijs, Ed.), Plenum, New York, 1992, Chap. 9.
10. A. Goodwin and B. M. Novak, *Macromolecules*, **27**, 5520 (1994).
11. H. H. Hörhold and J. Opfermann, *Makromol. Chem.*, **131**, 105 (1970).
12. G. Kossmehl, M. Härtel, and G. Manecke, *Makromol. Chem.*, **131**, 15 (1970).
13. R. J. Cotter and M. Matzner, *Ring-Forming Polymerizations*, Part A, Academic, New York, 1969, Chap. 1.
14. G. S. Kumar, *Azo Functional Polymers: Functional Group Approach in Macromolecular Design*, Technomic, Lancaster, Pennsylvania, 1992.
15. H. C. Bach and W. B. Black, *J. Appl. Polym. Sci., Appl. Polym. Symp.*, **22**, 799 (1969).
16. H. H. Bössler and R. C. Schultz, *Makromol. Chem.*, **158**, 113 (1972).
17. A. Ravve and C. Fitco, *J. Polym. Sci. A*, **2**, 1925 (1964).
18. A. A. Berlin and V. P. Parini, *Izv. Akad. Nauk SSSR Otd. Khim. Nauk.*, 1674 (1959) (*Chem. Abstr.*, **54**, 8715c 1960).
19. A. DeSouza Gomes and T. D. R. DeOliveira Cavalcanti, *J. Polym. Sci., Polym. Chem. Ed.*, **16**, 2671 (1978).
20. A. S. Hay, D. A. Bolon, K. R. Leimer, and R. F. Clark, *J. Polym. Sci. B*, **8**, 97 (1970).
21. D. M. White, *Polym. Preprints*, **12**(1), 155 (1971).
22. S. J. Havens and P. M. Hergenrother, *J. Polym. Sci., Polym. Lett. Ed.*, **23**, 587 (1985).
23. B. Tieke and G. Wegner, *Makromol. Chem.*, **179**, 2573 (1978).
24. G. N. Patel, R. R. Chance, E. A. Turi, and Y. P. Khanna, *J. Am. Chem. Soc.*, **100**, 6644 (1978).
25. G. N. Patel and N.-L. Yang, *J. Chem. Educ.*, **60**, 181 (1983).
26. W. Bracke, *J. Polym. Sci. A-1*, **10**, 975 (1972).
27. P. E. Cassidy, *Thermally Stable Polymers*, Marcel Dekker, New York, 1980, Chap. 2.

28. J. G. Speight, P. Kovacic, and F. W. Koch, *J. Macromol. Sci., Rev. Macromol. Chem.*, **C5**, 295 (1971).

29. G. C. Engstrom and P. Kovacic, *J. Polym. Sci., Polym. Chem. Ed.*, **15**, 2453 (1977).

30. S. A. Milosevich, K. Saichek, L. Hinchey, W. B. England, and P. Kovacic, *J. Am. Chem. Soc.*, **105**, 1088 (1983).

31. C.-F. Hsing, I. Khoury, M. D. Bezoari, and P. Kovacic, *J. Polym. Sci., Polym. Chem. Ed.*, **20**, 3313 (1982).

32. D. L. Gin, V. P. Cortincello, and R. H. Grubbs, *J. Am. Chem. Soc.*, **116**, 10934 (1994).

33. M. Rehahn, A.-D. Schluter, G. Wegner, and W. J. Feast, *Polymer*, **30**, 1060 (1989).

34. T. I. Wallow and B. M. Novak, *J. Am. Chem. Soc.*, **113**, 7411 (1991).

35. M. Bochmann and K. Kelly, *J. Chem. Soc., Chem. Commun.*, 532 (1989).

36. J. P. Critchley, G. J. Knight, and W. W. Wright, *Heat-Resistant Polymers*, Plenum, New York, 1983: (a) Chap. 4; (b) Chap. 2.

37. D. M. Gale, *J. Appl. Polym. Sci.*, **22**, 1955, 1971 (1978).

38. W. F. Gorham, in *Addition and Condensation Polymerization Processes* (N. A. J. Platzer, Ed.), Adv. Chem. Ser. 91, American Chemical Society, Washington, D.C., 1969, Chap. 42.

39. T. L. St. Clair and H. M. Bell, *J. Polym. Sci., Polym. Chem. Ed.*, **12**, 1321 (1974).

40. H. A. Carter, *J. Chem. Educ.*, **73**, 1160 (1996).

41. K. Niume, F. Toda, K. Uno, and Y. Iwakura, *J. Polym. Sci., Polym. Lett. Ed.*, **15**, 283 (1977).

42. M. Ueda, T. Kano, T. Waragai, and H. Sugita, *Makromol. Chem., Rapid Commun.*, **6**, 847 (1985).

43. J. E. Chandler, B. H. Johnson, and R. W. Lenz, *Macromolecules*, **13**, 377 (1980).

44. N. Grassie, I. G. Meldrum, and J. Gilks, *J. Polym. Sci. B*, **8**, 247 (1970).

45. W. J. Bailey, in *Step-Growth Polymerizations* (D. H. Solomon, Ed.), Marcel Dekker, New York, 1972, Chap. 6.

46. J. K. Stille, F. W. Harris, H. Mukamal, R. O. Rakutis, C. L. Schilling, G. K. Noren, and J. A. Reed, in *Addition and Condensation Polymerization Processes* (N. A. J. Platzer, Ed.), Adv. Chem. Ser. 91, American Chemical Society, Washington, D.C., 1969, Chap. 41.

47. R. J. Cotter and M. Matzner, *Ring-Forming Polymerizations, Part A*, Academic, New York, 1969, Chap. 3.

48. M. Hasegawa, *Adv. Polym. Sci.*, **42**, 1 (1982).

49. F. C. DeSchryver, N. Boens, and G. Smets, *J. Polym. Sci. A-1*, **10**, 1687 (1972).

50. S. S. Labana, *J. Macromol. Sci., Rev. Macromol. Chem.*, **C11**, 299 (1974).

51. W. L. Dilling, *Chem. Rev.*, **83**, 1 (1983).

52. Y. Musa and M. P. Stevens, *J. Polym. Sci. A-1*, **10**, 319 (1972).

53. N. Kardush and M. P. Stevens, *J. Polym. Sci. A-1*, **10**, 1093 (1972).

54. N. Yoda, in *Encyclopedia of Polymer Science and Technology* (H. F. Mark, N. G. Gaylord, and N. M. Bikales, Eds.), Wiley-Interscience, New York, 1969, pp. 630–653.

55. K. W. Leong, V. Simonte, and R. Langer, *Macromolecules*, **20**, 706 (1987).

56. A. J. Domb, E. Ron, and R. Langer, *Macromolecules*, **21**, 1925 (1988).

57. H. R. Kricheldorf and D. Lubbers, *Makromol. Chem., Rapid Commun.*, **11**, 83 (1990).

58. R. Subramanyam and A. G. Pinkus, *J. Macromol. Sci.-Chem.*, **A22**, 23 (1985).

59. A. J. Domb, *Macromolecules*, **25**, 12 (1992).

60. A. J. Domb, S. Anselem, R. Langer, and M. Maniar, in *Biomedical Polymers: Designed-to-Degrade Systems* (S. W. Shalaby, Ed.), Hanser, New York, 1994, Chap. 4.

61. E. M. Geniès, A. Boyle, M. Lapkowski, and C. Tsintavis, *Synthetic Metals*, **36**, 139 (1990).

62. N. Toshima and S. Hara, *Prog. Polym. Sci.*, **20**, 155 (1995).

63. R. Baum, *Chem. Engr. News*, April 19, 1993, p. 36.

64. B. C. Sherman, W. B. Euler, and R. R. Forcé, *J. Chem. Educ.*, **71**, A94 (1994).

65. F. Danusso and P. Ferruti, *Polymer*, **11**, 88 (1970).

66. J. F. Klebe, *J. Polym. Sci. A*, **2**, 2673 (1964).

67. E. Tsuchida and T. Tomomo, *J. Polym. Sci., Polym. Chem. Ed.*, **11**, 723 (1973).

68. T. Tomomo, E. Hasegawa, and E. Tsuchida, *J. Polym. Sci., Polym. Chem. Ed.*, **12**, 953 (1974).

69. G. B. Butler and S. H. Hong, *Polym. Preprints*, **27**(1), 122 (1986).

70. M. Tramantoni, L. Angiolini, and N. Ghedini, *Polymer*, **29**, 771 (1988).

71. M. Hauser, in *Ring-Opening Polymerization* (K. C. Frisch and S. L. Reegen, Eds.), Marcel Dekker, New York, 1969, Chap. 5.

72. T. Saegusa and S. Kobayashi, *Makromol. Chem., Macromol. Symp.*, **1**, 23 (1986).

73. T. Saegusa, Y. Nagura, and S. Kobayashi, *Macromolecules*, **6**, 495 (1973).
74. F. Gutman and L. E. Lyons, *Organic Semiconductors*, Wiley-Interscience, New York, 1967.
75. D. A. Seanor, *Adv. Polym. Sci.*, **4**, 317 (1965).
76. R. C. Schulz, D. Fleischer, A. Henglein, H. M. Bössler, J. Trisnadi, and H. Tanaka, *Pure Appl. Chem.*, **38**, 227 (1974).
77. S. Kanda and H. A. Pohl, in *Organic Semiconducting Polymers* (J. E. Katon, Ed.), Marcel Dekker, New York, 1968, Chap. 3.
78. T. Sulzberg and R. J. Cotter, *Macromolecules*, **2**, 146, 150 (1969).
79. T. Sulzberg and R. J. Cotter, *J. Polym. Sci. A-1*, **8**, 2747 (1970).
80. C. I. Simionescu, V. Bărboiu, and M. Grigoraş, *J. Macromol. Sci.-Chem.*, **A22**, 693 (1985).
81. S. R. Turner and M. Stolka, *Macromolecules*, **11**, 835 (1978).
82. J. L. Nash Jr., R. E. Thompson, J. W. Schwietert, and G. B. Butler, *J. Polym. Sci., Polym. Chem. Ed.*, **16**, 1343 (1978); J. W. Schweitert and G. B. Butler, ibid., **16**, 1359 (1978).
83. A. Eisenberg and M. King, *Ion-Containing Polymers: Physical Properties and Structure*, Academic, New York, 1977.
84. W. J. MacKnight and T. R. Earnest, Jr., *Macromol. Rev.*, **16**, 41 (1981).
85. A. Eisenberg (Ed.), *Ions in Polymers*, Adv. Chem. Ser. 187, American Chemical Society, Washington, D.C., 1980.
86. J. Economy and J. H. Mason, in *Ionic Polymers* (L. Holliday, Ed.), Applied Science, London, 1975, Chap. 5.
87. D. Dieterich, W. Kaberle, and H. Witt, *Angew. Chem. Int. Ed. English*, **9**, 40 (1970).
88. H. S. Egboh, A. Gaffar, M. H. George, J. A. Barrie, and D. J. Walsh, *Polymer*, **23**, 1167 (1982).
89. S. L. Hsu, H. X. Xiao, H. H. Szmant, and K. C. Frisch, *J. Appl. Polym. Sci.*, **29**, 2467 (1984).
90. J. C. Salamone and B. Snider, *J. Polym. Sci. A-1*, **8**, 3495 (1970).
91. G. Henrici-Olivé and S. Olivé, *J. Polym. Sci. A-1*, **10**, 625 (1972).
92. T. Tsutsui, *Devel. Ionic Polym.*, **2**, 163 (1986).
93. R. Jerome, R. Fayt, and T. Ouhadi, *Prog. Polym. Sci.*, **10**, 87 (1984).
94. J. Economy, J. H. Mason, and L. C. Wohrer, *J. Polym. Sci. A-1*, **8**, 2231 (1970).
95. H. Matsuda, *J. Appl. Polym. Sci.*, **23**, 2603 (1979).
96. H. Matsuda, *J. Polym. Sci., Polym. Chem. Ed.*, **12**, 455, 469 (1974).
97. H. Matsuda, *J. Appl. Polym. Sci.*, **20**, 995 (1976); **22**, 3371 (1978).
98. H. Matsuda, *J. Macromol. Sci.-Chem.*, **A9**, 397 (1975).
99. R. Kunin, *Ion Exchange Resins*, 2nd ed., Wiley-Interscience, New York, 1958.
100. M. J. Lysaght, in *Ionic Polymers* (L. Holliday, Ed.), Applied Science, London, 1975, Chap. 6.
101. H. Dautzenberg, W. Jaeger, J. Kötz, B. Philipp, C. Seidel, and D. Stscherbina, *Polyelectrolytes: Formation, Characterization and Application*, Hanser Gardner, Cincinnati, Ohio, 1994.
102. B. A. Bolto, *Prog. Polym. Sci.*, **9**, 89 (1983).
103. J. C. Salmone, L. Quach, A. C. Watterson, S. Krauser, and M. U. Mahmud, *J. Macromol. Sci.-Chem.*, **A22**, 653 (1985).
104. S. B. Clough, D. Cortelek, T. Nagabhushanam, J. C. Salamone, and A. C. Watterson, *Polym. Eng. Sci.*, **24**, 385 (1984).
105. C. Neculescu, S. B. Clough, P. Elayaperumal, J. C. Salamone, and A. C. Watterson, *J. Polym. Sci. C, Polym. Lett.*, **25**, 201 (1987).

Review Exercises

1. Write the repeating unit of the polymer formed. (Reactions without literature references are hypothetical.)

(a) HC≡CCH₂O—⟨benzene⟩—C(CH₃)(CH₃)—⟨benzene⟩—OCH₂C≡CH $\xrightarrow[\text{O}_2]{\text{Cu}_2\text{Cl}_2 \cdot \text{R}_3\text{N}}$

(A. S. Hay et al., *J. Polym. Sci. B*, **8**, 97, 1970)

(b)

(D. B. Sobanski et al., *J. Macromol. Sci.-Chem.* **A23**, 189, 1986)

(c)

(A. Fritz, *Polym. Preprints* **12**(1), 232, 1971)

(d)

(M. P. Stevens, *J. Polym. Sci., Polym. Lett. Ed.*, **22**, 467, 1984)

(e)

(ref. 27)

(f)

(S. Hashimoto and T. Yamashita, *J. Macromol. Sci.-Chem.*, **A23**, 295, 1986)

(g) $HNNH + CH_2{=}CHSO_2CH{=}CH_2 \longrightarrow$

(ref. 65)

(h) $HNNH + 2CH_2O + CH_3CH_2CH_2NO_2 \longrightarrow$

(ref. 69)

(i)

(ref. 72)

(j)

(ref. 42)

(k) $ClCH_2CH_2OCH_2CH_2Cl + H-N\underset{}{\bigcirc}N-H \longrightarrow$

(l) $H_2N-\underset{}{\bigcirc}-OCH_2CH_2CH_2O-\underset{}{\bigcirc}-NH_2 + OCH-CHO \longrightarrow$

(J. Bartulin et al., *Polym. Bull.*, **15**, 405, 1986)

(m) $O=\underset{}{\bigcirc}=O + Ph_3PCH_2-\underset{}{\bigcirc}-CH_2PPh_3 \xrightarrow[\text{DMF}]{C_6H_5Li}$

(K. Al-Junah and J. E. Fernandez, *Macromolecules*, **20**, 1177, 1987)

(n) $CH_3-\underset{}{\bigcirc}-CH_2-\underset{}{\bigcirc}-CH_2-\underset{}{\bigcirc}-CH_3 +$ $\longrightarrow$

(X. He et al. *Makromol. Chem., Rapid Commun.*, **9**, 191, 1988)

(o) $+ ClCH_2-\underset{}{\bigcirc}-CH_2Cl \xrightarrow{AlCl_3}$

(P. S. Patel and R. S. Patel, *J. Macromol. Sci.-Chem.*, **A23**, 1251, 1986)

(p) $+ BrCH_2CH_2Br \longrightarrow$

(q) $H_2N-\underset{}{\bigcirc\bigcirc}-NH_2 + CH_3-\overset{O}{\underset{}{C}}-\underset{}{\bigcirc}-\overset{O}{\underset{}{C}}-CH_3 \longrightarrow$

(r) $OCH-\underset{}{\bigcirc}-CHO + H_2N-\underset{}{\bigcirc}-NH_2 \longrightarrow$

(Z. Hui and A. Gandini, *Eur. Polym. J.*, **28**, 1461, 1992)

(s) $CH_3\overset{O}{\underset{}{C}}\overset{O}{\underset{}{C}}CH=CH\overset{O}{\underset{}{C}}\overset{O}{\underset{}{C}}CH_3 \xrightarrow{\Delta}$

(A. J. Domb et al., *J. Polym. Sci., Part A: Polym. Chem.*, **29**, 571, 1991)

(t) $+ OCH-\underset{}{\bigcirc}-CHO \xrightarrow{Base}$

(L. G. Hudson and M. P. Stevens, *J. Polym. Sci., Part A: Polym. Chem.*, **33**, 71, 1995)

2. Give the structure of (a) the monomers to synthesize the Diels–Alder polymer

and (b) the repeating unit of the polymer formed by Diels–Alder homopolymerization of

(ref. 45)

3. Suggest alternative syntheses of (a) polymer **25** (reaction 17.50); (b) the conjugated polymer given in reaction (17.14). [*Hint*: Consider reaction (17.31).]

4. Considering the proposed mechanism of benzene polymerization (Scheme 17.1), which monomer should give the higher molecular weight polymer, C_6H_6 or C_6D_6 (assuming identical reaction conditions)? Explain. (C.-F. Hsing et al., *J. Polym. Sci., Polym. Chem. Ed.*, **21**, 457, 1983)

5. Suggest plausible mechanisms for reactions (17.30), (17.31), and (17.63).

6. Considering reaction (17.38), write the probable repeating unit obtained on addition polymerization of

7. Rewrite reactions (17.43) through (17.45) using compounds **20** and **21** in place of **17**.

8. It is well known that benzocyclobutene undergoes thermally induced electrocyclic ring opening:

Predict the result of heating a bismaleimide with

(L.-S. Tan and F. E. Arnold, *J. Polymer Sci A: Polym. Chem.*, **26**, 3103, 1988)

9. Styrene undergoes the Diels–Alder reaction with good dienophiles at elevated temperatures. Two equivalents of dieneophile react per equivalent of benzene. The reaction shown for maleic anhydride is as follows:

Predict the product obtained by reacting 1,1-diphenylethylene with the bismaleimide given in exercise 2(n). (W. N. Emmerling and M. L. Hallensleben, *Eur. Polym. J.*, **13**, 179, 1977)

10. Predict the outcome if polystyrene containing dissolved bismaleimide were to be photolyzed. (M. P. Stevens, in *Ultraviolet Light-Induced Reactions of Polymers*, S. S. Labana, Ed., ACS Symp. Ser. 25, 1976, p. 64)

11. In the presence of catalytic amounts of a stable cation radical (R$\overset{+}{\cdot}$), the following 2 + 2 cycloaddition reaction occurs:

Predict the result of similarly reacting

(N. L. Bauld et al., *Macromolecules*, **29**, 3661, 1996)

12. A polyamine has been synthesized by ADMET polymerization of $(CH_2=CHCH_2CH_2)_2NPh$. Give the repeating unit of the polymer. (J. D. Portmess and K. B. Wagener, *J. Polym. Sci., Part A: Polym. Chem.*, **34**, 1353, 1996)

13. Charge-transfer polymers can be prepared by cationic or by anionic polymerization of

Write the repeating unit in each case. (J. L. Nash, Jr., et al., *J. Polym. Sci., Polym. Chem. Ed.*, **16**, 1343, 1978)

14. Predict the outcome of reacting a carboxyl-terminated telechelic polydiene with $Ti(OAc)_4$. (G. Broze et al., *J. Polym. Sci., Polym. Lett. Ed.*, **21**, 237, 1983)

15. (a) As the molecular weight of a barium-neutralized carboxyl-terminated polydiene increases, the stress and modulus at constant strain increase. With zirconium-neutralized polymer, however, the opposite is observed; that is, stress and modulus decrease at constant strain. Suggest a reason. (b) Predict the stress–strain behavior of barium-, aluminum-, and zirconium-neutralized polymer if the molecular weight of the telechelic polymer is kept constant. (M. R. Tant et al., *Polymer*, **27**, 1815, 1986)

18

Natural Polymers

18.1 Introduction

The macromolecular chemistry described in the previous seventeen chapters reflects an impressive range of research and development built on a foundation established by the brilliance and foresight of a number of great scientists. When one delves into the fascinating realm of natural polymers, however, one cannot escape the conclusion that our own knowledge of polymer synthesis is, at best, rudimentary. Consider, for instance, that complex chemical reactions involving natural polymers within the human body were prerequisite to the intellectual processes that led to the pioneering work in synthetic polymer chemistry. An entire field of chemistry—biochemistry—is very much involved with these natural polymers; hence this chapter serves only as a necessarily brief introduction to the topic, with particular emphasis being placed, where pertinent, on their application or modification for commercial use.

Because of the physiological activity of natural polymers, three major classifications have occupied most of the literature in this field: polysaccharides, proteins, and polynucleotides. These are treated in Sections 18.3, 18.4, and 18.5 of this chapter. They are listed in Table 18.1. In keeping with our emphasis on commercial applications, this table also gives some end uses that go beyond the obvious biochemical significance of these polymers.

In addition to the three major classifications, there are a variety of natural organic polymers (Table 18.2), many of which have been employed commercially for a long time. Not included in Table 18.2 are microbial polyesters, which were covered in Chapter 12. We begin our discussion with a brief look at these "miscellaneous" natural polymers. Also included are synthetic polymers derived from the naturally occurring *tall oil.*

18.2 Miscellaneous Natural Polymers

18.2.1 Rubber

Rubber[1-5] is the most important and widely used natural polymer from the industrial standpoint. It was used for centuries by the Mayans in the Western Hemisphere before it was introduced into Europe by Columbus. The Mayans obtained the material from a tree they called *caoutchouc*

TABLE 18.1. Physiologically Active Natural Polymers

Polymer	Description	Representative Commercial Applications
Polysaccharides	Polyacetals and/or polyketals of monosaccharides	Foodstuffs, plastics, fibers, structural materials, adhesives, coatings
Proteins	Polyamides of α-amino acids	Foodstuffs, fibers, pharmaceutical agents
Polynucleotides (nucleic acids)	Polyesters of phosphoric acid and nucleosides (ribose or 2-deoxyribose with attached heterocyclic amine bases)	Genetic engineering (production of pharmaceutical agents, agricultural chemicals, etc.)

TABLE 18.2. Miscellaneous Natural Organic Polymers

Polymer	Description	Representative Commercial Applications
Rubber	Polyterpene having a 1,4-polyisoprene structure	Elastomers (principally tires), coatings
Lignin	Network aromatic/aliphatic oxygenated polymer	Fuel, phenolic adhesives, oil-well drilling mud components
Humus	Similar to lignin	Boiler scale remover, pigment extenders, emulsifiers
Coal	Very complex; similar to lignin, but also containing fused ring structures	Fuel, coal tar chemicals, synfuels
Kerogen	Complex, similar to coal	Fuel
Asphaltenes (bitumens)	Cycloaliphatic and aromatic fused-ring system	Highway surfacing, roofing, flooring
Shellac	Polyester of long-chain acids and hydroxy acids	Coatings, laminating resins, inks, sealing wax
Amber	Fossilized terpenoid resin	Decorative items, jewelry

TABLE 18.3. Principal Types of Natural Rubber

Name	Isoprene Structure	Principal Uses
Hevea	cis-1,4	Tires, footwear, weather stripping, shock absorbers, sports accessories
Guayule	cis-1,4	Latex gloves
Gutta percha	trans-1,4	Wire coatings, textile belting, varnishes, orthopedic devices
Balata	trans-1,4	Wire coatings, textile belting, varnishes
Chicle	cis- and trans-1,4	Chewing gum

("weeping tree"), a term still used to denote the polymer in many countries. It was Joseph Priestley who coined the term *rubber* when he noted the caoutchouc could be used to rub out pencil marks.

Rubber is a polyterpene synthesized naturally by the enzymatic polymerization of isopentenylpyrophosphate. Its repeating unit (**1**) is the same as that of 1,4-polyisoprene. Isoprene is, in fact, a major degradation product of rubber, being identified as such in the early 1860s.[6] The chemistry of diene polymers is covered in detail in earlier chapters and will not be repeated here; however, some of the structural features of the various forms of rubber, which are listed in Table 18.3, are of interest.

$$\left[CH_2C{=}CHCH_2 \right] \overset{CH_3}{}$$

1

The principal form of natural rubber, which consists of 97% *cis*-1,4-polyisoprene, is Hevea. It is obtained by tapping the bark of a tree (*Hevea brasiliensis*) that grows wild in South America and is cultivated in other parts of the world. Nonisoprene constituents, including proteins, are also present in small amounts in Hevea rubber, which give the natural polymer properties distinct from those of synthetic *cis*-1,4-polyisoprene. Hevea-type rubber is also found in a variety of shrubs and small plants, including milkweed and dandelion. Guayule, one of the most important shrubs, grows well in the arid climates of northern Mexico and the southwestern United States. There is much current interest in commercial cultivation of guayule (*Parthenium argentatum*) because of a potentially life-threatening allergy caused by proteins in Hevea rubber. The problem first came to light in the 1980s following increased use of latex gloves to prevent transmission of human immunodeficiency virus (HIV) and hepatitis B. Guayule rubber contains hypoallergenic proteins; that is, they do not trigger an allergic reaction. High-yield varieties of guayule are now being developed using tissue culture techniques.[7]

Almost all natural rubber is harvested as a latex consisting of about 32 to 35% rubber and about 5% of other compounds, including fatty acids, sugars, proteins, sterols, esters, and salts. Guayule rubber is an exception, being obtained by pulping and parboiling the plant before refining. The cellulosic crop residue (bagasse) is a potential source of fermentation alcohol. Rubber is of very high molecular weight (averaging about 1.5 million) and is amorphous, although it becomes randomly crystallized at lower temperatures.

The latex is converted to foam rubber by mechanical aeration followed by vulcanization. Rubber gloves and balloons are usually made by coating latex on forms before vulcanization. Most latex is coagulated (for example, with acetic acid) and used in bulk form. Most Hevea rubber (about 65%) is used in tire manufacture, but it is also found in a host of commercial products (Table 18.3). All these utilize rubber in vulcanized form. One of the few applications of unvulcanized rubber is in the form of *crepe* which, because of its excellent abrasion resistance, is used for shoe soles.

Gutta percha and balata, having a *trans*-1,4-polyisoprene structure, are much harder, less soluble, and more crystalline than Hevea, which shows the importance of stereochemistry as it relates to polymer properties. Chicle is unusual in that it contains a mixture of *cis*- and low-molecular-weight *trans*-1,4-polyisoprene. All three are obtained in latex form from trees indigenous to Southeast Asia or Mexico and South and Central America. Gutta percha, balata, and chicle command a very small commercial market compared with the ubiquitous Hevea rubber. Until ionomers captured the market, gutta percha was the material of choice for golf-ball covers.

18.2.2   Lignin, Humus, Coal, and Kerogen

Wood consists almost entirely of three materials: the polysaccharides cellulose and hemicelluloses, and lignin. Lignin[8–10] is the "cement" holding the cellulose fibrils together and providing much of wood's dimensional stability. Constituting about 25 to 30% of wood, lignin is an extremely abundant polymer that has yet to reach its potential with regard to polymer applications. At the present time most of the lignin produced in pulping operations is burned as fuel at the pulping site.

The structure of lignin varies according to source, but an approximation of a segment of softwood lignin (Figure 18.1) illustrates its complexity.[11] The molecular weight of native lignin is presumed to be very high, but because separation from cellulose inevitably leads to degradation, it is impossible to tell how high. Because it contains a preponderance of activated benzene rings, degraded lignin reacts readily with formaldehyde, which has led to limited commercial development in the area of plywood adhesives (Chapter 14). Lignin sulfonates obtained from wood pulping are also used as adhesives, asphalt extenders, and oil-well drilling mud additives. Efforts have continued over the years to modify lignin for possible use as engineering plastics.[8,12] Reaction with propylene oxide, for example, yields hydroxypropyl derivatives that have been converted to thermosetting polyurethanes.

Humus and coal are related in structure to lignin. Humus[13,14] constitutes the organic component of soils and natural waters that is relatively resistant to biodegradation. It is also found in fossil fuels. Humus is presumed to originate from lignin, but unlike lignin it contains significant numbers of carboxyl groups. For convenience, humus is classified according to whether it is water soluble (*fulvic acid*), base soluble but water insoluble (*humic acid*), or base insoluble (*humin*). All three have essentially the same structural fea-

FIGURE 18.1. Representative section of the softwood lignin. [From Freudenberg,[10] courtesy of the American Association for the Advancement of Science.]

tures, but they differ in molecular weight and degree of functionality. Humic materials chelate metals and hydrogen bond strongly to proteins and polysaccharides. Because humus is an excellent scavenger of metals, and because it plays a role in soil drainage and movement of water and nutrients, humus is of considerable interest to environmentalists. Industrially, humic materials find use as emulsifiers, pigment extenders, and boiler scale removers.

Coal is an extremely complex polymeric material whose structure varies widely with source.[15] Although it contains structural features analogous to those of humus and lignin, which is to be expected given its derivation from vascular plants, coal also contains clusters of fused aromatic rings linked with hydroaromatic and alicyclic units. Oxygen-, nitrogen-, and sulfur-containing heterocyclic units are also present. Dispersed in this porous network structure is a complex mixture of substances called *vitrinite*. Coal is used as a fuel directly, or as a feedstock for the manufacture of liquid and gaseous fuels.

Related to coal, but varying widely in structure according to source, is *kerogen*, the organic constituent of oil shale.[16] Thermal cracking of kerogen to low-molecular-weight compounds represents a potential alternative to petroleum-based fuels.

FIGURE 18.2. Hypothetical structure of a petroleum asphaltene. [From Speight and Moschopedis,[17] copyright 1981. Reprinted by permission of the American Chemical Society.]

$(C_{79}H_{92}N_2S_2O)_3$
mol wt. 3449

18.2.3 Asphaltenes

Also called *bitumens*, asphaltenes[17,18] are resinous materials widely used in highway construction as aggregate binders, as binders for roofing and flooring compositions, and for waterproofing buildings. They occur in natural deposits, but are obtained principally from the residue of petroleum distillation. Asphaltenes consist primarily of polynuclear aromatic and cycloaliphatic ring systems with aliphatic side chains, and vary in molecular weight from a few thousand to several hundred thousand. Although the structure varies considerably with source, a hypothetical lamellar-type structure has been proposed[17] (Figure 18.2) on the basis of spectroscopic and x-ray analysis.

18.2.4 Shellac

A number of natural resins have been used for varnishes and molding compounds, with *shellac* being the best known. Shellac[19–21] is secreted by the lac insect (*Kerriar paca*), found in southern parts of Asia. The resin, which is collected by scraping the branches of shellac-encrusted trees, consists of a very complex mixture of crosslinked polyesters derived from hydroxy acids, principally aleuritic acid (9,10,16-trihydroxyhexadecanoic acid, **2**). Gas chromatographic analysis of the products of chemical degradation of shellac has also shown the presence of several saturated and unsaturated long-chain aliphatic acids together with other hydroxy-substituted acids and nonaliphatic compounds.[22]

OH

HO⌒⌒⌒⌒⌒⌒⌒⌒CO$_2$H

OH

2

Following a long history of use in Asia, shellac was introduced into Europe in the tenth century. Phonograph records were made of shellac composites until the late 1940s, when vinyl polymers captured the market. Blends of shellac with other resins, especially phenol–, urea–, and melamine–formaldehyde, find occasional use. Given the fact that about 150,000 insects are needed to produce one pound of resin, it is hardly surprising that shellac has been largely supplanted with synthetic polymers.

18.2.5 Amber

Amber is the generic name for all fossil resins[23,24] secreted in prehistoric times by certain types of evergreens, now extinct. It consists of complex mixtures of sesquiterpenoids, diterpenoids, and triterpenoids, that have undergone polymerization and molecular reorganization during fossilization. Phenolic units may also be present. Baltic coast deposits, which contain about 8% succinic acid, are often called *succinite*. Amber is amorphous, and its infrared spectrum closely resembles those of nonvolatile rosins from extant pine trees. *Copal*, a fossil resin found in Africa, is similar to amber, but generally softer and geologically younger.

Interesting from the biological perspective is the fact that amber often serves as a repository for a variety of extinct species of insects.

18.2.6 Tall Oil–Derived Polymers

Tall oil constitutes about 3% of dry wood and is an important byproduct of wood pulping. (The name is derived in part from the Swedish *Tallolja*, which combines "tall pine" and "oil.") The oil is not polymeric, but it has been adapted for polymer synthesis. Tall oil represents one of the three classifications of *naval stores*, an archaic term that has its roots in early shipbuilding when pitch and rosin were used for caulking.[26] The other two classifications are *gum* and *wood* naval stores, the former collected from cuts in tree bark, the latter extracted from shredded tree stumps. Turpentine is another major constituent of naval stores.

Tall oil is made up of two distinct components: *rosin* and *tall oil fatty acids*. Rosin is primarily a mixture of 20-carbon fused-ring monocarboxylic acids of which abietic acid (**3**) and levopimaric acid (**4**) are representative. Varnishes made from rosin and drying oils were developed in ancient times. Esterification of rosin with glycerol or pentaerythritol leads to products known as *ester gums* that are used in adhesives and lacquers.[27]

Petroleum shortages in recent years have led to renewed interest in rosin as a renewable polymer feedstock.[28,29] Diels–Alder reaction of **4** with maleic anhydride (18.1), for example, results in an acid–anhydride, **5**, which can be converted to polyamideimide[29] (**6**) by reaction with diamine (18.2).

Tall oil fatty acids, which are separated from rosin by distillation, consist mainly of 18-carbon acids, particularly oleic and linoleic acids. Oxidative cleavage of the latter gives nonanedioic (azeleic) acid, used in production of nylon 69 (Chapter 13). Epoxidized linoleic esters are used as vinyl plasticizers. Treatment of linoleic acid with an activated clay results in a complex series of reactions involving isomerization to a conjugated diene and free radical and Diels–Alder dimerization and trimerization. The products, called *dimer acids* and *trimer acids*,[30] are used as curing agents for epoxy and phenolic resins. Polyamides of dimer acids act as printing ink binders and thixotropic agents.

$$4 + \quad \longrightarrow \quad \tag{18.1}$$

$$5 + H_2N-Ar-NH_2 \quad \longrightarrow \quad \tag{18.2}$$

18.3 Polysaccharides

18.3.1 *Cellulose*

In terms of chemical structure, cellulose[31–33] (**7**) is among the simplest of the natural polymers in that it consists of a single repeating unit, D-glucose, linked through carbons 1 and 4
by β linkages.* This means that the ring substituents other than hydrogen, including the
bonds linking the glucose rings together, are all equatorial.

7

Cellulose is found widely in nature, being the major constituent of the cell wall of plants
and averaging about 50% in a typical wood. It is the most abundant organic compound on
the planet. Cellulose is also the principal constituent of a variety of natural fibers that occur
as seed hairs surrounding the seeds of certain plants (for example, cotton), as the fibrous inner bark of others (bast fibers), and as the fibrous constituent of certain leaf stalks (leaf
fibers). The amount of cellulose in the fibers varies according to source and is usually associated with such materials as water, wax, pectin, proteins, lignin, and mineral substances.
The approximate percentage of cellulose in these various fibers and the principal commercial uses are given in Table 18.4. The DP of cotton cellulose is about 15,000, compared with
about 10,000 for wood cellulose. Separation of the latter from lignin results in a decrease in
DP to about 2600.[34]

Certain types of bacteria also synthesize cellulose, which can be obtained as a continuous film by cultivating the bacteria in a glucose solution. Cellulose is also secreted by such
marine chordates (tunicates) as the sea squirt.

The polymer, known as *native cellulose* (to distinguish it from regenerated cellulose), is
linear, highly crystalline, and characterized by a very high degree of intramolecular and intermolecular hydrogen bonding. As a result, cellulose is not thermoplastic but decomposes
before melting. Despite its large number of hydroxyl groups, cellulose is insoluble in water

---

*The monosaccharide D-glucose exists in two cyclic forms, α and β. In the β form, all hydroxyl substituents and
the hydroxymethyl group are in the equatorial position, whereas in the α form, the hydroxyl at C-1 (the hemiacetal
hydroxyl group) is axial.

β-D-Glucose   α-D-Glucose

TABLE 18.4. Cellulosic Fibers[a]

Fiber	Approximate % Cellulose	Major Uses
Seed hair fibers		
Cotton	85–90	Textiles
Kapok	65	Thermal and acoustic insulation; mattress, cushion, and life preserver stuffing
Coir[b]	40	Upholstery, sacking, matting, cordage
Bast fibers		
Flax	80–90	Textiles (linen)
Hemp	78	Twine, rope, netting, sacking
Jute[b]	65	Textiles, cordage, sacking, linoleum backing
Ramie	100	Paper (special types), gas mantles, cordage
Leaf fibers		
Manila hemp[c]	65	High-strength paper, muslin, cordage, rope
Sisal hemp[c]	65	Rope, twine

[a] Data from J. R. Scott and W. J. Roff, *Handbook of Common Polymers*, CRC Press, Cleveland, 1971, pp. 140–142.

[b] Coir and jute contain 40 and 10% lignin, respectively.

[c] Hemp contains about 22% pectin.

and most other common solvents, although it will succumb to certain solvent mixtures. Solutions of metal complexes such as copper(II)-ammonia[35] (*Schweitzer's reagent*) or copper(II)-diamine[36] dissolve cellulose, presumably by forming complexes of copper(II) with the hydroxyl groups of the polymer. Among the solvent systems of more recent vintage[37,38] are LiCl-dimethylacetamide, dimethyl sulfoxide–paraformaldehyde, amine oxides, and phosphoric acid.

Cellulose swells in water and especially in concentrated base. The base-swollen polymer, known as *alkali cellulose* or *soda cellulose*, is used to prepare regenerated cellulose, as described in the next section. The process of treating cotton with aqueous base, and then washing out the base, is known as *mercerization*. Mercerized cotton has a higher luster than natural cotton, is less dense, and is somewhat less crystalline. It also has improved dyeability.

18.3.2 Regenerated Cellulose

Some important commercial products are prepared from cellulose by first dissolving the polymer, then precipitating it from solution. The regenerated cellulose is different from native cellulose because extensive degradation takes place during the process of dissolution and the final product is usually less crystalline.

The most widely used regeneration method is called the *xanthate process* or *viscose process*.[39] Alkali cellulose, usually prepared from wood pulp (that is, wood separated from lignin by treatment with alkali or bisulfite solution), is allowed to age, during which time the molecular weight is reduced. The exact role of the alkali is not known with certainty, but it is believed that sodium ions are held interstitially, while hydroxide ions are bound by hydrogen bonding. The reduction in molecular weight probably arises mainly from oxidative degradation. Addition of carbon disulfide then converts cellulose (Cell–OH) to the xanthate (**8**) (18.3). Not all the OH groups are converted, just enough to make the polymer soluble in aqueous alkali. The basic solution, known as *viscose*, is orange in color, is viscous, and has an unpleasant odor arising from volatile sulfurous byproducts of the xanthation reaction.

$$\text{Cell—OH} + \text{CS}_2 + \text{NaOH} \longrightarrow \underset{\mathbf{8}}{\text{Cell—O—}\overset{\displaystyle \overset{\text{S}}{\|}}{\text{C}}\text{—S}^-\text{Na}^+} + \text{H}_2\text{O} \qquad (18.3)$$

The viscose is allowed to age further to allow a series of complex reactions to occur that results in an increase in viscosity. The solution is then extruded as a filament into a bath containing dilute sulfuric acid, sodium bisulfite, and zinc sulfate to hydrolyze the xanthate back to cellulose (18.4). The filaments are stretched to orient the molecules and increase crystallinity. Fibers prepared this way are known as *viscose rayon*, a product useful for textiles and tire cord. When viscose is extruded through a slit die into an acid bath, a film of regenerated cellulose known as *cellophane* results. Cellophane is plasticized by washing it in a bath containing glycerol during processing.

$$\text{Cell—O—}\overset{\displaystyle \overset{\text{S}}{\|}}{\text{C}}\text{—S}^-\text{Na}^+ \xrightarrow{\text{H}^+} \text{Cell—OH} + \text{CS}_2 + \text{Na}^+ \qquad (18.4)$$

Regenerated cellulose is also prepared from solutions in Schweitzer's reagent (the *cuprammonium process*) by hydrolysis with acid, and by saponifying cellulose acetate.

18.3.3 Derivatives of Cellulose

A large number of cellulose derivatives have been synthesized,[40–44] including esters of inorganic and organic acids (apart from xanthate esters), ethers, and graft copolymers, although relatively few have achieved commercial importance. Those that have include the nitrate, acetate, propionate, and butyrate esters and the methyl, ethyl, and hydroxyethyl ethers. In addition, crosslinking reactions have been widely used for treating textiles.

18.3.3.1 Cellulose Esters

As was mentioned in Chapter 1, cellulose nitrate was one of the very first commercial polymers. It was developed originally during the last century and eventually found use in the manufacture of explosives (*guncotton*), photographic film, lacquers, antiseptic film, textile fibers, and some molding applications (for example, as an ivory substitute for billiard balls). Cellulose nitrate plasticized with camphor was known by the trade name Celluloid.

The polymer is also known as *nitrocellulose*, but this is a misnomer because it contains no nitro groups. Cellulose nitrate is prepared from cellulose and nitric acid (18.5), usually in the presence of sulfuric acid, although phosphoric or acetic acids may also be used. The degree of nitration varies according to application. Cellulose containing 10 to 11% nitrogen is used mainly for lacquers or plastics, whereas that containing about 13.5% nitrogen (guncotton) is used for explosives. (Completely nitrated cellulose contains 14.14% nitrogen.)

$$\text{Cell—OH} + \text{HONO}_2 \rightleftharpoons \text{Cell—ONO}_2 + \text{H}_2\text{O} \qquad (18.5)$$

Cellulose nitrate is an extremely flammable material and not particularly stable; hence it has been replaced in most applications by other polymers in more recent years. Some work has also been done on cellulose sulfate and phosphate, the major concern being to improve dyeability of cellulosic fibers in the former case and to reduce flammability in the latter.

Organic esters are much more important at the present time. Cellulose acetate of varying degrees of substitution is prepared by first treating cellulose with acetic acid, then with acetic anhydride (18.6) and a mineral acid catalyst. The fully acetylated product is subsequently reacted with water to achieve the desired degree of substitution. Mild hydrolysis gives an almost completely acetylated cellulose known as *triacetate*. Further hydrolysis gives an acetone-soluble product called *secondary acetate*. Both are used in plastics and fiber applications, in the former case in combination with appropriate plasticizers. As a textile fiber, cellulose acetate is called *acetate rayon*. The $\overline{\text{DP}}$ of cellulose acetate is much lower than that of the cellulose feedstock (averaging about 200–250) because of cleavage of glucosidic linkages by the acidic esterification catalyst.

$$\text{Cell—OH} + (\text{CH}_3\text{CO})_2\text{O} \xrightarrow{\text{H}^+} \text{Cell—O} \overset{\displaystyle \text{O} \atop \displaystyle \|}{\text{C}} \text{CH}_3 + \text{CH}_3\text{CO}_2\text{H} \qquad (18.6)$$

Cellulose propionate and cellulose butyrate, or, more commonly, acetate–propionate and acetate–butyrate mixtures, are the only other commercially important cellulose esters. The mixed esters are generally more soluble and have better impact resistance than cellulose acetate. They are widely used in molding compositions, motion picture film, and coatings.

Some miscellaneous longer chain esters have been prepared, also esters of unsaturated acids (for example, methacrylic and crotonic), halogenated acids, and dibasic acids. The unsaturated esters are of interest from the standpoint of crosslinking and graft copolymerization.

18.3.3.2 Cellulose Ethers

Cellulose ethers[41] are prepared by pretreating cellulose with base (alkali cellulose), then reacting with a halogen compound (18.7) or epoxide (18.8). Methyl sulfate is also used for preparing methyl ethers. Some of the cellulose ethers of interest are methyl ($R = CH_3$), ethyl ($R = CH_2CH_3$), hydroxyethyl ($R = CH_2CH_2OH$), hydroxypropyl ($R = CH_2CH(OH)CH_3$), carboxymethyl ($R = CH_2CO_2H$), aminoethyl ($R = CH_2CH_2NH_2$), and benzyl ($R = CH_2C_6H_5$).

$$\text{Cell—OH} + \text{NaOH} + \text{RCl} \longrightarrow \text{Cell—OR} + \text{NaCl} + \text{H}_2\text{O} \qquad (18.7)$$

$$\text{Cell—OH} + \overset{\displaystyle \text{O} \atop \displaystyle \diagup \diagdown}{\text{CH}_2 — \text{CH}_2} \xrightarrow{\text{NaOH}} \text{Cell—OCH}_2\text{CH}_2\text{OH} \qquad (18.8)$$

Ethylcellulose is the most widely used, principally in plastics applications similar to those of cellulose acetate. Methylcellulose is water soluble* and is used as a food-thickening agent

---

*Methylcellulose is unusual in that it is much more soluble in cold water than in hot. Some other cellulose derivatives exhibit similar inverse solubility–temperature properties.

and as an ingredient in certain adhesive, ink, and textile-finishing formulations. Car-boxymethyl cellulose, hydroxyethyl cellulose, and hydroxypropylcellulose are also used in adhesives and textile finishes and as emulsifiers (for example, in latex paints.) Hydro-xypropylcellulose sandwiched between two water-insoluble films has found use in the manufacture of degradable bottles. When the outer layer is peeled off, the hydroxypropylcellulose dissolves rapidly, thus reducing the solid-waste problem normally associated with nonreturnable bottles.

18.3.3.3 Graft Copolymers

A large number of graft copolymers of cellulose have been prepared,[45] and although there has been little commercial exploitation, there is considerable interest from the standpoint of modification of cellulosic fibers. Vinyl grafts[46a] can be initiated with the usual free radical sources, including peroxides, redox initiators, radiant energy, or even mechanical rupturing (mastication). Cellulose derivatives can also be grafted. Cellulose acetate grafted with vinyl acetate using a cobalt-60 γ-ray source is one example of a graft copolymer that has achieved some commercial interest.

Free radical grafting can also be accomplished by incorporating active chain transfer groups into the cellulose backbone.[46b] Mercaptoethylcellulose (**9**), for example, undergoes grafting in the presence of a vinyl monomer under conditions of vinyl polymerization (18.9).

$$\text{Cell—OCH}_2\text{CH}_2\text{SH} \xrightarrow[\text{peroxide}]{\text{CH}_2=\text{CHR}} \text{Cell—OCH}_2\text{CH}_2\text{SCH}_2\text{CH}\text{ᴡᴡ} \qquad (18.9)$$
$$\underset{\mathbf{9}}{} \qquad\qquad\qquad\qquad\qquad \underset{R}{|}$$

An example of nonvinyl grafting includes the reaction of cellulose with ethylene oxide (18.10) to give a product marketed under the trade name Ethylose.

$$\text{Cell—OH} + n\overset{\displaystyle O}{\overset{\displaystyle /\!\backslash}{\text{CH}_2\text{CH}_2}} \longrightarrow \text{Cell—O}\text{+CH}_2\text{CH}_2\text{O+}_{n-1}\text{CH}_2\text{CH}_2\text{OH} \qquad (18.10)$$

18.3.3.4 Crosslinked Cellulose

In recent years the treatment of cotton with crosslinking agents to impart crease-resistant or "durable press" properties has become a major industry.[46c,47] Most of those presently in use are hydroxymethyl derivatives of amides, prepared from formaldehyde and amide. Some representative examples include dimethylolurea (**10**), 1,3-bis(hydroxymethyl)-2-imidazolidinone (**11**) (known simply as dimethylolethyleneurea), dimethylol uron (**12**), and dimethylol alkyl carbamate (**13**). The reaction of **10** with cellulose is given (18.11) in simplified form because there is probably some polymerization of the urea and formaldehyde in addition to reaction with cellulose. The hydroxymethyl compounds are water soluble and are applied to the fabric from aqueous solution using an acid catalyst.

$$\begin{array}{cc} \overset{\displaystyle O}{\overset{\displaystyle \|}{\text{HOCH}_2\text{NHCNHCH}_2\text{OH}}} & \qquad \overset{\displaystyle O}{\overset{\displaystyle \|}{\text{HOCH}_2\text{—N}\diagdown\diagup\text{N—CH}_2\text{OH}}} \\[2ex] \mathbf{10} & \qquad\qquad \mathbf{11} \end{array}$$

$$HOCH_2—N \overbrace{\qquad}^{O} N—CH_2OH$$

12

$$ROC—N(CH_2OH)_2$$

13

$$Cell—OH + 10 + HO—Cell \longrightarrow Cell—OCH_2NHCNHCH_2O—Cell \qquad (18.11)$$

Other crosslinking agents include diepoxides, epichlorohydrin (which behaves as a diepoxide), aldehydes, acetals (prepared from formaldehyde and polyol), and tetraoxane (the cyclic tetramer of formaldehyde) which undergoes ring opening in the presence of acid to yield polyoxymethylene crosslinks (**14**).

$$Cell—O\text{-}[CH_2O]_n\text{-}Cell$$

14

18.3.4 Starch

From the standpoint of chemical structure, starch[48] differs from cellulose in two major ways: the glucose rings are linked together through carbons 1 and 4 by α rather than β linkages, and considerable chain branching occurs through carbon 6. But, like cellulose, complete hydrolysis yields D-glucose.

Starch occurs naturally as minute granules in the roots, seeds, and stems of numerous types of plants, including corn, wheat, rice, millet, barley, and potatoes; it constitutes the main carbohydrate reserve of the plants. It consists of two polysaccharides, *amylose* and *amylopectin*, which can be separated according to differences in solubility. Amylose (**15**) is mainly linear in structure, with molecular weight ranging from about 30,000 up to 1 million, although upper limits of 200,000 to 300,000 are more common. Apart from the α linkage (axial C—O bond at carbon 1), amylose therefore resembles cellulose.

15

Amylopectin is highly branched through carbon 6 and has molecular weights of over 1 million. It is believed to consist of chains of 20 to 25 glucose units linked through carbons 1 and 4, as in amylose, but with the chains connected to each other through the 1,6 linkage.

A partial structure (**16**) for amylopectin is given. There is usually about three times as much amylopectin as amylose in natural starch, although much higher proportions of either occur in certain plants.

16

Because of its highly branched structure and permeability, starch is not suitable for plastics or fiber applications as is the case with cellulose. Unlike cellulose, however, it is useful as a food because animals produce the enzymes necessary to catalyze hydrolysis of α linkages. Apart from its utility in the food and fermentation industries, starch has also found some use in adhesive formulations (for example, flour paste) and as a sizing or glazing agent in the paper and textile industries.

18.3.5 Derivatives of Starch

As might be expected, the types of modifications applied to cellulose have also been applied to starch. Starch nitrate is a useful explosive, and starch acetate, which forms clear films, has found commercial use as a paper and textile sizing agent. Amylose triacetate forms fibers similar to those of cellulose acetate, but they are not as strong and offer no special advantages. Ether derivatives of starch, particularly hydroxyethyl ethers, have been used as adhesives, and an allyl ether has been used as an air-drying coating.

A large number of graft copolymers of starch have also been reported.[46b] The above-mentioned allyl ether derivative, for example, has been reacted with styrene under free radical conditions to give styrenated starch, and other vinyl monomers have been reacted directly by the same methods described for cellulose. There is considerable interest in graft copolymers of starch because of their potential as biodegradable packaging materials and agricultural mulches (see Chapter 4, Section 4.7). A practical application of starch crosslinking using epichlorohydrin is the treatment of rice granules to make them more resistant to breaking down in canned soups.

18.3.6 Other Polysaccharides

A number of other polysaccharides exist in various natural sources,[49] some related in structure to cellulose or starch, others having uniquely different structures. Most important from the industrial standpoint is *chitin*,[50-53] (**17**), which is like cellulose except for an acetylated amino group at carbon 2. Chitin, the second most abundant polysaccharide, is found in the shells of crustacea, in the horny exoskeletons of insects and spiders, and in some types of fungi. It is also present in massive amounts in marine organisms, especially plankton. Chitin's molecular weight is estimated to be about 1–2 million. Far less common is *chitosan*, the deacetylated version of chitin (NH_2 at carbon 2). Both chitin and chitosan are available commercially as film and fiber. They are used in foods and cosmetics because of their gel-forming properties, and in wastewater treatment because they chelate heavy metal ions and flocculate proteins. Because they are biocompatible, they are also finding increasing application in the biomedical field.[54]

17

A number of polysaccharidelike materials are obtainable from seaweed or algae. *Alginic acid* (**18**), a polymer similar to cellulose, is linked through carbons 1 and 4 and has a carboxyl function attached to carbon 5. The polymer also has some units with the configuration reversed at carbon 5.

18

Alginic acid, which is found in certain types of brown seaweed, is used (as the sodium salt) in adhesive and sizing formulations and in some fiber applications. (One interesting use is in the manufacture of water-soluble disposable military parachutes.) *Carrageenin* and *agar*, obtainable from red seaweed or red algae, have similar structures. The former consists of two separable fractions, one (**19**) having galactose units linked through carbons 1 and 3 or 1 and 4 with sulfate groups at carbon 2; the other (**20**) having galactose units linked 1,4 with an ether group linking carbons 3 and 6. Agar is similar except that it has fewer sulfate groups and the 1,4-linked ether-containing fraction is derived from L-galactose rather than the D-isomer.

19

20

Related in structure to starch, or more particularly the amylopectin form of starch, is *glycogen*, the reserve carbohydrate of animals. Glycogen is more highly branched than amylopectin and is partially bound to proteins. Other polysaccharides include *inulin*, a polymer found in certain plants, that has fructose rings linked together with a glucose unit at one end of the chain; some natural gums such as *gum arabic* and *gum tragacanth* that have complex structures that yield several monosaccharides on hydrolysis; *dextran*, a high-molecular-weight polysaccharide synthesized from sucrose or starch by certain bacteria, consisting of glucose units linked through carbons 1 and 6 by α-linkages, and with chain branching through carbons 1 and 3; and *hemicelluloses*, which are found with cellulose in the cell walls of plants. Typical ingredients of hemicelluloses are *xylan* (like cellulose, but with the CH_2OH group replaced with H); *mannan*, which gives mannose on hydrolysis; *araban*, composed of arabinose units; and *galactans*, consisting of a complex mixture of monosaccharides. Hemicelluloses are amorphous and of much lower molecular weight than cellulose, having DP's of only a few hundred; nevertheless, they represent one of the three major constituents of wood.

18.4   Proteins

18.4.1   Amino Acids, Polypeptides, and Proteins

Proteins are the basic building blocks of animals and are, therefore, of primary interest to biochemists.[55,56] They are the major constituent of skin, nerve tissue, tendons, muscle, hair, and blood, and they exist in all living cells. Because proteins are also the principal constituent of enzymes and many hormones, they help control body functions. The study of proteins is fundamental to the field of biochemistry; therefore, we shall discuss only some of the basic principles of protein structure and synthesis.

$$\left[NH-\underset{\underset{R}{|}}{CH}-\underset{\overset{O}{\parallel}}{C} \right]$$

21

Proteins are polyamides (**21**) with α-amino acids being the monomer units. They are far more complex than polysaccharides because a large number of α-amino acids exist naturally which can form an almost limitless number of sequential arrangements. Unlike polysaccharides, however, proteins are monodisperse. Protein chemists refer to the amide group linking amino acids together as a *peptide* group. *Dipeptides* contain two amino acids, *tripeptides* three, and so on. The terms *polypeptide* and protein are often used interchangeably, although many protein chemists reserve the former for poly(α-amino acid)s having molecular weights below 10,000. Others consider chains having fewer than 50 amino acids to be polypeptides.

The 20 amino acids most commonly found in proteins are listed in Table 18.5 under three subheadings; neutral, basic, and acidic. Basic amino acids contain more than one *basic* amino group, acidic more than one carboxyl group. Because of the complex nature of proteins, each amino acid may be referred to by a mnemonic three-letter code (Ala for alanine, etc.). The more recent one-letter code, given in parentheses in the table, is gaining popularity.

The amino acids of mammalian proteins have three things in common. First, the amino group that forms the peptide linkage is always located on the α-carbon. Second, with the exception of glycine, which contains no stereogenic carbon, they are all optically active and all have the L configuration (**22**), related structurally to L-glyceraldehyde (**23**). And, third, the free amino acids do not exist in the form shown in Table 18.5; instead they exist principally as *dipolar ions* (zwitterions) (**22**).

$$\underset{\textbf{22}}{\underset{R\diagup\overset{|}{C}\diagdown\underset{+}{NH_3}}{\overset{CO_2^-}{|}}{H}} \qquad \underset{\textbf{23}}{\underset{HOCH_2\diagup\overset{|}{C}\diagdown OH}{\overset{CHO}{|}}{H}}$$

Certain conventions are followed in writing peptide structures. For example, for glutathione (**24**), a tripeptide found in living cells, both the structural and abbreviated formulas are written with the free amino group, or *N-terminal amino acid*, to the left, and the free carboxyl, or *C-terminal amino acid*, to the right. In naming peptides, the ending "-yl" is used for all amino acids except the C-terminal one, and the amino acids are named in order from the N-terminal one. Thus **24** is glutamylcysteinylglycine.

$$\underset{\underset{CO_2H}{|}}{H_2NCHCH_2CH_2} \overset{\overset{O}{\parallel}}{C} NH\underset{\underset{CH_2SH}{|}}{CH} \overset{\overset{O}{\parallel}}{C} NHCH_2CO_2H$$

or   Glu—Cys—Gly

24

Nonvinyl Polymers

TABLE 18.5. Common Amino Acids $H_2N—CH—CO_2H$
$$\underset{R}{|}$$

Name	R	Rotation	Abbreviation
Neutral			
Alanine	$—CH_3$	(+)	Ala (A)
Asparagine	$—CH_2\overset{\overset{O}{\|}}{C}NH_2$	(−)	Asn (N)
Cysteine[a]	$—CH_2SH$	(−)	Cys (C)
Glutamine	$—CH_2CH_2\overset{\overset{O}{\|}}{C}NH_2$	(+)	Gln (Q)
Glycine	$—H$		Gly (G)
Isoleucine	$—\underset{\overset{\|}{CH_3}}{C}HCH_2CH_3$	(+)	Ile (I)
Leucine	$—CH_2CH(CH_3)_2$	(−)	Leu (L)
Methionine	$—CH_2CH_2SCH_3$	(−)	Met (M)
Phenylalanine	$—CH_2—$⟨◯⟩	(−)	Phe (F)
Proline[b]	⟨N⟩$—CO_2H$ (with NH)	(−)	Pro (P)
Serine	$—CH_2OH$	(−)	Ser (S)
Threonine	$—\underset{}{\overset{\overset{OH}{\|}}{C}}HCH_3$	(−)	Thr (T)
Tryptophan	(indole structure)	(−)	Try (W)
Tyrosine	$—CH_2—$⟨◯⟩$—OH$	(−)	Tyr (Y)
Valine	$—CH(CH_3)_2$	(+)	Val (V)
Basic			
Arginine	$—CH_2CH_2CH_2NH\underset{\overset{\|}{NH_2}}{C}=NH$	(+)	Arg (R)
Histidine	$—CH_2—$(imidazole structure)	(−)	His (H)
Lysine	$—CH_2CH_2CH_2CH_2NH_2$	(+)	Lys (K)

continued

TABLE 18.5. (continued)

Name	R	Rotation	Abbreviation
Acidic			
Aspartic acid	—CH$_2$CO$_2$H	(+)	Asp (D)
Glutamic acid	—CH$_2$CH$_2$CO$_2$H	(+)	Glu (E)

[a] Cystine, the oxidized form of cysteine, contains two cysteine units linked by a sulfur–sulfur bond.
[b] Full structure given.

Proteins are classified into two major groups, *fibrous* and *globular*, and the shapes of each can be roughly deduced from the names. Fibrous proteins are held together as fibrils by hydrogen bonds. They are insoluble in water, and this insolubility, together with the strong intermolecular forces, makes them ideally suited for their role as structural materials of animal tissue. Typical of the fibrous proteins are *keratin* (hair, nails, feathers, horn), *collagen* (connective tissue), *fibroin* (silk), and *myosin* (muscle).

Globular proteins are folded and more compact than fibrous proteins. Their shape arises primarily from a combination of intramolecular secondary bonding forces, and most of the polar groups are directed to the outside, thus rendering them soluble in water or dilute salt solutions. Because of their solubility, globular proteins are more concerned with body functions that require mobility of the protein molecules. Enzymes, for example, are globular proteins, as are many hormones. Hemoglobin and egg albumin are other examples.

Some proteins, known as *conjugated proteins*, contain a nonpeptide structure associated with the polypeptide portion. The nonpeptide structure is called the *prosthetic group*, and it usually is extremely important from the standpoint of biological activity. Hemoglobin, for example, consists of a peptide moiety (globin) and an iron–porphyrin prosthetic group (heme, **25**).

25

It is the heme that forms a reversible complex with oxygen, allowing oxygen transport from the lungs to other parts of the body. Other conjugated proteins, called *nucleoproteins*, are found in living cells in combination with nucleic acids (Section 18.5).

18.4.2 *Protein Structure*

The sequence of amino acids in proteins is referred to as the *primary structure*. To determine the primary structure of any polypeptide or protein, the types of amino acids present as well as the quantity of each must first be determined. This is accomplished by complete

FIGURE 18.3. Alpha helix.

hydrolysis, followed by separation and identification of the constituents of the hydrolyzate. This, in itself, is a monumental task, considering the complexity of proteins; however, gas and liquid chromatographic techniques and separations by ion exchange have been developed and refined into automatic amino acid analyzers that do the job rapidly and accurately. The reader is referred to biochemistry texts for methods of determining amino acid sequences.

To understand the physiological action of proteins or the catalytic action of enzymes, it is necessary to know more than just the amino acid sequence. The shape of the molecule is also important. Like all molecules that exhibit conformational flexibility, proteins tend to form stable geometries, generally those that allow the greatest degree of hydrogen bonding. The conformation of a polypeptide or protein chain is referred to as the *secondary structure*.[57]

X-ray analysis, notably by Linus Pauling, has shown that a right-handed helical arrangement, called an *alpha helix* (Figure 18.3), is probably the most important secondary structure. Such an arrangement allows room for the bulky R groups present in most amino acids and is stabilized by intramolecular hydrogen bonding between the amide nitrogen atoms and the carbonyl oxygens. A short section of the peptide chain of α-keratin, the main constituent of hair, wool, and nails, is shown in Figure 18.4. The repeat distance—that is, the distance between amino acid residues along the axis of the helix—is 1.5 Å.

Proteins that have predominantly small R groups (for example, silk fibroin) can assume a configuration that is essentially an extended chain, but with some contraction so that crowding between the R groups of one chain and those of a neighboring chain is minimized, and the chains can then lie side by side in a sheetlike arrangement. Such an arrangement is referred to as a *pleated sheet*, also as the *beta arrangement*. A schematic representation of the chain association of silk fibroin with a repeat distance of 7.0 Å is shown in Figure 18.5.

The shape of a protein molecule is not given completely by the secondary structure. Sections of a peptide chain may exhibit some irregularity, helical sections may be folded in a manner that allows hydrogen bonding between sites separated by some distance along the chain, and sections of chains may be linked chemically through the sulfur–sulfur bonds of cystine groups or by *salt bridges* between carboxyl and ammonium groups. This overall three-dimensional structure of a protein molecule is known as the *tertiary structure*, and it is this structure that allows certain molecules, such as enzymes, to fit together with other

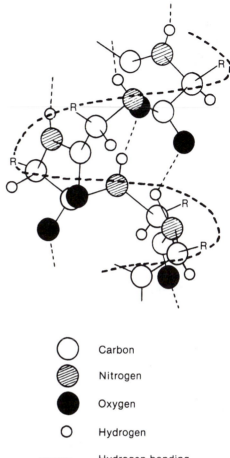

◯ Carbon

◨ Nitrogen

● Oxygen

○ Hydrogen

----- Hydrogen bonding

FIGURE 18.4. Section of an alpha helix of α-keratin.

FIGURE 18.5. Pleated sheet structure (β-arrangement) of silk fibroin.

molecules in a way that gives rise to such remarkably efficient and specific catalytic activity. When the tertiary structure is disrupted, a process called *denaturation*, the protein loses its physiological activity.

In addition to tertiary structure, proteins may exhibit a *quaternary structure*; that is, they may be associated with other protein molecules or with nonprotein groups, as is the case with conjugated proteins. The determination of tertiary and quaternary structure is a formidable task requiring the measurement of thousands of x-ray diffraction patterns of protein crystals and calculations necessitating the use of computers. The first to be determined were those of myoglobin and hemoglobin by Kendrew and Perutz at Cambridge University (Nobel Prize 1962). Tertiary structures of dozens of proteins are now known, and certain structural patterns have emerged. For example, glutamic acid, methionine, alanine, and leucine are commonly found in α-helices, whereas valine, isoleucine, phenylalanine, and tyrosine are more often found in pleated sheets. Glycine, proline, and aspartic acid occur frequently in hairpin loops. From a knowledge of primary structure, it is thus possible to predict with some measure of confidence the location of various secondary structures in proteins.

18.4.3 Synthesis of Polypeptides and Proteins

Heating a mixture of α-amino acids would yield a polypeptide, but the distribution would be random. To make a polypeptide having a specific sequence it is necessary to protect one of the functional groups of the amino acid by converting it to a derivative, and to react each protected amino acid in the proper sequence.[58,59] When the first peptide linkage is formed, the protecting group is removed, and the next protected amino acid unit is reacted. The two steps—formation of the peptide linkage and removal of the protecting group—are repeated until the desired peptide chain is formed. Given the complexity of proteins, the number of reactions involved can run into the hundreds. The most common protecting group is *t*-butyloxycarbonyl (BOC), formed by reaction of *t*-butyloxycarbonyl chloride with amino acid (18.12).

$$
\begin{array}{cc}
\overset{\displaystyle O}{\overset{\displaystyle \|}{}}\quad \overset{\displaystyle R}{\overset{\displaystyle |}{}} & \overset{\displaystyle O}{\overset{\displaystyle \|}{}}\quad \overset{\displaystyle R}{\overset{\displaystyle |}{}} \\
(CH_3)_3COCCl + H_2NCHCO_2H & \longrightarrow \underbrace{(CH_3)_3COCNHCHCO_2H}_{BOC}
\end{array}
\qquad (18.12)
$$

Techniques have been developed that allow automation of the reaction sequence, notably the *Merrifield solid-phase peptide synthesis*,[60] (R. B. Merrifield, Nobel Prize 1984). The Merrifield method uses an insoluble crosslinked polymer for binding the peptide chain as each amino acid unit is added. Byproducts and excess reagents are then removed following each addition by simply washing the insoluble polymer with an appropriate solvent. Typically, polystyrene microgel (see Chapter 3, Section 3.8) is chloromethylated on some of the benzene rings, as described earlier in Chapter 9 (reaction 9.5). The chloromethylated rings then serve as initiation sites for the peptide. The sequence of steps in polypeptide synthesis, outlined in Scheme 18.1, is as follows:

SCHEME 18.1. Merrifield solid-phase peptide synthesis.

1. BOC-protected amino acid is attached to the insoluble polymer Ⓟ via an ester group.
2. The protecting group is removed under conditions mild enough to preclude cleavage from the polymer support.
3. The second BOC-protected amino acid is added, and a strong dehydrating agent such

as dicyclohexylcarbodiimide (DCC, **26**) is added to promote formation of the peptide group. The DCC is hydrated to dicyclohexylurea (**27**) in this step.

4. Steps 2 and 3 are repeated with each successive amino acid until the desired sequence is obtained.

5. Finally, the polypeptide or protein is removed from the polymeric support.

The entire sequence of reactions can be carried out in a single reaction vessel fitted with a fritted disk filter for drawing off the byproducts, solvents, and excess reagents. Modern instruments incorporating microprocessor technology that carry out the synthetic steps in a programmed sequence are now available commercially. The instruments are similar in design to that shown later in this chapter (Figure 18.12) for polynucleotide synthesis. In an early application of the Merrifield method, bovine ribonuclease, an enzyme containing 124 amino acids, was synthesized in six weeks in 17% overall yield via 369 reactions involving 11,931 individual steps, with no isolation of intermediate products.[61]

An occasional problem with the Merrifield method is that impurities are sometimes difficult to remove, in which case a more traditional solution-phase synthesis may be preferred. Even so, the same general sequence of steps using protected amino acids is followed.

18.4.4 Wool, Silk, Collagen, and Regenerated Protein

The history of wool and silk for use in clothing dates back some 5000 years. Wool[62] is obtained mainly from sheep by scouring the sheep hair with alkaline soap solution to remove fats and greases. Different breeds of sheep produce wools with properties that vary considerably in elasticity, strength, color, and shrinkage, and wools are usually classified as fine, medium, long, cross-bred, and carpet, according to their source. Other animals, including goats (mohair and cashmere wools) and camels or members of the camel family (llama, alpaca, vicuna), are also sources of wool. Silk[63,64] is secreted by silkworms as two filaments encased in a proteinaceous gum. Other species, such as gypsy moth caterpillars and spiders, also secrete silk, but only the silkworm variety has a long history of commercial use. There is much current interest, however, in spider silk,[65] in particular the dragline silk secreted by the Panamanian golden orb weaver (*Nephila clavipes*).[66] Dragline silk, used by the spider to fashion the framework of its web and to suspend itself, has better strength properties even than the high-strength aramid Kevlar (Chapter 3, Section 3.7). The spider spins the silk from a 30% aqueous solution contained within a gland through one of several spinnerets. Other types of silk, each with a unique function, are extruded through the other spinnerets. Recombinant DNA techniques have now been mobilized to make bacteria produce dragline silk for commercial exploitation.

TABLE 18.6. Composition of Silk and Wool[a]

Amino Acid		Approximate Grams Amino Acid per 100 g Protein	
Type	*Name*	*Silk (Fibroin)*	*Wool (Keratin)*
Aliphatic	Glycine	41.2	5.5
	Alanine	33.0	4.3
	Serine	16.2	10.6
	Valine	3.6	5.7
	Threonine	1.55	7.15
	Leucine ⎤ Isoleucine ⎦	2.0	12.6
Aromatic	Phenylalanine	3.35	4.1
	Tyrosine	11.4	5.5
Sulfur containing	Cystine	0.2	13.0
	Methionine	0	0.55
Heterocyclic	Proline	0.7	6.8
Acidic	Aspartic acid	2.75	6.8
	Glutamic acid	2.15	14.5
Basic	Histidine	0.4	1.2
	Arginine	1.0	9.8
	Lysine	0.5	3.3

[a]Data from *Handbook of Common Polymers*, Chemical Rubber Co., Cleveland, 1971.

There is considerable variation in the amino acid composition of wool and silkworm silk, as shown in Table 18.6. It can be seen, for example, that silk consists mainly of four amino acids—glycine, alanine, serine, and tyrosine—the first three of which have small R groups (H, CH_3, and CH_2OH, respectively), whereas wool has a more even distribution. Furthermore, wool contains a much higher proportion of cystine, indicative of more chain coiling through disulfide groups. As was pointed out in Section 18.4.2, it is the relative absence of very bulky R groups in silk that is responsible for its assuming the extended-chain beta configuration, whereas wool (keratin) exhibits an alpha-helical secondary structure. (Stretched wool also has an extended-chain beta arrangement.)

Numerous chemical modifications of silk and wool have been reported.[46d,67,69] Wool containing less than 1% of grafted nylon 610 has considerable shrink resistance. The grafting is accomplished commercially by interfacial polymerization. The wool is first treated with a 1% aqueous solution of hexamethylenediamine plus inorganic base, then with a 2 to 3% hydrocarbon solution of sebacoyl chloride. The graft is assumed to form by reaction between acid chloride end groups and free protein amino groups. Wool has been similarly grafted with polyurethane and polyurea by treatment with diamine followed by diisocyanate and bischloroformate, respectively (see Chapter 13). Treatment with epoxide leads to polyether grafts. Vinyl polymers have also been grafted onto wool and silk—for example, by radiation grafting.

A number of reagents have also been used to crosslink protein fibers. If the cystine disulfide crosslinks of wool are reduced to thiol, then recrosslinked with dihalide (18.13), the treated wool exhibits improved resistance to acids, bases, oxidizing and reducing agents, and moths and carpet beetles. Other reagents used to crosslink wool and silk include formaldehyde, diepoxides, and epichlorohydrin. Wool fabrics have also been flameproofed with phosphorus-modified melamine–formaldehyde resins.

$$\{-SH + Br(CH_2)_x Br + HS-\} \longrightarrow \{-S(CH_2)_x S-\} \qquad (18.13)$$

Collagen,[70,71] the principal fibrous protein of animal tissue, is found in skin, bone, tendon, and cartilage. It owes its high strength to a "triple helix" structure consisting of three helical peptide chains twisted into a cablelike configuration. Collagen from skin and hides is converted to leather by treatment with tannic acid or heavy metal salts. On heating, the triple helix of collagen unravels, and, when cooled, the proteins assume a new tertiary structure. This denatured collagen is called *gelatin*. Crosslinked gelatin is a potential source of biodegradable plastics.[72]

Plastics and fibers have been manufactured from regenerated protein obtained from a number of sources,[73] including milk (casein), maize (zein), ground nuts (arachin), and soya beans (glycinin). The predominant amino acid of these proteins is glutamic acid (particularly in zein). Arginine is also present in significant amounts in arachin, and isoleucine in the other three.

The regenerating process involves dispersing the protein in dilute sodium hydroxide solution, followed by extrusion through spinnerets into an acid bath to form fibers. The protein is also crosslinked with formaldehyde to harden it and reduce its solubility. As a plastic, regenerated protein is usually used for indoor decorative applications (knobs, buttons, handles, etc.) because of its poor weatherability. Uncrosslinked casein is also used to improve adhesion of viscose rayon tire cord to rubber. Regenerated protein fibers are usually used in combination with wool, cotton, or synthetic fibers. Many of the chemical modifications applied to wool and silk have also been carried out with regenerated proteins.[46d]

18.5   Nucleic Acids

18.5.1   Nucleic Acid Structure

It was mentioned earlier that nucleoproteins are proteins found in living cells in combination with nucleic acids. It is the nucleic acids that direct the synthesis of proteins in the cell, and it is the elucidation of nucleic acid structure and function, upon which is based our knowledge of the chemical basis of heredity, that has been among the most exciting developments of biochemical research in recent years.[74,75]

Nucleic acids are polyesters of phosphoric acid and sugars, to each of which a heterocyclic amine (referred to as a *base*) is attached (**28**). There are two principal types of nucleic acids—one containing the sugar D-ribose (**29**), and the other D-2-deoxyribose (**30**), having one less hydroxyl group. Nucleic acids containing D-ribose are known as *ribonucleic acids* (RNA), and those containing D-2-deoxyribose are known as *deoxyribonucleic acids* (DNA). The points of

28

29   30

attachment to phosphoric acid are the hydroxyl groups at carbons 3 and 5 for both sugars. Bases are attached to the sugars at carbon 1, replacing the hydroxyl group at that position. A sugar with a base attached is called a *nucleoside*, and these represent the products of hydrolysis of nucleic acids. The ester of phosphoric acid and a nucleoside is known as a *nucleotide*; hence nucleic acids are also called *polynucleotides*. The structures of RNA and DNA may thus be depicted by the three nucleotide portions of each chain shown in Figure 18.6.

RNA   DNA

FIGURE 18.6. Sections of RNA and DNA chains.

The heterocyclic bases found in RNA and DNA are all derivatives of pyrimidine (**31**) or purine (**32**). The most commonly occurring pyridimine derivatives, each of which is attached to the sugar at the position-1 nitrogen atoms, are *cytosine* (2-hydroxy-4-aminopyrimidine) (**33**), *5-methylcytosine* (**34**) (somewhat less prevalent), and *thymine* (5-methyl-2,4-dihydroxypyrimidine) (**35**), found in DNA; and *cytosine* and *uracil* (2,4-dihydroxypyrimidine) (**36**), found in RNA. These bases are shown in the *isomeric keto form*, rather than the hydroxy form, because this is how they exist in the nucleic acids. There are two purine derivatives occurring commonly in both DNA and RNA: *adenine* (6-aminopurine) (**37**) and *guanine* (2-amino-6-hydroxypurine) (**38**). Guanine also exists, as shown, in the keto form. Each is attached to the sugar at the position-9 nitrogen atom.

31   32

33   34   35   36

37   38

Some nucleosides and nucleotides occur as free compounds as well as constituents of nucleic acids, and frequently they play important biochemical roles. Nucleosides consisting of ribose and a purine base are named *adenosine* and *guanosine* (for the bases adenine and guanine, respectively). The corresponding deoxyribose nucleosides are *deoxyadenosine* and *deoxyguanosine*. The ribose–pyrimidine bases are *cytidine* and *uridine* (corresponding to cytosine and uracil, respectively). The deoxyribose–pyrimidine nucleosides are *deoxycytidine* and *deoxythymidine* (for cytosine and thymine, respectively). With the last name the deoxy prefix is sometimes omitted because thymine occurs *only* in combination with deoxyribose.

The corresponding nucleotides are named as phosphates of the nucleosides. Compound **39**, for example, is adenosine 5'-monophosphate (AMP). (The 5' indicates the position on the sugar where phosphoric acid is attached.) Adenosine diphosphate (ADP) and adenosine triphosphate (ATP) are also found in cells, and it is hydrolysis of the phosphoric anhydrides, an exothermic process, that provides energy for numerous biochemical processes.

Monophosphates, diphosphates, and triphosphates of the other nucleosides are also present in cells, although ATP is most important from the standpoint of energy transfer.

39

Apart from the structural differences described above, RNA and DNA also differ in molecular weight, secondary structure, and biochemical function. DNA is a very-high-molecular-weight polymer, in some instances in the hundreds of millions. (The DNA of the bacterium *Escherichia coli* is about 2.5 *billion*!) Chemical analysis of the bases present in DNA shows a remarkable regularity in that the number of adenine bases is equal to the number of thymine bases, and the number of guanines equals the number of cytosines. It was this regularity, together with supporting evidence based on x-ray studies by Wilkins and Franklin, that led Watson and Crick (Nobel Prize, together with Wilkins, 1962) to postulate the double-helix structure of DNA. DNA exists as two separate right-handed helical chains wound around each other and held in place by hydrogen bonding between *base pairs*, as depicted in Figure 18.7. Depicted also is the unraveling of the double helix with concomitant formation of complementary DNA chains, the process that occurs when DNA molecules reproduce (*replicate*). The base pairs are adenine (A)–thymine (T) and guanine (G)–cytosine (C). They extend perpendicularly toward the center with the deoxyribose–phosphate chains on the outside of the helixes. Such an arrangement accounts for the regularity of the double helix, because both base pairs allow approximately the same distance between deoxyribose units (Figs. 18.8 and 18.9). Attempts to construct models containing other base pairs lead to distortion of the regular double helix. Not all DNA has the double-helix structure. Single-stranded DNA is also known, as is a less regular, left-spiraling form, but the double helix is most important. The elucidation of the overall structural picture of DNA is still a monumental task because of the sheer size of the molecules. Certain types of DNA are catenated (see Figure 18.10).[76,77]

While a substantial amount of information is available on the secondary structure of DNA, less is known about RNA. It is clear, however, that RNA is much lower in molecular weight than DNA; also that there are three main types of RNA found in living cells, known as *ribosomal RNA* (rRNA), *transfer* or *soluble RNA* (tRNA), and *messenger RNA* (mRNA). These have molecular weights of about 1 million, 25,000, and 500,000, respectively. These various forms of RNA are associated with proteins in the cell. Some viruses, particularly plant viruses, consist of RNA associated with protein. Tobacco mosaic virus, for example, contains about 95% protein and 5% RNA, with the protein forming a protective sheath around the RNA. Although some double-chain RNA molecules are known to occur in certain viruses, RNA is primarily single strand.

Nucleic acids represent the chemical basis of heredity. DNA, the chromosome of the cell, contains the information necessary to duplicate a new cell, and this information is con-

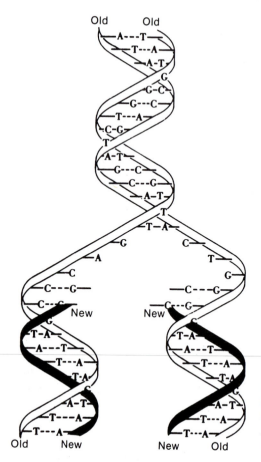

FIGURE 18.7. Schematic representation of the DNA double helix showing unraveling and formation of complementary chains. [From R. J. Light, *A Brief Introduction to Biochemistry*, Benjamin, 1968, reprinted by permission.]

tained in the sequence of bases or groups of nucleotides (genes). When DNA undergoes replication (Figure 18.7), each separated strand acts as a template for a complementary strand; thus new double-helix structures are formed with the same base pairing as in the original. Enzymes (DNA polymerase) catalyze the synthesis of the new chains. Sections of the DNA also serve as templates for RNA synthesis, and it is the various forms of RNA that are responsible for assembling the amino acids into the proper sequence for protein synthesis. The

FIGURE 18.8. Adenine–thymine base pair.

Cytosine

SUGAR

Guanine

SUGAR

|← 10.7 Å →|

FIGURE 18.9. Guanine–cytosine base pair.

recognition factor that allows RNA to select a particular amino acid from the amino acid pool is contained in the base sequence of the RNA molecules (the primary structure). The determination of the primary structure of nucleic acids and how nucleic acids direct the synthesis of proteins (the genetic code) are detailed in modern biochemistry texts.

18.5.2 Nucleic Acid Synthesis

A knowledge of the genetic code has obvious ramifications. For example, if the base sequence of the polynucleotide responsible for synthesis of a particular protein is known, the amino acid sequence of the protein can be deduced. This is, in fact, becoming the method of choice for determining primary protein structure.

FIGURE 18.10. Electron micrograph of catenated circular DNA molecules.[76] [Courtesy of Prof. Jerome Vinograd, California Institute of Technology. Reprinted by permission from *Nature*, **216**, No. 5116, 647–52. Copyright © 1967. Macmillan Journals Limited.]

Genetic engineering technologies were developed in the 1970s that allowed the modification of host cell DNA by insertion of new polynucleotide fragments. The new DNA produced is called *recombinant DNA*. Thus, *E. coli* bacteria can be induced to produce insulin or interferon or some other pharmacologically useful polypeptide by inoculation with the DNA sequence necessary for that polypeptide's synthesis. These developments have led to a demand for improved methods of polynucleotide synthesis.[78,79] Commercial polynucleotide sequencers (called "gene machines") are now available that use solid-phase synthesis analogous to that of the Merrifield method for proteins.[80] More functional groups need protecting, however, which makes polynucleotide synthesis more complicated.

The solid support may be a crosslinked polymer or, more commonly, a controlled-pore glass. Typical protecting groups, shown in Figure 18.11 are dimethoxytrityl (DMT) for the carbon-5 OH group of deoxyribose (designated 5′ OH) and phosphoramidite for carbon-3 OH (3′ OH) having variable R groups. Adenine and guanine bases are usually protected by benzoyl groups, guanine by an isobutyryl group. Thymine needs no protecting group. Synthesis begins with the first protected nucleotide attached to the solid support at the 3′ OH, as shown in Scheme 18.2. In step 1, the trityl group is removed with trichloroacetic acid. Step 2 involves reaction with the second protected nucleotide using an activating agent such as tetrazole (**40**), which serves to protonate the phosphoramidite nitrogen. In step 3 the trivalent phosphite ester is oxidized to phosphate. Detritylation of the newly added nucleotide sets the stage for reaction with the next nucleotide. When the chain is at the desired length, it is removed from the support with aqueous ammonia, which also removes the protecting groups.

40

A typical polynucleotide synthesizer[81] is shown in Figure 18.12. The basic components of the machine are a reaction vessel (mounted on the front of the instrument), a programmable controller, solvent and reagent delivery systems, and numerous valves.

FIGURE 18.11. Representative protecting groups used in solid-phase polynucleotide synthesis. (DMT = dimethoxytrityl; R = CH_3, CH_2CH_2CN, etc.; R′ = CH_3, $CH(CH_3)_2$, ⌬O, etc.)

SCHEME 18.2. Solid-phase polynucleotide synthesis. (Ⓢ = solid support; B/P = protected base; DMT = dimethoxytrityl)

FIGURE 18.12. Automated polynucleotide synthesizer. [Courtesy of Dr. Leon Barstow, Protein Technology.]

18.6 Conclusion

We began this book on a practical note—textile fibers and plastic bottles. We end with a glimpse of the macromolecular chemistry of life. With our rapidly expanding knowledge of the chemistry of polynucleotides, we have entered an age of genetic engineering, with all the ethical questions that go with it. The synthesis of artificial genes capable of functioning within living cells is already a reality.[82] Yet we have a long way to go before we are able to duplicate the efficiency and selectivity that characterize polymer processes in living systems. Consider the orb-weaver spider producing fibers from an aqueous solution at room temperature that are stronger than any made by industry under far less environmentally benign conditions.

Compared with the chemistry of life, nylon or polyester fibers and plastic bottles lose much of their glamour. But if nothing else, such contrasts serve to remind us of the subtlety and beauty of nature.

References

1. F. R. Eirich (Ed.), *Science and Technology of Rubber*, Academic, New York, 1978.
2. A. D. Roberts (Ed.), *Natural Rubber Science and Technology*, Oxford University Press, New York, 1988.
3. G. B. Kauffman and R. B. Seymour, *J. Chem. Educ.*, **67**, 422 (1990).

4. Y. Tanaka, *Prog. Polym. Sci.*, **14**, 339 (1989).
5. E. A. MacGregor and C. T. Greenwood, *Polymers in Nature*, Wiley, New York, 1980, Chap. 7.
6. J. C. F. Williams, *J. Chem. Soc.*, **15**, Part 10 (1862).
7. K. Cornish and D. J. Siler, *Chemtech*, **26**(8), 38, (1996).
8. W. G. Glasser and S. Sarkenan (Eds.), *Lignin: Properties and Materials*, ACS Symp. Ser. 397, American Chemical Society, Washington, D.C. 1989.
9. I. A. Pearl, *The Chemistry of Lignin*, Marcel Dekker, New York, 1967.
10. K. Freudenberg and A. C. Neish, *The Constitution and Biosynthesis of Lignin*, Springer-Verlag, New York, 1968.
11. K. Freudenberg, *Science*, **148**, 595 (1965).
12. D. Feldman, M. Lacasse, and L. M. Beznaczuk, *Prog. Polym. Sci.*, **12**, 271 (1986).
13. M. Schnitzer and S. U. Khan, *Humic Substances in the Environment*, Marcel Dekker, New York, 1972.
14. F. J. Stevenson, *Humus Chemistry: Genesis, Composition, Reactions*, Wiley-Interscience, New York, 1982.
15. L. Gorbaty and K. Ouchi (Eds.), *Coal Structure*, Adv. Chem. Ser. 192, American Chemical Society, Washington, D.C., 1981.
16. P. F. Dickson, in *Kirk-Othmer: Encyclopedia of Chemical Technology*, Vol. 16 (H. F. Mark, D. F. Othmer, C. G. Overberger, and G. T. Seaborg, Eds.), Wiley-Interscience, New York, 1981.
17. J. G. Speight and S. E. Moschopedis, in *Chemistry of Asphaltenes* (J. W. Bunger and N. C. Li, Eds.), Adv. Chem. Ser. 195, American Chemical Society, Washington, D.C., 1981, pp. 1–15.
18. E. W. Sheu and O. C. Mullins (Eds.), *Asphaltenes: Fundamentals and Applications*, Plenum, New York, 1995.
19. P. K. Bose, Y. Sankaranarayanan, and S. C. Sen Gupta, *Chemistry of Lac*, Indian Lac Research Institute, Rachi, India, 1963.
20. E. Hicks, *Shellac*, Chemical Publishing, New York, 1961.
21. S. Maiti and M. D. S. Rahman, *J. Macromol. Sci., Rev. Macromol. Chem. Phys.*, **C26**, 441 (1986).
22. W. W. Christie, F. D. Gunstone, H. G. Prentice, and S. C. Sen Gupta, *J. Chem. Soc., Suppl.* **1**, 5833 (1964).
23. J. H. Langenheim, *American Scientist*, **78**(1), 16 (1990).
24. J. H. Langenheim, *Science*, **163**, 1157 (1969).
25. G. G. Larsson, *Baltic Amber*, Scandinavian Science Press, Klampenbory, Denmark, 1978.
26. Y. Jen and E. E. McSweeney, in *Applied Polymer Science*, 2nd ed. (R. W. Tess and G. W. Poehlein, Eds.), ACS Symp. Ser. 285, American Chemical Society, Washington, D.C., 1985, Chap. 48.
27. R. L. Whistler and J. N. BeMiller (Eds.), *Industrial Gums*, 3rd ed., Academic, New York, 1992.
28. S. Maiti, S. Das, M. Maiti, and A. Ray, in *Polymer Applications of Renewable-Resource Materials* (C. E. Carraher, Jr., and L. H. Sperling, Eds.), Plenum, New York, 1983, p. 129.
29. S. Maiti, S. S. Ray, and A. K. Kundu, *Prog. Polym. Sci.*, **14**, 297 (1989).
30. E. C. Leonard (Ed.), *The Dimer Acids*, Humko Sheffield Chemical, Memphis, Tenn., 1975.
31. R. M. Brown Jr. (Ed.), *Cellulose and Other Natural Polymer Systems: Biogenesis, Structure, and Degradation*, Plenum, New York, 1982.
32. R. D. Gilbert, *Cellulosic Polymers: Blends and Composites*, Hanser Gardner, Cincinnati, Ohio, 1994.
33. M. Yalpani (Ed.), *Industrial Polysaccharides*, Elsevier, New York, 1987.
34. J. D. Wilson and J. K. Hamilton, *J. Chem. Educ.*, **63**, 49 (1986).
35. S. M. Hudson and J. A. Cuculo, *J. Macromol. Sci., Rev. Macromol. Chem. Phys.*, **C18**, 1 (1980).
36. K. F. Gadd, *Polymer*, **23**, 1867 (1982).
37. C. L. McCormick and T. C. Shen, in *Macromolecular Solutions* (R. B. Seymour and G. S. Stahl, Eds.), Pergamon, New York, 1982, pp. 101–107.
38. A. S. Turbak, R. B. Hammer, R. E. Davies, and H. L. Hargert, *Chemtech*, **10**, 51 (1980).
39. E. Kline, in *Cellulose and Cellulose Derivatives*, 2nd ed. (E. Ott, H. M. Spurlin, and M. W. Griffin, Eds.), Wiley-Interscience, New York, 1954, Part 2, pp. 959–1018.
40. A. Hebeish and J. T. Guthrie, *The Chemistry and Technology of Celuluose Copolymers*, Springer-Verlag, New York, 1981.
41. N. M. Bikales and L. Segal, *Cellulose and Cellulose Derivatives*, Wiley, New York, 1971.

42. D. J. McDowall, B. S. Gupta, and V. T. Stannett, *Prog. Polym. Sci.*, **10**, 1 (1984).
43. B. Rånby and C. Rodehed, in *Polymer Science Overview* (G. A. Stahl, Ed.), ACS Symp. Ser. 175, American Chemical Society, Washington, D.C., 1981, pp. 253–264.
44. B. P. Morin, I. P. Breusova, and Z. A. Rogovin, *Adv. Polym. Sci.*, **42**, 139 (1982).
45. D. N.-S. Hon (Ed.), *Graft Copolymerization of Lignocellulosic fibers*, ACS Symp. Ser. 187, American Chemical Society, Washington, D.C., 1982.
46. E. M. Fettes (Ed.), *Chemical Reactions of Polymers*, Wiley-Interscience, New York, 1964: (a) K. Ward, Jr., and A. J. Morak, Chap. 5; (b) N. G. Gaylord and F. S. Ang, Chap. 10; (c) W. A. Reeves and J. D. Guthrie, Chap. 16; (d) R. E. Whitfield and W. L. Wasley, Chap. 6.
47. J. G. Frick, *Chemtech*, 100 (1971).
48. R. L. Whistler and E. F. Paschall (Eds.), *Starch, Chemistry and Technology*, Vols. 1 and 2. Academic, New York, 1965.
49. R. L. Whistler and C. L. Smart, *Polysaccharide Chemistry*, Academic, New York, 1953.
50. R. A. A. Muzzarelli, *Chitin*, Pergamon, Oxford, 1977.
51. G. A. F. Roberts, *Chitin Chemistry*, Macmillan, Houndsmills, 1992.
52. M. G. Peter, *J. Macromol. Sci.—Pure Appl. Chem.*, **A32**, 629 (1995).
53. G. Skjak-Braeks, T. Anthonsen, and P. Sandford (Eds.), *Chitin and Chitosan*, Elsevier, London, 1989.
54. D. L. Kaplan, B. J. Wiley, J. M. Mayer, S. Arcidiacono, J. Keith, S. J. Lombardi, D. Ball, and A. L. Allen, in *Biomedical Polymers: Designed-to-Degrade Systems* (S. W. Shalaby, Ed.), Hanser, New York, 1994, Chap. 8.
55. A. G. Walton, *Polypeptides and Protein Structure*, Elsevier, New York, 1981.
56. R. E. Dickerson and I. Geis, *The Structure and Action of Proteins*, Harper & Row, New York, 1969.
57. R. D. B. Fraser and T. P. MacRae, *Conformation in Fibrous Proteins and Related Synthetic Polypeptides*, Academic, New York, 1973.
58. M. Bodansky, A. Bodansky, and B. M. Trost, *The Practice of Peptide Synthesis*, 2nd ed., Springer-Verlag, New York, 1994.
59. P. Lloyd-Williams, F. Albericio, and E. Giralt, *Chemical Approaches to the Synthesis of Peptides and Proteins*, CRC, Boca Raton, Florida, 1997.
60. R. B. Merrifield, *J. Am. Chem. Soc.*, **85**, 2149 (1963).
61. B. Gutte and R. B. Merrifield, *J. Am. Chem. Soc.*, **91**, 501 (1969).
62. P. Alexander and R. F. Hudson, *Wool, Its Chemistry and Physics*, Reinhold, New York, 1954.
63. F. Lucas, J. T. B. Shaw, and S. G. Smith, *Adv. Protein Chem.*, **13**, 107 (1958).
64. E. Iizuka, *J. Appl. Polym. Sci., Appl. Polym. Symp.*, **41**, 163, 173 (1985).
65. F. Vollrath, *Sci. Am.*, **266**(3), 70 (1992).
66. M. A. Colgin and R. V. Lewis, *Chem. Ind.*, No. 24, 1009 (1995).
67. H. A. J. Battaerd and G. W. Tregear, *Graft Copolymers*, Wiley-Interscience, New York, 1967, pp. 238–239.
68. K. Arai, in *Block and Graft Copolymerizations*, Vol. 1 (R. J. Ceresa, Ed.), Wiley-Interscience, New York, 1973, Chaps. 7 and 8.
69. D. Kaplan, W. W. Adams, B. Farmer, and C. Viney (Eds.), *Silk Polymers: Materials Science and Biotechnology*, ACS Symp. Ser. 544, American Chemical Society, Washington, D.C., 1994.
70. K. H. Gustavson, *The Chemistry and Reactivity of Collagen*, Academic, New York, 1956.
71. A. Veis, *The Macromolecular Chemistry of Gelatin*, Academic, New York, 1964.
72. W. Zhao, A. Kloczkowski, J. E. Mark, B. Erman, and I. Bahar, *J. Macromol. Sci.—Pure Appl. Chem.*, **A33**, 525 (1996); *Chemtech*, **26**(3), 32 (1996).
73. J. H. Collins, *Casein Plastics and Allied Materials* (Plastics Monograph), London, 1952.
74. S. M. Hecht (Ed.), *Bioorganic Chemistry: Nucleic Acids*, Oxford University Press, New York, 1996.
75. L. B. Townsend and R. S. Tipson (Eds.), *Nucleic Acid Chemistry*, Wiley-Interscience, New York, 1986.
76. B. Hudson and J. Vinograd, *Nature*, **216**, 647 (1967).
77. H. L. Frisch and D. Klempner, *Adv. Macromol. Chem.*, **2**, 149 (1970).
78. G. R. Pettit, *Synthetic Nucleotides*, Vol. 1, Van Nostrand, New York, 1972.
79. K. L. Agarwal, A. Yamazaki, P. J. Cashion, and H. G. Khorana, *Angew. Chem. Int. Ed. English*, **11**, 451 (1972).
80. J. Van Brunt, *Chemtech*, **17**, 186 (1987).

81. L. E. Barstow and C. M. Groginsky, *Res. Dev.*, **27**(2), 184 (1985).
82. H. G. Khorana, in *Bioorganic Chemistry*, Vol. 7, Academic, New York, 1978, p. 351.

Review Exercises

1. What properties of asphaltenes make them suitable for road surfacing? What singular advantage does asphaltene have over synthetic polymers with similar properties?

2. Explain why phenol–formaldehyde resins are particularly good adhesives for wood. Select an aromatic ring from the lignin structure given in Figure 18.1 and illustrate its reaction with formaldehyde to form a phenolic-type resin.

3. Besides tall oil, name some other nonpolymeric natural products that have been used in polymer manufacture.

4. Suggest a way to make a polyesterimide from rosin and other readily available chemicals. (S. Das and S. Maiti, *J. Macromol. Sci.-Chem.*, **A17**, 1177, 1982)

5. The synthesis of "dimer acid" from isomerized linoleic acid is discussed in Section 18.7. Show how isomerized linoleic acid might undergo Diels–Alder dimerization to form dimer acid. (See Chapter 12, structure **36**, for linoleic acid.) (J. S. Cowan, *J. Am. Oil Chem. Soc.*, **39**, 534, 1962)

6. Castor oil is a triglyceride containing about 90 percent ricinoleic acid,

$$CH_3(CH_2)_5CH(OH)CH_2CH=CH(CH_2)_7CO_2H$$

   [Explain how castor oil might be converted to (a) a thermosetting polyurethane (Z. S. Petrovic and D. Fajnik, *J. Appl. Polym. Sci.*, **29**, 1031, 1984); (b) an IPN (M. Patel and B. Suthar, *J. Polym. Sci. A, Polym. Chem.*, **25**, 2251, 1987, or M. A. Linné et al., in *Rubber-Modified Thermoset Resins*, C. K. Rien and J. K. Gillham, Eds., Adv. Chem. Ser. 208, American Chemical Society, Washington, D.C., 1984).

7. Myrcene is a monoterpene that occurs naturally in hops and verbena and is formed by thermal rearrangement of β-pinene, a major constituent of turpentine. What repeating units would most likely be formed if myrcene were subjected to free radical polymerization? (J. L. Cawse et al., *J. Appl. Polym. Sci.*, **31**, 1963, 1986)

8. Unsaturated polyesters (Chapter 12) have been prepared from the Diels–Alder adduct of myrcene (exercise 7) with maleic anhydride. Show the structure of the adduct. Apart from the fact that myrcene is a renewable resource, what advantage might this anhydride offer? (D. W. Abate and M. P. Stevens, *J. Appl. Polym. Sci.*, **46**, 389, 1992)

9. Write repeating units of each of the following: (a) Mannan, one of the constituents of hemicelluloses, which consists primarily of mannose units linked β(1 → 4). (Mannose differs from glucose in its configuration at carbon 2.) (b) A water-soluble paper, popular with intelligence agencies and bookmakers (marketed under the trade name Dissolvo), which consists primarily of sodium carboxymethylcellulose. (Assume complete substitution.) (c) Dextran. (See description in Section 18.3.6.) Show chain branches in addition to the repeating unit.

10. Using the $\overline{DP}$ data given in this chapter, compare the number average molecular weights of "triacetate" and its cellulose feedstock. (Assume acetylation is quantitative.)

11. Cellulose acetate is spun into fibers from a viscous 95:5 acetone:water solution (25–30% solids). Apart from the obvious cost advantage, why might the acetone be diluted with water? (If the answer is not immediately obvious, you might want to review Chapter 2, Section 2.2.)

12. A more rot-resistant form of cellulose has been marketed. It is prepared by reaction of alkali cellulose with acrylonitrile. Write a reaction for this process. (*Text. Res. J.*, **25**, 58, 1955; **26**, 67, 1956)

13. Write a reaction showing the grafting of cellulose with $ClCH_2CH\overset{\displaystyle S}{\overbrace{}}CH_2$. (T. L. Ward et al., *J. Appl. Polym. Sci.*, **13**, 607, 1969)

14. When doing physical exertion in very cold weather, is it better to wear undergarments made of cotton or polypropylene? Explain.

15. Given the fact that baseballs are made largely of tightly wound wool fibers, explain why it is more difficult to hit home runs in very humid weather.

16. The molecular weight of spider silk is in the range of 2 to 3×10^5 g/mol. Prior to spinning, the proteinaceous polymer is stored as an aqueous solution in an abdominal sac. As the web is spun, the silk becomes insoluble. Speculate on the structural changes taking place during spinning. (A discussion of this topic may be found in R. F. Foelix, *Biology of Spiders*, Harvard University Press, Cambridge, Mass., 1982, Chap. 5.)

A P P E N D I X

A

COMMONLY USED

POLYMER ABBREVIATIONS

ABS	Acrylonitrile–butadiene–styrene copolymer
BMI	Bismaleimide
CA	Cellulose acetate
CAB	Cellulose acetate butyrate
CAP	Cellulose acetate propionate
CN	Cellulose nitrate
CP	Cellulose propionate
EPDM (or EPD)	Ethylene–propylene–diene monomer copolymer
EPR (or EPM)	Ethylene–propylene rubber
ETFE	Ethylene–tetrafluoroethylene copolymer
EVA	Ethylene–vinyl acetate copolymer
HDPE	High-density polyethylene
HIPS	High-impact polystyrene
LCP	Liquid crystal polymers
LDPE	Low-density polyethylene
LLDPE	Linear low-density polyethylene
MDPE	Medium-density polyethylene
MF	Melamine–formaldehyde polymer
PAA	Poly(acrylic acid)
PAEK	Polyaryletherketone
PAI	Polyamideimide
PAN	Polyacrylonitrile
PAS	Polyarylsulfone
PB	Poly(1-butene)
PBI	Polybenzimidazole
PBO	Polybenzoxazole
PBT	Poly(butylene terephthalate)
PC	Polycarbonate

PCT	Poly(cyclohexanedimethylene terephthalate)
PCTFE	Polychlorotrifluoroethylene
PDMS	Polydimethylsiloxane
PE	Polyethylene
PEEK	Polyetheretherketone
PEI	Polyetherimide
PEK	Polyetherketone
PEKK	Polyetherketoneketone
PEN	Poly(ethylene 2,6-naphthalenedicarboxylate)
PEO	Poly(ethylene oxide)
PES	Polyethersulfone
PET	Poly(ethylene terephthalate)
PF	Phenol–formaldehyde polymer
PHB	Poly(2-hydroxybutyrate)
PHVB	Poly(3-hydroxybutyrate–valerate)
PI	Polyimide
PIB	Polyisobutylene
PMMA	Poly(methyl methacrylate)
POD	Polyoxadiazole
POM	Polyoxymethylene
POP	Polyoxypropylene
PP	Polypropylene
PPO*	Poly(phenylene oxide)
PPS	Poly(phenylene sulfide)
PS	Polystyrene
PTFE	Polytetrafluoroethylene
PVA	Poly(vinyl alcohol)
PVAC (or PVAc)	Poly(vinyl acetate)
PVB	Poly(vinyl butyral)
PVC	Poly(vinyl chloride)
PVDC	Poly(vinylidene chloride)
PVDF	Poly(vinylidene fluoride)
PVF	Poly(vinyl fluoride)
SAN	Styrene–acrylonitrile copolymer
SBR	Styrene–butadiene rubber
SMA	Styrene–maleic anhydride copolymer
SIR	Styrene–isoprene rubber
TPE	Thermoplastic elastomer
TPI	Thermoplastic polyimide
UF	Urea formaldehyde polymer
UHMWPE	Ultrahigh-molecular-weight polyethylene
UP	Unsaturated polyester
VLDPE	Very-low-density polyethylene

*Occasionally written as PPE, poly(phenylene ether).

A P P E N D I X

POLYMER

LITERATURE

There has been a tremendous proliferation of texts, monographs, and journals dealing with polymer chemistry in recent years. Books covering almost every specialized area of polymers have been published, including bound volumes of symposia proceedings. Because polymer science is such an active field of research and development, such works become dated rather quickly. This appendix is limited to publications of an ongoing nature (journals, advances, continuing series) and to handbooks or encyclopedic compilations. Publishers should be contacted for electronic versions.

B.I   Encyclopedias and Yearbooks

Comprehensive Polymer Science (G. Allen and J. C. Bevington, Eds.), 7 vols., Pergamon, New York, 1989. 1st and 2nd supplements published in 1992 and 1996, respectively.

Concise Encyclopedia of Polymer Processing and Applications (P. J. Corish, Ed.), Pergamon, 1991.

Concise Encyclopedia of Polymer Science and Engineering (J. I. Kroschwitz, Ed.), Wiley, New York, 1990.

Encyclopedia of Composite Materials and Components (M. Grayson, Ed.), Technomic, Lancaster, PA, 1983.

Encyclopedia of Composites (S. M. Lee, Ed.), 6 vols., VCH, New York, 1989–1990.

Encyclopedia of Polymer Science and Engineering, 2nd ed. (H. F. Mark, N. M. Bikales, C. G. Overberger, G. Menges, and J. I. Kroschwitz, Eds.), 17 vols., plus supplement and index, Wiley-Interscience, New York, 1985–1990. [The first edition was published between 1964 and 1967 as *Encyclopedia of Polymer Science and Technology* (H. F. Mark, N. G. Gaylord, and N. M. Bikales, Eds.), in 16 volumes plus a supplement.]

Encyclopedia of PVC, 2nd ed. (L. I. Nass, Ed.), 3 vols., Marcel Dekker, New York, 1992.

Kirk-Othmer: Encyclopedia of Chemical Technology, 3rd ed. (M. Grayson and D. Eckroth, Eds.), 24 vols. plus supplement and index. Wiley-Interscience, New York, 1978–1984.

(The 4th edition began publication in 1991.) Although a more general reference, contains excellent reviews on most polymer topics.

Modern Plastics Encyclopedia (J. Agranoff, Ed.), McGraw-Hill, New York; published annually. Contains a wealth of information on properties and processing of commercially important plastics.

Polymeric Materials Encyclopedia, 12 vols. (J. C. Salamone, Ed.), CRC, Boca Raton, FL, 1996.

Polymer Yearbook (R. A. Pethrick, Ed.), Harwood, New York, 1984 (first issue); published annually. A concise reference work covering new developments in polymer science, polymer literature, conferences, trade names, and so on.

Rosato's Plastics Encyclopedia and Dictionary, D. Rosato, Hanser Gardner, Cincinnati, 1992.

Treatise on Adhesion and Adhesives (R. L. Patrick, Ed.), Marcel Dekker, New York; multi-volume work published irregularly.

B.2 Handbooks

Of those listed, *Polymer Handbook*, edited by Brandrup and Immergut, is the best one-volume source of physical data, reaction constants, nomenclature, and other information about polymers.

Additives for Plastics Handbook, J. Murphy, Elsevier, New York, 1996.

Adhesives Technology Handbook (A. H. Landrock, Ed.), Noyes, Park Ridge, NJ, 1986.

Atlas of Polymer Morphology, A. E. Woodward, Hanser Gardner, Cincinnati, 1989.

Blow Molding Handbook (D. Rosato and D. V. Rosato, Eds.), Hanser, Munich, 1988.

Chitin Sourcebook: A Guide to the Research Literature, E. R. Pariser, Wiley, New York, 1989.

CRC Handbook of Chromatography: Polymers, C. G. Smith, N. E. Skelly, and R. A. Solomon, CRC, Boca Raton, FL, 1982.

CRC Handbook of Polymer-Liquid Interaction Parameters and Solubility Parameters, A. F. M. Barton, CRC, Boca Raton, FL, 1990.

Dictionary of Composite Materials Technology (S. M. Lee, Ed.), Technomic, Lancaster, PA, 1989.

Electrical and Electronic Properties of Polymers: A State of the Art Compendium (J. I. Kroschwitz, Ed.), Wiley, New York, 1988.

Encyclopedic Dictionary of Commercial Polymer Blends (L. A. Utracki, Ed.), Chemtec, Toronto-Scarborough, 1994.

Engineered Materials Handbook, ASM, Metals Park; Vol. 1, *Composites* (1987); Vol. 2, *Engineering Plastics* (1988).

Flammability Handbook for Plastics, 4th ed., C. J. Hilado, Technomic, Lancaster, PA, 1990.

Handbook of Adhesive Technology (A. Pizzi and K. L. Mittal, Eds.), Marcel Dekker, New York, 1994.

Handbook of Coatings Additives (L. J. Calbo, Ed.), Marcel Dekker, New York, Vol. 1, 1987; Vol. 2, 1992.

Handbook of Common Polymers (W. J. Roff and J. R. Scott, Eds.), CRC, Boca Raton, FL, and Butterworth, London, 1971.

Handbook of Composites, 2nd. ed. (G. Lubin, Ed.), Van Nostrand Reinhold, New York, 1982.

Handbook of Conducting Polymers (T. A. Skotheim, Ed.), 2 vols., Marcel Dekker, New York, 1986.

Handbook of Conductive Organic Molecules and Polymers (H. S. Nalwa, Ed.), 4 vols., Wiley, New York, 1997.

Handbook of Corrosion Resistant Coatings (D. J. De Renzo, Ed.), Noyes, Park Ridge, NJ, 1986.

Handbook of Elastomers: New Developments and Technology (A. K. Bhowmick and H. L. Stevens, Eds.), Marcel Dekker, New York, 1988.

Handbook of Fiber Science and Technology, 9 vols., Marcel Dekker, New York, 1983–1984.

Handbook of Organic Coatings: A Comprehensive Guide for the Coatings Industry, R. B. Seymour and H. F. Mark, Elsevier, New York, 1990.

Handbook of Plastics Compounds, Elastomers and Resins (M. Ash and I. Ash, Eds.), VCH, New York, 1991.

Handbook of Plastics Test Methods, 3rd ed. (R. P. Brown, Ed.), Wiley, New York, 1989.

Handbook of Polymer Composites for Engineers (L. Hollaway, Ed.), Technomic, Lancaster, PA, 1994.

Handbook of Polymer Degradation (S. H. Hamid, M. B. Amin, and A. G. Maadhah, Eds.), Marcel Dekker, New York, 1992.

Handbook of Polymeric Foams and Foam Technology (D. Klempner and K. C. Frisch, Eds.), Hanser Gardner, Cincinnati, 1992.

Handbook of Polymer Science and Technology (N. P. Cheremisinoff, Ed.), 4 vols., Marcel Dekker, New York, 1989.

Handbook of Polymer Synthesis (H. R. Kricheldorf, Ed.), 2 vols., Marcel Dekker, New York, 1992.

Handbook of Polyolefins (C. Vasile and R. B. Seymour, Eds.), Marcel Dekker, New York, 1993.

Handbook of Pressure Sensitive Adhesion Technology, 2nd. ed (D. Satas, Ed.), Van Nostrand Reinhold, New York, 1989.

Handbook of Rubber Technology (W. Hofmann, Ed.), Hanser, Munich, 1988.

Handbook of Thermoplastic Elastomers (B. M. Walker and C. P. Rader, Eds.), Van Nostrand Reinhold, New York, 1988.

Handbook of Thermoset Plastics (S. H. Goodman, Ed.), Noyes, Park Ridge, NJ, 1986.

Industrial Synthetic Resins Handbook (E. W. Flick, Ed.), Noyes, Park Ridge, NJ, 1985.

International Plastics Flammability Handbook (J. Troitzsch, Ed.), Hanser Gardner, Cincinnati, 1990.

International Plastics Handbook, 2nd ed., H. Saechtling, Hanser, Munich, 1987.

Mold Making Handbook for the Plastics Engineer (K. Stoeckhert, Ed.), Hanser, Munich, 1983.

Nylon Plastics Handbook (M. I. Kohan, Ed.), Hanser Gardner, Cincinnati, 1995.

Physical Properties of Polymers Handbook (J. E. Mark, Ed.), AIP, Williston, VT, 1996.

Plastics Additives Handbook, 2nd ed. (R. Gächter and H. Müller, Eds.), Hanser, Munich, 1987.

Plastics Engineering Handbook of the Society of the Plastics Industry, Inc., 4th ed. (J. Frados, Ed.), Van Nostrand Reinhold, New York, 1976.

Plastics Technical Dictionary, A. M. Wittfoht, Hanser Gardner, Cincinnati, 1992.

Plastics Technology Handbook, 2nd ed., M. Chanda and S. K. Roy, Marcel Dekker, New York, 1992.

Polymer Handbook (J. Brandrup and E. H. Immergut, Eds.), 3rd ed., Wiley-Interscience, New York, 1989.

Polymers: An Encyclopedic Sourcebook of Engineering Properties (J. I. Kroschwitz, Ed.), Wiley, New York, 1987.

Polymer Science—A Materials Science Handbook (A. D. Jenkins, Ed.), North-Holland, Amsterdam, 1971.

Polymer Science Dictionary, M. S. M. Alger, Elsevier, New York, 1989.

Polyurethane Handbook, 2nd ed. (G. Oertel, Ed.), Hanser Gardner, Cincinnati, 1993.

Reinforced Plastics for Commercial Composites, Source Book (G. Shook, Ed.), ASM, 1986.

Rubber Technology Handbook, W. Hofmann, Hanser Gardner, Cincinnati, 1989.

Whittington's Dictionary of Plastics, 3rd ed. (J. F. Carley, Ed.), Technomic, Lancaster, PA, 1993.

B.3 Continuing Series

The following listing gives the title and publisher and, in many instances, the first year of issue. Reviews published as part of a polymer journal are listed in Section B.4.

Advances in Colloid and Interface Science (Elsevier, Amsterdam, 1967)
Advances in Dendritic Macromolecules (JAI, Greenwich, CT, 1994)
Advances in Organic Coatings Science and Technology (Technomic, Lancaster, PA, 1979)
Advances in Polymer Science (Springer-Verlag, New York, 1958)
Advances in Polymer Technology (Wiley, New York. Formerly *Advances in Plastics Technology*)
Advances in Urethane Ionomers (Technomic, 1995)
Advances in Urethane Science and Technology (Technomic, Lancaster, PA, 1971)
Current Topics in Polymer Science (Hanser, New York, 1987)
Current Trends in Polymer Science (Research Trends, Poojapura, India)
Developments in Ionic Polymers (Elsevier, New York)
Developments in Oriented Polymers (Elsevier, New York)
Developments in Polymer Degradation (Elsevier, New York)
International Polymer Processing (Hanser, Munich, 1986)
International Polymer Science and Technology (PAPRA, Shawbury, UK, 1974)
Macromolecular Chemistry: Specialist Periodical Report (Chemical Society, London, 1979)
Macromolecular Synthesis (Wiley, New York, 1963)
Organic Coatings: Science and Technology (Marcel Dekker, New York, 1983)
Plastics Engineering (Marcel Dekker, New York, 1981)
Reviews in Macromolecular Chemistry (Marcel Dekker, New York, 1968)

B.4 Journals

A large number of polymer journals are published worldwide. They may be classified as general or specialized; that is, they deal with a specific area of polymer science. The compilation given here has the following characteristics:

Format	Title (publisher, country of publication, number of issues per year, language, date first published).
Language code	C, Chinese; E, English; F, French; G, German; I, Italian; J, Japanese; P, Polish; R, Russian.

The date first published refers to the original date regardless of title. (If titles have changed, former titles follow the entry.) In some instances, publication is irregular. Data are taken largely from *Ulrich's International Periodicals Directory*, 35th ed., Bowker, New York, 1997; in some instances data are incomplete.

Note: This is not an exhaustive listing; a good many trade journals dealing with plastics, rubber, fibers, coatings, and adhesives are omitted. In addition, other more general journals such as *Journal of the American Chemical Society* or *Bulletin of the Chemical Society of Japan*, occasionally contain papers dealing with polymers. Some journals (e.g., *Journal of Thermal Analysis, Journal of Rheology, Journal of Materials Chemistry*) are included because polymers represent a significant portion of the coverage.

B.4.1 General Journals

Acta Polymerica (Akad. Verlag, Germany, 12, EGR, 1950) (formerly *Faserforschung und Textiltechnik*)

Die Angewandte Makromolekülare Chemie (Hüthig und Wepf Verlag, Switzerland, 9, EFG, 1967)

Bulletin of High Polymers/Gaofenzi Tongxun (China Publ. Centre, China, 6, C)

Chinese Journal of Polymer Science (Science Press, Beijing, China, 4, E, 1983) (formerly *Polymer Communications*, 1983–1985)

Colloid and Polymer Science (Steinkopff, Germany, 12, E, 1906) (formerly *Kolloid Zeitschrift für Polymere*)

Designed Monomers and Polymers (VSP, Netherlands, 4, E, 1998)

European Polymer Journal (Pergamon, USA, 12, EFGI, 1965)

International Journal of Polymeric Materials (Gordon and Breach, USA, 16, E, 1971)

Journal of Applied Polymer Science (Wiley, USA, 16, E, 1956)

Journal of Applied Polymer Science, Applied Polymer Symposia (Wiley, USA, irreg., E, 1965)

Journal of Macromolecular Science: Part A—Pure and Applied Chemistry (Marcel Dekker, USA, 12, E, 1966) (formerly *Journal of Macromolmolecular Science: Part A—Chemistry*, 1966–1991)

Journal of Macromolecular Science: Part B—Physics (Marcel Dekker, USA, 12, E, 1966)

Journal of Macromolecular Science: Part C—Reviews in Macromolecular Chemistry and Physics (Marcel Dekker, USA, 4, E, 1966) (formerly *Journal of Macromolecular Chemistry: Reviews in Macromolecular Chemistry*, 1966–1975)

Journal of Polymer Materials (Oxford and IBH, India, 4, E, 1984)

Journal of Polymer Science, Macromolecular Reviews (Wiley, USA, 1963, discont. 1981) (formerly *Macromolecular Reviews*, 1963–1967)

Journal of Polymer Science, Part A: Polymer Chemistry (Wiley, USA, 17, 1946) (formerly *Journal of Polymer Science*, 1946–1962; *Part A (General Papers)*, 1963–1965; *Part A-1 (Polymer Chemistry)*, 1966–1972; *Polymer Chemistry Edition*, 1972–1986) (A *Polymer Letters* section was published separately until 1991, when it was integrated into *Parts A* and *B*)

Journal of Polymer Science, Part B: Polymer Physics (Wiley, USA, 17, EFG, 1962) (formerly *Journal of Polymer Science, Part A-1, Polymer Physics*, 1966–1972; *Polymer Physics Edition*, 1972–1986)

Journal of Polymer Science, Symposia Proceedings (Wiley, USA, irreg., E, 1963) (formerly *Journal of Polymer Science, Part C, Polymer Symposia*, 1963–1972)

Kobunshi/High Polymers (Soc. Polym. Sci. Japan, 12, J, 1952)

Kobunshi Ronbunshu (Soc. Polym. Sci. Japan, 12, J, 1944) (formerly *Kobunshi Kagaku*)

Macromolecular Chemistry and Physics (Hüthig & Wepf., Switzerland, 12, EFG, 1947) (formerly *Die Makromolekülare Chemie*, 1947–1993)

Macromolecular Chemistry—Theory and Simulations (Hüthig & Wepf., Switzerland, 6, E, 1975) (formerly *Die Makromolekülare Chemie, Supplement*, 1975–1993)

Macromolecular Rapid Communications (Hüthig & Wepf., Switzerland, 12, EFG, 1947) (formerly *Die Makromolekülare Chemie, Rapid Communications*, 1947–1993)

Macromolecular Reports (Marcel Dekker, USA, 8, E, 1991) (integrated into *Journal of Macromolecular Science—Pure and Applied Chemistry*, 1997)

Macromolecular Symposia (Hüthig & Wepf., Switzerland, 12, EFG, 1980) (formerly *Die Makromolekülare Chemie, Macromolecular Symposia*, 1947–1993)

Macromolecules (Am. Chem. Soc., USA, 26, E, 1968)

Polimery (Inst. Chemii Przemyslowej, Poland, 12, P, 1956) (formerly *Tworzywa*, 1956–1958; *Tworzywa, Guma, Lakiery*, 1959–1960; *Tworzywa Wieloczasteczkowe*, 1961)

Polymer (Butterworth, UK, 26, E, 1960) (*Polymer Communications* was published separately until 1992, when it was integrated into *Polymer*)

Polymer Applications/Kobunshi Kako (High Poly. Publ. Assoc., Japan, 12, J, 1951)

Polymer Bulletin (Springer Verlag, USA, 12, E, 1978)

Polymer Contents (Elsevier, UK, 12, E, 1984)

Polymer International (Elsevier, UK, 12, E, 1969) (formerly *British Polymer Journal*, 1969–1987)

Polymer Journal (Soc. Polym. Sci., Japan, 12, E, 1970)

Polymer News (Gordon and Breach, UK, 12, E, 1970)

Polymer Preprints (Am. Chem. Soc., Div. Polym. Chem., USA, 2, E, 1960)

Polymer Science Library (Elsevier, Netherlands, irreg., E, 1979)

Polymer Science Russia (Maik Nauka-Interperiodica, Russia, 12, E) (English translation of *Vysokomolekularnye Soedineniya*; formerly, until 1992, *Polymer Science USSR*)

Progress in Colloid and Polymer Science (Springer-Verlag, USA, irreg., E, 1909)

Progress in Polymer Science (Elsevier, UK, 8, E, 1967)

Trends in Polymer Science (Elsevier, UK, 12, E, 1993)

Vysokomolekularnye Soedineniya, Seriya A i B (Maik Nauka, Russia, 12, R, 1959) (formerly published separately as *Seriya A* and *Seriya B*, 1959–1993)

B.4.2 Specialty Journals

Additives for Plastics (Yarsely Res. Labs., UK, 12, E, 1971)
Additives for Polymers (Elsevier, UK, 12, E, 1971)
Advanced Composite Materials (VSP, Netherlands, 4, E, 1991)
Advanced Composites Bulletin (Elsevier, UK, 12, E)
Advanced Materials (VCH, USA, 12, E, 1989)
Bioactive and Compatible Polymers (Technomic, USA, 4, E, 1986)
Biomaterials (Butterworth, UK, 24, E, 1980)
Biomedical Materials (Elsevier, UK, 12, E, 1975) (formerly *Biomedical Polymers*, 1978–1988)
Biopolymers (Wiley, USA, 12, E, 1961)
Carbohydrate Polymers (Elsevier, UK, 6, E, 1981)
Cellular Polymers (Elsevier, UK, 6, E, 1982)
Celluluose (Chapman and Hall, UK, 4, 1994)
Chemistry of Materials (Am. Chem. Soc., USA, 12, E, 1989)
Colloids and Surfaces (Elsevier, Netherlands, 24, E, 1980)
Composite Interfaces (VSP, Netherlands, 6, E, 1993)
Composite Structures (Applied Science, UK, 4, E, 1982)
Composites (Butterworth, UK, 4, E, 1969)
Composites and Adhesives Newsletter (T-C Press, USA, 4, E, 1984)
Composites Science and Technology (Elsevier, UK, 12, E, 1968)
Computational Polymer Science (PRA, USA, 4, E, 1992)
Computational and Theoretical Polymer Science (Elsevier, UK, 4, E, 1996)
High Performance Plastics (Elsevier, UK, 12, E, 1983)
High Performance Polymers (IOP, UK, 4, E, 1989)
International Journal of Adhesion and Adhesives (Butterworth, UK, 4, E, 1980)
International Journal of Biological Macromolecules (Butterworth, UK, 8, E, 1971)
International Journal of Plasticity (Pergamon, USA, 6, E, 1985)
International Journal of Polymer Analysis and Characterization (Gordon and Breach, USA, 4, E, 1995)
International Journal of Polymeric Materials (Gordon and Breach, UK, 16, E, 1971)
Journal of Adhesion (Gordon and Breach, UK, 16, E, 1969)
Journal of Adhesion Science and Technology (VSP, Netherlands, 12, EG, 1987)
Journal of Bioactive and Compatible Polymers (Technomic, USA, 4, E)
Journal of Biomaterials Science: Polymer Edition (VSP, Netherlands, 12, E, 1988)
Journal of Cellular Plastics (Technomic, USA, 6, E, 1965)
Journal of Coatings Technology (Fed. Socs. Coating Technol., USA, 12, E, 1922) (formerly *Journal of Paint Technology; FSPT Official Digest*)
Journal of Composite Materials (Technomic, USA, 24, E, 1967)
Journal of Composites Technology and Research (ASTM, USA, 4, E, 1979)
Journal of Controlled Release (Elsevier, Netherlands, 15, E, 1984)

Journal of Elastomers and Plastics (Technomic, USA, 4, E, 1969) (formerly Journal of Elastoplastics)

Journal of Environmental Polymer Degradation (Plenum, USA, 4, E, 1993)

Journal of Inorganic and Organometallic Polymers (Plenum, USA, 4, E, 1991)

Journal of Materials Chemistry (Royal Society of Chemistry, UK, 12, E, 1991)

Journal of Membrane Science (Elsevier, Netherlands, 26, E, 1977)

Journal of Polymer Engineering (Freund, UK, 4, E) (formerly *Polymer Engineering Reviews*)

Journal of Reinforced Plastics and Composites (Technomic, USA, 18, E, 1982)

Journal of Rheology (Wiley, USA, 6, E, 1957) (formerly *Transactions of the Society of Rheology*)

Journal of Thermal Analysis (Magyar Tudomanyos Akad., Hungary, 12, EPG, 1969)

Journal of Thermoplastic Composite Materials (Technomic, USA, 6, E, 1988)

Journal of Vinyl and Additive Technology (SPE, USA, 4, E, 1979) (formerly *Journal of Vinyl Technology*, 1979–1995)

Kunststoffe (Hanser, Germany, 12, E, 1911) (Engl. ed. of *Kunststoffe, German Plastics*)

Kunstoffe-Plast Europe (Hanser, Germany, 12, EG) (formerly, until 1996, *Kunstoffe, German Plastics*)

Kunststoffe—Plastics (Vogt-Schild, Switzerland, 12, EG, 1953)

Mechanics of Composite Materials (Engl. transl. of *Mekhanika Kompozitnykh Materialov*) (Plenum, USA, 6, E, 1965) (formerly *Polymer Mechanics*)

New Polymeric Materials (VSP, Netherlands, 4, E, 1987)

Plasmas and Polymers (Plenum, USA, 4, E, 1995)

Plastics and Rubber Processing and Applications (Elsevier, UK, 4, E, 1981)

Plastics Engineering (SPE, USA, 12, E, 1945) (formerly *SPE Journal*)

Polymer Composites (SPE, USA, 6, E, 1980)

Polymer Degradation and Stability (Elsevier, UK, 12, E, 1979)

Polymer Engineering and Science (SPE, USA, 24, E, 1961) (formerly *SPE Transactions*)

Polymer Gels and Networks (Elsevier, UK, 6, E, 1992)

Polymeric Materials (Gordon and Breach, UK, 4, E, 1971)

Polymeric Materials Science and Engineering (Am. Chem. Soc., Div. Polym. Mat. Sci. Eng., USA, 2, E) (formerly *Organic Coatings and Plastics Chemistry Preprints*)

Polymer Networks and Blends (Chemtec, Canada, E, 1991)

Polymer Photochemistry (Elsevier, UK, 6 per volume, E, 1981)

Polymer-Plastics Technology and Engineering (Marcel Dekker, USA, 4, E, 1967) (formerly *Journal of Macromolecular Science, Part D: Reviews in Polymer Processing and Technology*, 1967–1974)

Polymer Process Engineering (Marcel Dekker, USA, 3, E, 1983)

Polymer Reaction Engineering (Marcel Dekker, USA, 4, E, 1992)

Polymer Recovery (Kluwer, Netherlands, 4, E, 1993)

Polymer Recycling (RAPRA, UK, 4, E, 1995)

Polymers and Polymer Composites (RAPRA, UK, 8, E, 1988)

Polymers for Advanced Technologies (VCH, UK, 6, E, 1990)

Polymer Testing (Elsevier, UK, 6, E, 1980)

Polymers, Paint and Color Journal (Fuel and Metallurgical Journals, UK, fortnightly, E, 1879)

Progress in Organic Coatings (Elsevier, Switzerland, 12, EFG, 1972)
Proteins—Structure, Function, and Genetics (Liss, USA, 12, E, 1986)
Reactive and Functional Polymers (Elsevier, Netherlands, 9, E, 1982) (formerly *Reactive Polymers, Ion Exchangers, Sorbents*, 1982–1991; *Reactive Polymers*, 1991–1995)
Reinforced Plastics Elsevier, UK, 12, E, 1956)
Rheologica Acta (Steinkopff, Germany, 6, E, 1958)
Rubber Chemistry and Technology (Am. Chem. Soc., Rubber Div., USA, 5, E, 1928)
SAMPE Journal (SAMPE, USA, 6, E)
Sen'i Gakkaishi (Soc. Fiber Sci. and Tech., Japan, 12, J, 1944)
Textile Institute Journal (Text. Inst., UK, 6, E, 1910)
Textile Research Journal (TRI, USA, 12, E, 1930)
Textile Review/Sen'i Kogyo Zasshi (Text. Journ. and Book Publ. Co., Japan, 6, J, 1909)
Urethane Plastics and Products (Technomic, USA, 12, E, 1971)
Vinyls and Polymers (Inst. Polym. Chem., Japan, 12, J, 1961)

SOURCES OF LABORATORY
EXPERIMENTS IN
POLYMER CHEMISTRY

C.1 Laboratory Manuals

The following books were written specifically as laboratory manuals or could be used as such:

G. F. D'Alelio, *Experimental Plastics and Synthetic Resins*, Wiley-Interscience, New York, 1956.

D. Braun, H. Cherdron, and W. Kern, *Techniques of Polymer Syntheses and Characterization*, Wiley-Interscience, New York, 1972.

E. A. Collins, J. Bares, and F. W. Billmeyer, Jr., *Experiments in Polymer Science*, Wiley-Interscience, New York, 1973.

E. M. McCaffery, *Laboratory Preparation for Macromolecular Chemistry*, McGraw-Hill, New York, 1970.

E. M. Pearce, C. E. Wright, and B. K. Bordoloi, *Laboratory Experiments in Polymer Synthesis and Characterization*, Educational Modules for Material Science and Engineering Project, Pennsylvania State University, University Park, 1982.

S. H. Pinner, *A Practical Course in Polymer Chemistry*, Pergamon Press, Oxford, 1961.

W. R. Sorenson and T. W. Campbell, *Preparative Methods of Polymer Chemistry*, 2nd ed., Wiley-Interscience, New York, 1968.

C.2 Supplementary Sources

C.2.1 Laboratory Experiments

H. S. Finlay, *Experiments in Polymer Chemistry*, Shell International Petroleum, London, undated (Booklet containing 11 experiments.)

Macromolecular Syntheses, Wiley-Interscience, New York. (Continuing series. Detailed procedures for specific polymers, 1963ff.)

F. Rodriguez, *Principles of Polymer Systems*, 3rd ed., Hemisphere, New York, 1989. (Appendix 4 lists several experiments.)

S. R. Sandler and W. Karo, *Polymer Syntheses*, 2nd ed., Academic, New York, 3 vols., 1992–1996. (Examples of synthetic procedures.)

C.2.2 General Analytical Techniques

D. Braun (Ed.), *Simple Methods for the Identification of Plastics*, 2nd ed., Hanser, Cincinnati, 1987

T. R. Crompton, *Chemical Analysis of Additives in Plastics*, 2nd ed., Pergamon Press, Oxford, 1971.

J. Haslam, H. A. Willis, and D. C. M. Squirrell, *Identification and Analysis of Plastics*, 2nd ed., Butterworth, London, 1972. (Also contains a collection of IR spectra of polymers.)

G. M. Kline, *Analytical Chemistry of Polymers*, Wiley-Interscience, New York, Part 1, 1959; Parts 2 and 3, 1962.

J. F. Rabek, *Experimental Methods in Polymer Chemistry*, Wiley-Interscience, New York, 1980. (A superb compilation of practical information dealing with almost every physicochemical method for analysis and characterization of polymers.)

C.3 Journal Articles

The following experiments or demonstrations were published in *Journal of Chemical Education* between 1950 and 1997. (Authors, volume, page number, and year given.)

C.3.1 Synthesis

"Anionic polymerization of vinyl monomers." A. Zilkha, M. Albeck, and M. Frankel, **35**, 345 (1958).

"The aqueous ring-opening metathesis polymerization of furan–maleic anhydride adduct." T. Viswanathan and J. Jethmalani, **70**, 165 (1993).

"Classroom demonstrations of polymer principles: Part 2. Polymer formation." F. Rodriguez, L. J. Mathias, J. Kroschwitz, and C. E. Carraher, Jr., **64**, 886 (1987).

"The dead-end polymerization experiment. A modified treatment." D. J. T. Hill and J. H. O'Donnell, **61**, 881 (1984).

"A demonstration of the cuprammonium rayon process." G. B. Kauffman and M. Karbassi, **62**, 878 (1985).

"Emulsion polymerization and film formation of dispersed polymeric particles." G. W. Ceska, **50**, 767 (1973).

"Expanded polystyrene: An experiment for undergraduate students." A. Felgenbaum and D. Scholler, **64**, 810 (1987).

"Formation of titanium gels from titanium alkoxides: A demonstration of kinetically controlled polymerization." M. R. Mueller and J. D. Britt, **71**, 890 (1994).

"Free radical polymerization of acrylamide." E. F. Silversmith and W. C. Wolsey, **69**, 763 (1992).

"Free radical polymerization of styrene. A radiotracer experiment." R. J. Mazza, **52**, 476 (1975).

"Free radical polymerization using the rotating sector method." S. J. Moss, **59**, 1021 (1982).

"Generation of poly(vinyl alcohol) and arrangement of structural units." C. E. Carraher, Jr., **55**, 473 (1978).

"Interfacial polymerizations: Microscale polymer laboratory experiments for undergraduate students." R. G. Lewis, M. Choquette, E. H. Darden, M. P. Gilbert, D. Martinez, C. Myhaver, K. Schlichter, R. Wondenberg, and K. Zawistowski, **69**, A215 (1992).

"Kinetics of condensation polymerization." E. L. McCaffery, **46**, 59 (1969).

"A kinetic study of the adiabatic polymerization of acrylamide." R. A. M. Thomson, **63**, 362 (1986).

"Laboratory methods for the preparation of polymer films." E. B. Mano and L. A. Durao, **50**, 228 (1973).

"Low-temperature polymerization." L. T. Jenkins, **33**, 231 (1956).

"Modifications in the synthesis of caprolactam and nylon 6." J. E. Malmin, **57**, 742 (1980).

"Nylon 6—a simple, safe synthesis of a tough commercial polymer." L. J. Mathias, R. A. Vaidya, and J. B. Canterberry, **61**, 805 (1984).

"The nylon rope trick." P. W. Morgan and S. L. Kwolek, **36**, 182 (1959). (See also G. C. East and S. Hassell, **60**, 69, 1983.)

"Organometallic catalyzed synthesis and characterization of polyethylene. An advanced laboratory experiment." D. E. Tranbuehl, T. V. Harris, A. K. Howe, and D. W. Thompson, **52**, 261 (1975).

"Polyaniline—A conducting polymer: Electrochemical synthesis and electrochromic properties." B. C. Sherman, W. B. Euler, and R. R. Force, **71**, A94 (1994).

"Polymer preparations in the laboratory." G. M. Lampman, D. W. Ford, W. R. Hale, A. Pinkers, and C. G. Sewell, **56**, 626 (1979).

"Polymerization kinetics. Dead-end radical polymerization." E. Senogles and L. A. Woolf, **44**, 157 (1967).

"Polymerization of ethylene at atmospheric pressure." A. Zilkha, N. Calderon, J. Rabani, and M. Frankel, **35**, 344 (1958).

"Polymer synthesis in the undergraduate organic laboratory." W. R. Sorenson, **42**, 8 (1965).

"Polystyrene—a multistep synthesis." S. H. Wilen, C. B. Kemer, and I. Waltcher, **38**, 304 (1961).

"Preparation and crosslinking of an unsaturated polyester." M. P. Stevens, **44**, 160 (1967).

"Preparation and properties of polybutadiene (jumping rubber)." B. Z. Shakhashiri, G. E. Dirreen, and L. G. Williams, **57**, 738 (1980).

"Preparation and solution polymerization of diacetylenes." B. A. Hathaway and A. M. Scates, **74**, 111 (1997).

"Preparation of a polysulfide rubber." G. R. Pettit and G. R. Pettit III, **55**, 472 (1978).

"The preparation of 'bouncing putty.' " D. A. Armitage, M. N. Hughes, and A. W. Sindern, **50**, 434 (1973).

"Preparation of terephthaloyl chloride. Prelude to erzatz nylon." W. C. Rose, **44**, 283 (1967).

"A revision of a general chemistry experiment on an inorganic polymerization–condensation reaction." R. J. Palma, Sr., and E. Sargent, **67**, 614 (1990).

"Ring-opening and ring-forming polymerizations. An organic and polymer laboratory experiment." L. J. Mathias and R. Viswanathan, **64**, 639 (1987).

"A safe and novel polymerization of styrene." B. R. Flynn, **68**, 1038 (1991).

"Superabsorbent polymers." F. L. Buchholz, **73**, 512 (1996).

"Synthesis and a simple molecular weight determination of polystyrene." D. W. Armstrong, J. N. Marx, D. Kyle, and A. Alak, **62**, 705 (1985).

"Synthesis and characterization of a conducting polymer." R. K. Bunting, K. Swarat, D. Yan, and D. Finello, **74**, 421 (1997).

"Synthesis of a phenol–formaldehyde thermosetting polymer." Y. Bedard and B. Riedl, **67**, 977 (1990).

"Synthesis of a superabsorbent polymer." C. M. Garner, M. Nething, and P. Nguyen, **74**, 95 (1997).

"Synthesis of caprolactam and nylon 6." C. E. Carraher, Jr., **55**, 51 (1978).

"Synthesis of poly(β-alanine) and β-alanine." C. E. Carraher, Jr., **55**, 668 (1978).

"An undergraduate experiment in homogeneous catalysis: Synthesis of phenylethoxycarbene tungsten pentacarbonyl and polymerization of norbornene and phenylacetylene." D. Villemin, **64**, 183 (1987).

C.3.2 Properties

"Biodegradable films based on partially hydrolyzed corn starch." H. Sommerfeld and R. Blume, **69**, A151 (1992).

"Classroom demonstrations of polymer principles: Part 1. Molecular structure and molecular mass." F. Rodriguez, L. J. Mathias, J. Kroschwitz, and C. E. Carraher, Jr. **64**, 72 (1987).

"Classroom demonstrations of polymer principles: Part 2. Polymer formation." F. Rodriguez, L. J. Mathias, J. Kroschwitz, and C. E. Carraher, Jr., **64**, 886 (1987).

"Classroom demonstrations of polymer principles: Part 3. Physical states and transitions." F. Rodriguez, L. J. Mathias, J. Kroschwitz, and C. E. Carraher, Jr., **65**, 352 (1988).

"Classroom demonstrations of polymer principles: Part 4. Mechanical properties." F. Rodriguez, **67**, 784 (1990).

"Classroom demonstrations of polymer principles: Part 5. Polymer fabrication." F. Rodriguez, **69**, 915 (1992).

"Computer simulation of vinyl polymer tacticity." R. Radiotis and G. R. Brown, **72**, 133 (1995).

"Conformation of macromolecules." D. H. Napper, **67**, 305 (1969).

"Converting sunlight to mechanical energy. A polymer example of entropy." L. J. Mathias, **64**, 889 (1987).

"The crosslinked structure of rubber." L. H. Sperling and T. C. Michael, **59**, 651 (1982).

"Curing of unsaturated polyester: Network formation." R. F. Toorkey, K. C. Rajanna, and P. K. Sai Prakash, **73**, 372 (1996).

"Demonstrating rubber elasticity." F. Rodriguez, **50**, 764 (1973).

"Demonstration—ordered polymers." P. H. Mazzocchi, **50**, 505 (1973).

"Demonstration of solvent differences by visible polymer swelling." J. H. Ross, **60**, 169 (1983).

"Dependence of molecular weight of polystyrene on initiator concentration." P. Ander, **47**, 233 (1970).

"The dependence of strength in plastics upon polymer chain length and chain orientation." R. D. Spencer, **61**, 555 (1984).

"Differential scanning calorimetry study of the cross-linking of styrene and an unsaturated polyester. The chemistry of canoe manufacture." J. Vebrel, Y. Grohens, A. Kadmiri, and E. W. Gowling, **70**, 501 (1993).

"Easily built low-cost apparatus to measure polymer conductivity." J. Gruber, H. de Santana, I. Gruber, and W. Gazotti, **74**, 418 (1997).

"Electric birefringence: A simple apparatus for determining physical parameters of macromolecules and colloids." H. H. Trimm, K. Parslow, and B. R. Jennings, **61**, 1114 (1984).

"Expanded polystyrene." A. Feigenbaum and D. Scholler, **64**, 810 (1987).

"The gelation of polyvinyl alcohol with borax. A novel class participation experiment involving the preparation and properties of a 'slime.' " E. Z. Casassa, A. M. Sarquis, and C. H. Van Dyke, **63**, 57 (1986). (See also A. M. Sarquis, **63**, 60, 1986.)

"The gelation of poly(vinyl alcohol) with $Na_2B_4O_7$: Killing slime." K. W. McLaughlin, N. K. Wyffels, A. B. Bentz, and M. V. Keenan, **74**, 97 (1997).

"Gelatin as a physically crosslinked elastomer." G. V. S. Henderson, Jr., D. O. Campbell, V. Kuzmicz, and L. H. Sperling, **62**, 269 (1985).

"Growth and observation of spherulites in polyethylene." F. W. Billmeyer, P. H. Geil, and K. R. van der Weg, **37**, 460 (1960).

"Introducing plastics in the laboratory: Synthesis of a plasticizer, dioctyl phthalate, and evaluation of its effects on the physical properties of polystyrene." A. Casper, J. Gillois, G. Guillerm, M. Savignac and L. Vo-Quang, **63**, 811 (1986).

"Kinetic and thermodynamic study on the sorption of liquids by polymer films." T. M. Aminabhavi and S. B. Harogoppad, **68**, 343 (1991).

"A low-cost, small-scale polymer mixer." S. P. Nunes, P. Alves, Jr., and F. Galembeck, **67**, 982 (1990).

"Macrospherulites of poly(ethylene oxide)." F. E. Bailey, Jr., and J. V. Koleski, **50**, 761 (1973).

"Measurement of diffusivity of organic liquids through polymer membranes: A simple and inexpensive laboratory experiment." U. Shanthamurthy and T. M. Aminabhavi, **67**, 82 (1990).

"Molecular behavior of elastomers." E. J. Etzel, S. J. Goldstein, H. J. Panabaker, D. G. Fradkin, and L. H. Sperling, **63**, 731 (1986).

"Molecular motion in polymers." L. H. Sperling, **59**, 942 (1982).

"Monomer reactivity ratios." K. I. Ekpenyong, **62**, 173 (1985).

"Plastic density determination by titration." R. Bruzan and D. Baker, **5**, 397 (1993).

"Polydiacetylenes." G. H. Patel and N.-L. Yang, **60**, 181 (1983).

"Polymeric fast ion conductor: A physical chemistry experiment." W. Guo, B. M. Fung, and R. E. Frech, **66**, 783 (1989).

"Polymer mechanical properties via a new laboratory tensile tester." T. C. Gilmer and M. Williams, **73**, 1062 (1996).

"Polymer photophysics: A negative photoresist." F. B. Bramwell et al., **56**, 541 (1979).

"A polymer-solution 'thermometer': A demonstration of the thermodynamic consequences of specific polymer-solvent interactions." P. L. Rubin, **58**, 866 (1981). (See also J. C. Norman, **61**, 1094, 1984.)

"Predictions of transport properties of permeants through polymer films." L. N. Britton, R. B. Ashman, T. M. Aminabhavi, and P. E. Cassidy, **65**, 368 (1988).

"Producing fibers of poly(vinyl alcohol)." M. C. Sherman, **69**, 883 (1992).

"Propagating fronts of polymerization in the physical chemistry laboratory." J. A. Pajman, W. W. West, and J. Simmons, **74**, 727 (1997).

"Protein denaturation." M. Pickering and R. H. Crabtree, **58**, 513 (1981).

"Protein denaturation and tertiary structure." J. S. Barton, **63**, 367 (1986).

"A quick, simple demonstration to distinguish between HD and LD polyethylene." K. E. Kolb and D. K. Kolb, **63**, 417 (1986).

"Reactivity ratios from copolymerization kinetics." W. A. Mukatis and T. Ohl, **49**, 367 (1972).

"Rubber elasticity." J. P. Byrne, **71**, 532 (1994).

"Rubber elasticity—a physical chemistry experiment." M. Bader, **58**, 285 (1981).

"Some properties of poly(methyl methacrylate) studied by radiation degradation: An interdisciplinary student experiment." D. J. T. Hill and J. H. O'Donnell, **58**, 174 (1981).

"Stress-strain behavior of rubber." C. B. Arends, **37**, 41 (1960).

"Stretched elastomers: A case of decreasing length upon heating." S. B. Clough, **64**, 42 (1987).

"Viscoelasticity of cheese." Y. S. Yang, J. S. Guo, Y. P. Lee, and L. H. Sperling, **63**, 1077 (1986).

"Viscometric determination of the isoelectronic point of a protein." J. E. Benson, **40**, 468 (1963).

"Viscosity and the shapes of macromolecules." J. L. Richards, **70**, 685 (1993).

"The viscosity of polymeric fluids." J. E. Perrin and G. C. Martin, **60**, 516 (1983).

"Visualization of protein denaturation by chemical modification of sulfhydryl groups." A. Parody-Morreale and C. Baron, **63**, 1003 (1986).

C.3.3 Characterization

"Analysis of ethylene–vinyl acetate copolymers: A combined TGA/FTIR experiment." K. R. Williams, **71**, A195 (1994).

"Antioxidants in plastics: An instrumental analysis project." W. H. Chan, K. S. Lam, and W. K. Yu, **66**, 173 (1989).

"Determination of acrylonitrile/methyl methacrylate copolymer composition by infrared spectroscopy." K. I. Ekpenyong and R. O. Okonkwo, **60**, 429 (1983).

"The determination of number-average molecular weight." K. R. Williams and U. R. Bernier, **71**, 265 (1994).

"Evaluation of a viscosity–molecular weight relationship." L. J. Mathias, **60**, 422 (1983).

"Experiments illustrating metal–insulator transitions in solids." (Includes conducting properties of polypyrrole and polyaniline), S. W. Keller and T. E. Mallouk, **70**, 855 (1993).

"Fiber identification." J. Allan, **67**, 256 (1989).

"Gel filtration chromatography." J. A. Hurlbut and N. D. Schonbeck, **61**, 1021 (1984).

"Identification of polymers in university class experiments." H. J. M. Bowen, **67**, 75 (1990).

"Identifying plastics by density." K. E. Kolb and D. K. Kolb, **68**, 348 (1991).

"Identifying polymers through combustion and density." A. A. Blumberg, **5**, 399 (1993).

"Illustration of M_n and M_w in chain-growth polymerization using a simplified model. An undergraduate polymer chemistry laboratory exercise." D. M. Snyder, **69**, 422 (1992).

"Light scattering by polymers." G. P. Mathews, **61**, 552 (1984).

"Method for separating or identifying plastics." K. E. Kolb and D. K. Kolb, **68**, 348 (1991).

"A new viscometer design for polymer samples." J. R. Khurma, **68**, 69 (1991).

"Plasticizers in PVC. A combined IR and GC approach." W. H. Chan, **64**, 897 (1987).

"Polymerization evaluation by spectrophotometric measurements." J. Dunach, **62**, 450 (1985).

"Polymerization kinetics and viscometric characterization of polystyrene." J. H. Bradbury, **40**, 465 (1963).

"Polymer molecular weight distribution." D. R. Smith and J. W. Raymonda, **49**, 577 (1972).

"A polymer viscosity experiment with no right answer." L. C. Rosenthal, **67**, 78 (1990).

"Polymers and the visible spectrum. An 'enlightening' activity." E. H. Levine, **69**, 122 (1992).

"Polystyrene kinetics by infrared." H. Andres-Henry, **71**, 357 (1996).

"Quantification of polystyrene acylation by IR spectroscopy: An experiment on polymer IR analysis." E. G. Ramirez Lopez and M. Albores-Velasco, **74**, 551 (1997).

"Quantitative analysis by FTIR: Thin films of copolymers of ethylene and vinyl acetate." L. J. Mathias, M. G. Hankins, C. M. Bertolucci, T. L. Grubb, and J. Muthiah, **69**, A217 (1992).

"Rapid identification of thermoplastic polymers." H. Cloutier and R. E. Prud'homme, **62**, 815 (1985).

"Separation of surface active compounds by foam fractionations." R. M. Skomoroski, **40**, 470 (1963).

"A simple and inexpensive student viscometer." L. G. Daignault, D. C. Jackman, and D. P. Rillema, **67**, 81 (1990).

"A simple, discovery-based laboratory exercise: The molecular mass determination of polystyrene." G. A. Slough, **72**, 1031 (1995).

"A simple, inexpensive molecular weight measurement for water-soluble polymers using microemulsions." L. J. Mathias and D. R. Moore, **62**, 545 (1985).

"A simpler small scale method for the identification of plastics." G. E. Anderson, **73**, A173 (1996).

"A simplified oxygen-flask combustion procedure for polymer analysis." D. R. Burfield and S.-C. Ng, **61**, 917 (1984).

"Size-exclusion chromatography: Separating large molecules from small." D. J. McLoughlin, **69**, 993 (1993).

"Spectrofluorimeters as light-scattering apparatus: Applications to polymer molecular weight determination." M. A. Mougan, A. Coella, A. Jover, F. Meijide, and J. V. Tato, **72**, 284 (1995).

"Textile fiber identification: An organic-polymer laboratory." R. L. Flachskam, Jr., and N. W. Flachskam, **68**, 1044 (1992).

"Undergraduate organic and polymer lab experiments that exemplify structure determination by NMR." T. Viswanathan, F. Watson, and D. T. C. Yang, **68**, 685 (1991).

"The use of opto-electronics in viscometry." R. J. Mazza and D. H. Washbourn, **59**, 1067 (1982).

"The vinyl acetate content of packaging film: A quantitative infrared experiment." K. N. Allpress, B. J. Cowell, and A. C. Hurd, **58**, 741 (1981).

C.3.4 Reactions

"Acid hydrolysis of nylon 66." W. F. Berkowitz, **47**, 536 (1970).

"Hydrolysis of latex paint in dimethyl sulfoxide." J. A. Vinson, **46**, 877 (1969).

"Immobilization of enzymes in polymer supports." H. D. Conlon and D. M. Walt, **63**, 368 (1986).

"A laboratory introduction to polymeric reagents." L. G. Wade, Jr., and L. M. Stell, **57**, 438 (1980).

"Polymer crosslinking and gel formation without heating." J. H. Ross, **54**, 110 (1977).

"Polymer photooxidation—an experiment to demonstrate the effect of additives." N. S. Allen and J. F. McKellar, **56**, 4 (1979).

"Silicate-PVA polymers." B. A. Burke, **65**, 895 (1988).

"Sulfonation of polystyrene: Preparation and characterization of an ion exchange resin in the organic laboratory." A. E. Holboke and R. P. Pennell, **66**, 613 (1989).

"Uses of a vinylpyridine polymer in undergraduate organic synthesis." D. Getman, D. Hagerty, G. Wilson, and W. F. Wood, **61**, 550 (1984).

C.3.5 Miscellaneous

"A bag of slime: A novel lab procedure." D. A. Katz, **71**, 891 (1994).

"Chemistry of paint." G. Schuerman and R. Bruzan, **66**, 327 (1989).

"Competency based modular experiments in polymer science and technology." E. M. Pearce, C. E. Wright, and B. K. Bordoloi, **57**, 375 (1980). (See also Pearce, Wright, and Bordoloi in Section C.1.)

"Disposable macromolecular model kits." I. Nicholson, **46**, 671 (1969).

"Disposable models for the demonstration of configuration and conformation of vinyl polymers." H. Kaye, **48**, 201 (1971).

"Fun—audience participation polymerization demonstrations." J. Ihde, **67**, 264 (1990).

"Happy and unhappy balls: Neoprene and polynorbornene." G. B. Kauffman, S. W. Mason, and R. B. Seymour, **67**, 198 (1990).

"Happy polymer party." M. P. Tarazona and E. Saiz, **67**, 238 (1990).

"The laboratory for introductory polymer courses." L. J. Mathias, **60**, 990 (1983). (Suggestions for procedures.)

"Lecture demonstration of polymer structure using polarized light." F. Rodriguez, **46**, 456 (1969).

"Model of the alpha helix configuration in polypeptides." T. A. Whalen, **33**, 136 (1956).

"Models for linear polymers." P. W. Morgan, **37**, 206 (1960).

"Models illustrating the helix-coil transition in polypeptides." H. J. G. Hayman, **41**, 561 (1964).

"Models of the polypeptide α-helix and of protein molecules." G. Gorin, **41**, 44 (1964).

"Polyethylene and pipecleaner models of biological polymers." H. B. Pollard, **43**, 327 (1966).
"Polymer chemistry in the elementary physical chemistry course." J. D. Ferry, **36**, 164 (1959).
"Polymer models." C. E. Carraher, Jr., **47**, 581 (1970).
"A polymer pollution solution classroom activity." T. L. Helser, **73**, 843 (1996).
"Polymers in the physical chemistry laboratory. An integrated experimental program." G. L. Hardgrove, D. A. Tarr, and G. L. Miessler, **67**, 979 (1990).
"Rayon from dryer lint: A demonstration." M. A. Kropp, **74**, 401 (1997).
"Simple models for polymer stereochemistry." F. Rodriguez, **45**, 507 (1968).
"Using an electronic spreadsheet in the design of exercises for a polymer laboratory course." W. W. McGee and G. Mattson, **70**, 756 (1993).

INDEX